Disaster and Risk Research: GADRI Book Series

Disaster and Risk Research: GADRI Book Series is published under the auspices of the Global Alliance of Disaster Research Institutes (GADRI). The global status of disaster research reflects the major strides made in the disaster sciences. These volumes present the forefront of disaster research, including key scientific findings, methodologies, policy recommendations and case studies. Whilst disaster risk needs to be managed in an integrated manner, persistently isolated applications of knowledge, practice and policy are falling short of ensuring disaster-resilient societies. Responding to this deficit calls for measurement, tools, techniques and institutional structures that can realistically support comprehensive risk assessment and management across multiple hazard landscape. As such, disaster research is now faced with a multi-disciplinary, multi-stakeholder challenge. Contributions to this series therefore address many varied and critical opportunities to advance the subject area. A cross-cutting vision shared across the Disaster and Risk Research volumes is to improve the future of scientific and technological guidance with clearly identifiable roadmaps to ensure human safety and security. The Global Alliance of Disaster Research Institutes was established in March 2015, directly after the United Nations World Conference on Disaster Risk Reduction (WCDRR 2015) in Sendai, Japan, based on the belief that a multi-institutional alliance can strengthen disaster research and its influences around the world. GADRI has a mandate to support the implementation of the Sendai Framework for Disaster Risk Reduction 2015-2030 and is a member of the Scientific and Technical Advisory Group (STAG) of the United Nations Office for Disaster Risk Reduction (UNDRR). In addition, GADRI provides a platform for scientific communities from different disciplines, backgrounds and countries, helping them share their knowledge, findings and views. This approach yields more holistic and farther-reaching insights, which can contribute to further steps in effective disaster risk management.

Hirokazu Tatano · Paul Kovacs · Wilma James
Editors

Proceedings of the 6th Global Summit of the Global Alliance of Disaster Research Institutes

Towards GADRI Objectives of Achieving a Sustainable Disaster-Resilient World

Springer

Editors
Hirokazu Tatano
Disaster Prevention Research Institute
Kyoto University
Uji, Kyoto, Japan

Paul Kovacs
Institute for Catastrophic Loss Reduction
Western University
Toronto, ON, Canada

Wilma James
Disaster Prevention Research Institute
Kyoto University
Uji, Kyoto, Japan

ISSN 2524-5961 ISSN 2524-597X (electronic)
Disaster and Risk Research: GADRI Book Series
ISBN 978-981-96-6608-9 ISBN 978-981-96-6609-6 (eBook)
https://doi.org/10.1007/978-981-96-6609-6

This Springer imprint is published by the registered company Springer Nature Singapore Pte Ltd.
The registered company address is: 152 Beach Road, #21-01/04 Gateway East, Singapore 189721, Singapore

Message by the Secretary-General of GADRI; and the Chair of the Board of Directors of GADRI

Welcome to the 6th Global Summit of GADRI!

We are pleased to welcome you to the 6th Global Summit of GADRI at the Disaster Prevention Research Institute (DPRI), Kyoto University, Uji Campus.

Through the Summit, we intend to promote the current status of disaster prevention science and technology in various countries around the world and share knowledge and evidence-based results across various global platforms. Through our joint efforts, we will discuss the contribution and implementation targets of the Science and Technology Roadmap to implement the goals and priorities of the Sendai Framework.

The prolonged worldwide pandemic and cascading risks have taught us that the conventional approach to disaster risk planning and management is insufficient for the development of sustainable and resilient communities. A worldwide pandemic underscored the importance of integrating the following key areas:

(i) Disaster risk should not be treated in isolation but should be integrated with health, climate change, and environmental risks.
(ii) DRR objectives and vision should be integrated with Sustainable Development Goals to foster a resilient world.
(iii) Short-term DRR objectives need to be integrated with a long-term vision and plans for a resilient society.

In 2022, the Board of Directors of GADRI established five committees to implement global research networks, develop research roadmaps and plans, build the capacities of research institutes, share information and engage in collaborative research, and advocate for organizations. These committees are being directed toward the three key areas through integrated research and development.

The Panel Discussion Session One therefore explores the GADRI objectives of achieving a sustainable and resilient Society Against Disasters facilitated by the committees as follows:

- Committee on Networking.
- Committee on Science & Technology Roadmap.
- Committee on Institutional Capacity Building.
- Committee on Data and Information Sharing.
- Committee on Advocacy.

Panel Discussion Session Two takes the committee themes further to discuss some of the new challenges for action by GADRI:

- Big Science for DRR: Large-Scale Experimentation.
- Sustainable DRR: Integrating Climate Acton, SDGs, Field DRR (experience) Sharing – New Challenges for action by GADRI.
- Gender and Inclusivity in DRR Policy and Practice.

- Putting Health into Disaster Risk Reduction (DRR) and Recovery.
- Young Scientists Session on Youth DRR—Role of Youth and Young Professionals in Data Knowledge Sharing in Disaster Risk Management.

We will share the progress, strategies and challenges and discuss how to play an important role as GADRI. Through dedicated and vigorous discussion sessions, we intend to influence widespread decision-making processes.

Leading up to the summit, we conducted a survey among all 212 members of GADRI to:

(i) Learn more about the activities of member institutes.
(ii) Understand collectively the progress and achievement of member institutes towards GADRI's objectives (https://gadri.net/about/Aims/).
(iii) Make inputs as GADRI's collective contributions towards the Sendai Midterm Review. The survey form was distributed electrically to all GADRI institutions, and the results will be shared during the 6th Global Summit.

The survey results will be shared during the Opening Ceremony of the summit, with the intention to then share these further with the UNDRR office.

We hope that your contributions and debates will expand well beyond this summit and that GADRI activities will continue to provide and function as evidence-based science in response to urgent issues in the world.

We wish you every success during the 6th Global Summit of GADRI.

Hirokazu Tatano
Secretary-General, GADR; and Professor, DPRI, Kyoto University, Japan

Paul Kovacs
Chair, Board of Directors of GADR; and Executive Director, ICRL, Western University, Canada

Acknowledgements

Wilma James: james.wilmatheonesta.6r@kyoto-u.ac.jp

The Global Alliance of Disaster Research Institutes (GADRI) conveys its gratitude and deep appreciation to all those who were directly and indirectly involved in the successful organization of the 6th Global Summit of the Global Alliance of the Disaster Research Institutes (GADRI) at the Disaster Prevention Research Institute (DPRI), Kyoto University, Japan, from 15 to 17 March 2023. We value your work and contributions to the 6th Global Summit of GADRI. It also appreciates Kyoto University and the Disaster Prevention Research Institute (DPRI), Kyoto University for hosting the GADRI Secretariat at the DPRI, Kyoto University without whose financial and administrative support GADRI could not have executed its targeted objectives and activities.

Much appreciation is also expressed to all keynote speakers and panelists for taking time to deliver keynote addresses and participating in panel presentations and also for their much-valued insights and contributions to open discussion sessions. GADRI acknowledges with thanks the support and courtesy rendered by the United Nations office for Disaster Risk Reduction (UNDRR) and the contributions made by its members.

Deep appreciation and gratitude are also expressed to the members of the Board of Directors of GADRI for their tireless and continued contributions and support for GADRI activities. GADRI also thanks all of its members for their participation and contributions towards the success of the 6th Global Summit of GADRI and also for their continued contributions to the targets of the Science and Technology Roadmap for the implementation of the goals and priorities of the Sendai Framework.

A special thank goes to the GADRI Secretariat Committee members who tirelessly contributed to the execution of the conference and many students who ensured that the sessions were running successfully.

Deep thank you and appreciation to Prof. Adnrew Collin who invested his time and effort in editing the Proceedings of the 6th Global Summit of GADRI.

Welcome Greetings by the Director of the Disaster Prevention Research Institute (DPRI), Kyoto University, Japan

Distinguished guests, ladies and gentlemen, colleagues,

Good morning!

Welcome to Kyoto and to the DPRI, Kyoto University Uji Campus.

After a dormant four years of lockdown under the corona virus pandemic, and as the surrounding mountains are springing to life and the cherry blossoms are waiting to share its glory, I am happy to welcome you all to the 6th Global Summit of GADRI at the DPRI, Uji Campus of the Kyoto University.

I would also like to thank all of you for making time to attend the 6th Global Summit of GADRI in person, especially, Ms. Mami Mizutori for taking time off from her busy schedule to be with us; Prof. Nagahiro Minato, President of Kyoto University who made it a point to attend the summit and spend time with the Board members of GADRI and others; and Ms. Atsuko Matsumura, Mayor of Uji City to whom we are grateful for supporting the activities of DPRI, Kyoto University at all times. Thank you.

All of you have traveled from far and near to be here with us on this occasion. Seeing all of you here, brings a welcome constancy to our campus and our research world, as once again, we are able to interact in person, with our colleagues at other institutions, and renew our connections and collaborations. What a pleasant change it makes.

Prior to moving forward with my remarks, I would like to take this opportunity to express our condolences to the victims and families of the devastating earthquakes that took place in Turkey and Syria last month.

Our colleague, Prof. Mustafa Erdik from the Boğaziçi University, Turkey, is here among us today. Later in the morning, we will have the chance to listen to more information on the earthquakes from Prof. Erdik directly.

I learned that despite the challenges posed by the pandemic and in the face of perpetual changes in various boarder restrictions, the 6th Global Summit of GADRI has brought together nearly 200 participants from 40 states and economies. The programme of the conference consists of plenary sessions by distinguished speakers from varied disciplines, group discussion sessions, and much more. The summit themes cover current global agendas ranging from present to future disaster risks; mitigating climate change and increasing resilience through DRR objectives with the SDGs; and systemic risks and emerging future challenges. No doubt you will be engaged in very interesting and stimulating deliberations.

The late afternoons will be devoted to panel group discussions. Group discussion session 1 specifically targets the five-fold objectives of GADRI, carefully formed by its Board of Directors and embedded in the Charter of GADRI. We can take this occasion to discuss progress, strategies and challenges with research of our respective institution's activities and discuss how to play an important role as GADRI. We can voice our opinions supported by evidence-based scientific results to influence decision-making processes in various issues in today's global agendas. I look forward to working with you in these endeavours.

I would also like to stress that the Disaster Prevention Research Institute (DPRI), Kyoto University, is very pleased to host the Secretariat of GADRI. It has paved the way for us to play a core role as an international research institute of excellence in disaster prevention studies.

I think everyone involved in disaster prevention research has a sense of mission. It is this sense of mission to "save people's lives, make people happy, and make them smile", which is engraved in our hearts that drives us to continue to fulfil our mission. I, as the Director of DPRI, believe that this is very much at the heart of our activities. I would like to proceed with this from the perspective of not only the current generation, but also pass it on to the next generation, and the generations to come.

I look forward to interacting with you during the coming three days of the 6th Global Summit of GADRI.

I wish you every success in the 6th Global Summit of GADRI.

Eiichi Nakakita
Director
Disaster Prevention Research Institute (DPRI)
Kyoto University

Welcome Address by the President of the Kyoto University

It is my greatest pleasure, and honour, to welcome you all to the 6th Global Summit of the Global Alliance of Disaster Research Institutes (GADRI) at the Kyoto University, Uji Campus, Disaster Prevention Research Institute (DPRI).

I would like to begin by expressing our sincerest sympathies and condolences to the victims of the devastating earthquakes that occurred in the southern border of Turkey and Syria last month. Kyoto University, including the DPRI and together with GADRI, will do our utmost best to provide support for all those affected by this disaster. Japan has also experienced such devastating earthquakes in the past and suffered from serious damage and loss that is caused. We can therefore empathize with the pain and suffering of those affected by this terrible catastrophe.

Each time, in the midst of such a tragedy, we are starkly reminded of our vulnerability unforeseen hazards and disasters whether natural, man-made or caused by in relation to infectious diseases such as the Covid-19 pandemic. We are shocked and saddened by the statistics describing the victims and mortality rates, and when we see the devastation caused to infrastructure and other basic services and essentials. The need to improve and expand the utilization of early warning systems, facilitate early responses, and promote disaster awareness is urgently felt once again. The Sendai Framework, Paris Agreement on Climate Change, SDGs, and other global agendas, unequivocally draws our attention to the importance of promoting sustainable development and increasing disaster resilience and encourages us to focus and unite our efforts, with the common goal of leaving no one behind.

Those aims are in full alignment with Kyoto University's most fundamental institutional goal, which is expressed in the university Mission Statement as the pursuit of harmonious coexistence within the world's human and ecological community. The Kyoto University celebrated its 125th Anniversary last year, and we continue to pursue that goal through our pioneering institutional traditions of academic freedom and "teaching through research". It is an approach that, throughout the university's history, has enabled its scholars to pursue particularly original and innovative research, leading to groundbreaking achievements that continue to be acknowledged by the most respected international academic awards and accolades.

As the university continues to develop and refine its approach in step with the changing times, the importance of diversity as a key component of a vibrant and innovative education and research environment has never been more apparent. In particular, actively fostering the younger generation of researchers and increasing the ratio of female researchers are two current key priorities.

In this context, I am delighted to see that this summit will benefit from the contributions of over 20 female scholars among its keynote and panel discussion participants and that the program includes discussion sessions dedicated to "Gender and Inclusivity in DRR Policy and Practice" and "Young Scientists Session on Youth and DRR" in the programme of the 6th Global Summit of GADRI. These important themes are embedded in the Agenda of the Sendai Framework; they are very much at the core of Kyoto University's vision to promote diversity in education and research.

The presence of GADRI has become the foundation of an international research network to systematically promote disaster risk prevention and contribute to global agendas, on science and technology roadmaps and action plans. As a community of scientists, together, we can synergistically combine our research activities, resources, and knowledge to support our communities with evidence-based outcomes and implementable real-world solutions. We have already seen the importance of evidence-based scientific findings and how they can influence and support urgent decision-making for policy and practice. By combining our efforts, we can optimize our resources to confront the challenges we face and achieve a substantial reduction to disaster risk, thereby reducing the loss of lives and livelihoods, and helping to "make our communities safer and more resilient to disasters".

The programme of the 6th Global Summit of GADRI promises to keep all of you participants fully engaged throughout the coming three days. I look forward to learning the summit's outcomes, and more importantly, to the resulting actionable decisions that will ensure a safer future for the world.

I wish you all the very best for a productive and rewarding three days.

Nagahiro Minato
President
Kyoto University

Photo 3: Ms. Mami Mizutori, Assistant Secretary-General and Special Representative of the Secretary-General for Disaster Risk Reduction in the United Nations Office for Disaster Risk Reduction

Opening Remarks by Ms. Mami Mizutori, Assistant Secretary-General and Special Representative of the Secretary-General for Disaster Risk Reduction in the United Nations Office for Disaster Risk Reduction

6th Global Summit of the Global Alliance of Disaster Research Institutes (GADRI)

Wednesday15 March 2023

Disaster Prevention Research Institute, Kyoto University, Uji Campus, Kyoto, Japan

Ladies and Gentlemen,

Good morning. I would like to extend my warmest welcome to you to the 6th Global Summit of GADRI.

First of all, please allow me to join in expressing my condolences to the people and the Government of Türkiye and the Government of Syria.

I would like to start by acknowledging Prof. Hirokazu Tatano, Secretary-General of GADRI, and this session's co-chairs, Prof. Ana Maria Cruz and Dr. Paul Kovacs.

I would also like to thank our hosts, Prof. Nagahiro Minato, President of Kyoto University, and the Honorable Ms. Atsuko Matsumura, Mayor of Uji City, and Prof. Eiichi Nakakita, Director of the Disaster Prevention Research Institute of Kyoto University.

This Summit takes place at a time of rapid global change, where the impacts of complex risks are cascading across sectors and across borders.

That is why I welcome the focus areas of this year's GADRI summit which aptly capture the priorities we see for reducing global risks.

The summit's first area of focus recognizes that disaster risk should not be treated in isolation but should be integrated across sectors.

We could not agree more as while sectors are siloed in government and research, risks are not siloed.

To address this, we are calling for a comprehensive approach to risk management. An approach that accounts for all hazards and leverages all disciplines for risk mitigation.

The Sendai Framework for Disaster Risk Reduction identifies this as its second Priority of Action, which is to have resilient and strong risk governance. Without such an approach, countries would be blind to the full spectrum of risks and their responses would not be as effective.

The second area of focus is around integrating risk reduction into sustainable development.

The countries that are the furthest behind in achieving the Sustainable Development Goals are the Least Developed Countries. These countries also account for 70% of climate-related disaster mortality.

That is why I came to Japan directly from the LDC5 conference in Doha, Qatar, where we welcomed the integration of Disaster Risk Reduction into the entirety of the Doha Programme of Action.

Without a risk-informed approach to development, it would be impossible for LDCs to graduate to middle-income status, and then not to be pushed back again, let alone achieve the SDGs.

The third focus area is around the need to integrate short-term objectives with long-term plans for resilience.

In the 2022 Global Assessment Report for Disaster Risk Reduction, we identified short-termism as a barrier to effective risk reduction.

Indeed, the prevalence of short-termism in decision-making is one of the reasons why we predicted that the number of medium- to large-scale disasters would grow by 2030 to reach 560 a year – or one and a half disasters per day.

Through the implementation of the UN Secretary-General's Early Warnings for All Initiative, which we, the United Nations Office for Disaster Risk Reduction (UNDRR), are co-leading with WMO, we hope to support countries not only to address immediate risks but to also strengthen their ability to build long-term resilience.

I am glad to be here today among those who understand the challenges we face and are dedicated to helping the world to develop scalable solutions.

I look forward to learning of this Summit's outcomes and recommendations, I wish you a successful Summit, and I am most grateful for your commitment to building resilience for all.

Thank you very much.

Photo 4: Ms. Atsuko Matsumura, the Mayor of Uji City

Speech by Ms. Atsuko Matsumura, the Mayor of Uji City

Thank you for that kind introduction.

I am Atsuko Matsumura, the Mayor of Uji city.

I would like to congratulate related organizations on holding face-to-face and on a grand scale, the 6th Global Summit of Research Institutes for Disaster Risk Reduction, today. Also, welcome to Japan and to Uji city. As a representative of our city, I welcome all of you from the bottom of my heart.

In recent years, various natural disasters such as floods, droughts, heat waves, and cold waves have been reported on a global scale, and while the ability to respond to disasters is being emphasized again, on the 6th of last month, a quake struck Turkey and Syria.

More than 52,000 (fifty two thousand) people have died, and more than two million people are still in great need of help.

Japan has experienced the Great Hanshin-Awaji Earthquake and Great East Japan Earthquake, and we know how hard to live as an evacuee that will continue for a long time ahead. So, I do hope their sooner restoration and revival.

About ten years ago, here in Uji city as well, we had unexpectedly heavy rain and suffered a lot of damage, including partial burst and flooding of rivers. At that time, the Kyoto University Disaster Prevention Research Institute offered expert advice and lent us their support. And I am very grateful that they are still cooperating with disaster prevention and disaster mitigation training for citizens.

We feel they are very reliable, as is the government where the research institutes participating here are located.

Against a backdrop, I have heard that the theme of the summit, which will be held from today, will be "the realization of a sustainable and disaster-resilient world."

In this unstable social situation, this theme is on point. And I've heard that it will be held face-to-face for the first time in four years due to the influence of COVID-19.

I believe that this face-to-face meeting not only facilitates the sharing of results and knowledge among institutes, but also strengthens networks among them. Furthermore, it is my strong hope that this summit will deepen research of disaster prevention, and the results of this research will serve as an impetus to support disaster-stricken areas and promote disaster prevention measures in the government. And I believe that the importance of the Global Alliance of Disaster Research Institutes and of this summit will grow much more.

You have a brochure about Uji city for tourists. I hope you will enjoy Uji in spring, which is full of history and culture and where cherry blossoms are in bloom.

In closing, I wish the Global Alliance of Disaster Research Institutes' prosperity and success in the 6th Global Summit of Research Institutes for Disaster Risk Reduction.

Congratulations and thank you for your attention.

March 15th, 2023 (two thousand twenty three)

Atsuko Matsumura

Mayor of Uji city

Contents

Panel Group Discussion I: Towards GADRI Objectives of Achieving a Sustainable and Resilient Society Against Disasters

Panel Group Discussion II: New Challenges for Action by GADRI

Regional Alliances

Selected Presentation Papers (Arranged By Country Order)

Summary Report of Activity of the 6th Global Summit of GADRI

Report by the Secretary-General, GADRI 6th Global Summit of Global Alliance of Disaster Research Institutes: Towards GADRI Objectives of Achieving a Sustainable Disaster-Resilient World

Hirokazu Tatano, Andrew Collins, and Wilma James(✉)

Disaster Prevention Research Institute (DPRI), Kyoto University, Uji Campus, Kyoto, Japan
James.wilmatheonesta.6r@kyoto-u.ac.jp

Fig. 1. Group Photo of the Sixth Global Summit of the Global Alliance of Disaster Research Institutes (GADRI), Disaster Prevention Research Institute (DPRI), Kyoto University, Japan held from 15 to 17 March 2023.

The 6th Global Summit of GADRI: Towards GADRI Objectives of Achieving a Sustainable Disaster-Resilient World was successfully convened at the DPRI, Kyoto University, Uji Campus, Kyoto, Japan from 15 to 17 March 2023. The biennially held

H. Tatano et al. (Eds.): GSRIDRR 2023, DRRGADRIBS, pp. 3–13, 2025.
https://doi.org/10.1007/978-981-96-6609-6_1

Fig. 2. Opening Ceremony Session Chairs, Prof. Paul Kovacs, and Prof. Ana Maria Cruz.

Global Summits of GADRI bring together participants engaged in disaster risk reduction or disaster science from multidisciplinary research fields (Fig. 1 and 2).

The Opening Ceremony session of the 6th Global Summit of GADRI was chaired by Prof. Ana Maria Cruz, DPRI, Kyoto University; and Prof. Paul Kovacs, Chair of the Board of Directors of GADRI; and Founder and Executive Director, Institute for Catastrophic Loss Reduction at Western University, Canada. The following dignitaries participated in the session:

- Prof. Eiichi Nakakita, Director, DPRI, Kyoto University, Japan
- Prof. Nagahiro Minato, President, Kyoto University, Japan
- Ms. Mami Mizutori, Special Representative of the Secretary-General for Disaster Risk Reduction, UNDRR, Switzerland
- Ms. Atsuko Matsumura, Mayor, Uji City, Kyoto, Japan

On behalf of all members of the summit, the members of the Opening Ceremony took the opportunity to express condolences to the people and governments of Turkey and Syria and to state solidarity at this very challenging time of the earthquake that occurred on 6 February 2023 (Fig. 3).

The Secretary-General of GADRI, Prof. Hirokazu Tatano presented an overview of the Global Alliance of Disaster Research Institutes (GADRI) and of the history of the Summit (Fig. 4).

Fig. 3. Distinguished guest speakers at the Opening Ceremony – (L) – Prof. Eiichi Nakakita, Prof. Nagahiro Minato, Ms. Mami Mizutori, and Ms. Atsuko Matsumura.

Dr. Genta Nakano, DPRI, Kyoto University shared the results of the pre-conference survey undertaken among the members of GADRI. The survey was responded to by 87 institutes of GADRI. Dr. Nakano presented in detail the information of the survey. A complete and a detailed report on the survey results can be found in Chapter 2.

The three plenary sessions of the summit featured keynote speeches by ten prominent speakers from around the world to address multidimensional global crises including from climate change to the recent earthquakes in Southern Turkey, the global COVID-19 pandemic, systemic risks and emerging future challenges.

Plenary Session I: Listening to various stakeholders on integration of disaster risks to present and future disasters.

The session was chaired by Prof. Virginia Murray, Head, Global Disaster Risk Reduction, UK Health Security Agency, UK; and Prof. Rajib Shaw, Keio University, Japan (Fig. 5).

- Keynote 1 - The state of Disaster Risk Reduction and the Transformations Needed to take the World from Risk to Resilience, Ms. Mami Mizutori, UNDRR, Switzerland
- Keynote 2 - What Happened on 6 February 2023 Earthquakes in Türkiye, Prof. Mustafa Erdik, Emeritus Professor, Dept. Earthquake Engineering, Kandilli Observatory and Earthquake Research Institute, Boğaziçi Üniversitesi, Türkiye

Fig. 4. Presentation on GADRI Activities by Prof. Hirokazu Tatano, Secretary-General, GADRI; Professor, DPRI, Kyoto University, Japan.

- Keynote 3 - Climate Science for Action, Prof. Masahide Kimoto, President, National Institute for Environmental Studies (NIES); and Professor Emeritus, The University of Tokyo, Japan
- Keynote 4 - The Potential for Recovery to Support Transformative Improvement in Disaster Risk Reduction, Prof. Paul Kovacs, Chair, Board of Directors of GADRI; and Executive Director, Institute for Catastrophic Loss Reduction (ICLR), Western University, Canada

Plenary Session II: Visions to mitigate climate change and increase resilience through DRR objectives with SDGs.

The session was chaired by Prof. Yuichi Ono, IRIDeS, Tohoku University, Japan; and Dr. Tom De Groeve, Head, European Commission, Joint Research Centre (EC-JRC), Italy.

- Keynote 5 - The Climate Crisis is a Health Crisis - Climate Change and Public Health - links to the Sendai Framework, the Paris Agreement and the SDGs, Prof. Virginia Murray, Head, Global Disaster Risk Reduction, UK Health Security Agency, UK
- Keynote 6 - Digital Twinning of Communities for Disaster Risk Reduction: Climate Adaptive Solutions, Prof. John van de Lindt, Harold H. Short Endowed Chair Professor, Co-Director, Center for Risk-Based Community Resilience Planning, Colorado State University, USA

Fig. 5. Plenary Session one Chairs: (L) Prof. Rajib Shaw and Prof. Virginia Murray.

- Keynote 7 - Addressing Climate Change and Disaster Risks in the Context of Sustainable Development in Southern Africa, Dr. Chipo Mudavanhu, Vice-President, AADRI; and Senior Lecturer, Bindura University of Science Education, Zimbabwe

Plenary Session III: Systemic risks and emerging future challenges.

- Keynote 8 – presented online - Risk Communication and Governance in a Post-Truth Environment, Prof. Ortwin Renn, Retired Scientific Director, International Advanced Science Studies (IASS), Germany

The online presentation of the keynote by Prof. Ortwin Renn was chaired by Prof. Charles Scawthorn, Visiting Professor, University of California, Berkeley, USA; and Ms. Ritsuko Yamazaki-Honda, Deputy Director, National Research Institute for Earth Science and Disaster Resilience (NIED), Japan.

The following two keynotes were delivered on Day 3 of the summit and were chaired by Prof. Wei-Sen Li, Secretary-General, National Science and Technology Center for Disaster Reduction (NCDR), Taiwan; and Prof. Gretchen Kalonji, Dean, Institute for Disaster Management and Reconstruction (IDMR), Sichuan University, China.

- Keynote 9 - Anticipation, Integration and Impact – Three Elements for Tackling Future Risks, Dr. Tom De Groeve, Head, European Commission, Joint Research Centre (EC-JRC), Italy

- Keynote 10 - Disaster Science: Updating Theory to Guide Disaster Risk Reduction in the Future, Prof. David Alexander, Institute for Risk and Disaster Reduction (IRDR), University College London, UK

At the end of each plenary session, the keynote speeches were followed by a Q&A session moderated by the respective chairpersons. A few of the keynote speakers have submitted their full reports to be included in the proceedings.

During the GADRI Regional Alliances Session, representatives of the regional alliances were given a chance to showcase collaboration and research workflows and opportunities. The session was chaired by Prof. Kaoru Takara, Professor Emeritus, DPRI, Kyoto University; and Prof. Peter Sammonds, Director, UCL Institute for Risk and Disaster Reduction, UK (Fig. 6).

Fig. 6. GADRI Regional Session Chairs (L) Prof. Peter Sammonds and Prof. Kaoru Takara.

- Activities of Natural Disaster Research Council in Japan, Prof. Makoto Okumura, Natural Disaster Science Council, Japan
- North American Alliance of Hazards and Disaster Research Institutes (NAAHDRI) Activities: Fostering Interdisciplinary Research and Community Engagement to Better Confront Increasing Disasters under the Changing Climate - Dr. Guirong (Grace) Yan, Director, CHM&CR, Missouri University of Science and Technology, USA - on Behalf of Prof. Lori Peek, Director, NHC, University of Colorado Boulder, USA

- African Alliance for Disaster Research Institutes (AADRI) – Prof. Desmond Manatsa, President, AADRI; and Executive Dean, Faculty of Science and Engineering, Bindura University of Science, Zimbabwe
- South Asia Alliance of Disaster Research Institutes (SAADRI) – Prof. Mahua Mukherjee, Secretary-General, SAADRI; and Professor, IIT Roorkee, India
- UK Alliance for Disaster Research (UKADR) – Prof. Andrew Collins, Advisor, UKADR; and Disaster and Development Network, Northumbria University Newcastle, UK
- 3rd World Bosai Forum, March 2023 – Prof. Yuichi Ono, Director, IRIDeS, Tohoku University, Japan (Fig. 7 and 8).

Fig. 7. Presentation on UK Alliance for Disaster Research (UKADR) – Youth Engagement by Dr. Mark Ashley Parry, Disaster and Development Network, Northumbria University Newcastle, UK.

The two panel discussion sessions covered the sub-themes of the conference through ten parallel sessions. Forty-three experts in various fields shared their experiences, views and opinions during the two panel discussion sessions.

Panel Session I concentrated on the five Committees of GADRI - Towards GADRI Objectives of Achieving a Sustainable and Resilient Society Against Disasters.

- I-A: Committee on Networking - Fostering Networking to Enhance Global DRR
  Chair: Charles Scawthorn
  Co-chairs: Ritsuko Yamazaki-Honda - Kenji Tanaka
  Supported by Tetsuya Sumi
  Panelists: Rajib Shaw, Ronan Sato, Jean-Paul Pinelli, Hiroyuki Goto
- I-B: Committee on Science and Technology Roadmap
  Chair: Hirokazu Tatano
  Co-chairs: Toshio Koike, and Selim Gunay

Fig. 8. Presentation of the Third World Bosai Forum held at Tohoku University between 10 to 13 March 2023 by Prof. Yuichi Ono, Director, IRIDeS, Tohoku University, Japan.

Supported by Yuki Matsushi, and Maria Camila Suarez Paba
Panelists: Toshio Koike, Qunli Han, Yuichi Ono, Mahua Mukherjee, Guirong Grace Yan, Selim Gunay

- I-C: Committee on Institutional Capacity Building
  Chair: Wei-Sen Li
  Co-chairs: Desmond Manatsa and Gretchen Kalonji
  Supported by Subhajyoti Samaddar, Tetsuya Takemi, Kenji Kawaike
  Keynote: Prof. Reini Wirahadikusumah, Rector, Institute of Technology Bandung (ITB), Indonesia (online)
  Panelists: Prof. Reini Wirahadikusumah (online); Bill Ho (online); Ailsa Holloway (online), G. Mukwada (online), Dimiter Velev (online)
- I-D: Committee on Data and Information Sharing
  Chair: Andrew Collins
  Co-chairs: Tom De Groeve, Philipp Ulbrich, Hiroyuki Goto,
  Panelists: Norio Maki, Jean-Paul Pinelli, Rebecca Richardson, Hiroyuki Goto
- I-E: Committee on Advocacy
  Science Technology and Innovation linked to Advocacy
  Chair: Rajib Shaw
  Co-chair: Irasema Alcantara-Ayala, Paul Kovacs
  Supported by Masamitsu Onishi

Panelists: Antonia Loyzaga (video), Joy Pereira (video), Takeshi Komino, Sayaka Irie

Panel Session II focused on New Challenges for Action by GADRI in the areas of Big Science for DRR; Sustainable DRR covering Climate Change, SDGs, and Field DRR; Gender and Inclusivity in DRR Policy and Practices; Putting Health into DRR and Recovery; and Young Scientists Session on Youth and DRR.

- II-A: Big Science for DRR: Large-scale Experiment
  Chair: Nobuhito Mori, DPRI, Kyoto University, Japan
  Co-chair: Shinichi Matsushima, DPRI, Kyoto University, Japan
  Panelists: Nelson Pulido, Kojiro Suzuki, Kentaro Tabata
- II-B: Sustainable DRR: Integrating climate action, SDGs and Field DRR and Data (experience) Sharing
  New Challenges for Actions by GADRI
  Chair: Tetsuya Takemi, DPRI, Kyoto University, Japan
  Co-chair: John van de Lindt, Colorado State University, USA
  Panelists: Weqiang Ma, Fatima Akter, Mahua Mukherjee, Novvria Sagita
- II-C: Gender and Inclusivity in DRR Policy and Practice
  Chair: Peter Sammonds, UCL Institute for Risk and Disaster Reduction, UK
  Co-Chairs: Ana Maria Cruz, and Guirong Grace Yan
  Panelists: Punam Yadav, Miwako Kitamura, Shigeo Tatsuki, Ana Maria Cruz
- II-D: Putting Health into Disaster Risk Reduction (DRR) and Recovery
  Chair: Virginia Murray
  Co-Chair: Andrew Collins, Ryoma Kayano of WHO Kobe Office, Genta Nakano
  Panelists: Ryoma Kayano, Genta Nakano, Virginia Murray
- II-E: Young Scientists Session on Youth and DRR
  The Role of Youth and Young Professionals in data and knowledge sharing in disaster risk management
  Chair: Gretchen Kalonji, IRDR, Sichuan University
  Co-Chair: Wei-Sen Li, NCDR, Huan Liu, DPRI, Kyoto University, Japan
  Panelists: Mizan Bisri, Ryo Tsuchida, Nuraini Rahma Hanifa (online), Ranit Chatterjee, Maria Camila Suarez Pablo, Chipo Mudavanhu

Panel discussion session results were shared during the presentation sessions. All chairs of the panel sessions submitted full reports on key questions and outcomes which are included in this proceedings.

The poster session attracted over 50 abstracts. The prospective presenters were encouraged to specifically highlight their respective institutes' research activities, achievements, implementations in support of the Sendai Framework Priority Areas (Fig. 9).

In addition to the above, the poster session provided a platform for young and upcoming scientists to share their research achievements, activities and future plans. There were 33 presenters who joined the 6th Global Summit of GADRI in-person to present their work. Six of the presenters won the best abstract award of a GADRI Scholarship to participate in the summit (Fig. 10).

The Networking Session on Seed and Needs brought together 13 institutions who shared their institutions' research aspirations and challenges.

Fig. 9. Participants at the Poster Session.

In addition, booth space was provided for GADRI member institutes to showcase their respective institute activities.

At the closing session, the conference draft final outcomes and resolution was tabled for discussion and inputs called for from the audience. A draft document of outcomes and resolutions was compiled with important points contributed by all keynote speakers and panel discussion groups. The document of outcomes and resolutions was finalized by Prof. Andrew Collins with the support of Prof. Paul Kovacs and Prof. Hirokazu Tatano. This was later shared by e-mail with all participants of the summit.

The final resolution document is included in these proceedings.

A Closing Banquet celebrated the success of the 6th Global Summit of GADRI. Prof. Kyoko Inagaki, Vice President, Kyoto University, Japan; and Mr. Daisaku Kadokawa, Kyoto City Mayor, Japan delivered greetings to the participants. There was a unique performance of the taiko-drums by the Kyoto Tachibana High School students.

The GADRI Secretariat awarded 24 full and partial scholarships to various scholars out of which 12 were awarded to female scholars.

The 6th Global Summit of GADRI was attended by 185 participants out of which there were 67 females. The participants represented 74 institutes in 30 economies (Fig. 11).

Fig. 10. Booth by the National Science and Technology Center for Disaster Reduction (NCDR), Taiwan.

Fig. 11. Group Photo of the Global Banquet, Rihga Royal Hotel, Kyoto, Japan on 17 March 2023.

Contributions of Disaster Research Institutes to the Sustainable Disaster-Resilient World -Results of the Questionnaire Survey-

Genta Nakano[1](✉), Paul Kovacs[2], and Hirokazu Tatano[1]

[1] Disaster Prevention Research Institute (DPRI), Kyoto University, Uji, Japan
nakano.genta.8n@kyoto-u.ac.jp
[2] Institute for Catastrophic Loss Reduction (ICLR), Western University, London, Canada

Abstract. The Global Alliance of Disaster Research Institutes (GADRI) is a collaborative platform of 212 disaster research institutes worldwide. The questionnaire survey was conducted targeting member institutes with the aim of identifying the benefits and gaps of GADRI membership and the contributions to the Sendai Framework for Disaster Risk Reduction. The survey responses from 85 institutes shows that more than 5,000 professors and 30,000 students are involved in the disaster related research and education. Gender ratio of professors by region indicates the strong tendency in Japan that the professor positions are dominated by male. While GADRI acted as a node for networking member institutes and contributed to promote collaborative research and exchange among researchers, and making one-voice of member institutes, further needs to improve the accessibility to different types of disaster related data were identified. Majority of the institutes participated in the national, regional and local disaster risk reduction platforms to contribute in particular to drought and flood modeling, capacity building, early warning and cultural heritage management.

Keywords: disaster research institutes · Sendai Framework for Disaster Risk Reduction · GADRI · questionnaire survey

1 Introduction

The Global Alliance of Disaster Research Institutes (GADRI) is a collaborative platform to make a collective action of disaster research institutes worldwide to promote the disaster-related research, share knowledge, promote networking and contribute to disaster risk reduction (DRR). The alliance was established in March 2015 through the discussion held at the 2nd Global Summit of Research Institutes for Disaster Risk Reduction. GADRI set up the following five objectives to be achieved (GADRI, undated).

- To establish a global research network
- To provide a research road map, and plans and organization of disaster research groups

P. Kovacs—Chair, Board of Directors of GADRI; and Executive Director.

H. Tatano et al. (Eds.): GSRIDRR 2023, DRRGADRIBS, pp. 14–24, 2025.
https://doi.org/10.1007/978-981-96-6609-6_2

- To promote capacity development of disaster research institutes and enhance researcher and student exchange
- To promote exchange and sharing of data and information for scientific research across the globe
- To serve as an advocacy organization to speak with one voice in an effort to influence decision-making processes.

GADRI also aims to provide scientific support for the effective implementation of global agendas such as the Sendai Framework for Disaster Risk Reduction, the Sustainable Development Goals (SDGs) and the Paris Agreement for climate change adaptation and mitigation. As of December 2022, 212 disaster research institutes worldwide have joined GADRI and it consists of 83 institutes (39%) in Asia excluding Japan, 37 institutes (17%) in North and South America (Americas), 37 institutes (17%) in Europe, 33 institutes (16%) in Japan, 12 institutes (6%) in Africa, 10 institutes (5%) in Oceania.

GADRI organizes the biennial global summit and its 6th summit was held from 15 to 17 March 2023 at Kyoto University, Japan. The summit gave a specific focus on confirming the progress of GADRI's five objectives and the contributions of GADRI to the Sendai Midterm review of Sendai Framework. This paper introduces the questionnaire survey results held for the GADRI member institutes to understand collectively the progress, achievement and gap of GADRI's objectives and Sendai Framework.

2 Methodology

The questionnaire consisted of three sections. Section 1 covered general information of each member institute such as the location of the institute, the number of faculty members and students, the total expenditure for the academic year 2021, and the research areas covered by each institute. Section 2 asked the questions related to the GADRI's objective such as if the research networks have been expanded, if participation in GADRI enhanced the accessibility to the different types of scientific data, if the capacity has increased as a result of participation in GADRI and so on. Section 3 asked about the GADRI's contributions towards Sendai Midterm review. If the institute is involved in the national, state and local platform for disaster risk reduction in order to utilize the scientific research outcomes, in what research disciplines they contribute to the platforms, what type of project the institute carries out to directly apply the scientific outcomes for the society and so on. The survey form has been sent via email to the member institute and each member institute responded in the period between November 2022 and January 2023.

3 Survey Results

3.1 Results Based on the General Information

The survey was responded by 85 institutes, which is account for 40% of all the member institutes. The proportion of the responding institute by region is 24 institutes (28%) in Europe, 22 institutes (26%) in Asia, 20 institutes (24%) in Americas, 13 institutes (15%) in Japan, 3 institutes (4%) in Oceania, 3 (4%) in Africa. In these responding institutes, there are 5,456 professors including associate professors, junior associate professors/lecturers and assistant professors are involved in the disaster related research. In terms of the gender ratio, 2,387 are female and 2,735 are male as Fig. 1 shows. 334 are not specified as some institutes just reported the total number of professors.

Fig. 1. Number of professors and gender ratio

Table 1 shows the gender ratio by region of professors including associate, junior associate professors/lecturers, and assistant professors. There are more male professors in Africa, Americas and Japan. Especially the gender gap in Japan is significant compared to other regions and it reflects clearly that Japan ranked at 116 in the Global Gender Gap Index 2022 rankings (World Economic Forum, 2022). On the other hand, it is almost equal ratio of female/male in Europe. Responding institutes in Europe included Sweden that ranked 5 in above index, Switzerland ranked 13 and UK ranked 22 (World Economic Forum, 2022) and equal opportunity of participation towards the DRR research is ensured. Even though the gender gap still exists in Asian countries according to the Index, number of female professors is larger than that of male. It may be a result of institutional efforts to ensure the female participation in DRR research, however, this data is self-reported by each institute and minor errors in typing may have happened. As a reference information, 67 female and 118 male have participated in 6th summit of GADRI.

Table 1. Gender ratio by region of professors including associate, junior associate professors/lecturers and assistant professors

Region	Total	Female	Male	Other
Africa	395	117	278	0
Americas	759	277	378	0
Asia	2,561	1,445	1,197	0
Europe	907	488	418	0
Japan	830	58	462	0
Oceania	4	2	2	0
Total	**5,456**	**2,387**	**2,735**	**0**

Figure 2 shows the number of students that belong to the responding 85 institutes. In total, 25,942 undergraduate course students study about the disaster related topics and its gender ratio is 6,999 female students, 10,167 male students and other, which did not identify themselves neither female or male is 1. Sum total of female, male and other does not match with the total number in the graph because some institutes only reported the total number without specifying the gender ratio. Number of male students tend to be larger than that of female not only for undergraduate course students, but also for Master's course and Doctoral course students. There are in total 5,220 Master's course students, out of which 952 are female and 2,286 are male. For Doctoral course students, there are 635 female, 1,065 male, 3 others and in total 2,591 are conducting the disaster related research.

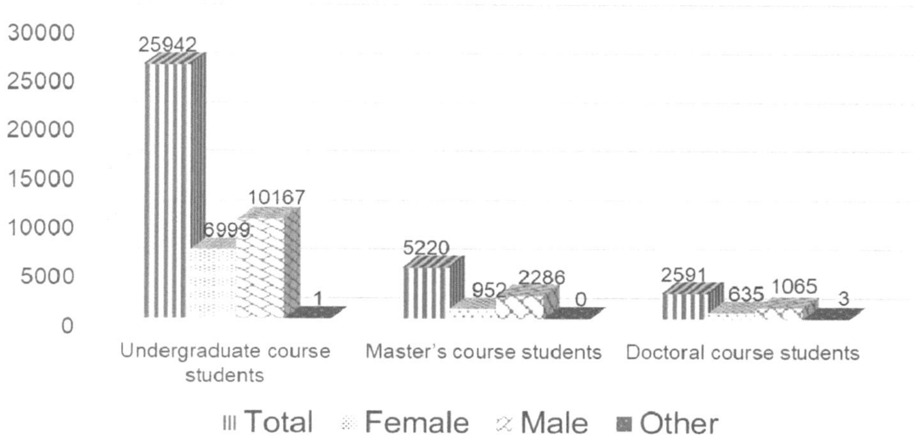

Fig. 2. Number of students and gender ratio (85 institutes)

Figure 3 shows the research areas that each institute cover to get overall ideas of disciplines that institutes contribute to in order to reduce the disaster risk. Basically, each institute cover multi-disciplines so that the number of covered research area that was responded by each institute were summed up. As a result, 37% account for hazard-related area such as geophysics, volcanology, hydrology, meteorology, climatology and so on. 19% account for engineering-related topics such as civil, coastal, environmental, mechanical engineering and so on. 43% is about social science related research such as economy, governance, sociology, psychology education and so on. Only 1% account for the health-related topic such as epidemiology and public health.

The amount of expenses also describes the scale of contributions to DRR. Thus, total expenses of academic year 2021 was asked to each institute and 48 responded to the question. Sum up total of 48 institutes is USD 916 million and its regional proportion is shown in Fig. 4. The number after the percentage in the Figure indicates the number of institutes responded. For example, 12 institutes responded to this question in Asia and it accounts for 44% of total amount. Even though the number of responding institutes is

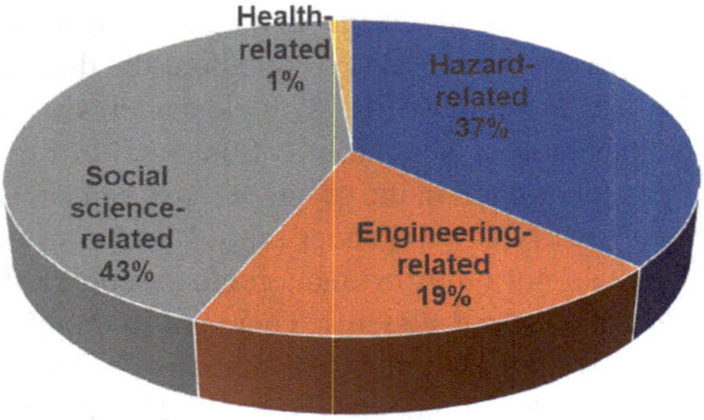

Fig. 3. Research areas that each institute cover

larger in Americas and Europe than Asia, nearly half of the expenses are occupied by Asia.

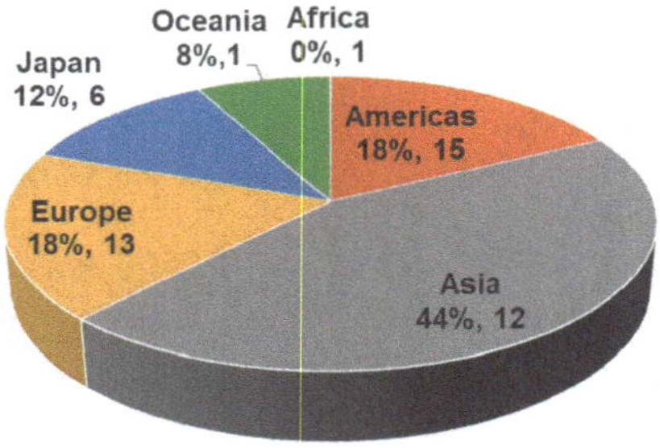

Fig. 4. Total expenses in USD by region

3.2 Progress, Achievement and Gap of GADRI's Objective

GADRI has five-fold objectives as mentioned in the introduction. Since its establishment, the alliance has carried out the actions such as organizing biennial global summits for networking, lecture series by prominent researchers, sending a voice of member institutes to United Nations' related initiatives, publishing the GADRI book series and so on to achieve the objectives.

Networking is one of the core-activities of GADRI and at the same time it contributes to the progress of Sendai Framework as it calls for coordinated efforts by the scientific community to deepen the understanding of disaster risks, promote evidence-based implementation, and transfer and disseminate the scientific knowledge and technologies in support of decision-making processes (UN, 2015). Since GADRI's establishment in 2015, 82 institutes of 86 institutes responded that the networks has expanded. Their networks expanded in different geographical scale as 8 institutes responded that the institute is connected to more researchers in their own country, 13 institutes responded to be connected more in their country and region, and 61 institutes were connected more not only in their country and region but also global level. Figure 5 indicates the important evidence that the networking has brought concrete actions. Total 31 institutes responded that the joint research with other institutes was developed through networking, 36 institutes responded that the interaction with researchers from other institutes was increased, and 15 institutes responded that the opportunities to interact with prominent researchers from other institutes increased.

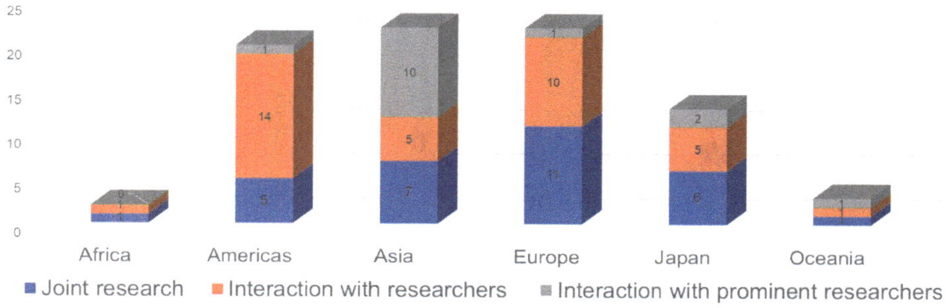

Fig. 5. Expansion of the institutional research network since 2015

GADRI also aims to promote the development of research roadmaps and plans of member institute. 26 institutes do not have institutional research plan, 14 institutes have the plan but is not published, and 43 institutes have the published multi-year research plan including long term goals. Apart from the development of research roadmaps and plans, the survey asked if institute's research plan was influenced by the global agendas including the Sendai Framework, Paris Agreement on Climate Change and SDGs. In fact, 60 institutes that account for 71% of responding institutes take into considerations three global agendas of Sendai Framework, climate change and SDGs as shown in blue bars in Fig. 6. Fourteen institutes rather focused on the Sendai Framework and 7 institutes focused on local and national issues that implies that the plan is not influenced by the global agendas.

Capacity building of the member institutes is also the important role of GADRI. Sixty-three institutes which is account for 74% of responding institutes considered that the capacity has increased since 2015 and increase in budget and creation of new positions for staff members including young generations were raised as the main reason of capacity development. Even though the GADRI offers the lecture series by prominent researchers and networking as above mentioned, these were not mentioned as the contributing factors for capacity building by responding institutes.

The fourth GADRI's objective is the enhancement of the accessibility to the different types of scientific data. Global Assessment Report 2022 (UNDRR, 2022) pointed out the lack in quality data that support the data-based decision-making for risk reduction and the enhancement of the accessibility to the data especially in developing countries are the urgent issues. GADRI also seeks to make the quality data available and usable to the member institutes, however, 51 institutes of 79 responding institutes answered that the GADRI participation does not enhance the accessibility so far. The institutes

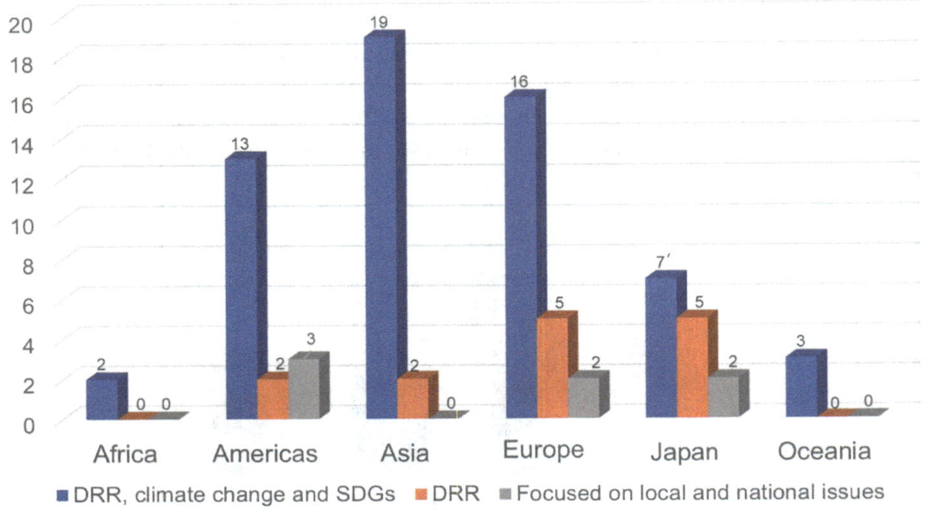

Fig. 6. Influence of the global agendas to the research plan

look for the accessible data and information such as building damage data, engineering and environmental assessment software, high resolution global climate model data, high resolution satellite image data, social vulnerability data and so on. On the other hand, 28 institutes answered that the accessibility has increased as they have access to climate change related data, comparative information among regions, data about potential partners and other networks, disaster map, remote sensing products, hydrological and meteorological data, information on research and scientific projects led by other institutes and seismological data for earthquake early warning shared by GADRI partners from time to time. So, these fact shows that there is a clear gap that only limited institutes have been able to be benefited from the accessibility and this has not widely recognized by the majority of the member institutes. It was also asked that if institute recognizes that GADRI makes one-voice of participating institutes to influence to and contribute to the decision-making process of DRR towards local and national governments as well as international organizations. 70 out of 87 institutes answered positive to this question.

3.3 Contributions to Sendai Midterm Review

The year 2023 that 6th global summit of GADRI is held coincides with the year for midterm review to confirm the progress of Sendai Framework. Especially the target E of the framework promotes the establishment and strengthening of national and local DRR platforms (UNDRR, undated) and involvement of scientific sectors are highly encouraged to reflect the DRR research for policy-making and decision-making. In fact, 74 of 87 institutes are involved in the national, regional and local DRR platforms in order to utilize the scientific research outcomes. Figure 7 shows its regional proportions of responding institutes.

Sendai Framework has four priorities of actions namely, Priority 1: understanding disaster risk, Priority 2: Strengthening disaster risk governance, Priority 3: Investing

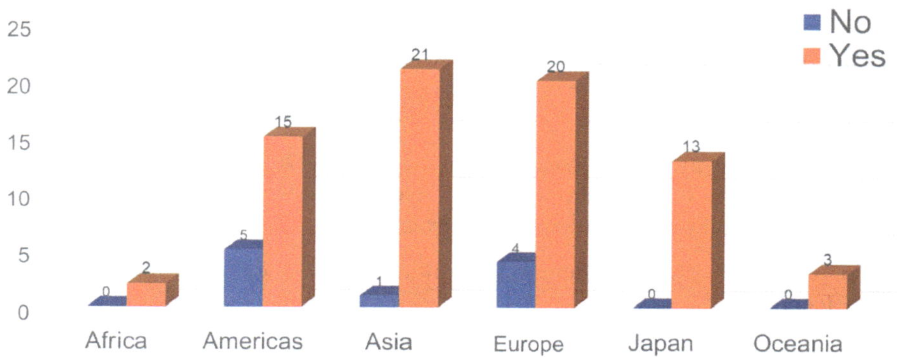

Fig. 7. Involvement of member institutes to the national, regional and local DRR platforms by region

in disaster risk reduction for resilience, Priority 4: Enhancing disaster preparedness for effective response and to "Build Back Better" in recovery, rehabilitation and reconstruction (UN, 2015). The survey asked how greater global progress can be achieved with respect to the four priorities for action set out in the Sendai Framework and followings are some major opinions from member institutes.

- Data sharing to ensure equitable access to disaster-related data at both regional and international levels
- Collaboration among researchers as well as multi-stakeholders including local front-line emergency manager to apply the scientific outcomes to the decision making
- Needs to foster "Facilitators," who can integrate and translate interdisciplinary scientific and indigenous knowledge for all stakeholders to cooperate in building social consensus
- Mechanisms for technology and knowledge transfer to developing countries
- Focus on community-based solutions and natural-based solutions
- Attention is needed on governance and financing issues. Needs more knowledge and frameworks in these fields to speed up the pace of battling the transboundary issues

The survey also asked what priorities for global action that emerged since the Sendai Framework has been identified based on the research and experiences of institutes. 56 institutes responded to this question with descriptive answer. Most of the comments were written with a few words so that it prevents from doing the details analysis. However, frequently utilized terms are "capacity", "local", "governance", "management" "early" and many institutes finds the necessary actions to promote the local level disaster risk management and capacity building as well as the strengthening of early warning. Some of the representative answers were extracted below.

- Policy recommendations build on local actions
- DRR knowledge is now important to be taken to the grassroot level, to the local communities including younger generations of different age groups.
- Early warning system and community empowerment for DRR
- Ease of access to relevant information from robust evidence

- Good progress on early warning and risk knowledge, also increasing bridging of gap between climate adaptation and disaster prevention, but now a strong focus is needed to understand systemic nature of risk and increase citizen awareness.
- Systemic risk governance

37 member institutes also contribute to the United Nations related international scientific committee of DRR and/or elaboration of United Nations related international report on DRR. Followings are some of the examples.

- Mid Term Review Sendai Framework Working Group-UN Major Group on Science and Technology
- Integrated Research on Disaster Risk (IRDR)
- IPCC Sixth Assessment Report
- UN Global Assessment Report on Disaster Risk Reduction (GAR)
- UN Global Risk Assessment Framework (GRAF)
- UNISDR Global Scientific and Technology Advisory Group (G-STAG)
- Making Cities Resilient 2030 (MCR2030)
- European Science and Technology Advisory Committee (ESTAG)
- WHO Health Emergency and Disaster Risk management Research Network
- UN Framework Convention on Climate Change (UNFCCC)
- UNESCO-IHP Regional Steering Committee for Asia and the Pacific (RSC-AP)
- UNCCD Science Policy Interface
- WMO/UNDRR Early Warning for All
- INFORM Annual Report
- International Flood Initiative (IFI)
- Asian Water Cycle Initiative (AWCI)

The survey also asked to the member institutes to provide examples of how research contribute to advancing disaster risk reduction and resilience. Descriptive answers were given by 64 institutes with total 1,984 words. In order to clarify the overall trend of examples, co-occurrence of the words has been explored with KH Coder, a quantitative content analysis software as shown in Fig. 8. It gives a broad idea that the institutes contribute to drought and flood modeling, city level capacity building, earthquake early warning, management of cultural heritage, risk management strategy etc.

Some of the concrete research examples raised by the institutes are listed below.

- Advancing resilience measures and understanding of the probability and vulnerability to floods and droughts. We are also collaborating to a great deal with actors on all levels, from local communities to global organizations.
- Decision making model for local government in flood risk preparedness and management
- Engagement with local government and agencies, publications targeted at practitioners. For example, developed the flood prediction technology for small-medium size rivers, consideration of "Short-time Calculation", "Low price", "Simple" and "Having necessary accuracy".
- It is developed earthquake hazard and risk models, software tools and data sets that are used by thousands of researchers, risk managers, engineers, governments and insurance companies, to assess and manage earthquake risk, including developing

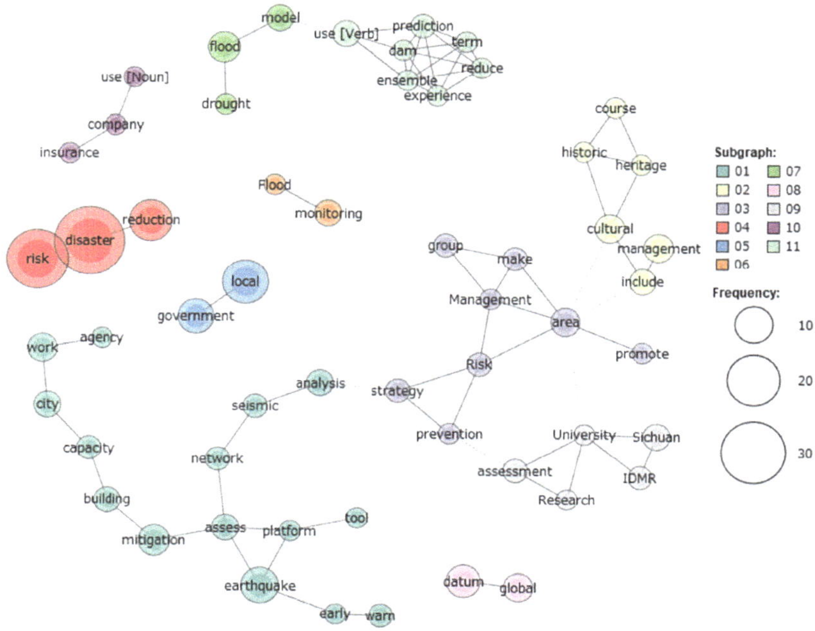

Fig. 8. Co-occurrence analysis of the contributions by member institutes

mitigation measures, building codes, earthquake alerts, building retrofits, insurance and reinsurance, catastrophe bonds.

- Landslide hazard and risk assessment, development of early warning systems for landslides, EO data for mapping and monitoring ground deformations
- Our institute has been working towards capacity building in the area of disaster risk management of cultural heritage as part of the UNESCO Chair Programme on Cultural Heritage and Risk Management
- The term 'action data', is targeted not on global commonalities for all-encompassing global policies but on uncovering the particular, distinct experiences unique to the communities themselves. Action data refers to data that effects change, this sort of data creates a voice that can be operational at all levels of human engagement in disaster prevention and response and forms a central pillar of the strategy for achieving the Sendai Framework

4 Conclusion

This paper introduced the questionnaire survey results that was held at the occasion of the 6th global summit of GADRI. It aimed to identify the progress, achievement and gap to the GADRI's five-fold objectives and contributions to the Sendai Midterm review of Sendai Framework. The survey was answered by 85 institutes, which is account for 40% of all the member institutes. In these institutes, there are 5,456 professors are advancing the DRR related research and gender ratio of professors in Europe and Asia is reaching to the equality while it shows the strong tendency in Japan that the professor positions are dominated by male. More than 30,000 students including undergraduate, Master's and Doctoral courses study about DRR related topics and conduct research. Sum up total of the annual expenses in the academic year 2021 responded by 48 institutes reached up to USD 916 million. These basic data imply the scale of contributions to the DRR and resilience of member institutes.

The survey also identified that GADRI functioned as a node of networking member institutes and contributed to promote joint research and exchange among researchers. Majority of the responding institutes take into considerations the three global agendas of Sendai Framework, climate change and SDGs to establish the institutional research plan. The institutes also recognizes that GADRI makes one-voice of participating institutes to influence to and contribute to decision-making process of DRR towards local and national governments as well as international organizations. While it is evident that GADRI brings benefit to the member institutes, the gaps exist. 51 institutes of 79 responding institutes feel that the GADRI participation does not enhance the accessibility to different types of disaster related data. In addition, GADRI offers the lecture series by prominent researchers and networking opportunities, these were not recognized by member institutes as a part of capacity building. Number of participants to the lectures by prominent researchers and networking opportunities are still limited so that it requires further advertisement and easy-to-attend method to these opportunities.

The member institutes are also strongly involved in the application of scientific research outcomes to the society. 74 of 87 institutes contribute to the national, regional and local DRR platforms. Especially, drought and flood modeling, capacity building, early warning and management of cultural heritage are some of the principal ways of contributions. In order to make the greater global progress to the Sendai Framework, institutes call for further data sharing initiatives to ensure the equitable access, needs of facilitators who can integrate and translate interdisciplinary scientific and indigenous knowledge to build social consensus, further community-based DRR and governance are raised. It is important to periodically carry out this type of survey targeting member institutes to clarify the temporal and spatial progresses for further collective actions as GADRI to achieve the sustainable disaster resilient worlds.

References

GADRI: Global Alliance of Disaster Research Institutes (undated), Aims. https://gadri.net/about/Aims/. Accessed 15 Apr 2023

World Economic Forum: Global Gender Gap Report (2022). https://www3.weforum.org/docs/WEF_GGGR_2022.pdf. Accessed 15 Apr 2023

UN: United Nations, Sendai Framework for Disaster Risk Reductions 2015–2030 (2015). https://www.undrr.org/publication/sendai-framework-disaster-risk-reduction-2015-2030. Accessed 1 Sept 2021

UNDRR: United Nations Office for Disaster Risk Reduction, Global Assessment Report on Disaster Risk Reduction (2022). https://www.undrr.org/media/79595/download. Accessed 15 Apr 2023

UNDRR: United Nations Office for Disaster Risk Reduction (undated), Sendai Framework Indicators. https://www.preventionweb.net/sendai-framework/sendai-framework-indicators. Accessed 1 Sept 2021

Plenary Sessions

Ms. Mami Mizutori, Special Representative of the Secretary-General (SRSG) for Disaster Risk Reduction, United Nations Office for Disaster Risk Reduction (UNDRR)

Wilma James(✉)

DPRI, Kyoto University, Uji Campus, Kyoto, Japan
James.wilmatheonesta.6r@kyoto-u.ac.jp

Ms. Mami Mizutori, Special Representative of the Secretary-General (SRSG) for Disaster Risk Reduction, United Nations Office for Disaster Risk Reduction (UNDRR).

Keynote Remarks (15 min).

Event: 6th Global Summit of GADRI "Plenary Session I: Listening to Various Stakeholders on Integration of Disaster Risks to Present and Future Disasters".

Time: 11:30- 12:30, Wednesday 15 March 2023.

Location: DPRI, Kyoto University, Uji Campus, Kyoto.

Distinguished Chairs, Professors Virginia Murray and Rajib Shaw,

Esteemed fellow speakers, Professor Masahide Kimoto and Professor Paul Kovacs, Ladies and Gentlemen, I am delighted to be with you again; this time to share the perspective of the UN Office for Disaster Risk Reduction on the current state of disaster risk management and what we can do together to strengthen long-term resilience.

Specifically, I am going to share with you some of the findings from the Midterm Review of the implementation of the Sendai Framework, and also share with you some highlights of the 2022 UN Global Assessment Report on Disaster Risk Reduction.

But at the outset, I would like to thank GADRI for this comprehensive survey. Many of the survey findings relate to my remarks today and it has given everyone a clear perspective of what GADRI is.

So first and foremost, I would like to share with you some of our thoughts on the current risk landscape. In 2019, when the GAR2019 was launched at the Global Platform it declared that the era of hazard-by-hazard risk reduction was over. This declaration was to show that the risk landscape is very complex, that risk is systemic, it crosses borders, and that risks are not siloed. Our approach to risk management was then very much siloed, and this is perhaps still true today.

Very soon after that, we all came to understand the meaning of systemic risk, cascading impact, and compounded disasters. The Sendai Framework was pioneering in this respect. When the Sendai Framework was adopted, there was a very strong feeling that biological hazards and the pandemics/epidemics caused by them would be the next big disaster, but not many governments put them into their strategies.

Since the beginning of the pandemic, uncertainty has become stronger. We are living in a world where we have to deal with not only systemic risk and complex risks, but also

H. Tatano et al. (Eds.): GSRIDRR 2023, DRRGADRIBS, pp. 27–30, 2025.
https://doi.org/10.1007/978-981-96-6609-6_3

a lot of uncertainty and we need to adapt our approaches to this. As a result, conventional risk management and crisis response systems are not working. We need to cope with this reality, but our systems are not there yet.

Most global systems of governance have been designed to address individual risks. This needs to change. It's not only governments, as I mentioned in my opening remarks. Research may also be a bit siloed. The private sector is also siloed. The United Nations system is siloed. We are all siloed and we need to change this.

If we continue in this way, the aims of the Sendai Framework will not be achieved. The SDGs will also not be achieved. Many of us are starting to feel pessimistic about this. The feeling that we can do it by 2030 is waning. We need to course correct now, and this is what the Sendai Framework midterm review process is all about.

What is the Sendai Framework midterm review telling us? It is encouraging that 72 member states have conducted a voluntary national review. As you all know the Sendai Framework is not a legal framework so UN Member States do not have to conduct voluntary reviews, but more than 70 countries did. This, I think is the first encouraging point. In doing so, even in cases where the reviews were not comprehensive or complete, many Member States were able to understand where they were. It is also very encouraging that 25 UN agencies conducted their own voluntary reviews. In 2015, and even in 2018 when I took on this role, the UN system was not rallying around disaster risk reduction or prevention. However, with the onset of Covid-19, and also with the climate emergency, many UN agencies, including both the development agencies and humanitarian agencies UNDP, FAO, WHO, UNHCR, UNICEF among others, they are all understanding that if they do not embed disaster risk reduction at the center of their activities, and if they do not take risk informed decisions our work as a system will not be sustainable.

We also had more than 25 reviews from various stakeholders. I know that this community was very engaged with it. As the Sendai Framework says, we need to have whole-of-society implementation of the Sendai Framework and the fact that so many stakeholders conducted reviews was also very encouraging. I take this as a big step forward. The Sendai Framework was the poor cousin of all the 2015 agreements. Many people did not know about it. Many people did not know about the UN Office for Disaster Risk Reduction. This has changed as we experienced so many tragic events recently.

What are the findings of the midterm review? There are some areas of great progress. In terms of risk governance, in 2015 only 55 countries said they had national strategies for disaster risk reduction. Now 125 countries say they have a national strategy for disaster risk reduction. I think this is very positive. We always say that a failure to plan is a plan for failure. Certainly the content of these plans is important, but it is important progress to see so many countries now have DRR plans. 110 countries now report that they have established a system to account for loss and damage. This is also very important. If you do not know what you are losing, you cannot make policies to improve your systems. Now 155 countries are submitting their data to the Sendai Framework Monitor. There is a difference in the amount of data that they submit, but in one way or another 155 countries are contributing data. This is also very positive.

I would like to share with you some of the specific findings and recommendations that are coming out from the midterm review process.

I invite you to read the full report for the complete findings, but I will focus now on sharing some of the report's recommendations under the four Priorities for Action of the Sendai Framework.

First, under Priority 1, to better 'understand disaster risk,' the report recommends that high-quality data on disaster risk reduction be made available for planners and practitioners. There has been a lot of discussion at this conference about data. The important thing is to translate this information into risk informed policies. This part is still lacking. This means improving the standards for the collection and analysis of official risk data, and broadening the application of risk assessments. The goal is for risk data to be accessible, interoperable, and integrated into decision-making at all levels. Risk-informed decision making is the key to achieving the SDGs.

Under Priority 2, 'strengthening disaster risk governance', at the national level we are doing very well but we are not doing as well at the local level. The number of countries which report that they have cities with local disaster risk reduction strategies is still very low. The reality is that during the time of Covid-19, response to Covid was tested at the local level. Prevention must be strong at the local level.

At the national level, many governments are not yet taking a whole of government approach. In many countries including in the global north, disaster risk management is still a matter of the disaster risk management agency. This is not enough. If we do not take risk informed decisions in every governmental sector, it is not going to work. We saw the effects of this during the pandemic in the education system, in the agriculture system, in the health system of course. Even as 90% of all disasters are now climate related, DRR and climate are not as comprehensively integrated as they should be.

Priority 3 is about 'investing in disaster risk reduction,' and it presents the most challenges. The report calls for a reimagining of the fundamental relationship between the economy, environment and society to better finance risk reduction and de-risk investments. We do not see a sufficient flow of money going to resilience. We are talking about many things, including national budgets, international cooperation. We need to make sure there is a flow of money going toward building resilience. If money keeps going to creating new risks then we are failing.

Finally, under Priority 4 which includes better preparedness and building back better, preparedness is improving. This is shown in the fact that reduction of mortality is the only global target that is succeeding. There is a trend of reduction in global mortality, if you take out the number of people who died from Covid-19. There are many reasons, but the development of early warning systems in more countries explains this in part. However, when it comes to building back better, we have not seen evidence that countries are using the opportunity of recovery to build resilience. This is because the review found that significant opportunities are being missed to build back better from disasters.

This is just a summary of some of the recommendations from the report. The full report will come out at the end of this month, and it will include best practices from the Member States' national reviews. On 18 and 19th of May, the President of the General Assembly will convene a high-level meeting in New York. The political declaration will be adopted by UN Member States. Yesterday, the consultation among the Member States started. It is an intense process. The Member States care about the outcome, and feel the need to come up with a political declaration that can guide us through the next seven years.

Next, I would like to share with you highlights from the 2022 GAR report, which along with the midterm review findings, can guide us to build long-term resilience.

The report asks a central question: how can decision-making systems evolve to better address current and future risks? One of the determinations of the GAR2022 is that we need to apply metrics that reflect economic, planetary and societal health. We are not sufficiently measuring the right data for decision-making and the major drivers of risk.

The GAR 2022 also identified the importance of how we communicate about risk. We have seen bad examples of risk communication during Covid-19, illustrating that many lives can be lost as a result. I recently participated in a risk communication conference in Romania. All the participating governments gave examples of how difficult it was for them to communicate about the risk of covid and to encourage their citizens to protect themselves against infodemics. How we communicate about risk is very important, and when we do this we need to think about how our minds work. Cognitive research shows how decisions regarding disasters are often influenced by short-term thinking, and unfounded expectations that risk is not near us either in relation to geography or time. That is how people want to perceive risk. We need to find ways that people can associate the risk to their lives and surroundings and be able to communicate.

The GAR2022 also found that our governance systems and financial systems are not fit for service on prevention and building resilience. We need to be able to come up with a system of governance that is more participatory and more transparent. We always say we need to involve all stakeholders. I saw in the survey that many of you feel like you are involved in the conversation in your region your country, and you're a part of the national platform. But I think that this is something we need to strengthen even further. We always say vulnerable people need to be protected but we need an approach that recognizes why they are vulnerable, and that their voices are not integrated into disaster risk reduction policies and other policies.

We are in the process of coming up with a special report of the GAR, which is looking at the linkages between the Sendai Framework and the SDGs. We hope to publish it this year because in September there will be an SDG summit. The next GAR 2025 will be launched at the Global Platform 2025. This will be an important Global Platform – the first since the MTR process.

I would like to conclude by thanking you for your commitment to your community and caring for the resilience of your people. As we saw in the survey we must caring for others as well, beyond our countries and communities. We always talk about the importance of the interface between science and policy. We always talk about the importance of basing policies on evidence and science. I do truly believe that now when we are challenged by so many risks and so many people who want to influence and counteract the evidence, the importance of your community, wisdom and experience have never been as important as today.

I once again thank you for your commitment to disaster risk reduction and look forward to learning of the outcomes of your deliberations.

Thank you.

Important Features of the February 6, 2023 Earthquakes in Türkiye

M. Erdik[1](✉), M. B. D. Tümsa[2], A. Pınar[1], and E. Altunel[3]

[1] Kandilli Observatory and Earthquake Research Institute, Department of Earthquake Engineering, Boğaziçi University, İstanbul, Turkey
erdikm@gmail.com
[2] Turkish Earthquake Foundation, Istanbul, Turkey
[3] Engineering Faculty, Department of Geology, Eskişehir Osmangazi University, 26040 Eskişehir, Turkey

Abstract. On Feb. 6, 2023, a magnitude Mw7.8 earthquake struck in south-central Türkiye, followed by a magnitude Mw 7.7 event about 9 h later. The paper provides a general view of the important and relevant characteristics of the tectonics, seismicity, strong ground motion, near-fault and basin effects, liquefaction, building codes, and damages to buildings and infrastructure with appropriate references to field reports, related papers, and presentations. The financial losses and earthquake insurance applications are also encompassed. The paper further elaborates on the reasons behind extensive damages and associated casualties highlighting the non-code-compliance issue. Important findings, conclusions, and suggestions for the ways and means of reducing earthquake losses in future earthquakes are also included. The paper also provides a comprehensive reference to the investigations, studies, papers, reports, articles from the media, and presentations.

Keywords: February 2023 turkey earthquakes · strong ground motion · earthquake damage

1 Introduction

Major earthquakes, associated with the East Anatolian fault system, struck south-central Turkiye on February 6, 2023 (Schmitt et al., 2023). The Kahramanmaraş Earthquake Sequence consisted of the M7.8 Kahramanmaraş event at 04:17 AM local time, M6.7earthquake near the M7.8 event (11 min after the M7.8 event), and, the M7.7 Elbistan event at 01:24 PM local time (within 9 h). These earthquakes were felt as far away as Syria, Cyprus, Greece, Jordan, Lebanon, Iraq, Georgia, Armenia, Egypt, and Israel. The following 3-month aftershock activity encompassed 3 events with M6-M7 and 49 events with M5-M6 (Seismological Bulletins, 2023). M7.8 earthquake that ruptured more than 300 km fault with fault offsets more than 4 m, was physically very similar to the 1906 San Francisco earthquake, where over 80% of the city was destroyed.

Figure 1 shows the epicenters of these two earthquakes and the other important earthquakes of this century, together with the number of lives lost, on a base map that

H. Tatano et al. (Eds.): GSRIDRR 2023, DRRGADRIBS, pp. 31–67, 2025.
https://doi.org/10.1007/978-981-96-6609-6_4

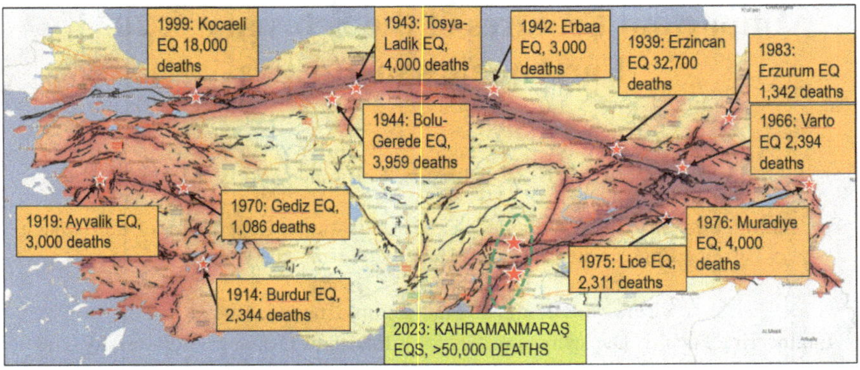

Fig. 1. The epicenters of the February 6 (Kahramanmaraş) earthquakes and the other important earthquakes of this century, together with the number of lives lost.

includes fault lines and the earthquake hazards indicated with shades of red color (EERI Buildings Team, 2023).

The earthquakes were well recorded both in near-fault conditions and at far-field conditions by more than 500 stations of the Turkish Strong Motion Network. A unique set of strong motion records was recorded that will provide data for the analysis of directivity, near-fault, and basin effects. As of May 2023, on the basis of governmental announcements, the death toll from the Türkiye earthquakes has reached 50,783, with more than 107,204 injured and 297 missing. At least 15.73 million people and 4 million buildings were affected, 345,000 buildings were destroyed and more than 2 million residents in the affected provinces were evacuated to nearby provinces (Cumhuriyet Newspaper, 2023). The earthquakes caused about $103.6 billion of direct damage (TERRA, 2023; GEER Association report, 2023; TC-CSBB, 2023; USGS-Podcast, 2023; Erdik et al., 2023a).

This paper is intended to update the preliminary report (Erdik et al., 2023b; Feb 23, 2023) of the authors to provide correct information regarding the geo-tectonic and seismic characteristics of these earthquakes, a study of the strong ground motion. A review of the structural damage observed in the region with a discussion of the reasons for such extensive damage and consequential casualties as well as suggestions for the avoidance of such losses in future earthquakes.

2 Seismotectonics

The East Anatolian Fault Zone (EAFZ) is one of the major active tectonic structures of the Eastern Mediterranean region (Fig. 2a). It constitutes the southeastern boundary of the westward escaping Anatolian Block and accommodates the relative motion between the Anatolian Block and the Arabian Plate (e.g. McKenzie, 1972; Şengör et al., 1985; Dewey et al., 1973; Barka and Kadinsky-Cade, 1988; Barka and Reilinger, 1997). The EAFZ is a left-lateral strike-slip fault that is about 550 km in length and runs from Karlıova in the northeast to the Mediterranean Sea in the southwest. The northeastern termination of the EAFZ is in Karlıova (Fig. 2b) where it intersects with the right-lateral North Anatolian

Fault Zone (NAFZ). The fault zone between Karlıova and Çelikhan consists of NE-SW trending faults and there is no significant fault branch separating from the zone. In the southwest of Çelikhan, an E-W-trending left-lateral fault (Çardak-Sürgü Fault of Emre et al. (2013)) splits from the EAFZ and extends towards Göksun (Fig. 2b). Further southwest of Çelikhan, the EAFZ extends along a narrow corridor up to Pazarcık where almost N-S-trending Narlı Fault splits towards the south (Fig. 2b). While there is general agreement on the main trace of the East Anatolian Fault Zone (EAFZ) between Karlıova in the northeast and Türkoğlu in the southwest, the fault's continuation beyond Türkoğlu remains a topic of debate. Researchers propose the following three main perspectives on the southwestern extension of the EAFZ.

1. Intersection with the Dead Sea Fault Zone (DSFZ) near Türkoğlu:
   According to this view, the EAFZ intersects with the DSFZ near Türkoğlu, crossing the Amanos Mountains and extending into the İskenderun Bay. This interpretation is supported by several studies (e.g., McKenzie, 1970, 1972, 1976; Dewey et al., 1973; Şengör, 1980; Hempton, 1987; Jackson and McKenzie, 1984; Barka and Kadinsky-Cade, 1988; Kempler and Garfunkel, 1991; Westaway and Arger, 1996; Koçyiğit and Erol, 2001; Yönlü et al., 2017).
2. Extension along the Karasu Segment to the Amik Basin:
   This perspective suggests that the EAFZ extends along the Karasu Segment within the Karasu Valley, heading south to meet the DSFZ in the Amik Basin. Proponents of this view include Allen (1969), Arpat and Şaroğlu (1975), Kelling et al. (1987), Şaroğlu et al. (1992), Şengör et al. (1985), Över et al. (2004), and Duman and Emre (2013).
3. Termination near Türkoğlu:
   A third interpretation posits that the EAFZ ends near Türkoğlu without extending further southwest. In this scenario, it connects with the DSFZ via the Karasu Fault. This hypothesis is supported by Lovelock (1984), Muehlberger and Gordon (1987), Perinçek and Çemen (1990), Chorowicz et al. (1994), and Yürür and Chorowicz (1998).

Various criteria, including geological evidence (e.g., Arpat and Şaroğlu, 1975; Westaway and Arger, 1996; Yönlü et al., 2017), GPS measurements (e.g., McClusky et al., 2000; Reilinger et al., 2006), plate kinematic analyses (e.g., Lyberis et al., 1992; Yürür and Chorowicz, 1998), and seismological studies (e.g., Taymaz et al., 2004), indicate that the slip rate along the East Anatolian Fault Zone (EAFZ) ranges from 4 to 31 mm per year. The total displacement on the EAFZ varies depending on the features displaced by the fault. For instance, cumulative left-lateral offsets of up to 27 km have been estimated based on displaced geological markers (e.g., Arpat and Şaroğlu, 1975; Şaroğlu et al., 1992). A leftward shift of the Euphrates River southwest of Elazığ was identified as a left-lateral displacement by Arpat and Şaroğlu (1975), who estimated a cumulative offset of 15 km. Additionally, Yönlü et al. (2017) proposed that the Aksu River valley has been offset left laterally by approximately 16.5 ± 0.5 km along the EAFZ.

The EAFZ is the source of large earthquakes both during the instrumental period and historical times. The southwestern parts of the EAFZ were reactivated with two large earthquakes on February 6, 2023. During the first event (Mw 7.8), the Narlı Fault,

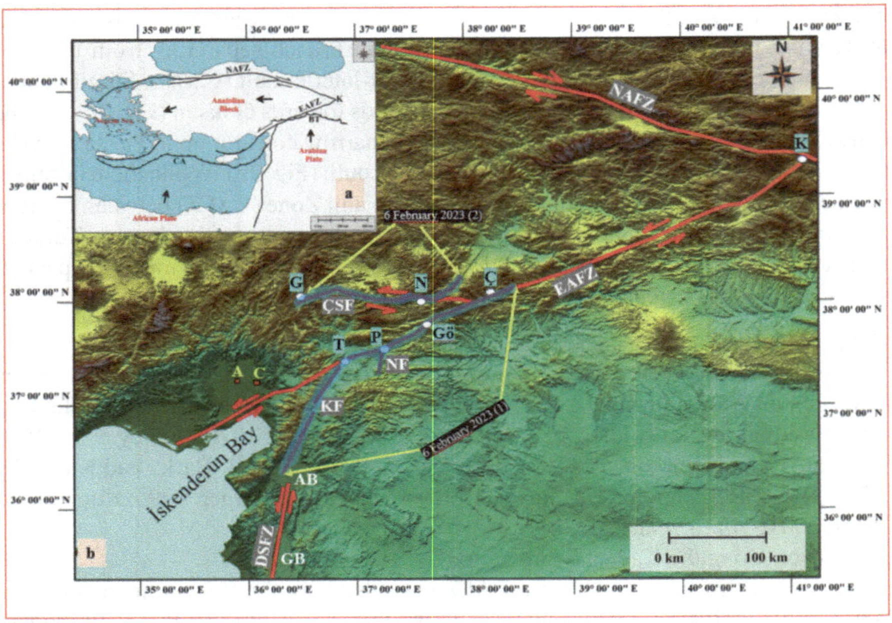

Fig. 2. (a) The seismotectonic framework of the Eastern Mediterranean region with the movement of Africa, Arabia, and Anatolia relative to Eurasia. (NAFZ: North Anatolian Fault Zone, EAFZ: East Anatolian Fault Zone, BT: Bitlis Thrust, CA: Cyprus Arc, K: Karlıova). (b) Map of the East Anatolian Fault Zone, showing the major active fault segments in red lines. Shaded lines show reactivated segments during the February 6 events. NAFZ: North Anatolian Fault Zone, EAFZ: East Anatolian Fault Zone, DSFZ: Dead Sea Fault Zone, ÇSF: Çardak-Sürgü Fault, NF: Narlı Fault, KF: Karasu Fault, K: Karlıova, Ç: Çelikhan, N: Nurhak, G: Göksun, Gö: Gölbaşı, P: Pazarcık, T: Türkoğlu, AB: Amik Basin, GB: Ghab Basin, A, and C in yellow are ancient cities of Anazarbus and Castabala, respectively. Map produced from SRTM Worldwide Elevation Data (1-arc-second resolution).

Türkoğlu-Çelikhan portion of the EAFZ and Karasu Fault ruptured (Fig. 2b). Field observations show that the surface rupture extends between Suvatlı village in the southwest (about 7 km south of the Hatay Airport in the Amik Basin) and around Mutlu village (about 15 km northeast of Çelikhan) in the northeast (Fig. 2b). The maximum horizontal displacement is about 3 m on the Narlı Fault, 6 m on the Türkoğlu-Çelikhan portion of the EAFZ, and 4 m on the Karasu Fault. The second event (Mw 7.7) took place on the Çardak-Sürgü Fault (Fig. 2b), the surface rupture extends approximately E-W between Göksun and east of Nurhak and it extends in NE-SW direction from this location further east. The maximum horizontal displacement is about 8 m on the E-W trending portion of the rupture. Before the 6 February 2023 earthquakes, some moderate-sized earthquakes (e.g. 1905, M:6.8 Malatya; 1971, Ms:6.7 Bingöl; 1986, Ms:6.0 Sürgü; 2003, Ms:6.4 Bingöl; 2010, M: 6.0 Karakoçan–Elazığ; 2020, M: 6.8 Elazığ earthquakes) occurred on the EAFZ further northeast of Çelikhan (Tan et al., 2008) but no surface rupture was reported.

Several surface rupturing events reportedly took place on the EAFZ in the 19th century (e.g. 1866 M:6.8, 1874 M: > 7.1, 1874 M:6.7, 1893 M:7.1 earthquakes) and Ambraseys (1989) claims that two-thirds of the fault zone was ruptured during the 19th-century earthquake activity. According to Ambraseys (1989), an earthquake took place in 1544 on the Çardak-Sürgü Fault, which splits from the EAFZ in the southwest of Çelikhan. Earthquakes in 1114 and 1513 likely occurred in the southwestern part of the fault zone and caused great damage in northern Syria and eastern Anatolia (Calvi, 1941; Soysal et al., 1981; Ambraseys, 1989; 2004). There is no account in historical documents about the magnitude of the 1114 event, but Ambraseys reports that the 1114 earthquake affected northern Syria and southeastern Turkey. Paleoseismological data suggest that the 1513 earthquake took place on the Türkoğlu - Pazarcık portion of the EAFZ while the 1114 event occurred northeast of Pazarcık (Yönlü 2017) and these two earthquakes were the last large events before the 6 February 2023 earthquakes.

The instrumental and historical earthquake activity of the region between Türkoğlu and İskenderun Bay is summarized by Yönlü et al. (2017) as follows. Several large or moderate-size historical earthquakes are reported in the area; for example, the 526 AD and 561 AD earthquakes caused great damage to the ancient cities of Anazarbus and Castabala and these major cities were abandoned after these events. A destructive earthquake also took place in 1269 and more than 60,000 died in the Adana Basin. However, there is insufficient information in the literature to attribute these events to a specific fault segment. On the other hand, based on paleoseismic studies in the southwest of Türkoğlu, Yönlü et al. (2017) revealed that there was a large earthquake after 425–570 AD that could be attributed to one of the 526 AD, 561 AD, or 1269 historical events. Regarding the instrumental period, the 27 June 1998 Ceyhan earthquake (M:6.2) is the largest event in the north of İskenderun Bay. The epicenter location and the source mechanism of the mainshock are well constrained by Aktar et al. (2000) providing clear evidence of pure left-lateral strike-slip motion. Considering the Amik Basin area, there is no surface rupturing event in the instrumental period, but some historical earthquakes are involved in surface faulting. The 1408 earthquake (M = 7.5) took place at the northern end of the DSFZ and was associated with about 20-km-long surface rupture (Ambraseys and Melville, 1995; Sbeinati et al., 2005a; Akyüz et al., 2006). The 13 August 1822 (M = 7.4) event occurred in the Karasu Valley (Ambraseys 1989). Ambraseys (1989) reports that the 1872 April 3 earthquake (M = 7.2) occurred in the Amik Basin and generated heavy damage around the former Amik Lake.

3 Seismological Characteristics

3.1 Historical Earthquakes

A reliable study on the historical earthquakes in the region is prepared by an expert team (Sesetyan et al., 2023) in the time-window 1000–1900. A list of these historical events is provided in Table 1 and Fig. 3.

In addition to these historical events, the May 526 Antioch (today. Antakya) earthquake that created extensive damages, was followed by a fire that destroyed most of the buildings left standing after the earthquake, killing about 250,000 people (NCEI/WDS, 2022; Sbeinati et al., 2005b).

Table 1. List of the Historical Earthquakes (Şeşetyan et al., 2023)

Y	M	D	AREA	N MDP	LAT	LON	Mw	Unc	SOY (int)	SHETA Mw	SBE Mw	CFT Mw	TAN Mw	SES Mw	DUEM Mw
1003			al-Thughur	9			6.63	0.32	8					6.21	
1114	8	10	Antioch						9	8.0			8.1	6.77	
1114	11	13	Misis	2	36.954	35.600	6.63	0.32			7.4	7.8			6.3
1114	11	29	Antioch, Maraş	21	37.311	37.423	7.75	0.42			7.7			6.21	
1138	10	11	Atharib	10	35.844	36.680	7.18	1.13	8	7.4	7.4	7.5	7.4		
1269	5		Sis	8	36.936	36.265	7.22	0.46				6.4			6.3
1344	1	3	Aintab	9	37.265	37.907	7.55	0.43	7	7.6	6.8	7.3			
1514			Malatya, Adana	3	37.500	36.500			7				7.5		7.4
1544	1		Zeytun	4	38.034	36.995	7.00	0.75	8	6.4			6.8	6.21	6.8
1795			Maraş	1	37.577	36.928	6.63	0.32							
1822	8	13	Antakya	47	36.391	36.841	7.74	0.38	9		7.0			7.43	7.5
1872	4	3	Amik Golu	33	36.131	36.015	7.01	0.52	9		5.9			7.15	7.2
1893	3	2	Malatya	49	38.154	38.293	6.84	0.45		7.0			7.1	7.10	7.1

Y = year; M = month; D = day; Area = most damaged area; NMDP = number of macroseismic data points; LAT, LON = latitude, longitude of the epicentre (this study); Mw, Unc = Mw and uncertainty (this study); SOY (int) = Intensity by Soysal et al, 1981; SHETA Mw = Mw from Shebalin and Tatevossian (1987); SBE Mw = Mw from Sbeinati et al. (2005); CFT Mw = Mw from Guidoboni et al. 2019; TAN Mw = Mw from Tan et al. (2008); SES Mw = Mw from Sesetyan et al. (2013); DUEM Mw = Mw from Duman and Emre (2013).

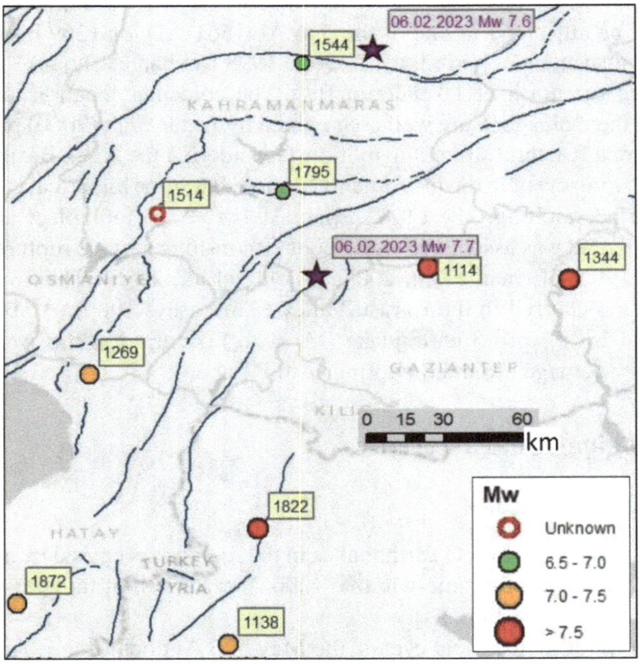

Fig. 3. Epicentral locations of the earthquakes.

3.2 Foreshocks, Mainshocks, and Aftershocks

The aftershock distribution associated with Kahramanmaraş Mw7.8 and Ekinözü Mw7.7 mainshocks indicates that the earthquake sequence was associated with ruptures on several fault segments (Fig. 4). Moreover, it is obvious from their elongations that each aftershock swarm corresponds to seismic activity associated with ruptures on fault segments having different fault geometries. The ruptures associated with such a complex fault network probably have contributed to generating the observed peak ground acceleration. Firstly, the multiple fault rupture nature of the mainshocks was obvious soon after examining the recorded complex near-field and far-field seismic waveforms (Erdik et al., 2023; Okuwaki et al., 2023; Peterson et al., et al., 2023; Zahradnik et al., 2023). Then, more detailed rupture extents were retrieved from the geodetic data showing non-uniform slip distribution along the rupture zone (Barbot et al., 2023; Hussein et al., 2023; Meng et. al., 2023). Furthermore, geological field observations carried out along the surface ruptures illuminated much finer details on the ruptured dynamics of large transform faults such as EAFZ (e.g., Karabacak et al., 2023).

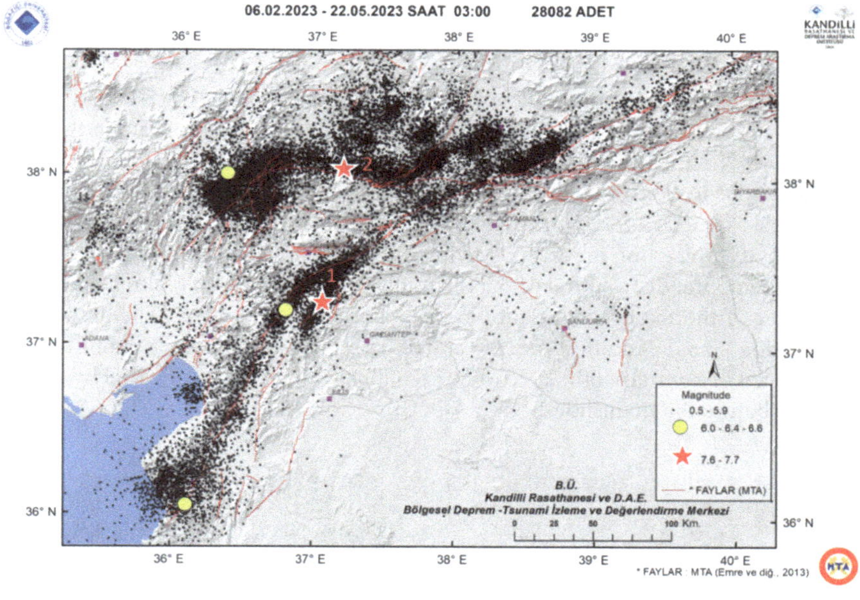

Fig. 4. Source rupture area of the February 6, 2023, major earthquakes and the associated aftershock activity between February 6 and May 22nd, 2023. The red star number-1 indicates the epicenter of the Mw7.8 event and the red star number-2 is the epicenter of the Mw7.7 Ekinözü earthquake. The yellow solid circles denote the location of M > 6 aftershocks and the black dots portray the smaller aftershock locations (Aftershock map by Kandilli Observatory and Earthquake Research Institute).

The aftershock lineament along the Narlı fault is marked by a red number 1 in Fig. 5. It is now a well-known fact that the ruptures during the February 6, 2023 earthquake sequence nucleated on Narlı off-fault located to the east of the NE-SW extending East Anatolian Fault Zone (EAFZ). The teleseismic modeling of the initial pulses of the P-waves yields a moment release equivalent to Mw6.8 moment magnitude for the rupture on the Narlı fault (Fig. 5). The modeling of the complex initials P-wave pulses of the teleseismic body waves requires two subevents to get a satisfactory fit between the observed seismograms and the synthetics. The first subevent corresponds to a rupture on the Narlı fault while the second subevent corresponds to a fault rupture on the EAFZ inferred by taking into account the NE-SW orientation of the nodal planes (Fig. 5). The moment tensor retrieved for the rupture on the Narlı fault shows predominantly strike-slip faulting with the normal faulting component. The left-lateral strike-slip motion with a significant normal faulting component of the 1st subevent is also retrieved from the deformations observed along the railroad (Fig. 5, right panel).

Spatial and temporal modeling yields a 5-s time difference between the two subevents. The rupture initiation point on the Narlı fault being located about 30 km away from the EAFZ requires super-shear rupture propagation velocity or a rupture triggered by P-wave propagation considering a seismic velocity close to Vp≈6 km/s. A supershear rupture velocity is currently one of the most debatable issues regarding the geodynamic modeling of rupture propagation associated with the Kahramanmaraş Mw7.8 and Ekinözü Mw7.7 mainshocks (Abdelmeguid et al., 2023; Melgar et al., 2023; Rosakis et al., 2023).

The NE-SW-oriented nodal plane in the focal mechanism of the second subevent is in-line with the strike of the EAFZ. Also, the focal mechanism of an earthquake of magnitude Mw4.2 that took place three days before the main shock (February 3, 2023, 11:05) is similar to the focal mechanism of the second subevent (Fig. 5). The Mw4.2 event was classified as a foreshock (Kusky et al., 2023). A more detailed analysis of the seismic activity preceding the Mw7.8 mainshock is handled by Kwiatek et al. (2023). Spatial and temporal distribution of the seismicity observed within 65 km distance from the epicenter of the mainshock identifies several clusters of events. The authors show that the earthquake preparation process was initiated approximately 8 months prior to the occurrence of February 6th, 2023, Mw7.8 mainshock and was followed by a seismic quiescence before the mainshock.

Fig. 5. Spatial and temporal modeling of the initial pulses of the teleseismic body waves of the Mw7.8 Kahramanmaras earthquake is achieved with two subevents. The location and focal mechanism of the foreshock on February 3, 2023, 11:05, Mw4.2 is also shown on the left panel. The left-lateral strike-slip faulting with a significant normal faulting component associated with the 1st subevent is also retrieved from the deformations observed along the railroad (right panel; photo taken by the ITU field investigation team lead by Prof. Serdar Akyüz).

Having occurred 9 h after the Kahramanmaraş M7.8 earthquake it was a matter of investigation whether the Ekinözü Mw7.8 earthquake was triggered by the first main-shock (Chen et al., 2023; Karabulut et al., 2023; Stein et al., 2023). All the studies estimated a significant stress increase at the epicentral area of the second mainshock resulting from the ruptures of the first mainshock. Erdik et al. (2023a) derived a detailed slip model for the first and second mainshocks. Incorporating the slip distribution model of February 20, 2023, major aftershock (Mw6.4) in Hatay a Coulomb stress change model was established (Fig. 6). The correlation between the M > 4 aftershock locations and the stress increase are shown as a red color background image is obvious. A significant stress increase is estimated along the eastern flank of Karasu Valley where the faults are left unbroken during the M7.8 mainshock.

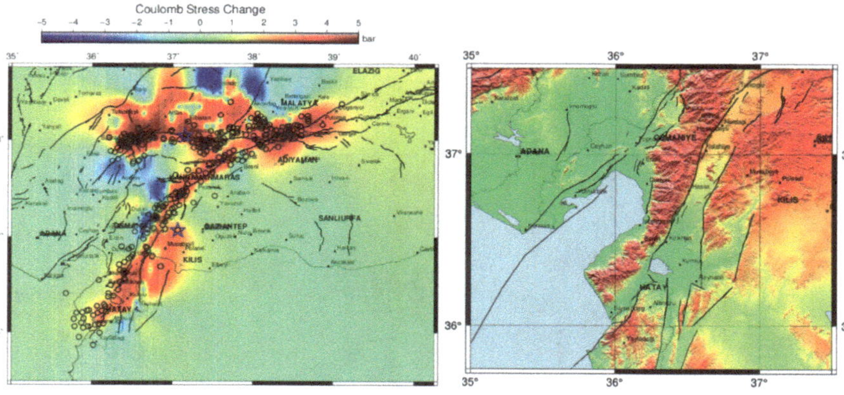

Fig. 6. Integrated Coulomb stress changes associated with the Kahramanmaraş Mw7.8 and Ekinözü Mw7.7 mainshocks including the effects of the largest aftershock Mw6.4 on February 20 near Hatay. The circles denote the aftershock location lager than M > 4 (left panel). Note that the aftershocks are taking place along the western flank of the Karasu valley extending between Hatay and Turkoğlu. The western and eastern flanks of Karasu valley have distinct morphological features (right panel).

4 Strong Ground Motion

The general information and seismic records of earthquakes are from AFAD, Turkey (https://tadas.afad.gov.tr/list-event). The AFAD database has a total of 402, 474, and 795 ground-motion records for earthquakes with magnitudes of Mw6.7, Mw7.7, and Mw7.8, respectively. It should be noted that AFAD has changed the contents of its web page several times. Reviews were performed to identify and remove records with non-usable data, which include noise-dominated, unclear event signals, early termination, late trigger, instrument malfunction, and spike records (such as GEER-EERI Report, 2023).

PGAs up to 1.2 g and PGVs up to 216 cm/s were recorded. The highest PGA values are recorded in Antakya, Hatay. High vertical ground motions were observed. The analyses of the excellent set of records obtained will serve to understand the attenuation, near-fault directivity, and basin amplification effects.

The horizontal peak ground acceleration (PGA, measured in cm/s^2) and peak ground velocity (PGV, measured in cm/s) associated with the magnitude-7.8 earthquake are illustrated in Fig. 7 and Fig. 8. The fault rupture is marked by white lines in these figures. Notably, the PGA and PGV values are significantly higher in Hatay Province, located to the southwest, likely due to the combined effects of basin amplification and rupture directivity. Similar visualizations for the magnitude-7.5 earthquake are presented in Fig. 9 and Fig. 10.

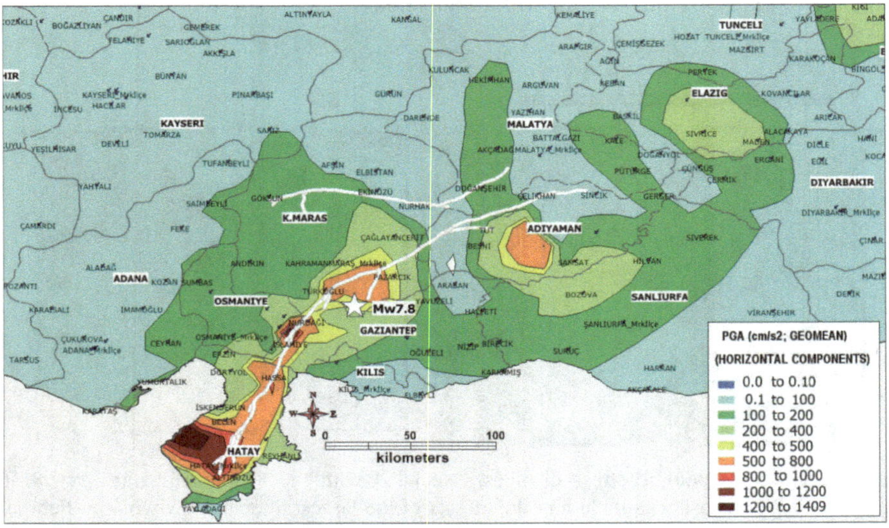

(a)

Fig. 7. Contour map of the peak ground acceleration (in units of centimeters per second squared; maximum values of north-south and east-west horizontal components) associated with the magnitude-7.8 earthquake. (Erdik et al., 2023b) Surface rupture taken from USGS.

A spatially interpolated geometric mean of the peak ground acceleration for the fault-parallel and normal components of the strong ground-motion observations for the M7.8 earthquake (Mai et al., 2023) is provided in Fig. 11.

As can be observed in these figures, in essence, the maximum levels of ground shaking were observed along the fault ruptures and in the Kahramanmaraş and Antakya-Hatay cities and regions. These high ground motion levels were reasons for the comparatively high damage and consequential casualties (Figs. 12 and 13).

In Fig. 7 and Fig. 8, the severity of the earthquake ground motion during the magnitude-7.8 event, The primary structural system for urban residential buildings affected by earthquakes typically consists of reinforced concrete frames with a symmetrical floor plan and unreinforced masonry infill walls. These infill walls are usually made of hollow clay tiles with insufficient mortar at the joints. Although not originally

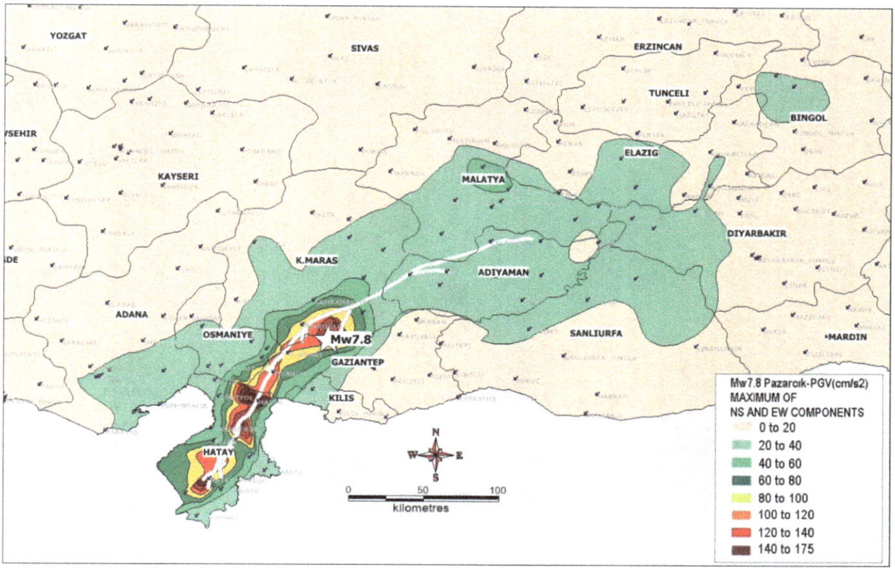

Fig. 8. Contour map of the peak ground velocity (in units of centimeters per second; maximum value of north-south and east-west horizontal components) associated with the magnitude-7.8 earthquake. (Erdik et al., 2023a) Surface rupture taken from USGS.

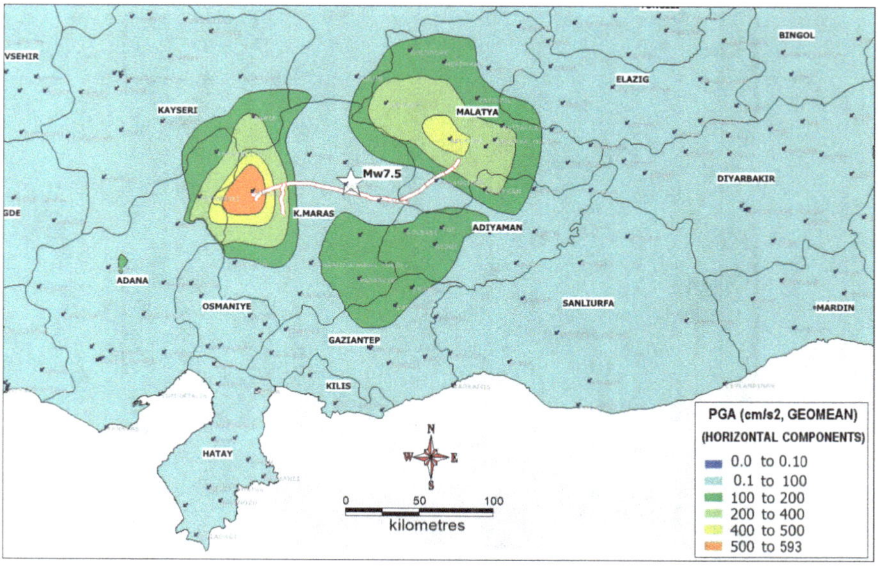

Fig. 9. Contour map of the peak ground acceleration (in units of centimeters per second squared; geomean values of north-south and east-west horizontal components) associated with the magnitude-7.5 earthquake. (Erdik et al., 2023b) Surface rupture taken from USGS.

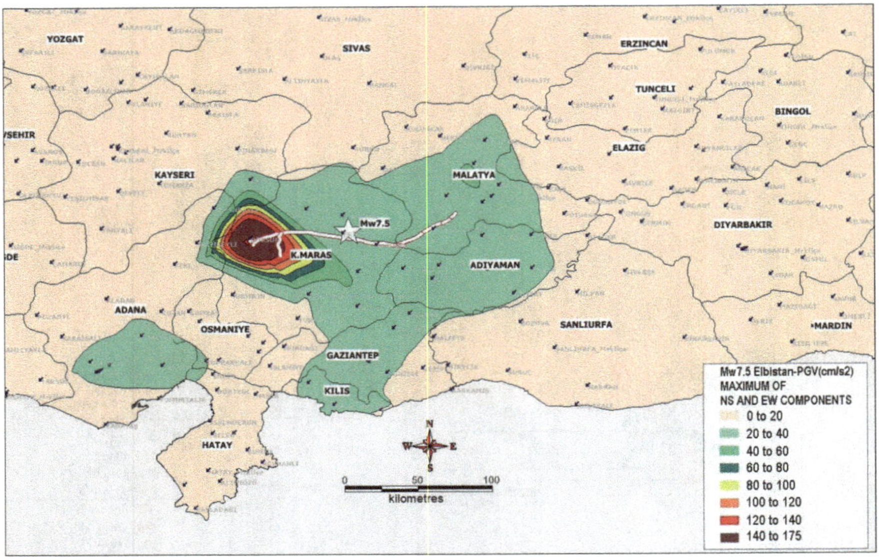

Fig. 10. Contour map of the peak ground velocity (in units of centimeters per second; maximum value of north-south and east-west horizontal components) associated with the magnitude-7.5 earthquake. (Erdik et al., 2023a) Surface rupture taken from USGS.

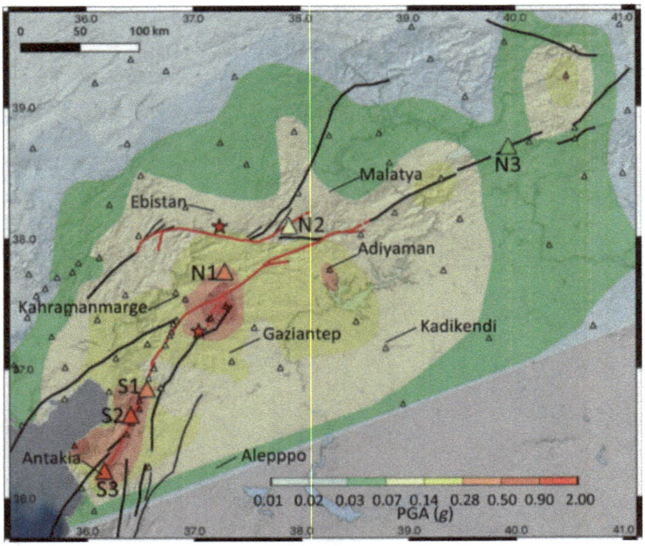

Fig. 11. The spatially interpolated geometric mean of the peak ground acceleration for the fault-parallel and normal components (Mai et al., 2023). Triangles represent local strong ground-motion seismic stations.

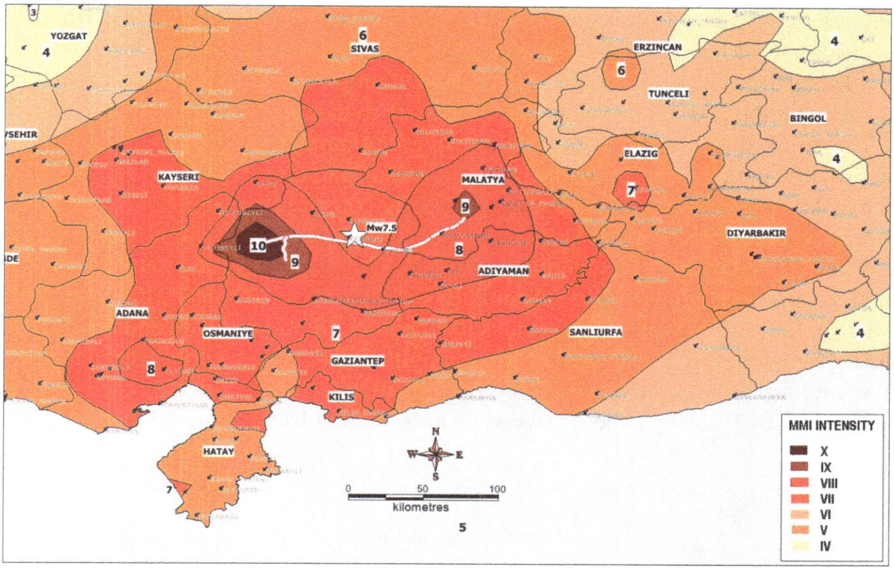

Fig. 12. Contour map of the MMI intensity associated with the magnitude-7.5 earthquake. Surface rupture taken from USGS.

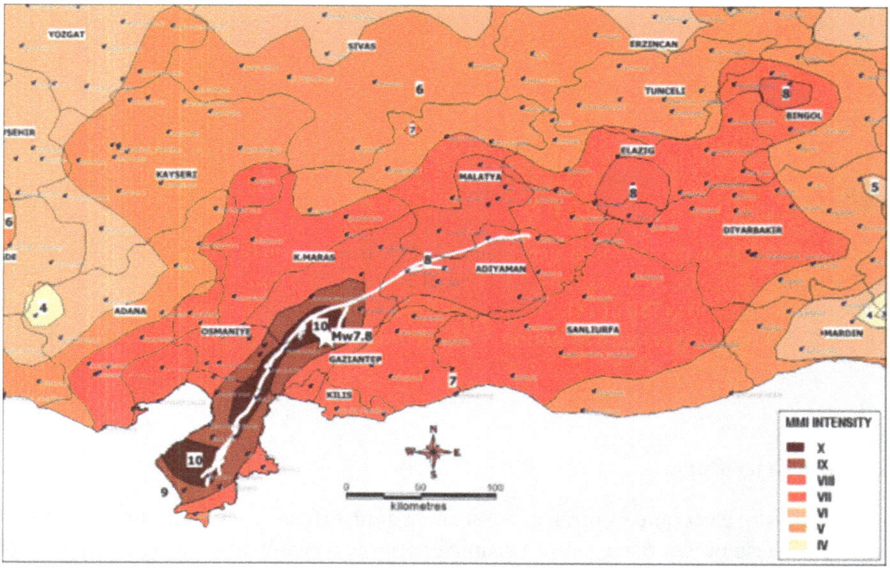

Fig. 13. Contour map of the MMI intensity associated with the magnitude-7.8 earthquake. Surface rupture taken from USGS.

designed for seismic resistance, they often serve as the first barrier against earthquakes and help control lateral drift. Once these infill walls collapse, the lateral resistance is provided by the reinforced concrete frames and, if present, by the shear walls, to transfer the lateral earthquake forces on the building to its foundation.

The Amik Basin is equipped with a strong motion network (Durukal et al., 2009) that provides data for analyzing the ground motion characteristics observed during this earthquake. The transfer function, determined by comparing records from stations on different geological substrates (Fig. 14), offers insights into site amplification effects. As expected, sediment layers amplify ground motion. Figure 14 illustrates the transfer function between stations within the Amik Basin (Stations #3123 and #2124) and nearby strong-motion stations on stiff substrates like bedrock (Stations #3136 and #3140). The transfer function reveals two to three times amplification of long-period ground motion. These findings are consistent with basin amplification factors reported by Campbell and Bozorgnia (2014) for $Z_{2.5}$ shear wave velocities ranging from 3.5 to 4.5 km/s.

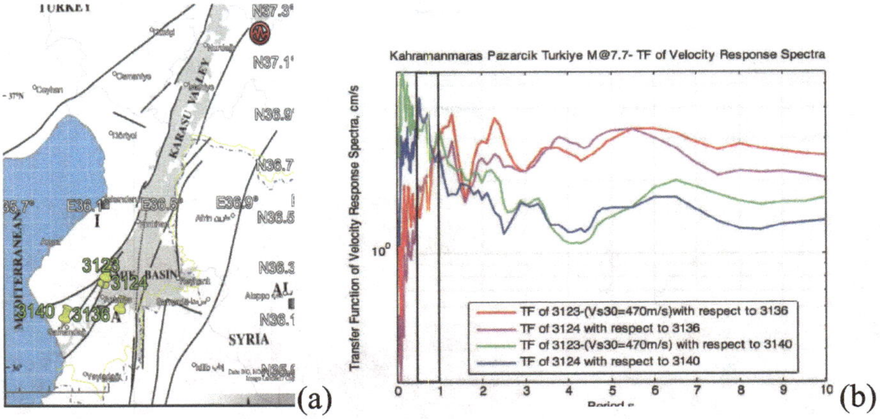

Fig. 14. (a) Map showing the location of stations 3136 (installed on bedrock) and 3140 (installed on basin sediments), as well as stations 3123 and 2124, which are both located in the Amik Basin. (b) The transfer function (TF) of velocity response spectra (horizontal component with geometric mean) with 5% damping relative to station 3136 (bedrock) and station 3140 (basin sediments).

4.1 Directivity Pulses

Large strike-slip earthquakes often generate near-fault effects, including rupture directivity pulses. These pulses occur when seismic energy is concentrated along the fault in the direction of rupture, resulting in a significant burst of motion. Such effects can be identified by analyzing velocity and displacement time histories from the earthquake (e.g., Fig. 15 and Fig. 16). These ground motions pose a severe threat to long-period structures such as tall buildings, long-span bridges, seismically isolated structures, and large liquid tanks, which must be designed to endure long-period ground motions (Somerville and

Graves, 1993). Understanding directivity pulses is essential for designing earthquake-resistant structures (Erdik et al., 2023a). Extracting information about pulse characteristics—including their shape, period, and amplitude—from strong motion records can support improvements in engineering codes and earthquake-resistant designs.

To investigate near-fault pulse-like ground motions, Wu et al. (2023) analyzed 100 sets of strong motion records from within 200 km of the causative fault of the M7.8 earthquake. These velocity waveforms exhibited bidirectional movement with multiple reflections due to rupture directivity effects. The velocity pulse amplitudes ranged from 30.5 to 219.8 cm/s (e.g., at Station #3138), with pulse periods varying between 2.3 and 14.5 s. While pulse amplitude decreases significantly with distance from the fault, the pulse period remains largely unaffected by fault distance. The observed pulse characteristics are consistent with the average levels observed in other earthquakes of similar magnitude.

Baltzopoulos et al. (2023) conducted another analysis of near-fault ground motions, examining 30 velocity time histories. These were rotated incrementally over 180 degrees to identify candidate pulse waveforms. Spectra from the ten stations closest to the fault reveal that all primary events produced motions that generally posed significant challenges to structures.

For the magnitude-7.8 earthquake, directivity pulses were observed in records from near-fault stations such as NAR and 4615, as shown in Fig. 15 and Fig. 16. To identify these pulses, the method of Baker (2007) was employed, which uses wavelet-based signal decomposition to isolate the largest velocity pulse in the ground motion record. Ground motions with extracted signals that are substantially larger than the remaining signal are classified as "pulse-like." The pulses presented in Fig. 17 are an updated interpretation of observations from Erdik et al. (2023b), emphasizing the near-fault effects associated with these earthquakes and highlighting the need for further study.

4.2 Comparison of Ground Motion-Based and Code-Based Spectra

The two major earthquakes were recorded by stations distributed across a wide area, enabling a comparison of the observed ground motion spectra with the spectra used to develop building codes. Building codes for earthquake-resistant structures are based on a 5% damping ratio, which serves as a design standard. To facilitate this comparison, the recorded ground motion spectra from the recent earthquakes were converted to 5%-damped horizontal response spectra. This analysis utilized data from four Stations: #3132, #3117, #3123, and #3145—situated in urban areas (Fig. 18). These converted spectra were compared with the site-specific design basis spectra outlined in Türkiye's 2018 Earthquake-Resistant Design Code (TBDY, 2018), which is derived from probabilistic seismic hazard assessments.

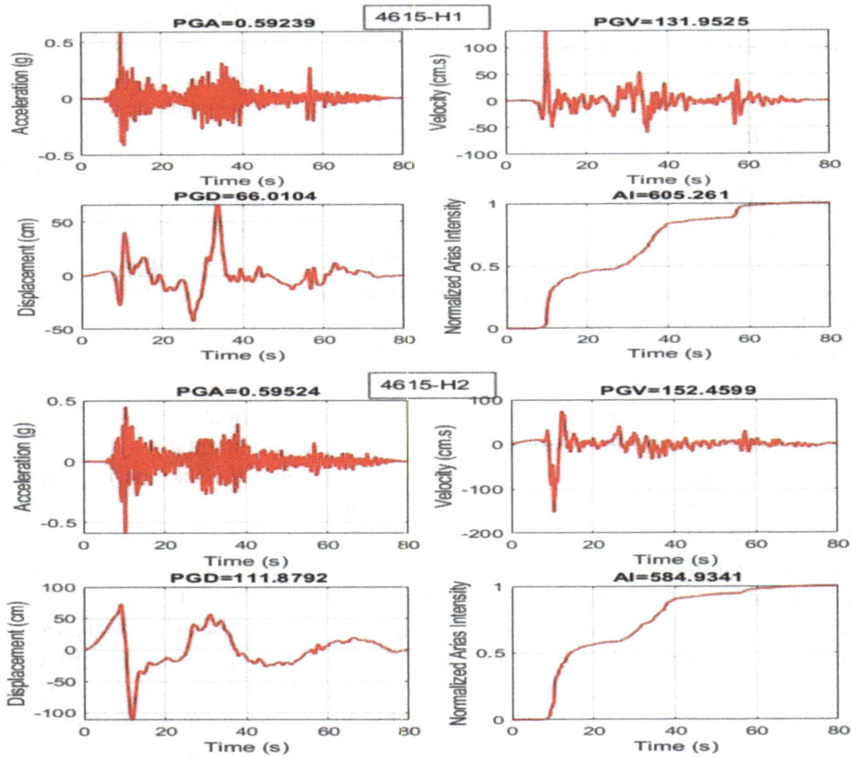

Fig. 15. The peak ground acceleration, velocity, and displacement time histories and normalized arias intensity associated with 4615 station from the M7.8 earthquake

The code-based spectra are designed for average return periods of 2,475 years (DD-1), 475 years (DD-2), 72 years (DD-3), and 43 years (DD-4), with residential buildings typically designed based on the 475-year return period spectrum.

Figure 19 illustrates these comparisons for the magnitude-7.8 earthquake. In Figs. 19a and 19b, the empirical ground motion spectra recorded at Stations #3145 and #3123 in Hatay Province exceeded the 2,475-year return period spectra for periods greater than 0.5 s. In contrast, Figs. 19c and 19d show that the spectra recorded at Stations #3132 and #3117, also in Hatay Province, more closely align with the code-based spectra. This indicates that, in some areas, the ground shaking exceeded the design parameters for buildings, even if constructed to code standards.

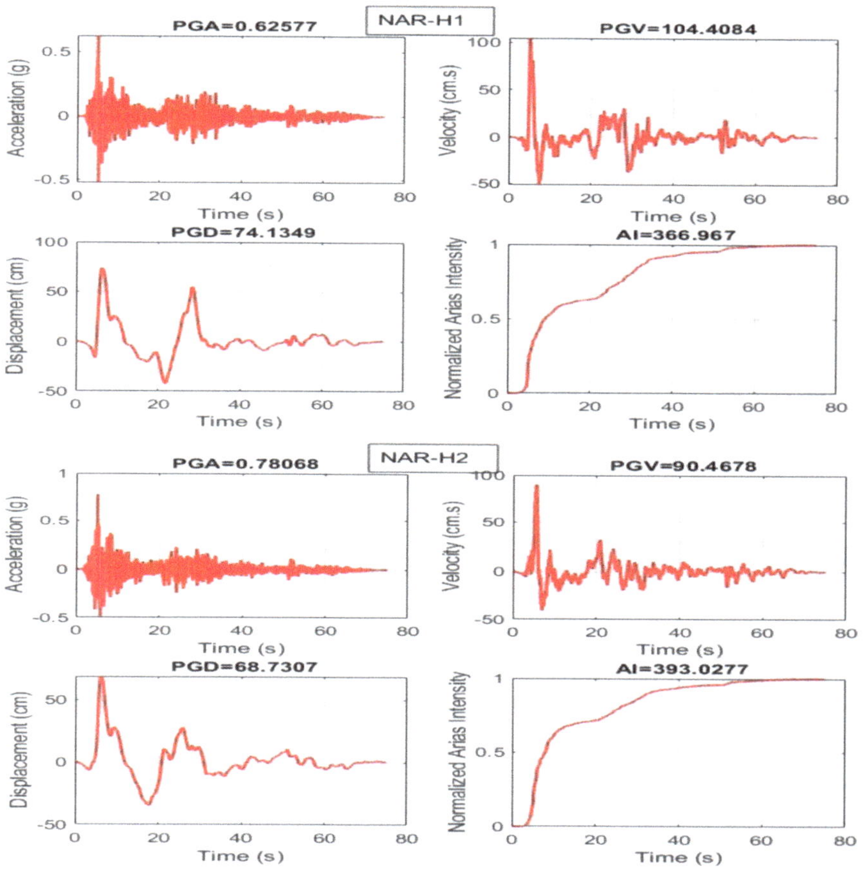

Fig. 16. The peak ground acceleration, velocity, and displacement time histories and normalized arias intensity associated with the NAR station from the M7.8 earthquake.

Figure 20 compares the original records, extracted pulse signals, and velocity response spectra of the residual records. The results show that the velocity spectrum of the residual record closely matches the original record in the short-period range but is significantly lower in the long-period range for periods greater than Tp*Tp*. The velocity spectrum of the pulse signal aligns with the original record in the long-period range, indicating that the long-period response in the original record is primarily influenced by the pulse signal.

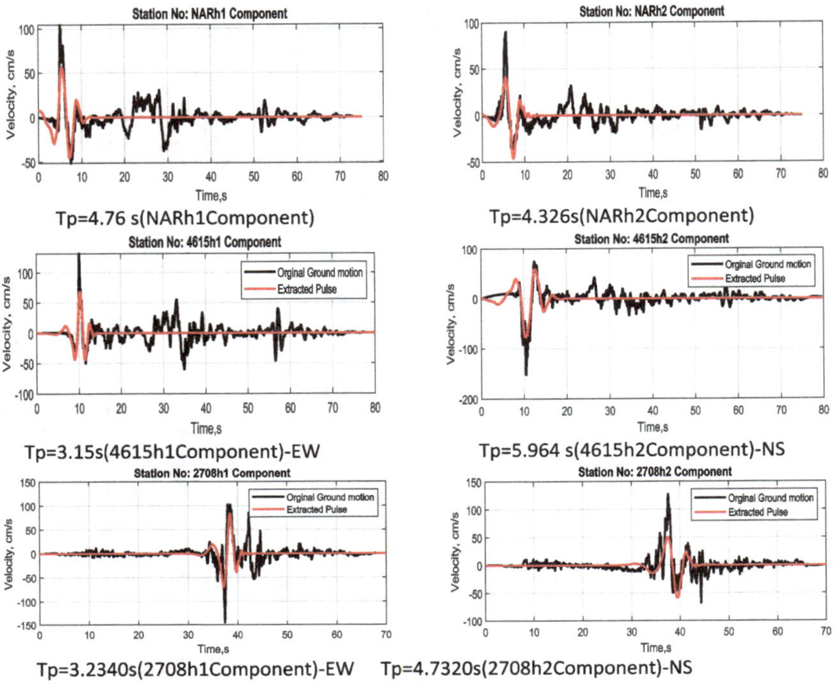

Fig. 17. Extracted near-fault velocity pulses obtained from three near-fault stations.

Interestingly, the velocity spectrum of the residual record for the 4651 station shows a peak at T = 3 s, suggesting the presence of a secondary pulse with a characteristic period of approximately 3 s. These secondary pulses were not extracted because the wavelet method used in this study identifies and extracts only the strongest pulse signal from each record. For other records, such as Station #2708, NAR, and #4615, the residual velocity spectra no longer exhibit long-period peaks after pulse extraction, indicating the absence of secondary pulses in these records.

5 Damages and Losses

Ten provinces, including the hard-hit Hatay, Kahramanmaraş, Gaziantep, and Adıyaman provinces, were heavily affected by building damage. State Bureau of Statistics indicates about 3.9 million housing units in these 10 provinces, out of which about 40% of them were built prior to the earthquake-resistant design code change in 2000. These 10 provinces provide 10% of the GDP in Türkiye, largely related to agriculture, industry, and service.

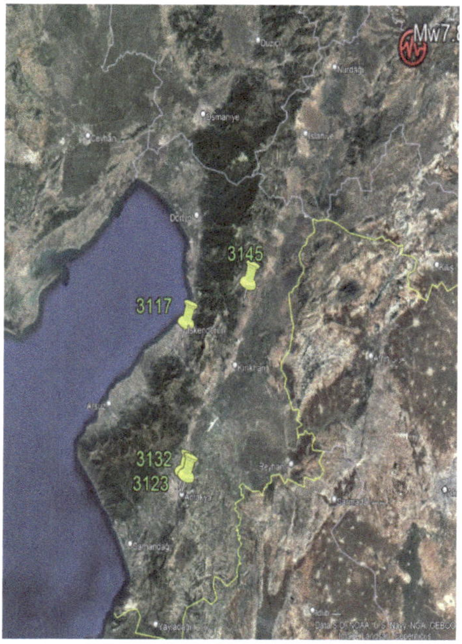

Fig. 18. Selected station locations for comparison of the acceleration response spectra with TBDY (2018)-based horizontal 5%-damped acceleration spectra. Credit: basemap Google Earth

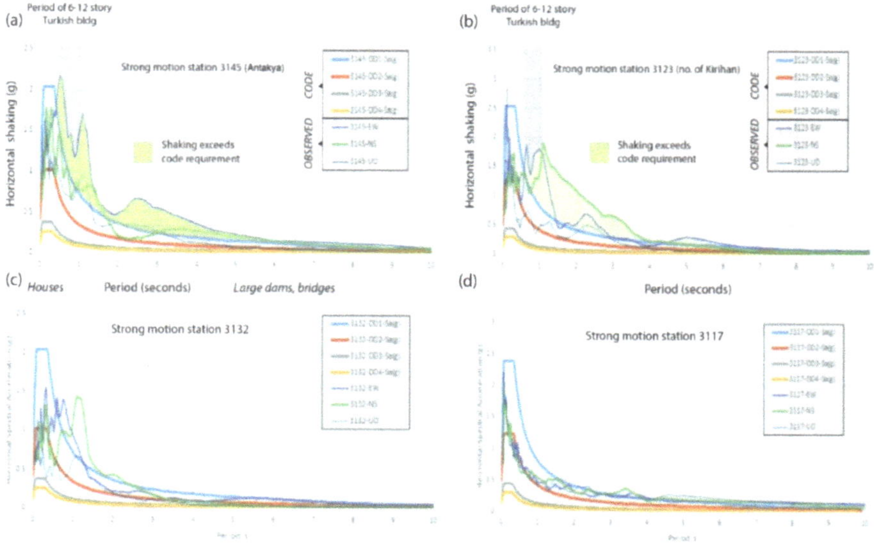

Fig. 19. Comparison of the 5%-damped acceleration response spectra (thin curves) with TBDY (2018) code-based spectra (bold curves) at the station of (a) 3145, (b) 3123, (c) 3132, and (d) 3117 (Erdik et al., 2023a).

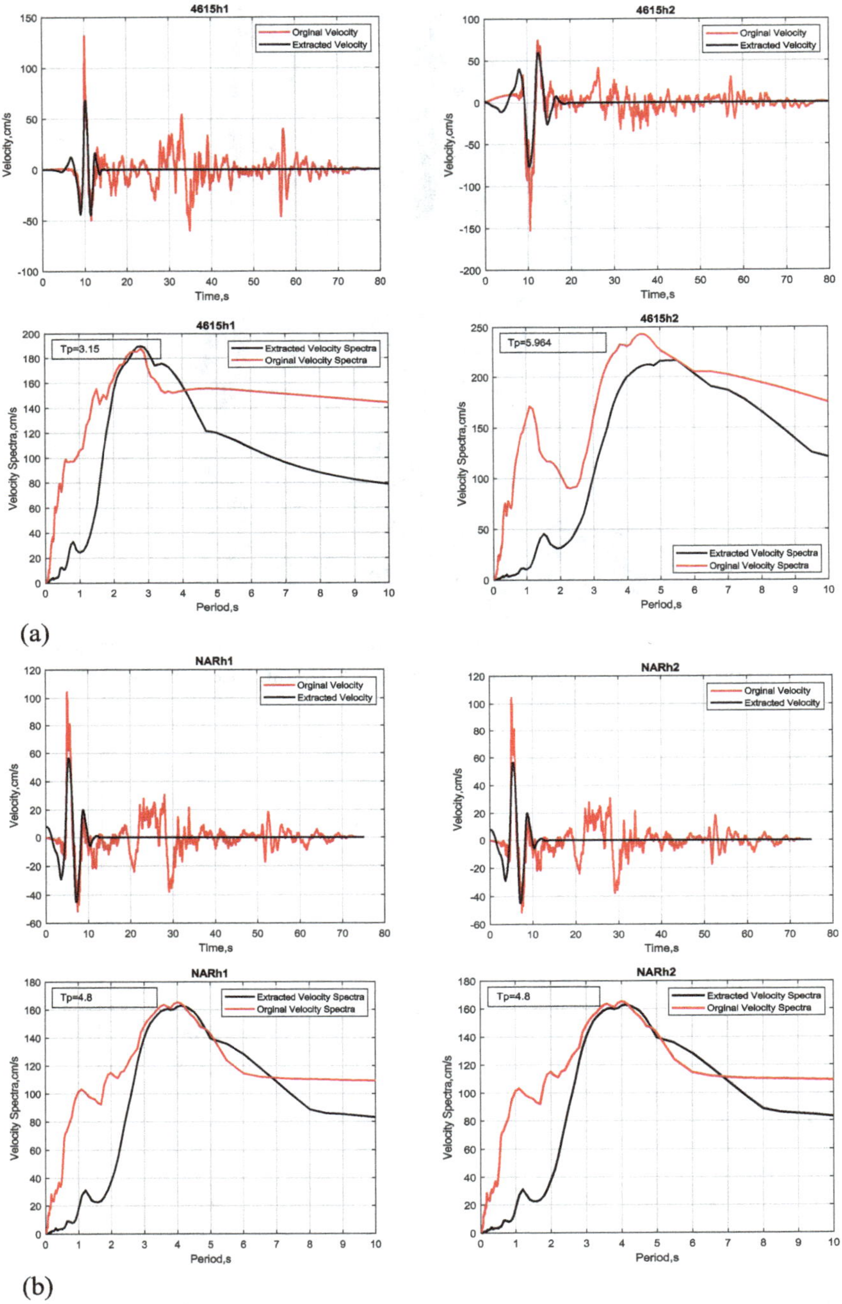

Fig. 20. Comparison of velocity response spectrum of original pulse recording with different pulse period Tp, extracted pulses, and 234 residual recordings. (a) record 4615; (b) record NAR; (c) record 2708

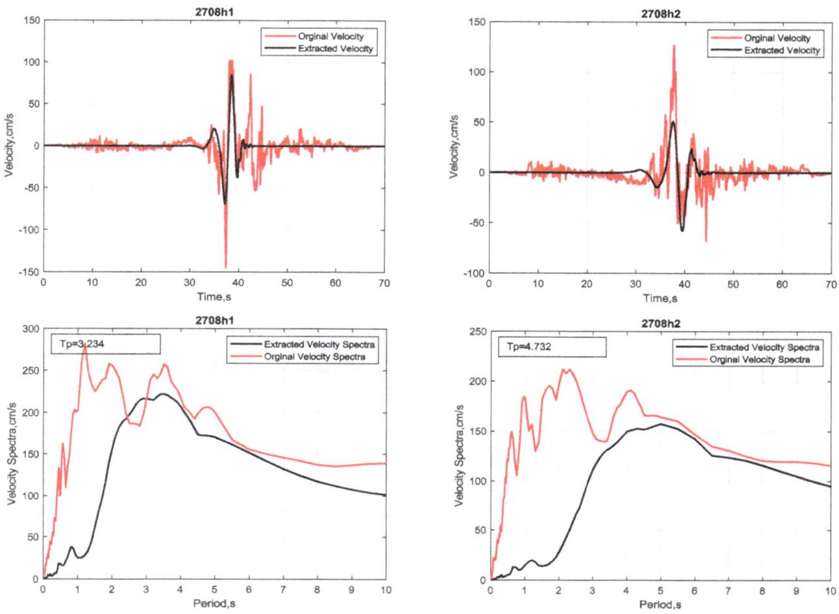

(c)

Fig. 20. (*continued*)

Figure 21 is provided (EERI Buildings Team, 2023, USGS-Podcast, 2023) to indicate the provinces affected by the February 6 earthquake sequence and the locations of Kahramanmaraş and Hatay provinces where the maximum damage and casualties took place. The MMI VII intensity contour matches approximately the PGA = 0.2 g contour provided in (EERI Buildings Team, 2023; GRADE, 2023) and approximately bounds the significantly affected region and encompasses the ruptured segments of the fault and the higher intensities that reached MMI IX –X and approximately bounds the significantly affected region.

Figure 22 portrays the distribution of the damaged buildings in the cities of Antakya (Hatay Province) and Kahramanmaraş on the basis of the proxy damage mapping using PALSAR-2 ScanSAR Imagery (Matsuoka et al., 2023). The figure shows that almost 70% of Antakya's and almost 50% of Kahramanmaraş's urban area were spotted by heavily damaged buildings.

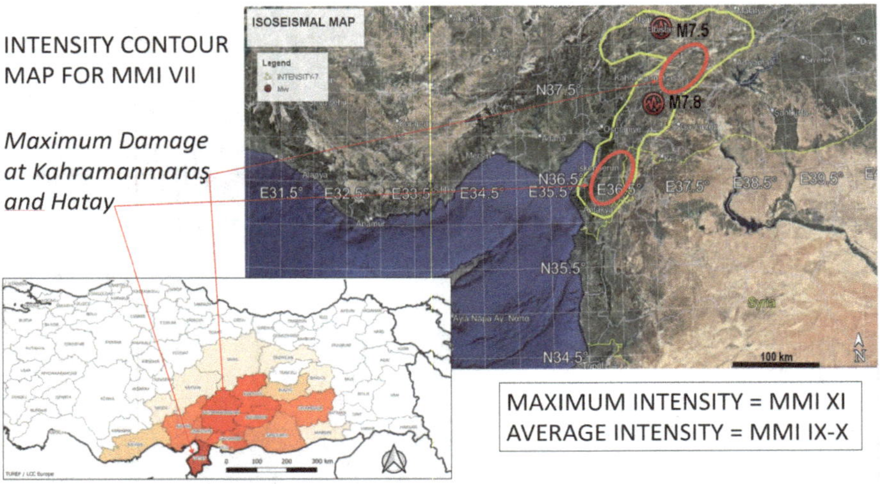

Fig. 21. MMI VII isoseismal contour (yellow lines) and the regions with heavy damage (red lines) (USGS-Podcast, 2023; GRADE, 2023).

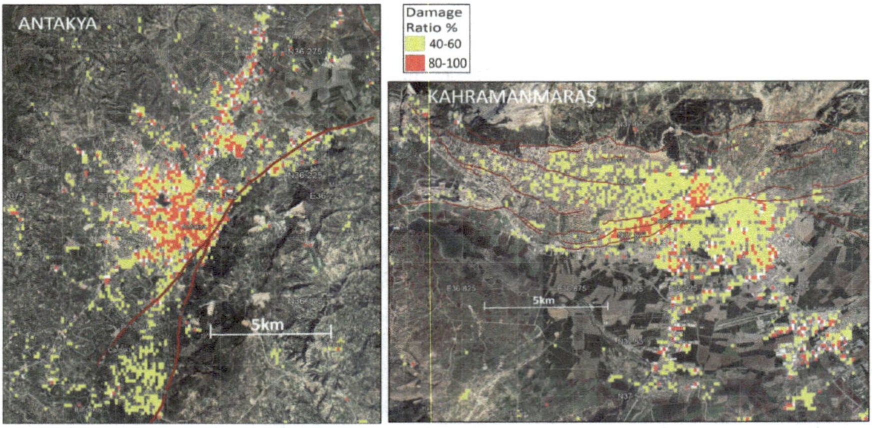

Fig. 22. Distribution of the damaged buildings in the cities of Antakya and Kahramanmaraş (Matsuoka et al., 2023). As shown in the legend, yellow and red pixels indicate the damage ratios in the orders of respectively 40%-60% and 80%–100%

6 Residential Buildings

The dominant structural system for residential buildings in urban areas of the affected region is reinforced concrete moment frames with unreinforced masonry infill walls. Approximately 40% of these structures were constructed after 2000, adhering to the 1998 building code with relatively adequate design and construction controls. Damage in these newer buildings is primarily observed in high-rise structures (10+ stories), likely due to design deficiencies and the impact of intense ground motion. In contrast, the majority of the severely damaged and collapsed buildings were constructed before 2000, based on the 1968 or 1975 building codes. These older structures were built with inadequate design and construction practices, resulting in non-ductile (brittle) concrete frames that are highly susceptible to total collapse during strong earthquakes. On the other hand, the majority of government buildings, tunnel-form buildings, and retrofitted buildings exhibited much better (code-level) earthquake performance.

According to field investigation reports (ITU, 2023; METU, 2023; AFAD, 2023; 2023 EEFIT Mission; GEER Association Report, 2023), in buildings that were either heavily damaged or collapsed, the load-bearing structural elements had low concrete quality, inadequate reinforcement, and/or poor detailing. The extensive inelastic action at the critical junctions in the buildings' structures caused the formation of a hinge at the connection of columns with the beams, thereby, especially in soft 1st story buildings, creating mechanisms that led to pancake-type collapses, where the floors pile up on each other as typically illustrated in Fig. 23. This is the worst type of collapse mechanism for life safety that should have been avoided.

Fig. 23. A typical pancake-type collapse of a residential apartment building in Kahramanmaraş (Anatolian Agency Photo)

RC frames with unreinforced masonry infill walls are common in Türkiye. In low levels of ground motion, these infill walls add to the stiffness and strength of the structure. However, in large ground motion levels, brittle failure of the infill walls can lead to the formation of soft stories during an earthquake, which may lead to pancake-type collapse.

The development stages of the earthquake-resistant design code in Türkiye are provided in Fig. 24. As can be seen the buildings were designed and constructed on the basis of the pre-1998 codes, where the designer had an opportunity to design cost-saving non-ductile reinforced concrete buildings (Sezen et al., 2023), are denoted as "low-code".

The 1998 code was similar to the US 1997 Uniform Building Code and, the currently applicable 2019 code has similar features to the ASCE 7-16 and EC8 Codes.

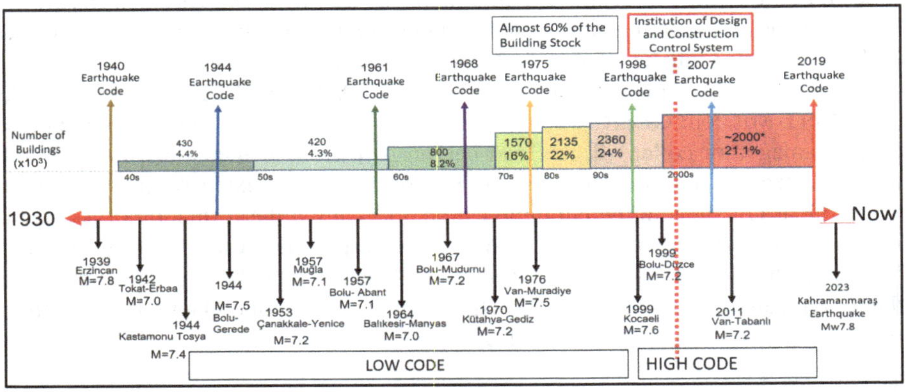

Fig. 24. Development stages of the earthquake-resistant design code in Türkiye

Figure 25 provides the breakdown of the damage states of the building stock (USGS-Podcast, 2023). The data is based on the results of the March 6 survey conducted by the Ministry of Environment, Urbanization, and Climate Change. Out of the 1,800,00 Buildings Inspected in the Affected Region: 54% No Damage, 27% Minor Damage, 3% Moderate Damage, 14% Major Damage + Collapse and, 780,000 Damaged (Minor Damage to Collapse) Buildings. 87% of Buildings with Major Damage-to-Collapse were Located in Adıyaman (11%), Kahramanmaraş (20%), Malatya (14%) and Hatay (42%). The non-conformity to earthquake-resistant design codes that resulted in non-ductile (brittle) buildings is clearly evident in the binary distribution of earthquake damage (Fig. 26). The total number of damaged buildings are about %25 higher as of May 15 but we believe that the ratio of damage states is not affected.

In the magnitude-7.8 earthquake, extensive liquefaction was observed at the shores of lakes in Gölbaşı (Adıyaman), in the İskenderun Port wharf area, and in the Antakya Amik Basin. Effects of liquefaction resulted in settled and tilted buildings disfigured railroad tracks and damaged embankments.

The prime reason for heavy damage and unreasonably high casualties is non-compliance with earthquake-resistant design codes. This issue is elaborated under the section NON-COMPLIANCE with the CODE.

6.1 Hospitals

Within the earthquake-affected area, there were 116 hospitals, and 24,000 beds (GEER Association Report, 2023). The hospitals ranged from one to ten stories in height, with many three or four stories. They included both fixed base and seismically isolated structures. Reportedly, some hospitals in the region, including Iskenderun State Hospital, a private hospital in Hatay, two private hospitals in Malatya, and another private hospital in Adıyaman received heavy damage. However, in general, observed structural damage

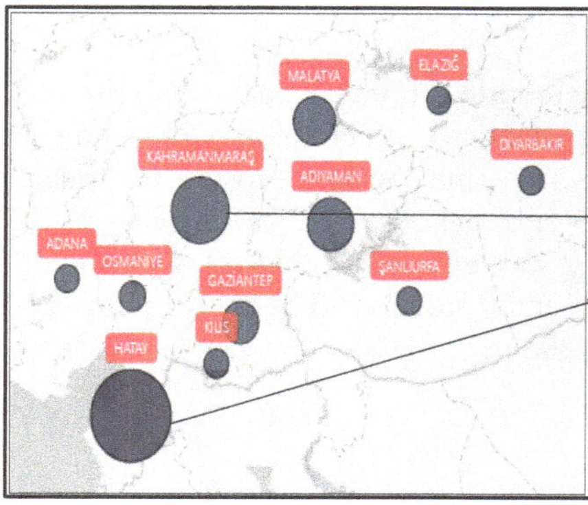

Damage States out of the 780,000 Damaged (Minor to Collapse) Buildings in the Affected Region:

CITY	# Damaged Buildings	# Minor Damage	# Medium Damage	# Major to Collapse
Adıyaman	59,000	49%	13%	38%
Kahramanmaraş	115,000	58%	6%	36%
Malatya	77,000	55%	7%	38%
Hatay	172,000	44%	6%	50%
Other Cities	357,000	74% - 85%	3% - 14%	3% - 23%
Total	780,000	62%	6%	32%

Fig. 25. Distribution of Damages States out of the 780,000 Damaged (Minor to Collapse) Buildings in the Affected Region (USGS-Podcast, 2023).

Fig. 26. Almost binary distribution of damage in Antakya (Anatolian Agency Photo)

in the lateral force-resisting system of the hospital buildings was minimal except for older buildings. On the other hand, however, non-structural damage was widespread and impeded their functionality (EERI Lifeline Team Report, 2023; GEER Association Report, 2023; NCEI/WDS, 2022; TC-CSBB 2023). Seismically isolated hospitals had no significant damage and were mostly functional after the events.

6.2 Schools

More than 20,000 educational buildings have shown very good earthquake performance. Only 3% collapsed and about 7% of buildings were heavily damaged (USGS-Podcast, 2023). Damaged school buildings are mostly stone masonry with or without retrofitting with RC elements (The 2023 EEFIT Mission, 2023; GEER Association Report, 2023).

6.3 Bridges/Viaducts

Damages on highway and railway bridges were reported either due to the fault rupture on the surface, by rock falls, or soil problems (liquefaction), which in some cases impeded transportation. Out of 1000+ bridges, only 15 had some damage. Most of the transport routes were operational within 2 days [24]. Most of the observed partial damages were due to: longitudinal movement of the bridges, "walking out" of thick elastomeric bearing pads, Abutment settlements, and, Leaning and residual drift of bridge piers (EERI Lifeline Team Report, 2023; EEFIT Mission, 2023).

6.4 Pipelines

The main hydrocarbon pipelines were damaged at the fault rupture crossings, landslides, and liquefaction. There were explosions and failures at about 20 different points of the natural gas transmission lines owned by Petroleum Pipeline Corporation (BOTAS) in the earthquake-affected region (EEFIT Mission, 2023). This damage reportedly has been remediated and their functionality is mostly resumed within one week. (USGS-Podcast, 2023; Erdik et al., 2023a).

6.5 Airports and Ports

Most airport buildings (Kahramanmaraş, Gaziantep, Hatay Airport) performed well (with some non-structural damage) - operations in some cases were impacted by other damage (runway damage in Hatay Airport) that was resolved within a week (USGS-Podcast, 2023; EERI-GEER Webinar, 2023).

The port of Iskenderun sustained liquefaction damage in the wharf area and experienced a significant fire the next day, making it unusable as an entry port for emergency supplies. Within a few days, the fire was suppressed, and the port was partially opened to service. (USGS-Podcast, 2023; Erdik et al., 2023a).

6.6 Dams

Out of about 150 dams, only a few of them had some damage. Embankment Dams provided reasonable performance (TC-CSBB 2023; EERI-GEER Webinar, 2023). The 60-m-high Sultansuyu embankment dam had some cracks that led to the controlled release of water through the spillway. Another dam in Syria, the 75-m-high Afrin embankment dam, experienced heavier damage (USGS-Podcast, 2023). In Berke Dam, there was minor vertical tilting of a concrete arch (EERI Lifeline Team Report, 2023).

6.7 Power Plants

There were 8 power plants in the earthquake-affected region. The Iskenderun coal plant and Afşin Plants A and B have received minor damage (EERI Lifeline Team Report, 2023; EEFIT Mission 2023). The damage were repaired and as of May 15, they are all operational.

6.8 Economic Impact

As reported by WB, the best preliminary estimate of direct damages from this event is \$34.1 billion. However, recovery and reconstruction costs will be much larger, potentially twice as large. UN's estimate of the total cost is \$100 billion. The Turkish Government's estimate of losses is 103.6 billion USD (USGS-Podcast, 2023; Erdik et al., 2023a).

Turkish Government estimate shows that the most prominent component of the burden imposed by the earthquake on the Turkish economy is the damage in housing units by 54.9% (56.9 billion USD). The second largest damage is the destruction of public infrastructure and damage to public service buildings (12.9 billion USD). The damage incurred by the private sector except for housing is estimated at (11.8 billion USD). This category includes the manufacturing industry, energy, communications, tourism, healthcare, education sectors, and damage to small tradespersons and houses of worship. Further, considering the losses to the insurance sector, revenue losses of tradespersons, and macroeconomic impacts, it is estimated that the total financial burden of the earthquake disaster for the country is (103.6 billion USD), which is equal to 9% of the GDP forecast for 2023 (TERRA, 2023).

About 500,000 new housing units are needed to sufficiently re-house the approximately 1.4 million people left homeless. The government plans to build 405,000 new urban housing units (in 3–4 story apartment buildings) and 83,000 rural houses. (USGS-Podcast, 2023; Erdik et al., 2023b).

6.9 Earthquake Insurance

Total insured losses are likely to top \$5 billion (Insurance day, 2023). Since 2000, earthquake insurance has been compulsory for private dwellings and is covered by the Turkish Catastrophe Insurance Pool (TCIP). With about 12 million policy holders, 54% penetration in the country, and a claims-paying capacity of approx. US \$6 Billion, TCIP is one of the largest compulsory earthquake insurance systems in the world. Insurance penetration is about 55% in the affected region. About 530,000 TCIP damage claims and a total payment reaching \$3 billion are estimated. The damage distribution of the claims is: 5% collapse, 6% heavy damage, 9% medium damage, and, 80% minor damage (DASK, 2023).

7 Non-compliance with the Code

7.1 Introduction

Compared to infrastructure (lifelines, dams, power plants) and governmental buildings (public offices, hospitals, educational buildings) the damage and casualties were almost totally associated with the private residential buildings and some of the commercial buildings and facilities. The reason behind this observation is the non-compliance with the code.

The primary structural system for urban residential buildings affected by earthquakes typically consists of reinforced concrete frames with a symmetrical floor plan and unreinforced masonry infill walls. These infill walls are usually made of hollow clay tiles with insufficient mortar at the joints. Although not originally designed for seismic resistance, they often serve as the first barrier against earthquakes and help control lateral drift. Once these infill walls collapse, the lateral resistance is provided by the reinforced concrete frames and, if present, by the shear walls, to transfer the lateral earthquake forces on the building to its foundation.

In Türkiye, many buildings have been designed to comply with modern building codes, but a significant number, particularly older structures, feature non-ductile concrete framing that is brittle and highly susceptible to total collapse during strong earthquakes, as demonstrated by the recent events.

Buildings constructed in accordance with modern codes sustained damage but generally did not collapse, likely avoiding fatalities. In contrast, as seen during the February 6 earthquakes, buildings that failed to meet code requirements were severely damaged, collapsed, and caused casualties.

This discrepancy highlights a clear pattern in earthquake damage, with a binary distribution: most buildings were either lightly damaged or severely damaged/collapsed, while relatively few experienced moderate damage. This distribution reflects a brittle structural response and indicates non-compliance with earthquake-resistant design standards.

The differing outcomes between similar buildings—those that survived with minor damage and those that failed—provide strong evidence that adherence to design codes and proper construction practices significantly reduces damage during strong earthquakes.

Field investigations reveal that, beyond general issues with code compliance, buildings that were severely damaged or collapsed often exhibited poor concrete quality, insufficient reinforcement, and inadequate detailing in their load-bearing structural elements.

Significant inelastic deformation occurred at critical connections between columns and beams, resulting in the formation of hinges at these junctions. This led to mechanisms causing pancake-type collapses, where floors stacked upon one another. Such collapses are the most catastrophic failure mode, often resulting in a high number of casualties.

The damage sustained by buildings built to code can be partly attributed to the severity of the earthquake ground motions, which at several locations exceeded the design basis ground motion levels. However, even under exposure to such high-level ground motions, the code-conforming buildings can have the reserve capacity to accommodate such extra demand, whereas the non-code-conforming buildings can certainly experience total collapse.

A substantial majority of the heavily damaged and collapsed buildings were built before the year 2000, designed following the 1968 or 1975 code with inadequate design and construction controls resulting in non-ductile (brittle) concrete frames prone to total collapse during strong earthquake shaking. New buildings constructed in compliance with 2007 and 2018 earthquake regulations reduced the destruction relatively, but the observed damages indicate that the measures taken did not go far enough.

To enforce and ensure code compliance, law No. 4708 on Building Inspection was enacted in 2001 for implementation in 19 pilot provinces. After 9 years of this pilot implementation, in 2011 the application of the law was enlarged to encompass the whole country. The law was amended in 2018 in line with the needs that emerged over time in order to ensure better implementation of inspection activities and the Ministry of Environment, Housing and Climate Change was empowered for selection of the Building Inspection firms. This amendment was introduced.

New provisions to ensure the independence and effectiveness of building inspections through the removal of the conflict of interest in the selection of Building Inspection firms by the contractor.

Following is a general overview of the reasons for extensive residential building damage.

7.2 Non-code-Compliant Structural Design

In all earthquake-resistant design codes, a proper, continuous load path is a design requirement to ensure that the loads must be transferred rationally from their point of origin to the final point of resistance.

The basic steps in the structural design of a building structure for acceptable seismic performance are as follows:

- The gravity- and seismic force-resisting systems are appropriate to the anticipated intensity of ground shaking.
- These systems are configured to produce a continuous, regular, and redundant load path so that the structure acts as an integral unit in responding to ground shaking.
- The mathematical model of the structure is analyzed under exposure to seismic motions and gravity forces to obtain internal forces and displacements.
- The members and connections are designed to have adequate strength, stiffness, and deformation capacities.

Turkish seismic design codes appear to define design force levels and other requirements that would suffice for earthquake-resistant design as revealed by the standing buildings. 1998, 2007, and 2018 codes include the up-to-date requirements of modern seismic design codes. Despite its limited coverage of ductility, irregularity, detailing, and capacity design principles, even the buildings designed and constructed properly according to the 1975 code are expected to provide life safety. Design errors stemming from non-compliance with these basic code regulations are believed to constitute the leading cause of heavier damage than expected in these earthquakes. The non-compliance problems are due to the incompetence of the designers, the use of improper analysis software or blind use of software, and inadequate or non-existent structural design control and assurance. Furthermore, the design problems cannot be detected by the current construction control system.

7.3 Soft Story

A typical feature of the buildings along the main roads is the presence of shops at the ground levels, where, to create wider space, infill walls are generally removed. At locations where the seismic demands are maximum, the removal of infill walls, coupled with weak columns and poor beam-column connections creates soft-story mechanisms that lead to partial collapse and even to progressive pancake type collapse failure due to soft first stories,

7.4 Vertical Load Bearing Elements

Heavy damage was observed in some of the building columns due to detailing deficiencies such as insufficient transverse reinforcement, use of smooth bars, absence of hooks, and inadequate concrete strength. The beam-column joints are the most vulnerable parts of the structural system, and the codes require damage to occur at the end of the beams, while the columns should remain intact (Strong column – relatively weak beam principle). In this way, the total collapse of the buildings would be prevented. However, the observations regarding earthquakes have revealed that most of the heavily damaged buildings have beams that are stronger than the columns, contrary to the above-mentioned principle.

7.5 Diaphragm Action

The common damages observed in buildings with hollow tile floors (thin slab and joists with hollow blocks in between) is due to the lack of appropriate diaphragm action needed for developing proper load transfer mechanisms to the vertical load-carrying structural elements. In the 2018 code, the use of such floors is heavily restricted to low seismicity regions and to low-rise buildings.

7.6 Infill Walls

The most widespread damage type is infill wall damage, consisting of interface cracking, in-plane diagonal cracking, and out-plane collapses. These infill walls should be considered in the design of these structures. Masonry infill walls are modeled as diagonal braces as per code requirements. Furthermore, the infill wall construction technique suggested in the latest seismic code in 2019 did not seem to be applied in any of the recent buildings.

7.7 Other Problems

More specific problems that are associated with the extensive damage to residential buildings are Short (captive) column effects; Pounding between buildings or between blocks of the same building separated by joints; Poor quality of the concrete and the reinforcing bars that do not follow Turkish Standards; Excessive corrosion levels causing loss of bond strength and cover spalling.

8 Conclusions and Suggestions for Risk Mitigation

The main reasons for extensive damage to residential buildings and life losses can be listed as follows:

- Many aging buildings (pre-2000) were constructed following the 1968 and 1975 codes, with inadequate design and construction control. (Brittle Structures)

- Severe shaking from the two shallow earthquakes. A large proportion of the buildings were demolished with the second shock of almost equal size (also, 7 Mw6+ aftershocks and, basin effects).
- Especially in basin sites the recorded spectra exceeded the code-based spectra in medium and long period ranges (Seismic Hazard Map revision for basins is needed)
- Post-2000 buildings performed as expected (i.e. collapse prevention). Some of the Post-2000 buildings were heavily damaged due to code compliance problems (mostly design issues)
- The type of infill wall construction sometimes prevented columns from responding as designed and exacerbated the damage (should be addressed in the codes)

  - For the mitigation of damage and life losses in future earthquakes the following measures can be suggested: Strengthening of the existing Building Design and Construction Inspection System Institution of Professional Engineering
  - Mandatory Professional Liability Insurance both for the designer and the constructor.
  - Requirement of 3rd Party Liability Insurance for the provision of service/operation permits to facilities with high occupant loads (hotels, shopping centers, etc.)
  - Provision of Earthquake Risk Report for house sales and rentals
  - Strengthening and improvement of the on-going urban renewal program with the incorporation of external retrofit methods, that do not require tenants to vacate their apartments for earthquake retrofit.

References

Abdelmeguid, M., Zhao, C., Yalcinkaya, E., Gazetas, G., Elbanna, A., Rosakis, A.: Revealing the dynamics of the Feb 6th, 2023 M7.8 Kahramanmaraş/Pazarcik Earthquake: near-field records and dynamic rupture modeling (2023, submitted)

AFAD: 06 Şubat 2023 Kahramanmaraş (Pazarcık ve Elbistan) Depremleri Saha **Çalişmalari** Ön Değerlendirme Raporu - AFAD Deprem Dairesi Başkanliği, 24 Şubat 2023, AFAD (2023)

Aktar, M., Ergin, M., Özalaybey, S., Tapirdamaz, C.: A lower-crustal event in the northeastern Mediterranean: the 1998 Adana Earthquake (Mw:6.2) and its aftershocks. Geophys. Res. Lett. **27**, 2361–2364 (2000)

Aktar, M.T., Yörük, A., Kaplan, H.: Kurtkulağı events in the Cilician Basin and its aftershocks. In: XXIV General Assembly of the European Seismological Commission, Proceedings and Activity Report, Athens, vol. 1, pp. 74–80 (1994)

Akyüz, S.H., Altunel, E., Karabacak, V., Yalçıner, C.C.: Historical earthquake activity of the northern part of the Dead Sea Fault Zone, southern Turkey. Tectono-Physics **426**, 281–293 (2006)

Allen, C.R.: Active Faulting in Northern Turkey. Contribution No. 1577, Division of Geological Sciences, California Institute of Technology, p. 32 (1969)

Altunel, E., et al.: Archaeological sites (tell and road) offset by the dead sea fault in the Amik Basin, Southern Turkey. Geophys. J. Int. **179**, 1313–1329 (2009). https://doi.org/10.1111/j.1365-246X.2009.04388.x

Ambraseys, N.N.: Earthquakes in the Eastern Mediterranean and the Middle East: a multidisciplinary study of 2000 years of seismicity, Cambridge, Cambridge, UK, p. 968 (2009). ISBN: 9780521872928

Arpat, E., Şaroğlu , F.: Türkiye'deki bazı önemli genç tektonik olaylar. TJK Bül. **18**, 91–101 (1975)
Arvanitakis, G.L.: Essai d'une statistique des templements de terre en Palestineet Syrie. Bull. Inst. Egypte **4**(4), 178–183 (1903)
Ambraseys, N.N., Melville, C.P.: Historical evidence of faulting in eastern Anatolia and northern Syria. Ann. Geophys. **38**(3–4), 337–343 (1995)
Ambraseys, N.N.: Temporary seismic, quiescence: SE Turkey. Geophys. J. **96**, 311–331 (1989)
Ambraseys, N.N.: The 12th century seismic paroxysm in the middle east: a historical perspective. Ann. Geophys. **47**(2), 1–26 (200)
Arpat, E., Şaroğlu, F.: Türkiye'deki bazı önemli genç tektonik olaylar: TJK. Bull. **18**, 91–101 (1975)
Baker, J.W.:. Quantitative classification of near-fault ground motions using wavelet analysis. Bull. Seismol. Soc. Am. **97**(5), 1486–1501 (2007)
Barka, A., Kadinsky-Cade, K.: Strike-slip fault geometry in Turkey and its influence on earthquake activity. Tectonics **7**(3), 663–684 (1988)
Baltzopoulos, G., Baraschino, R., Chioccarelli, E., Cito, P., Vitale, A., Iervolino, I.: Preliminary engineering report on ground motion data of the Feb. 2023 Turkey Seismic Sequence **3** (2023)
Barbot S., et al.: Slip distribution of the February 6, 2023 Mw 7.8 and Mw 7.6, Kahramanmaraş, Turkey earthquake sequence in the East Anatolian Fault Zone, Seismica (2023). https://doi.org/10.26443/seismica.v2i3.502
Barka, A., Reilinger, R.: Active tectonics of the Mediterranean region: deduced from GPS, neotectonic and seismicity data. Ann. Geofis. **XI**(3), 587–609 (1997)
Boore, D.M., Stewart, J.P., Seyhan, E., Atkinson, G.M.: NGA-West2 equations for predicting PGA, PGV, and 5% damped PSA for shallow crustal earthquakes. Earthq. Spectra **30**, 1057–1085 (2014)
Calvi, S.: Türkiye ve bazı Komşu Ülkelerin Deprem Kataloğu, Çeviren. Kumcu, E. (1979), İstanbul (1941)
Campbell, K., Bozorgnia, Y.: NGA-West2 ground motion model for the average horizontal components of PGA, PGV, and 5% damped linear acceleration response spectra (2014)
Chen, J., Zillio, L.D., Zhang, H., Yang, G., Shi, Y., Liu, C.: Decoding stress patterns of the 2023 Turkey-Syria earthquake doublet. Res. Square (2023). https://doi.org/10.21203/rs.3.rs-2922091/v1
Chen, G.: Report on pulse-like ground motions in the Feb. 2023 Turkey earthquakes. Earthquake Spectra **30**(3), 1087–1115 (2023)
Chorowicz, J., et al.: The Maraş Triple junction (Southern Turkey) based on digital elevation model and satellite imagery interpretation. J. Geophys. Res. **99**(B10), 20225–20242 (1994)
ÇŞİDB Turkish Government, Ministry of Environment, Urbanisation and Climate Change
Cumhuriyet Newspaper (2023). Accepted 26 May 2023. https://www.cumhuriyet.com.tr/siyaset/son-dakika-suleyman-soylu-acikladi-depremde-can-kaybi-50-bin-783-oldu-2074031
DASK: Doğal Afet Sigortaları Kurumu, 6 Şubat 2023 Kahramanmaraş Depremleri Raporu, 11. 05. 2023, ANKARA (2023). Accessed 26 May 2023. https://www.sigortamedya.com.tr/wp-content/uploads/2023/03/Kahramanmaras-Deprem-Raporu_Bulten_270323.pdf
Dewey, J.F., Pitman, W.C., III., Ryan, W.B.F., Bonnin, J.: Plate tectonics and the evolution of the Alpine system. Geol. Soc. Am. Bull. **84**, 3137–3180 (1973)
Djamour, Y., Vernant, P., Nankali, H.R., Tavakoli, F.: NW iraneastern Turkey present-day kinematics: results from the Iranian permanent GPS network. Earth planet. Sci. Lett. **307**(1–2), 27–34 (2011)
Duman, T.Y., Emre, Ö.: The East Anatolian Fault: geometry, segmentation and jog characteristics. Geol. Soc. (Lond.) Spec. Publ. **372** (2013)
Durukal, E.: Establishment of antakya basin strong ground motion monitoring system. Geophys. Res. Abstracts **11**, EGU2009-4326-1, EGU General Assembly (2009)

EEFIT Mission 2023: Kahramanmaras EQ Sequence, youtube. Accessed 26 May 2023. https://www.youtube.com/watch?v=yjeywsgkKtg
EERI Buildings Team: February 6, 2023, Kahramanmaraş, Türkiye, Earthquakes - EERI Buildings Team Performance of Buildings, 23 May 2023
EERI Lifeline Team Report: Impact of the Kahramanmaraş Earthquakes on Hospitals: A Preliminary Report from the EERI Lifelines Reconnaissance Team (2023). Accessed 26 May 2023. https://www.eeri.org/images/eeri2021/2023/2023am/proceedings-files/turkey-special-session/05-2023AM-Lifelines_Team.pdf
EERI-GEER Webinar: February 6 2023 Turkiye Events GEER Advanced Team Reconnaisance Briefing (2023). Accessed 26 May 2023. https://www.eeri.org/what-we-offer/webinars/15718-joint-eeri-geer-webinar-2023-kahramanmaras-turkey-earthquake-sequence-reconnaissance-briefing
Erdik, M., Tümsa, M.B.D., Pınar, A., Altunel, E., Zülfikar, A.C.: A preliminary report on the February 6, 2023 earthquakes in Türkiye, Temblor report (2023). https://doi.org/10.32858/temblor.297
Erdik, M., Şadan, B., Tüzün, C., Demircioglu-Tumsa, M.B., Ülker, Ö., Harmandar, E.: Near-fault earthquake ground motion and seismic isolation design. In: Cimellaro, G.P. (ed.) Seismic Isolation, Energy Dissipation and Active Vibration Control of Structures. WCSI 2022. LNCE, vol. 309. Springer, Cham (2023). https://doi.org/10.1007/978-3-031-21187-4_9
Ergin, M., Aktar, M., Eyidoğan, H.: Present-day seismicity and seismotectonic of the Cilician Basin: eastern mediterranean region of Turkey. Bull. Seismol. Soc. Am. **94**, 930–939 (2004). https://doi.org/10.1785/0120020153
GEER Association Report: February 6 , 2023 Türkiye Earthquakes: Report on Geoscience and Engineering Impacts, GEER Association Report 082 (2023). Accessed 26 May 2023. https://learningfromearthquakes.org/images/2023_02_06_nurdagi_turkey/GEER_2023_Turkey_Earthquake_FullReport_ReducedSize.pdf
Global Rapid Post-Disaster Damage Estimation (GRADE) Report, February 6, 2023 Kahramanmaraş Earthquakes, 20 February 2023
Guidoboni, E., Comastri, A.: Catalogue of earthquakes and tsunamis in the mediterranean area from the 11th to the 15th Century. Istituto Nazionale di Geofisica e Vulcanologia (2005)
Gülerce, Z., Kargıooğlu, B., Abrahamson, N.A.: Turkey-adjusted NGA-W1 horizontal ground motion prediction models. Earthq. Spectra **32**(1), 75–100 (2016). https://doi.org/10.1193/022714EQS034M
Hancılar, U., Tüzün, C., Yenidoğan, C., Erdik, M.: ELER software-a new tool for urban loss assessment. Nat. Hazards Earth Syst. Sci. (2010)
HAZUS 4.3 SP3 (2020): HAZUS 4.3 SP3 (2020). https://www.fema.gov/sites/default/files/2020-10/fema_hazus_earthquake_technical_manual_4-2.pdf
Hempton, M.R.: Constraints on Arabian plate motion and extensional history of the Red Sea. Tectonics **6**, 687–705 (1987)
Hessami, K., Nilforoushan, F., Talbot, C.J.: Active deformation within the Zagros mountains deduced from GPS measurements. J. Geol. Soc. **163**(1), 143–148 (2006)
https://hasar.6subatdepremi.org/
https://hasartespit.csb.gov.tr/
https://www.usgs.gov/news/featured-story/m78-and-m75-kahramanmaras-earthquake-sequence-near-nurdagi-turkey-turkiye
Hussain, E., Kalaycıoğlu, S., Milliner, C.W.D., Çakir, Z.: Preconditioning the 2023 Kahramanmaraş (Türkiye) earthquake disaster. Nat. Rev. Earth Environ. (2023). https://doi.org/10.1038/s43017-023-00411-2
İMO-İnşaat Mühendisleri Odası: Kahramanmaraş Pazarcik Ve Elbistan Depremleri ön Değerlendirme Raporu, 14 Şubat 2023 (2023)

Insuranceday web site. Accessed 26 May 2023. https://insuranceday.maritimeintelligence.informa.com/ID1144441/Turkey-earthquake-insured-loss-raised-to-44bn-Perils
ITÜ: 6 $ubat 2023 04.17 Mw 7,8 Kahramanmaras (Pazarcik, Turkoglu), Hatay (Kirikhan), 13.24 Mw 7,7 Kahramanmaras (Elbistan / Nurhak-Cardak) Depremleri On İnceleme Raporu, İstanbul Technical University (2023)
Jackson, J., McKenzie, D.: Active tectonics of the Alpine-Himalayan belt between western Turkey and Pakistan. Geophys. J. R. Astron. Soc. **77**, 185–264 (1984)
Jörg, W.: Türkei: Die Südküste von Kaunos bis Issos. München Artemis-Verlag, Zürih (1986). ISBN3-7608-0791-7
Karabacak, V., et al.: The 2023 Pazarcık (Kahramanmaraş, Türkiye) earthquake (Mw 7.7): implications for surface rupture dynamics along the East Anatolian Fault Zone. J. Geol. Soc. **180**(3), jgs2023–020 (2023). https://doi.org/10.1144/jgs2023-020
Karabulut, H., Güvercin, S.E., Hollingsworth, J., Konca, A.Ö.: Long silence on the East Anatolian Fault Zone (Southern Turkey) ends with devastating double earthquakes (6 February 2023) over a seismic gap: implications for the seismic potential in the Eastern Mediterranean region. J. Geol. Soc. **180**, jgs2023–021 (2023). https://doi.org/10.1144/jgs2023-021
Kelling, G., Gökçen, S.L., Floyd, P.A., Gökçen, N.: Neogene tectonics and plate convergence in the Eastern Mediterranean: new data from southern Turkey. Geology **15**, 425–429 (1987)
Kempler, D., Garfunkel, Z.: Northeast Mediterranean triple junction from a plate kinematics point of view. Bull. Tech. Univ. Istanb. **44**, 425–454 (1991). Special Issue
Kikuchi, M., Kanamori, H.: Inversion of complex body waves-III. Bull. Seism. Soc. Am. **81**, 2335–2350 (1991)
Koçyiğit, A., Erol, O.: A tectonic escape structure: Erciyes pull-apart basin, Kay-Seri, central Anatolia. Turkey. Geodin. Acta **14**(2001), 133–145 (2001)
Kusky, T.M., Erdin, B., Jiannan, M., Lu, W.: Twin earthquakes devastate southeast Türkiye and Syria: first report from the epicenters. J. Earth Sci. **XX**(XX), 1–6 (2023). https://doi.org/10.1007/s12583-023-1317-5. http://en.earth-science.net
Kwiatek, G., et al.: Months-long preparation of the 2023 MW 7.8 Kahramanmaraş earthquake, Türkiye. Res. Square (2023). https://doi.org/10.21203/rs.3.rs-2657873/v1
Lovelock, P.E.R.: A review of the tectonics of the northern middle east region. Geol. Mag. **121**(6), 577–587 (1984)
Lyberis, N., Tekin, Y., Chorowicz, J., Kasapoğlu, E., Gündoğdu, N.: The East Anatolian Fault: an oblique collisional belt. Tectonophysics **204**, 1–15 (1992)
Mai, P.M., et al.: The destructive earthquake doublet of 6 February 2023 in south central Türkiye and northwestern Syria: initial observations and analyses. Seismic Rec. **3**(2), 105–115 (2023)
Masson, F., et al.: Large-scale velocity field and strain tensor in Iran inferred from GPS measurements: new insight for the present-day deformation pattern within NE Iran. Geophys. J. Int. **170**(1), 436–440 (2007)
Matsuoka, M., Ogushi, F., Nojima, N., Karimzadeh, S.: Damage proxy mapping in large areas due to the 2023 Turkey-Syria earthquakes using PALSAR-2 ScanSAR imagery (2023). https://rslut.ir/scansar-palsar-2-damage-proxy-map-of-m7-8-and-m7-5-earthquakes-in-turkey-syria-2023/
McClusky, S.C., et al.: Global positioning system constraints on plate kinematics and dynamics in the eastern Mediterreanen caucasus. J. Geophys. Res. **105**, 5695–5719 (2000)
McKenzie, D.P.: Plate tectonics of the Mediterranean region. Nature **220**, 239–343 (1970)
McKenzie, D.P.: Active tectonics of Mediterranean region. Geophys. J. R. Astron. Soc. **30**, 109–185 (1972)
McKenzie, D.P.: The East Anatolian Fault: a major structure in eastern Turkey. Earth Planet. Sci. Lett. **29**, 189–193 (1976)
Melgar, D., et al.: Sub- and super-shear ruptures during the 2023 Mw 7.8 and Mw 7.6 earthquake doublet in SE Türkiye. Seismica **2**(3) (2023). https://doi.org/10.26443/seismica.v2i3.387

Meng, J., et al.: Surface deformations of the 6 February 2023 earthquake sequence, eastern Türkiye. Science **383**, 298–305 (2024). https://doi.org/10.1126/science.adj3770

METU: Preliminary reconnaissance report on February 6, 2023, Pazarcık Mw = 7.7 and Elbistan Mw = 7.6, Kahramanmaraş-Türkiye Earthquakes. In: Çetin, K.Ö., Ilgaç, M., Can, G., Çakır, E. (eds.) February 20, 2023, Report No: METU/EERC 2023-01, Earthquake Engineering Research Center, Middle East Technical University (2023)

Mousavi, Z., et al.: Global positioning system constraints on the active tectonics of NE Iran and the south Caspian region, earth planet. Sci. Lett. **377–378**, 287–298 (2013)

Muehlberger, R.W., Gordon, M.B.: Observations on the complexity of the East Anatolian Fault. Turkey. J. Struct. Geol. **9**, 899–903 (1987)

NCEI/WDS: National Geophysical Data Center / World Data Service (NGDC/WDS): NCEI/WDS Global Significant Earthquake Database. NOAA National Centers for Environmental Information (1972). "Comments for the Significant Earthquake". NOAA National Centers for Environmental Information (2022). https://doi.org/10.7289/V5TD9V7K. Accessed 29 Mar 2022

Bijelić, N., Lin, T., Deierlein, G.G.: Quantification of the influence of deep basin effects on structural collapse using SCEC cybershake earthquake ground motion simulations. Earthquake Spectra **35**(4), 1845–1864 (2019). https://doi.org/10.1193/080418EQS197M

Okuwaki, R., Yagi, Y., Taymaz, T., Hicks, S.: Multi-scale rupture growth with alternating directions in a complex fault network during the 2023 south-eastern Türkiye and syria earthquake doublet (2023). https://doi.org/10.31223/X5RD4W

Över, S., Özden, S., Yılmaz, H.: Late Cenozoic stress evolution along the Karasu Valley, SE Turkey. Tectonophysics **380**(2004), 43–68 (2004)

Palano, M.: On the present-day crustal stress, strain-rate fields and mantle anisotropy pattern of Italy. Geophys. J. Int. **200**(2), 969–985 (2015)

Palano, M., Imprescia, P., Agnon, A., Gresta, S.: An improved evaluation of the seismic/geodetic deformation-rate ratio for the Zagros Fold-and-Thrust collisional belt. Geophys. J. Int. **2018**(213), 194–209 (2017). https://doi.org/10.1093/gji/ggx524

Perinçek, D., Çemen, I.: The structural relationship between the East Anatolian and Dead Sea fault zones in southeastern Turkey. Tectonophysics **172**, 331–340 (1990)

Petersen, G.M., et al.: The 2023 SOUTHEAST Türkiye seismic sequence: rupture of a complex fault network. Seismic Rec. **3**(2), 134–143 (2023). https://doi.org/10.1785/0320230008

Peyret, M., et al.: Present-day strain distribution across the Minab-Zendan-Palami fault system from dense GPS transects. Geophys. J. Int. **179**(2), 751–762 (2009)

Reilinger, R., et al.: GPS constraints on continental deformation in the Africa–Arabia–Eurasia conti-nental collision zone and implications for the dynamics of plate interactions. J. Geophys. Res., Solid Earth **111** (B5), B05411, 52 (2006)

Reliefweb-UNDP 2023 100 DAYS LATER100 DAYS LATERA view from A view from Türkiye and Syria. https://reliefweb.int/report/turkiye/100-days-later-view-turkiye-and-syria. Accessed 26 May 2023

Rosakis, A., Abdelmeguid, M., Elbanna, A.: Evidence of early supershear transition in the Feb 6th 2023 Mw 7.8 Kahramanmaras, Turkey Earthquake From Near-Field Records (2023, submitted)

Şaroğlu, F., Emre, Ö., Kuşçu, I.: The East Anatolian fault zone of Turkey. Ann. Tecton. **6**, 99–125 (1992)

Sbeinati, M.R., Darawcheh, R., Mouty, M.: The historical earthquakes of Syria; an analysis of large and moderate earthquakes from 1365 B.C. to 1900 A.D. Ann. Geophys. **48**(3), 347–435 (2005). Accessed 22 Sept 2011

Sbeinati, M.R., Darawcheh, R., Mouty, M.: The historical earthquakes of Syria: an analysis of large and moderate earthquakes from 1365 B.C. to 1900 A.D. Ann. Geophys. **48**(3), 347–435 (2005). Accessed 22 September 2011

Schmitt, R., Herman, M., Barnhart, W., Furlong, K., Benz, H.: (2023). https://earthquake.usgs.gov/storymap/index-turkey2023.html. Accessed 26 May 2023
Seismological Bulletins: 2023 Kandilli Observatory and Earthquake Research Institute – Boğaziçi University. İstanbul, Turkey. http://www.koeri.boun.edu.tr/new/en
Şengör, A.M.C.: Türkiye'nin neotektoniğinin esasları, Türkiye Jeoloji Kurumu. Konf Ser. **40**(2) (1980)
Şengör, A.M.C., Görür, N., Şaroğlu, F.: Strike-slip faulting and related basin formation in zone of tectonic escape: Turkey as a case study. In: Biddle, K.T., Christie-Blick, N. (eds.) Strike-Slip Deformation, Basin Deformation and Sedi-mentation. In: Special Publication of the Society of Economic Paleontologists and Mineralogists, vol. 37, pp. 227–264 (1985)
Sesetyan, K., Stucchi, M., Castelli, V., Gomez Capera, A.A.: Kahramanmaras - Gaziantep Turkiye M7.7 Earthquake, 6 February 2023 (04:17 GMT+03:00) Large historical earthquakes of the earthquake-affected region: a preliminary report Bogazig University Kandilli Observatory and Earthquake Research Institute Department of Earthquake Engineering (2023). https://eqe.boun.edu.tr/sites/eqe.boun.edu.tr/files/kahramanmaras-gaziantep_earthquake_06-02-2023_large_hist_eqs_v1.pdf. Accepted 26 May 2023
Sezen, H., Elwood, K.J., Whittaker, A.S.: Evolution of seismic building design and construction practice in Turkey. Struct. Des. Tall Build. (2000)
Shelter Sector Türkiye ShelterCluster.org Coordinating Humanitarian Shelter
Shen, Z.-K., Wang, M., Zeng, Y., Wang, F.: Optimal interpolation of spatially discretized geodetic data. Bull. Seism. Soc. Am. **105**(4), 2117–2127 (2015)
Sieberg, A.: Erdbeben geographie. In: Gutenberg, B. (ed.) Handbuch der Geo-physik, vol. 4, pp. 775–812. Berlin (1932)
Somerville, P., Graves, R.: Conditions that give rise to unusually large long period ground motions. Struct. Des. Tall Build. **2**(3), 211–232 (1993)
Soysal, H., Sipahioğlu, S., Kolçak, D., Altınok, Y.: Türkiye ve Çevresinin Deprem Katalogu, MÖ 2100-MS 1900 (Earthquake Catalogue of Turkey and Surroundings, BC 2100-AD 1900). TÜB˙ITAK Project (No. TBAG 341) 99 s (1981)
StEER 2023 Türkiye Earthquake Sequence Initiated 6 February 2023 Released 29 March 2023
Stein, R.S., et al.: Interactions, stress changes, mysteries, and partial forecasts of the 2023 Kahramanmaraş, Türkiye, earthquakes. Temblor (2023). https://doi.org/10.32858/temblor.299
Tan, O., Tapırdamaz, M.C., Yörük, A.: The earthquake catalogues for Turkey. Turk. J. Earth Sci. **17**, 405–418 (2008)
Tavakoli, F., et al.: Distribution of the right-lateral strike-slip motion from the Main Recent Fault to the Kazerun Fault System (Zagros, Iran): evidence from present-day GPS velocities. Earth Planet. Sci. Lett. **275**(3–4), 342–347 (2008)
Taymaz, T., Westaway, R., Reilinger, R. (eds.): Active faulting and crustal de-formation in the Eastern Mediterranean Region. Tectonophysics **391**, 1–374 (2004)
TC-CSBB 2023: Kahramanmaraş ve Hatay Depremleri Raporu, Turkiye Cumhuriye, Cumhurbaşkanlığı , Strateji ve Bütçe Başkanlığı, March 2023. https://www.sbb.gov.tr/wp-content/uploads/2023/03/2023-Kahramanmaras-and-Hatay-Earthquakes-Report.pdf. Accepted 26 May 2023
TDV-DASK: Güncellenmiş Türkiye Deprem Tehlikesi, Haritasina Göre Dask Deprem Sigortasi Primlerinin Güncellenmesi (2015)
TDV-DASK: Güncellenmiş Türkiye Deprem Tehlikesi, Haritasina Göre Dask Deprem Sigortasi Primlerinin Güncellenmesi (2018)
TERRA: "Türkiye Earthquakes Recovery and Reconstruction Assessment", Turkish Government – Directorate of Strategy and Budget (2023). https://www.sbb.gov.tr/wp-content/uploads/2023/03/Turkiye-Recovery-and-Reconstruction-Assessment.pdf. Accessed 26 May 2023
T-RUPT: Olay Bazli Deprem Modeli: M7.8 Pazarcik, M7.7Ekinozu, T-RUPT, TURK-RE, Turkey (2023)

USGS-Podcast (2023). https://podcasts.apple.com/us/podcast/earthquake-science-center-seminars/id413770595. Accessed 26 May 2023

Wu, F., Xie, J., An, Z., Ch, L.: Pulse-like ground motion observed from the February 6, 2023 Turkish MW7.8 earthquake. Earthq. Sci. **36** (2023)

Wu, Y.M., Kanamori, H., Allen, R.M., Hauksson, E.: Determination of earthquake early warning parameters, TC and PD, for southern California. Geophys. J. Int. **2007**(170), 711–717 (2007). https://doi.org/10.1111/j.1365-246X.2007.03430.x

Yönlü, Ö., Altunel, E., Karabacak, V.: Geological and geomorphological evidence for the southwestern extension of the East Anatolian Fault Zone, Turkey. Earth Planet. Sci. Lett. **469**, 1–14 (2017)

Yürür, M.T., Chorowicz, J.: Recent volcanism, tectonics and plate kinematics near the junction of the African, Arabian and Anatolian plates in the Eastern Mediterranean. J. Volcanol. Geotherm. Res. **85**, 1–15 (1998)

Zahradnik, J., Turhan, F., Sokos, E., Gallovic, F.: Asperity-like (segmented) structure of the 6 February 2023 Turkish earthquakes (2023). https://doi.org/10.31223/X5T666

Zarifi, Z., Nilfouroushan, F., Raeesi, M.: Crustal stress map of Iran: insight from seismic and geodetic computations. Pure. appl. Geophys. **171**(7), 1219–1236 (2014)

Digital Twinning of Communities for Disaster Risk Reduction: Climate Adaptive Solutions, John van de Lindt, Harold H. Short Endowed Chair Professor, Co-director, Center for Risk-Based Community Resilience Planning, Colorado State University, USA

John van de Lindt and Wilma James(✉)

GADRI Secretariat, DPRI, Kyoto University, Uji Campus, Uji, Japan
james.wilmatheonesta.6r@kyoto-u.ac.jp

Abstract. As global temperatures continue to rise there is increasing evidence that coastal storms are also increasing in frequency and intensity. These phenomena coupled with the effects of Sea Level Rise (SLR) as polar ice caps melt presents increasing challenges for coastal communities worldwide and requires urgent action to combat climate change to ensure these communities are sustainable and resilient. Resilience-based decision-making support is integral to this process to ensure community stakeholders have quantifiable information to inform their actions which may include mitigations and/or adaptation strategies. Such quantification requires community-level models of physical infrastructure such as buildings, roadways, and water and power networks to be fully coupled with household, social institution (e.g. hospitals and schools), and economic models. The Interconnected Networked Community Resilience Modeling Environment (IN-CORE) platform is a computational environment developed the Center for Risk-Based Community Resilience Planning that consists of a suite of tools to analyze, measure, and improve the resilience of communities and includes these physical, social and economic components thereby enabling the development of "digital twins" of communities. IN-CORE provides a framework to examine alternative policy choices and financial investments to improve community resilience. Several illustrative examples of community-level models will be presented to illustrate several strategies for disaster risk reduction to climate change.

Keywords: Digital twinning · communities · disaster risk reduction · climate adaptive solutions · resilience

1 Presentation

In his keynote address, Prof. John van de Lindt emphasized the need to enhance resilience, particularly in the face of challenges like climate change and related sea level rise (SLR). He explained the process of measuring the resilience of entire cities or communities,

H. Tatano et al. (Eds.): GSRIDRR 2023, DRRGADRIBS, pp. 68–69, 2025.
https://doi.org/10.1007/978-981-96-6609-6_5

highlighting the importance of not only measuring but also understanding how we can change policies to adapt.

Addressing this uncertainty requires interdisciplinary modeling that includes the physical infrastructure, social institutions, households, and the economy. Prof. van de Lindt highlighted an open-source computational environment, or model, known as IN-CORE, which standard for the Interdependent Networked Community Resilience Modeling Environment, to predict damage at the individual building level, taking into account factors such as race/ethnicity, income, and other demographics. By working with census data and this model can facilitate socially equitable decisions related to community resilience planning.

Professor van de Lindt mentioned that U.S. Presidential Policy Directive 21 provides a guiding definition for understanding resilience as the capacity to prepare for, adapt to, withstand, and recover rapidly from disruptions.

Professor van de Lindt then emphasized the importance of adapting to changing conditions and climate, as well as the challenges in modeling recovery. Ultimately, the desire to improve community resilience and city resilience requires measuring what can, and cannot be seen, touched or felt. Modeling tangible aspects is relatively easier and incorporating unseen factors like the economy and its interactions can be complex. Prof. van de Lindt continued to explain that by integrating various models and data sources, they aim to capture the complexity of community resilience. Considering the five areas of community stability ensures an understanding of resilience, population dynamics, economic factors, social and physical services as well as governance. This multi-disciplinary approach is effective for modeling and planning.

Prof. van de Lindt continued to address crucial questions about resilience metrics and their implications. Defining what to measure and who decides those metrics are fundamental to assessing resilience accurately. Additionally, considering human decisions in resilience calculations underscores the importance of behavioral factors. Lastly, understanding how to utilize measured values effectively is key for implementing resilience strategies.

The Center for Risk-Based Community Resilience Planning, is funded by NIST, the National Institute for Standards and Technology, and is essential for understanding and mitigating hazards. By using complex models like IN-CORE, the users are able to generate data to calculate the real costs of hazards, including impacts on people and the economy. In the long term, this helps stakeholders make informed decisions regarding resilience and climate change.

Prof. van de Lindt emphasized the importance of modeling resilience before, during, and after the hazards like earthquakes and floods. While physical damage modeling is essential, it's only part of the picture. The true digital twin encompasses people, the economy, and many other factors.

Providing a comprehensive measurement of resilience in terms of stability areas is critical for providing optimal solutions that align with community goals. The ultimate goal to improving resilience is through effective communication, decision-making and actionable strategies.

Addressing Climate Change and Disaster Risks in the Context of Sustainable Development in Zimbabwe

Chipo Mudavanhu(✉)

Bindura, Zimbabwe
chipomuzenda@gmail.com

Abstract. Climate change poses a significant threat to Zimbabwe's sustainable development trajectory. Rising temperatures, erratic rainfall patterns, and increasingly frequent extreme weather events disrupt livelihoods, ecosystems, and progress towards achieving the Sustainable Development Goals (SDGs). This paper explores this vulnerability and emphasizes the importance of integrating Sustainable Development (SD), Climate Change Adaptation (CCA) and Disaster Risk Reduction (DRR) strategies. Key challenges in this integration include siloed policy frameworks, limited social equity considerations, and inadequate financing and capacity. It proposes solutions such innovative financing mechanisms, capacity building initiatives, and effective communication strategies for empowering communities. Furthermore, the paper explores the opportunities presented by SD, including the potential for green jobs, enhanced societal resilience, and greater social equity. Finally, it focuses on community participation in building long-term resilience and achieving the ambitious goals outlined in the SDGs. This paper offers a roadmap for Zimbabwe to navigate the complexities of climate change and build a more secure and sustainable future by integrating CCA, DRR, and SD.

Keywords: climate change · disaster risk reduction · sustainable development · adaptation · vulnerability

1 Introduction

Climate change presents a significant threat to countries across the globe, and Zimbabwe is no exception. Consistent with observations across sub-Saharan Africa such as by Funk et al., (2023), Zimbabwe confronts a multitude of challenges due to climate change. Rising temperatures have disrupted historical precipitation patterns, resulting in a hotter and drier climate (Chenje et al., 2017; Moyo et al., 2019). This shift has cascading consequences for Zimbabwe's agricultural productivity (Pedzisayi et al., 2022), water security (Makuvaro et al., 2020), public health (Maphosa et al., 2022), and infrastructure (Nyasha & Mukanganise, 2018). This paper analyzes the current challenges posed by climate change and disasters in Zimbabwe, exploring the impacts on various sectors, and investigating strategies for mitigating risks and promoting sustainable development.

H. Tatano et al. (Eds.): GSRIDRR 2023, DRRGADRIBS, pp. 70–78, 2025.
https://doi.org/10.1007/978-981-96-6609-6_6

2 Climate Change and Disaster Risks in Zimbabwe

Climate change poses significant challenges to Zimbabwe's sustainable development trajectory. Zimbabwe is highly vulnerable to the impacts of climate change, experiencing rising temperatures, erratic rainfall patterns, and an increase in the frequency and intensity of extreme weather events like droughts, floods, and cyclones. These events translate into disasters, causing widespread damage to infrastructure, livelihoods, and ecosystems, hindering progress towards poverty reduction (SDG 1) and environmental sustainability (SDG 13). For instance, recurrent droughts disrupt agricultural production, leading to food insecurity and malnutrition (SDG 2).

One of the most concerning aspects of climate change in Zimbabwe is the observed rise in average temperatures (Chenje et al., 2017). This warming trend disrupts historical precipitation patterns (Moyo et al., 2019), leading to a shift towards a hotter and drier climate. These changes have significant implications for weather patterns, with an increased likelihood of extreme weather events such as droughts, floods, and heatwaves (Funk et al., 2023). According to IPCC (2014a), the earth has experienced a general increase in temperature by about 0.6 °C over the past 1000 years and the mean annual temperature has increased by ~0.01 to 0.02 °C/year over the time period 1950–2002 (USAID Fact Sheet, 2015). The period from 1980 to date has been the warmest since Zimbabwe started recording its temperature. There is a shift to hotter days and nights with the hottest year being recorded in 2016. Several studies have shown that Zimbabwe's climate will be warmer than the 1961–1990. This was confirmed by Unganai (1996) who found an increasing trend in mean maximum temperatures of 0.1 °C per ten-year period between 1933 and 1993, with an overall increase of up to 0.8 °C in Zimbabwe. Furthermore, Mugabe et al. (2012) predicted a 1.5–2 °C increase in annual maximum temperatures for the period up to 2050 in most of the country. Thus, warming rates of 0.5–2 °C by 2030, 1–3.5 °C by 2070, and 3–4 °C by 2100 were projected (Engelbrecht and Bopape, 2009).

Erratic rainfall patterns have emerged as a hallmark of climate change in Zimbabwe. While long-term average rainfall totals may not exhibit a clear upward or downward trend, the variability has become more pronounced (Moyo et al., 2019). This is characterized by periods of exceptionally high rainfall, like those observed in the 1970s and 2000s, juxtaposed with extended dry spells that dominated the 1980s and early 1990s (Chenje et al., 2017). This increased variability translates to a growing uncertainty in the timing and amount of rainfall received during any given season, making agricultural planning and water resource management significantly more challenging. The IPCC (2022) has emphasized that these extreme weather events, including floods, droughts, and cyclones, are projected to become more frequent and intense as global temperatures continue to rise. This shift not only disrupts agricultural production but also increases the risk of weather-related disasters with cascading consequences for human and ecological well-being.

Household vulnerability in Zimbabwe is a multifaceted challenge shaped by a complex interplay of social, economic, and environmental factors (Moyo et al., 2019). Inequitable access to resources and means of production, including land and infrastructure, disproportionately burdens poor households (WEDO, 2023). Limited access to education further hinders a household's ability to adapt to changing circumstances. Gender inequality creates a situation where women, often with less decision-making power and resource access, face even greater vulnerability (Vincent et al., 2021). Furthermore, dependence on climate-sensitive resources such as rain-fed agriculture makes households particularly susceptible to erratic rainfall patterns and droughts. These vulnerabilities are further compounded by pre-existing health conditions, such as HIV/AIDS, which can be exacerbated by climate change impacts such as food insecurity and water scarcity, creating a vicious cycle (Makarau et al., 2019). Disease outbreaks that include cholera and dysentery, more likely during floods or droughts with poor sanitation, further threaten household well-being (Maphosa et al., 2022). However, vulnerability is not uniform. Studies by Vincent et al. (2021) highlight that the ability to diversify livelihoods can create opportunities even amidst challenges, suggesting a nuanced understanding of vulnerability across different contexts is crucial.

Climate change also amplifies macro-level vulnerabilities in Zimbabwe, impacting various socio-economic sectors through extreme weather events such as erratic rainfall and frequent droughts (Makuvaro et al., 2020). These vulnerabilities are further compounded by social, economic, and political processes that influence human settlements, agricultural practices, and natural resource management (Nhamo et al., 2019). For instance, recurrent droughts lead to water stress and scarcity, jeopardizing agricultural production. This not only threatens food security (SDG 2) but also human security, as water scarcity can exacerbate existing social tensions. Furthermore, the "perennially high incidence of malaria" in Zimbabwe, as documented by Makarau et al. (2019), poses a significant public health threat and deters tourism, a critical economic driver (Govere & Baudouin, 2020). This highlights the interconnectedness of climate change impacts, with vulnerabilities cascading across water, agriculture, health, and tourism sector. Addressing these challenges requires a multifaceted approach that integrates DRRstrategies, climate-smart development practices, and investments in social safety nets.

3 Addressing Climate Change and Disaster Risk Challenges Within the Framework of Sustainable Development

Disaster risk, climate change, and sustainable development (SD) are intricately linked, forming a complex challenge for nations worldwide. Climate change, as documented by the IPCC (2022), intensifies the frequency and severity of natural hazards, increasing disaster risk (Aitsi-Selmi et al., 2023). This jeopardizes progress towards achieving the SDGs (UNDRR, 2023). For instance, extreme weather events disrupt agricultural production (Bhat et al., 2021), jeopardizing food security (SDG 2) and livelihoods (SDG 1). Furthermore, damaged infrastructure from disasters hampers accessibility to clean water and good sanitation (SDG 6) and healthcare services (SDG 3). Therefore, building resilience to climate-related disasters is crucial for achieving sustainable development. Investing in early warning systems (Nyasha & Mukanganise, 2018) and climate-smart infrastructure (Aerts et al., 2014) can significantly reduce disaster risk, fostering a more sustainable future. Integrating DRR and CCA strategies within a SD framework offers a pathway to build resilience and safeguard development gains (Schipper & Vanessa, 2021).

Unlike traditional development models that prioritize short-term economic gains at the expense of the environment and social well-being, sustainable development, as outlined by the Brundtland Commission (1987), offers a holistic framework for addressing complex challenges such as climate change and DRR in Zimbabwe. SD emphasizes a "triple bottom line" approach, integrating economic development, social equity, and environmental protection (Moore, 2022). This integrated approach is crucial for Zimbabwe, as a siloed approach to climate change and disaster risk is demonstrably insufficient (Aitsi-Selmi et al., 2023). DRR as emphasized by the Sendai Framework for DRR (UNDRR, 2023), focuses on preparedness, early warning systems, and risk mitigation to reduce vulnerability to hazards. CCA complements DRR by aiming to adjust to the changing climate through strategies like developing drought-resistant crops and improving water management (Aerts et al., 2014). By integrating DRR and CCA within a SD framework, Zimbabwe can build resilience to climate-related disasters and achieve long-term prosperity.

Within the framework of SD, economic development takes on a new meaning, transcending mere growth. It focuses on building a robust and equitable economy that offers decent jobs and livelihoods for all, ensuring long-term prosperity without compromising the environment (Moore, 2022). To achieve SD, we need a transition to economic models that prioritize decoupling growth from resource depletion. Green jobs, particularly those in renewable energy, resource efficiency, and pollution control, can be a powerful driver of this shift. These green jobs not only create economic opportunities but also contribute to environmental protection, fostering a win-win scenario for both economic prosperity and environmental sustainability.

Social equity takes centre stage alongside environmental protection and economic development. It recognizes that social justice, human rights, and a reduction in poverty and inequality are fundamental for building long-term resilience in the face of climate change and disaster risks (Aitsi-Selmi et al., 2023). SD emphasizes that everyone can contribute to and benefit from a sustainable future, fostering a sense of shared responsibility. The UN's SDGs (UN, 2015) provide a blueprint for achieving this, highlighting the importance of addressing issues like gender equality, access to education, and decent work opportunities (SDG 5, 4, and 8 respectively). By promoting social equity, SD fosters a more inclusive and resilient society, better equipped to manage the challenges posed by climate change and disasters. This is because empowered and educated communities are better equipped to be prepared for, able to respond to, and recover from extreme weather events and other climate-related disasters (Wisner et al., 2020).

Environmental protection acts as the foundation for a healthy planet that can sustain life for generations to come. It prioritizes responsible resource management, biodiversity conservation, and climate change mitigation (Moore, 2022). This focus on environmental well-being is crucial in the face of climate change and its associated disaster risks. SD encourages practices that decouple economic growth from environmental degradation. The Paris Agreement exemplifies a global effort towards environmental protection (UNFCCC, 2015). By reducing the emission of greenhouse gases and promoting climate change mitigation strategies, the Paris Agreement aims to create a more resilient planet, thereby reducing climate induced disasters (Höhne et al., 2020).

4 Challenges in Integrating CCA, DRR and SD

The integration of CCA, DRR, and SD poses significant challenges, hindering progress towards a resilient future. One key obstacle lies in the siloed nature of policy frameworks. Existing policies often address these issues in isolation, neglecting the interconnectedness of climate risks, vulnerabilities, and development pathways (Aitsi-Selmi et al., 2023). This compartmentalization leads to missed opportunities for synergies and in some cases duplication of efforts. Furthermore, a lack of convergence between CCA and DRR strategies can create confusion and hinder effective risk management. DRR strategies may not adequately address slow-onset climate changes, while CCA efforts might fail to consider the immediate risks posed by extreme weather events (UNISDR, 2015). Additionally, achieving social equity within this framework remains a challenge. Traditional approaches often overlook the specific vulnerabilities faced by marginalized groups, such as women or those living in poverty (Aitsi-Selmi et al., 2023). Finally, limited financial resources and capacity constraints can further impede progress. Developing countries like Zimbabwe often lack the resources for essential adaptation measures, while capacity gaps in areas like infrastructure development and early warning systems hinder preparedness for disasters (World Bank, 2022). Overcoming these challenges requires a holistic approach that integrates SD, DRR, and CCA strategies, with a focus on social equity and community participation. Additionally, innovative financing mechanisms and capacity building initiatives are crucial for building long-term resilience in the face of climate change.

Striking a balance between economic growth and environmental protection is often a challenge. Businesses may be hesitant to adopt costly green technologies or resource-efficient practices that could initially impact profitability (Jakob et al., 2015). Bridging the technological gap is another hurdle. Developing countries may lack the resources or infrastructure to implement cutting-edge sustainable solutions (UNEP, 2023).

There is also a barrier to effective CCA in Zimbabwe lies in the disconnect between the generation and utilization of climate information. While data may be collected by various institutions, stakeholders at the community level often report that this information is presented in a complex and technical manner (Moyo et al., 2023). This technical jargon makes it difficult for local communities to understand and translate the data into concrete actions for adaptation. Furthermore, media coverage on climate change, particularly regarding long-term trends and their connection to daily weather events, is often limited (Manzungu et al., 2021). This lack of clear and accessible information is compounded by inadequate communication infrastructure in rural areas and limited capacity at the local level to utilize even simplified information (Moyo et al., 2023). This breakdown in knowledge sharing between data holders and communities weakens public awareness and engagement in crucial adaptation efforts. Effective communication strategies that bridge this gap are essential for empowering communities to become active participants in building resilience to climate change.

In addition, there is lack of convergence between DRR and CCA efforts. Existing policy frameworks, such as the Civil Protection Act of 1989, often address climate change and disaster risk in isolation, leading to a siloed approach that fails to capture the interconnected nature of climate-related risks (Aitsi-Selmi et al., 2023). Furthermore, the Civil Protection Act's limited focus on community participation in building resilience stands in stark contrast to best practices in disaster risk management (UNISDR, 2015). By failing to empower communities as active stakeholders in risk reduction processes, the Act hinders long-term preparedness and the sustainability of adaptation measures. A more holistic approach that integrates DRR and CCA strategies, with a strong emphasis on community engagement, is crucial for building robust climate resilience in Zimbabwe.

Integrating CCA, disaster risk reduction, and SD presents a complex challenge, impeding progress towards building resilient societies (UNFCCC). While all three domains are crucial for a sustainable future, fragmented approaches often lead to inefficiencies and missed opportunities for synergy (Wisner, et al, 2004). A key challenge arises from the inherent differences in timescales. CCA strategies focus on long-term trends like rising sea levels, while DRR prioritizes immediate preparedness for discrete events like floods. This creates potential tension in resource allocation, as evidenced by a recent study by Moore et al. (2024) which found that short-term funding cycles often hinder investments in long-term adaptation measures. Furthermore, fragmented institutional frameworks can significantly hinder progress. A 2023 report by the United Nations Office for DRR (UNDRR) emphasizes the need for improved inter-ministerial coordination to bridge the gap between policy and implementation (UNDRR, 2023). Finally, knowledge gaps and limited data sharing restrict the development of robust, evidence-based policies. A 2022 World Bank report highlights this challenge, noting it limits our ability to address the complex interplay between these crucial domains (World Bank, 2022). Overcoming these challenges necessitates a paradigm shift towards integrated approaches that leverage the strengths of each domain, fostering long-term resilience in a rapidly changing climate.

While achieving SD presents its own set of challenges, the opportunities it offers are substantial. Transitioning to a green economy unlocks a multitude of pathways for economic growth. The World Bank (2023) projects that a shift towards renewable energy and resource-efficient production systems can create millions of new jobs and unlock significant investment opportunities. Furthermore, social innovation directed at tackling environmental and social challenges can be a powerful driver of economic development (UNDP, 2023). Beyond economic benefits, SD fosters enhanced societal resilience. Integrating environmental considerations into infrastructure projects leads to more robust systems better equipped to withstand the impacts of climate change, such as extreme weather events (Biesbroek et al., 2013). Similarly, sustainable practices that conserve water resources and promote healthy ecosystems bolster a society's ability to cope with environmental shocks. Finally, SD promotes greater social equity. By ensuring access to clean water, sanitation, and sustainable livelihoods for all, particularly disadvantaged communities, it fosters a more just and equitable society (Amartya Sen, 1999).

Moving beyond top-down approaches, meaningful community participation is crucial for success. Local knowledge plays a vital role in identifying vulnerabilities and crafting context-specific solutions (Agrawal, 2008). Studies have shown that participatory processes foster a sense of ownership over solutions, leading to increased commitment and long-term sustainability (Schipper & Langston, 2015). Furthermore, collaboration strengthens social capital and builds trust between communities and authorities, enhancing communication and response during disasters (Adger et al., 2013). This shift towards empowering communities and leveraging their knowledge is essential for building resilience and achieving the ambitious goals outlined in the SDGs (UN, 2015).

In conclusion, addressing challenges within the framework of SD is critical for building resilience and achieving a sustainable future is an imperative. By integrating CCA, DRR, and SD strategies, we can create a more equitable and environmentally sound future for all. This approach offers a comprehensive roadmap for building a future where economic well-being, social equity, and environmental health go together. By embracing a united front under the banner of sustainable development, we can collectively navigate the complexities of the present and create a more prosperous and equitable future.

References

Aerts, J.C.J.H., Linham, M.M., Rogan, J.C.: Adapting to climate change: evaluating the effectiveness of flood risk management measures. Nat. Hazards **73**(2), 1043–1064 (2014)

Aitsi-Selmi, A., Leal Filho, W., Segnon, A.C.: The role of climate change in DRR in developing countries: a systematic review of the literature. Int. J. Disaster Risk Sci. **14**(1), 1–17 (2023)

Alder, J.R., Hostetler, S.W.: CMIP5 Global Climate Change Viewer. US Geological Survey (2013)

Berrone, P., Fosfuri, A., Gelabert, L.: Environmental and economic performance: evidence from a sample of S&P 500 firms. Nat. Sustain. **5**(2), 117–125 (2022)

Bhat, M.A., Wani, S.A., Dar, W.A.: Climate change and its impact on Agriculture in Kashmir: an overview. J. Appl. Nat. Sci. **13**(2), 242–248 (2021)

Brown, et al.: Climate Change Impacts, Vulnerability, and Adaptation in Zimbabwe. IIED Climate Change Working Paper Series (2012)

Brundtland Commission: Our Common Future. World Commission on Environment and Development. Oxford University Press (1987)

Chenje, C.M. (ed.): State of the Environment Report for Zimbabwe 2015. Harare: Ministry of Environment, Water and Climate (2017)

Chuma, T.: Population dynamics and climate change: a case study of Zimbabwe. Leadership for Environment and Development (LEAD), Harare, Zimbabwe (2012)

Davis, Hirji: Climate Change and Water Resources Planning, Development and Management in Zimbabwe: An Issues Paper. World Bank (2014)

Engelbrecht, F., Bopape, M.J.: Projections of Future Climate Change over Southern Africa. CSIR Natural Resources and the Environment Atmospheric Modelling (2009)

Food and Agriculture Organization (FAO): Climate Change and Adaptation in the Food and Agriculture Sector; Food and Agriculture Organization of United Nations (2003)

Intergovernmental Panel on Climate Change (IPCC): Climate Change 2022: Impacts, Adaptation, and Vulnerability. Contribution of Working Group II to the Sixth Assessment Report of the Intergovernmental Panel on Climate Change (2022). https://www.ipcc.ch/report/ar6/wg1/

IPCC: Climate Change: Impacts, Adaptation and Vulnerability: IPCC WGII AR5 (2014a)

Funk, C., et al.: Assessing recent climate change in sub-Saharan Africa using satellite remote sensing and weather models. Earth Syst. Sci. Data **15**(1), 891–913 (2023)

Govere, C., Baudouin, G.: Climate change and tourism in Sub-Saharan Africa: a review of key issues and challenges. Curr. Issue Tour. **23**(13), 1751–1772 (2020)

Höhne, O., Kuramochi, T., Rogelj, J.: The Paris Agreement: resolving the unfinished business of climate change. Climate Policy **20**(8), 1022–1035 (2020)

Makarau, A., Kudzayi, D., Gwatidzo, T.: The impact of climate change on mental health in Zimbabwe. Afr. J. Prim. Health Care Fam. Med. **11**(1), 1–6 (2019)

Makuvaro, E., Gwenzi, Y., Mandibaya, P.: Climate change and water security in southern Africa: a review of governance challenges and opportunities for transformation. Water Altern. **13**(2), 567–588 (2020)

Manzungu, E., Mushonga, B., Gwenzi, Y.: Climate change communication in rural Zimbabwe: a focus on knowledge, attitudes and practices. J. Nat. Conserv. **59**, 126024 (2021)

Maphosa, B., Bumhira, A., Mushonga, B.: Climate change and diarrhoeal diseases in Zimbabwe: a systematic review and meta-analysis. Int. J. Infect. Dis. JID **120**, 103–110 (2022)

Moore, B.: Sustainable Development: Combining Economic Prosperity, Social Equity, and Environmental Protection (2022). https://www.investopedia.com/sustainable-investing-4427774

Moyo, B.R., Nicholson, S.E., Davies, J.C.: Recent changes in agriculturally relevant rainfall characteristics in southern Africa. Int. J. Climatol. **39**(5), 2475–2488 (2019)

Moyo, B., Nyasha, J., Makonese, T.: Understanding local knowledge systems and perceptions of climate change in rural communities of Zimbabwe. J. Environ. Agric. Sustain. Dev. Countries **11**(2), 187–205 (2023)

Nhamo, S., Ndlela, T., Rurinda, J.: Vulnerability of rural livelihoods in southern Africa to climate change: the case of Guruve district Zimbabwe. Clim. Dev. **11**(3), 242–256 (2019)

Nyasha, J., Mukanganise, J.: CCA Strategies for Infrastructure Development in Zimbabwe. J. Sustain. Dev. **11**(2), 221 (2018)

Mugabe, F.T., Hachigonta, S., Sibanda, L.M., Thomas, T.S.: 'Zimbabwe'. In: Southern African Agriculture and Climate Change: A Comprehensive Analysis. IFPRI Monograph, Washington, DC: International Food Policy Research Institute (2012)

Mugandani, R., Wuta, M., Makarau, A., Chipindu, B.: Re-classification of agro-ecological regions of Zimbabwe in conformity with climate variability and change. Afr. Crop Sci. J. **20**(2), 361–369 (2012)

Paris Agreement to the United Nations Framework Convention on Climate Change (2015). https://unfccc.int/files/essential_background/convention/application/pdf/english_paris_agreement.pdf

Pedzisayi, R.W., Rurinda, J., Weisser, D.: CCA strategies for water security in southern Africa: a review. J. Hydrol. Reg. Stud. **38**, 101022 (2022)

Solomon, S, et al. (eds.): Climate Change 2007: The Physical Science Basis. Contribution of Working Group I to the Fourth Assessment Report of the Intergovernmental (2007)

Tadross, M., Jack, C., Hewitson, B.: On RCM-based projections of change in southern Africa summer climate. Geophys. Res. Lett. **32** (2005)

Unganai, L.S.: Historic and future climatic change in Zimbabwe. Clim. Res. **6**, 137–145 (1996)

United Nations Environment Programme (UNEP): The Green Jobs Initiative: Pathways to a Sustainable Future (2018). https://www.unep.org/events/workshop/green-jobs-future

United Nations (UN): Transforming Our World: The 2030 Agenda for Sustainable Development (2015). https://sdgs.un.org/goals

United Nations Framework Convention on Climate Change (UNFCCC): Paris Agreement (2015). https://unfccc.int/process-and-meetings/the-paris-agreement

United Nations Office for DRR (UNDRR): The Sendai Framework for DRR2015-2030 (2023). https://www.undrr.org/publication/sendai-framework-disaster-risk-reduction-2015-2030

USAID: Climate Change Information Factsheet, Zimbabwe, USAID (2015)

Vincent, K., Dougill, A.J., Fletcher, L., Rutten, M.: Climate change, forced migration, and livelihood insecurity in southern Africa. Climate Dev. **13**(2), 184–197 (2021)

Wisner, B., Pelling, M., Romero- Lankao, P.: Vulnerability: a multi-faceted perspective for disaster risk reduction. Prog. Hum. Geogr. **44**(1), 25–40 (2020)

Wisner, B., Blaikie, P., Cannon, T., Davis, I.: At Risk: Natural Hazards, People's Vulnerability and Disasters, 2nd edn. Routledge (2004)

United Nations Office for DRR (UNDRR). Global Report on DRR2023 (2023)
World Bank: World Development Report 2022: Finance for a Changing Climate (2022)
World Bank: Greening Growth: Investment Opportunities in Low-Carbon, Resilient Infrastructure (2023). [link to report - World Bank website]
Zimbabwe Metrological Services: Annual report. Harare, Zimbabwe (2012)

New Challenges for Risk Governance

Polycrisis and Systemic Risks

Ortwin Renn(✉)

Research Institute for Sustainability (RIFS) - Helmholtz Centre Potsdam, Potsdam, Germany
ortwin.renn@rifs-potsdam.de

Abstract. In recent years, the world has been exposed to several serious and simultaneously occurring crises. Risk researchers call them polycrises. First came Corona, then came the further manifestations of climate change, for example floods, draughts and forest fires, the war in Ukraine and Palestine, food crises in the world, inflation, galloping energy prices, and there are more every day. The hallmark of polycrises is the mutual amplification of nested, interconnected risks. Given these new context conditions and challenges, the traditional approach to risk assessment and management will not suffice to cope with the multiple facets of polycrisis. A new concept is required that addresses the systemic nature of contemporary risks. The paper will introduce such a new concept and explain the main features of a systemic risk framework that is designed to improve our understanding of polycrisis and provide better tools for managing and governing systemic risks.

Keywords: Systemic risks · polycrisis · risk governance · basic risk agents · complexity · uncertainty · ambiguity

1 Introduction

In the course of the last two decades, the world has faced an unprecedented convergence of severe crises occurring simultaneously, a phenomenon described by risk researchers as polycrises (Homer-Dixon et al. 2024; Homer-Dixon and Rockström 2022; Renn 2024). Events such as the COVID-19 pandemic, escalating climate impacts—including floods, droughts, and wildfires—ongoing conflicts like the war in Ukraine and Palestine, global food shortages, inflation, and soaring energy costs highlight the interconnected nature of these challenges. The defining feature of a polycrisis lies in how these interconnected risks amplify one another. For instance, the COVID-19 pandemic disrupted global supply chains, while the war in Ukraine led to grain supply shortages, creating cascading effects that exacerbate food insecurity worldwide, particularly in regions like India, China, Africa, and Europe (Trabucco 2022).

Every crisis—be it political, epidemiological, military, economic, or environmental—compels societies to rethink their assumptions about normalcy and imminent risks. One crisis, such as a pandemic, may shake societal foundations, while another, such

H. Tatano et al. (Eds.): GSRIDRR 2023, DRRGADRIBS, pp. 79–93, 2025.
https://doi.org/10.1007/978-981-96-6609-6_7

as a natural disaster, may trigger secondary crises like industrial accidents or chemical spills. At the same time, governance failures often hinder effective responses to these multifaceted challenges. These overlapping crises form a complex web of seemingly distinct issues that are, in reality, deeply intertwined (Lawrence et al. 2022). Their interconnectedness creates compounded impacts, resulting in global damage far exceeding the sum of individual crises.

The contemporary polycrisis landscape is characterized by several key attributes (WPKS2023; Renn 2024):

- Multiple crises unfolding simultaneously, each influencing and reinforcing the others.
- A lack of clear, optimal solutions that perform better across all criteria than alternative options.
- A pressing need for a systemic understanding of how social, natural, technological, and cultural domains interact.
- The imperative to identify and navigate conflicting goals while making resilient and ethically sound trade-offs.
- Recognition that single-focus optimizations often lead to excessive losses in other, equally critical areas.
- Transparent communication about conflicting goals and the trade-offs they entail.
- Inclusive decision-making processes that engage stakeholders and affected populations to develop collaborative solutions within limited timeframes.

Given these realities, traditional methods of risk assessment and management are insufficient to address the challenges posed by polycrises. A paradigm shift is necessary—one that embraces the systemic nature of modern risks. This paper proposes a novel systemic risk framework designed to enhance our understanding of polycrises and to offer improved strategies for managing and governing such interconnected risks. It seeks to answer a critical question: How can risk assessment and management be structured to effectively address the complexities of systemic risks and polycrises?

2 What are Systemic Risks?

Systemic risks refer to highly interconnected experiences of multiple hazards with effects that are difficult to assess in terms of scope, depth and time horizon, and which are associated with considerable knowledge and assessment problems due to the complexity, uncertainty and ambiguity of potential consequences (Renn 2014; 2016; Renn et al. 2020). The recent coronavirus pandemic can be seen as a prime example of a systemic risk displaying it transboundary nature and its complex web of interactions with other risks and human activities (Collins et al. 2020). The diagnosis of "delimited" trans-sectoral risks is a central element of systemic risks, in which the consequences of modern science and technology are interpreted from the perspective of "reflexive modernization", a theory formulated primarily by Ulrich Beck (for an overview of this literature, see Beck 2009; Adam et al. 2000; Rosa et al. 2014: 69ff.). According to the theory of reflexive modernization, the problematic nature of the risks generated by industrial and technological dynamics does not primarily lie in the health and ecological consequences that are at the heart of risk analysis and management within the professional risk community. Rather, it arises from the "social, economic and political side-effects of these side-effects" (Beck 1992). Applied to the recent pandemic, it is not so much the health damage caused by the pandemic as the indirect damage to the economy, politics and the social system that is increasingly dominating social discourse and creating conflicts and controversies (Collins et al. 2020).

What are systemic risks? In the financial sector, risks are described as systemic if the functionality of the financial market or a relevant part of this market is threatened when damage occurs. Or, in a nutshell: systemic risk refers to the possibility that a catastrophic event could affect the vital systems on which our society is based (Renn 2014). The risk behaves here like a pathogen. It also infects other systems or services which are vulnerable because of interdependencies (de Bandt et al. 2000: 10; Bubb and Zimmermann 2004). Particularly prominent are the effects of technical failures of or human attacks on critical infrastructures. They can be felt in the consequences of economic activity on the social and cultural experience of injustice and identity (Folke 2006; World Economic Forum 2013). Last but not least, the world experiences systemic risks in the area of infectious diseases and epidemics.

Several years before the financial crisis, the OECD had already set up a commission of experts to deal with systemic risks in all their breadth (OECD 2003). In addition to financial risks, which were already perceived as threatening at the time, the OECD experts selected global pandemics, flood disasters, nuclear accidents, terrorist attacks and global food poisoning as candidates for this new category of risks for their final report published in 2003 (OECD 2003: 96). The report had hardly any echo in the policy arenas and was widely ignored in the policy arenas of the world, yet it stimulated an intensive debate among risk scholars.

3 A New Perspective on Systemic Risks

Since the release of the OECD report, the concept of systemic risks has steadily gained prominence (Renn et al. 2022). Scholars have increasingly used the term *polycrisis* to describe the experience of intersecting risks (Lawrence et al. 2022; 2024). These interacting systemic risks defy conventional modeling approaches due to their boundary-crossing nature and the intricate causal structures they form, making them challenging—or even impossible—to model effectively. Consequently, traditional cause-and-effect models for risk assessment and management are losing their efficacy in a world characterized by interconnected risks. This is true on a global scale as well as within specific subsystems. To fully grasp the dynamics of risks spanning nature, technology, and society, they must be understood as evolving processes within complex systems. Rather than relying on linear datasets, the focus must shift to systemic interactions, where technical, natural, and socioeconomic factors intersect with social responses and discourses (Renn 2024).

Examples of systemic risks include global warming and its associated extreme events, zoonotic disease outbreaks, biodiversity loss, rising economic inequality, financial system instability, ideological extremism, cyberattacks, social and political unrest, and geopolitical imbalances (Schweizer 2021). Many of these risks have become more severe and disruptive than in previous decades, a phenomenon described in risk research as risk amplification (Kasperson et al. 1988). Additionally, the likelihood and intensity of these risks have been increasing at an accelerating pace.

Mapping the interdependencies among systemic risks often results in convoluted "spaghetti diagrams," which, while illustrating complex causal links, may miss critical connections or exaggerate weak ones. Such diagrams can also create the illusion of universal interconnection without identifying the most influential relationships. A more focused approach is to pinpoint key risk agents with significant harm potential and analyze their interactions.

This framework for systemic risks has been discussed in several publications (Renn et al. 2022, Renn 2024; SAPEA 2022; Schweizer et al. 2021). The core idea is that risks, regardless of their origin, manifest destructively through five principal agents: three physical agents—energy (in various forms), substances (e.g., toxic chemicals or pollutants), and biota (e.g., viruses, bacteria, fungi)—and two social agents—information (including money) and power (including violence). These agents interact dynamically in risk cascades and synergistic events. The sequence of a series of interconnected risks can in principle be modeled in a simple causal diagram starting with the risk triggers and ending with the consequences in different physical and social domains (see Fig. 1):

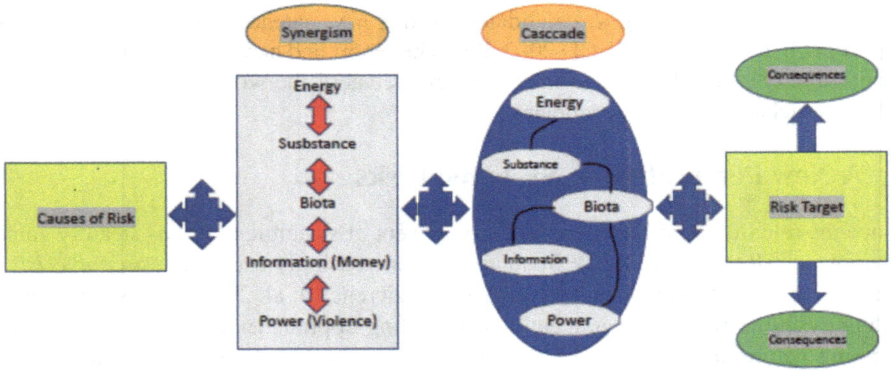

Fig. 1. Causal chain of cascading and interacting risks (adopted from Renn 2024)

Caused by some kind of trigger (events such as earthquakes or floods and activities such as polluting or warfare) one or more of these five agents are released. For example, an earthquake might release kinetic energy, triggering a chemical factory accident that spreads toxic substances. Alternatively, the close interaction of humans and animals might lead to a pandemic, which could fuel conspiracy theories and erode trust in public health governance, exacerbating the pandemic's impact. Similarly, financial rumors could alter company valuations, influencing wealth distribution. These interconnected events illustrate how systemic risks differ from traditional risk models, which typically examine the impact of a single agent on a specific target (IRGC 2018, 2019). In a polycrisis, multiple agents interact, amplifying or mitigating each other, necessitating a holistic approach to risk management.

Two primary pathways of interaction have been identified (Lawrence et al. 2022):

1. *Risk Cascades:* A sequential chain of events where one risk triggers another. For example, an earthquake might lead to a chemical spill, inadequate communication of the threat, loss of public trust in authorities, and subsequent hazards like climate change aggravation. Each agent—energy, substance, information, power—interacts in a chain, often through parallel sequences.
2. *Risk Synergism:* Simultaneous triggers release multiple agents that interact immediately and exacerbate potential harm. For instance, the interplay between climate change and geopolitical tensions might result in increased $CO2$ emissions from the use of liquid natural gas and eroded governmental credibility, fueling populist movements and undermining climate policy.

These pathways overlap, forming identifiable patterns of interaction that can sometimes be formalized into mathematical models or, where this is not feasible, translated into coherent narratives describing polycrisis cycles. Focusing on a limited set of agents enables better comparisons of these cycles and offers a foundation for developing an abstract, general taxonomy of polycrises. While these conceptual ideas are still in early stages of empirical testing, they hold promise for improving both the understanding and governance of polycrises.

Both, risk synergisms and cascades overlap and form specific patterns of interactions that can be identified, sometimes formalized in mathematical models and, if that is not possible, to be transferred into coherent narratives of a polycrisis cycle. The limited number of agents facilitates the comparison of different cycles and provides a powerful toolbox for developing a more abstract and generic taxonomy of polycrisis. At this point, these conceptual ideas have not been further tested or applied to empirical case studies. However, the approach provides some promising potential to improve our understanding of polycrisis and also our ability and capacity to govern polycrisis more effectively.

4 Characteristics of Systemic Risks

Systemic risks are now ubiquitous. It is therefore important to describe in more detail what makes a risk a systemic risk. The following characteristics serve this purpose (Renn 2014; 2016; Lucas et al. 2018; Renn et al. 2021):

- Systemic risks have a global or at least transregional impact (Reusswig 2011: 693f.; Lipsky 2009: 5). They can no longer be confined to a specific region. In this context, Ulrich Beck speaks of "unbounded" risks (Beck et al. 2004). Although such risks can be triggered locally, their effects are likely to spread to many other regions. A prime example of this is the emergence of a pandemic, such as the recent wave of COVID-19 disease. A new pathogen appears anywhere in the world. Tourists, business travelers or exported trade goods carry these pathogens around the world in a very short time. They can hardly be contained (Bornemann 2005).

- Systemic risks are closely interlinked with other risks and have an impact on different domains of the economy and society (Fischer-Kowalski and Weisz 1999). Their effects are linked to the impact chains of other activities and events, without this being apparent at first glance. The World Economic Forum's 2021 report on global challenges and risks offers many examples of such intuitively imperceptible links between risks (World Economic Forum 2021). For example, social inequality, negative trade balances, resistance to climate protection measures and even the rise of fundamentalist movements are closely intertwined.
- As a rule, systemic risks cannot be described by linear models of cause-and-effect chains, but often follow stochastic and non-linear cause-and-effect relationships (Bossel 2004: 385ff.). Identical causes do not lead to identical results, but to a range of consequences, all of which only occur with a certain probability. Non-linear systems describe another relationship that is difficult to access: they are characterized by the fact that marginal changes in one or more causes can lead to unexpectedly large swings in the effects. Such non-linearities are often referred to as the butterfly effect (Lorenz 1995). According to this, the flapping of a butterfly's wings in one area is said to trigger a change in the weather in another area. A much more realistic example is the overturning of a lake due to eutrophication (Goffman et al. 2006). If nutrients such as phosphates or nitrates are continuously introduced into a lake, this biotope can cope relatively well with these pollutants up to a certain concentration limit. There is even an increase in algae growth and an improvement in the food base for fish. It looks as if the lake is completely robust against these discharged pollutants. However, if this certain threshold is exceeded, the lake collapses relatively quickly due to a lack of oxygen. It is not possible to calculate in advance exactly where the threshold is located, as each lake must be viewed as a unique system of influence and effect chains. It is like a toggle switch that can be used to switch an effect on and off, such as a living pressure is present, which we cannot predetermine, the effect occurs immediately and the system changes abruptly. Any pressure that is insufficient to move the switch, on the other hand, has virtually no effect. Unfortunately, science does not know the exact pressure point of such a threshold for many systems (Renn 2016).
- The final characteristic of systemic risks is their underestimation by politicians and society (National Research Council 2010: 102ff.; Schweizer 2021) It is by no means the case that experts are unaware of these risks. They have generally been scientifically identified, analyzed and assessed. Some of these systemic risks, such as the impact of our energy supply on the climate, are even the focus of many scientific and political activities. The general public is also aware of these risks and classifies them as "relevant". However, in practice, it is clear that, unlike in the case of individual health risks, which the economy, government and society have been able to dramatically reduce over the last few decades, the effectiveness of risk-limiting and regulatory measures to date has been extraordinarily meagre (Randers 2012: 146).

Many systemic risks gain their particular explosiveness from the far-reaching, networked, non-linear and transnational effects that they cause on natural, economic, financial, social or political systems. The concept of "social amplification of risk" has provided important theoretical and empirical evidence of such indirect, secondary risk effects

(Kasperson et al. 1988, Renn et al. 1992). This analytical framework focuses on the effects that physical risks have at the social and institutional level. However, the concept of systemic risk goes beyond the amplification or mitigation of physical effects. It encompasses mechanisms according to which the various categories of damage interact with each other and influence the transnational distribution of risk (Renn et al. 2022).

In addition, systemic risks have a central symbolic and associative significance. Firstly, such risks are given symbolic connotations that exert a major influence on intuitive perception and evaluation (Slovic 1992, Boholm 1998, Jaeger et al. 2001: 101ff.; Siegrist et al. 2020). Secondly, risky activities are associated with certain interests, values and world views that are only loosely linked to the potential for physical harm but allow a close socio-political connection with current political disputes (Freudenburg and Pastor 1992, Luhmann 1986: 155ff.). Thirdly, cultural patterns of meaning are associated with risky activities, ranging from purely symbolic violations (e.g. of religious feelings) to combinations of harm and harmful situations (Douglas and Wildavsky 1982, Rayner 1990, Pidgeon 1998; Jaeger et al. 2001: 183ff.). These three constructive elements of risk perception are usually additionally socially reinforced by media coverage (Renn 2019: 107ff.).

Systemic risks are often accompanied by a loss of political trust, loss of legitimacy or stigmatization effects (Perrow 1990; Luhmann 1991, Sjöberg 2001). They call for new institutions and procedures that specifically address the interacting effects of technical-ecological transformations. This is associated with the need to develop new assessment processes that are tailored to the central characteristics of systemic risks such as increased complexity, uncertainty and ambiguity (see below).

Interact between physical consequences and their psychological, social, economic and cultural effects and require coordination with stakeholders and the affected public.

5 Three Key Dimensions of Systemic Risk

In order to better understand and assess systemic risks, it is helpful to divide the concept of risk into three dimensions. These three dimensions are: Complexity, Uncertainty and Ambiguity (Klinke and Renn 2021; Renn et al. 2011). Many of the risk problems addressed in the literature can be traced back to these three dimensions (cf. Bonß 1996, IRGC 2005, 2007). These three aspects go beyond the classic components of risk such as probability and extent of damage.

The problem of complexity should be addressed first. Complexity means something other than intricacy. "Complex" means that there are many intervening variables between cause and effect that either strengthen or weaken this relationship, so that it is not easy to draw conclusions about the triggering causes from the observed effect. Complexity refers to causal relationships that are difficult to identify and quantify. The reason for this can be interaction effects between a large number of causal factors, e.g. multiple synergies, or long delay times between causes and effects. These complex relationships require particularly sophisticated scientific investigations, as the cause-effect relationships are neither evident nor directly observable. In the case of highly non-linear relationships that can no longer be adequately captured by probabilistic techniques, complexity results in indeterminacy (i.e. a high degree of uncertainty, see below). Complex relationships are particularly prevalent in the case of health risks. Risk assessors and managers are dependent on model calculations, which in many cases can only claim hypothetical validity. These models are often controversial, even among experts. This uncertainty is in itself a problem of assessment and even more so of risk communication.

The second essential element of systemic risks refers to a high degree of uncertainty. Most risk assessments are based on the fact that there are rarely deterministic, i.e. fixed cause-effect chains in the nature of hazards. The same or similar exposures can lead to very different reactions in different individuals. Uncertainty includes, on the one hand, data measurement errors, model imprecisions and inter-individual variations expressed in statistical confidence intervals. On the other hand, it refers to indeterminacy and lack of knowledge, which can result in assessments not being possible or becoming too vague. The effects may only be modelled within certain system boundaries and, thus, system-wide, external influences and effects need to be disregarded. This second form of uncertainty is particularly prevalent in systemic risks because many events are mutually connected and difficult to assess in their interactions. They represent possible and contingent interpretation of a complex set of data. However, due to the complexity of the causal relationship patterns and the remaining uncertainties, ambiguities offer different possibilities to give meaning to the results and to define what they imply for risk management. One of the key tasks in the risk regulation process is therefore to define and explore the legitimate limits of a data-based interpretation.

Normative ambiguity refers to the moral acceptability of the risk. Is the risk acceptable or not? This classification requires value decisions that cannot be derived from system knowledge alone but requires a combination of system and orientation knowledge. The German Advisory Council on Global Change (WBGU 2000) has drawn up its own list of criteria and adopted it after intensive discussion:

- The two classic criteria of probability and likelihood of occurrence,
- Degree of remaining (un)certainty: This refers to the certainty with which the probabilities and the extent of damage can be estimated. As a rule, this certainty is expressed statistically by confidence intervals,
- Ubiquity defines the geographical range of potential damage (it refers to intragenerational equity),
- Persistence defines the temporal extent of potential damage (it relates to intergenerational equity),
- Reversibility describes the possibility of restoring the initial situation before the damage occurred (e.g. water purification, reforestation),
- Delay effect characterizes the time span between the original event and the actual consequences. The delay can be caused by physical, chemical or biological mechanisms,
- Mobilization potential is understood as the violation of individual, social or cultural interests and values. The mobilization potential is caused by social conflicts and psychological reactions of individuals or groups, who feel affected by the consequences of the risk. The mobilization potential can also stem from perceived inequities in the distribution of risks and benefits.

These criteria can be derived both from normative studies on the acceptability of decisions and from research studies on risk perception. They provide a rational basis for decision-making in order to assess and determine the acceptability of a risk. However, the integration and combination of the eight criteria into an overall judgment remains a domain of political prioritization (trade-offs). These priorities can and must be justified, but they cannot be derived from the results of the scientific risk analysis itself. The need to legitimize trade-offs in designing risk management options is particularly difficult and challenging when dealing with systemic risks.

The arbitrariness of justifications in the politicized risk debates is exacerbated by the uncertainty of the consequences themselves. Recently, there has been an increase in publications that address the handling of uncertainty under the headings of the precautionary principle and risk prevention and, above all, deal with the suppression of system-wide consequences and ignorance (fundamental: SAPEA 2019). In the context of the BSE crisis, the regulatory and marketing conflicts surrounding green genetic engineering and the difficult reform of chemicals policy, these authors often advocated strengthening precaution and participation in risk management. The basis for this demand is the observation that, from the point of view of legitimacy, the pressure to participate increases when decisions are made under uncertainty (Stirling 1998; 2003). Authors with a more empirical-analytical perspective argue that uncertainty is becoming more of a social issue and has also gained increasing practical significance in the political and administrative management of environmental and technological risks (Collingridge 1996, Japp 1996).

The literature on innovative methods of interest representation and citizen participation (Rowe and Frewer 2000, Renn and Schweizer 2020; OECD 2020) illustrates that procedures aimed at dealing with ambiguity are increasingly being developed and used at national and sub-national level in the environmental and technology sector. These participation processes are instruments for reflecting on and mediating different, often conflictual interpretations of risk, which are based on plural levels of knowledge, interests and value preferences. Systemic risks have a heightened ambiguity due to their transnational scope. This is also shown by previous experience with the corona crisis, where the lack of international cooperation in the distribution of vaccines, for example, has caused major conflicts among nations.

In view of different risk cultures and sometimes fierce transnational risk controversies in fields as diverse as nanotechnologies, energy transition, Industry 4.0 or pandemic control, some authors call for cross-border reflection, consultation and mediation processes not only in the administrative bodies provided for this purpose, but also in participatory, bottom-up procedures (Elstub 2010; Mansbridge 2012; Dryzek et al. 2019). The example of the coronavirus pandemic shows that cultural differences are usually not openly addressed in administrative bodies, but only have a latent effect, especially in controversies that are fought over scientific evidence and plausibility. However, they are often based on differences in cultural interpretation patterns (such as different concepts of naturalness and legitimacy of human intervention). Transnational participation procedures, in which the ambiguity of cross-border risks could be specifically addressed, have not yet been used. At the transnational and international level, participation is largely limited to conflict mediation, i.e. procedures for mediating conflicting interests through negotiation (Susskind and Babbitt 1992; Renn 2021). What will be increasingly in demand in the future is an integrated risk governance structure in which civil society actors play an important role in risk assessment beyond national boundaries (Schweizer and Renn 2019).

6 Conclusions

The world finds itself in a paradoxical situation: there is virtually no analysis of current developments in the economy and society that gives the global community a satisfactory or even good record with regard to the requirements of sustainable management of systemic risks. Successful handling of the concurrent transformation processes requires a robust knowledge base for assessing the likely consequences of each option for dealing with systemic risks. This primarily involves a systematic and inclusive weighing up of conflicting objectives that treats interests fairly and takes into account the pluralistic values that are present in contemporary societies. To this end, better integration of scientific expertise as well as the knowledge and concerns of the relevant stakeholders in deliberative negotiation and decision-making processes is essential. Comprehensive security concepts cannot be limited to the implementation of technical or statistical knowledge alone. Rather, it is important to localize the treatment of risks in a pluralistic environment of competing knowledge claims, different values and lifestyles and various technical contexts.

At the same time, modern risks to health and safety have become more complex, uncertain and ambiguous. This also makes it more difficult to identify and manage them. Therefore, an interdisciplinary and inclusive approach to risks is required from two sides: on the one hand, several concepts and perspectives compete for the appropriate treatment of systemic risks, which require an interdisciplinary approach; on the other hand, societal actors expect opportunities to actively participate in the selection and regulation of safety requirements. Significant reforms in the area of security behavior are essential so that the stakeholders affected by risks can better understand and master these complex interrelationships. Above all, a strategy of further education and training based on an interdisciplinary and systemic understanding of risk is required, as well as new deliberative procedures in order to constructively integrate the plurality of values, interests and concerns into decision-making processes.

Literature

Adams, W.C. (1986): Whose Lives Count? TV Coverage of Natural Disasters. *Communication,* 36 (2): 113-122.

Beck, U. (2009). *World at risk*. Polity Press: NewYork.

Beck, U. (1992). Modern society as a risk society. In: N. Stehr & R.V. Ericson (eds): *The Culture and Power of Knowledge. Inquiries into Contemporary Societies*. De Gruyter: Berlin and New York, pp. 199-214.

Beck, U.; Bonß, W. & Lau, C. (2004): Entgrenzung erzwingt Entscheidung: Was ist neu an der Theorie reflexiver Modernisierung? In: U. Beck & C. Lau (eds.): Entgrenzung und Entscheidung: Was ist neu an der Theorie reflexiver Modernisierung? Suhrkamp: Frankfurt/Main, pp. 13-62.

Boholm, A. (1998): Comparative Studies of Risk Perception: A Review of Twenty Years of Research. *Journal of Risk Research* 1(2): 135–163.

Bornemann, R. (2005): Influenza-Pandemien des 20. Jahrhunderts im Überblick – besteht die Möglichkeit einer Wiederholung der Pandemie von 1918? *GMDS/DAE-Jahrestagung 2005*, Freiburg, 14./15.9.2005. Manuskript. http://www.uni-bielefeld.de/gesundhw/ag2/infepi/avian-influenza.pdf.

Bonß, W. (1996): Die Rückkehr der Unsicherheit. Zur gesellschaftstheoretischen Bedeutung des Risikobegriffes. In: G. Banse (ed.): *Risikoforschung zwischen Disziplinarität und Interdisziplinarität.* Edition Sigma: Berlin, pp. 166-185.

Brown, K.; O'Neill, S. & Fabricius, C. (2013). Social Science Understandings of Transformation. *World Social Science Report 2013: Changing Global Environments.* Paris: ISSC and UNESCO.

Bubb, A. & Zimmermann, H. (2004): *Ansteckungsrisiken im Finanzsystem: Übersicht und aktuelle Problembereiche.* Forschungsbericht Nr. 9/04. Wirtschaftswissenschaftliches Zentrum (WWZ) der Universität Basel: Basel.

Collins, A.; Florin, M.-F. & Renn, O. (2020): COVID-19 risk governance: drivers, responses and lessons to be learned. *Journal of Risk Research,* 2, DOI: https://doi.org/10.1080/13669877.2020.1760332.

De Bandt, O. & Hartmann, P. (2000): *Systemic Risk: A Survey.* Working Paper Series/European Central Bank Nr. 35. Frankfurt/Main: European Central Bank.

Dryzek, J., André Bächtiger, Simone Chambers, Joshua Cohen, James N. Druckman, Andrea Felicetti, James S. Fishkin, David M. Farrell, Archon Fung, Amy Gutmann, Hélène Landemore, Jane Mansbridge, Sofie Marien, Michael A. Neblo, Simon Niemeyer, Maija Setälä, Rune Slothuus, Jane Suiter, Dennis Thompson & Mark E. Warren (2019). The Crisis of Democracy and the Science of Deliberation. *Science,* 363 (6432): 1144-1146. DOI: https://doi.org/10.1126/science.aaw2694.

Elstub, S. (2010): The Third Generation of Deliberative Democracy. *Political Studies Review,* 8: 291-307.

Fischer-Kowalski, M & Weisz, H. (1999): Society as a Hybrid Between Natural and Symbolic Realms. Toward a Theoretical Framework of Society-Nature Interaction. *Advances in Human Ecology,* 8: 215- 251.

Freudenburg, W.R. & Pastor, S.K. (1992): Public Responses to Technological Risk: Toward a Sociological Perspective. *Sociological Quarterly* 33 (3): 389-412.

Folke, C. (2006): Resilience: The Emergence of a Perspective for Social-Ecological System Analyses. *Global Environmental Change*, 16: 253-26.

Goffman, P.M.; Baron, J.S.; Blett, T.; Gold, A.J.; Goodman, I.; Gunderson, L.H.; Levinson, B.; Palmer, M.A.; Paerl, H.W.; Peterson, G.D.; Poff, N.L. & Rejeski, D:W. (2006): Ecological Thresholds: The Key to Successful Environmental Management or an Important Concept with No Practical Application? *Journal of Ecosystems*, 9 (1): 1-13.

Hillgartner, S. (1992): The Social Construction of Risk Objects: Or, How to Pry Open Networks of Risk. In: .F. Short & L. Clarke (eds): *Organizations, Uncertainties, and Risk.* Westview: Boulder, pp. 39-53.

Homer-Dixon, T. & Rockström, J. (2022): What Happens When a Cascade of Crises Collide? *The New York Times*, November 13.

IRGC, International Risk Governance Council (2005): *White Paper on Risk Governance. Towards an Integrative Approach.* International Risk Governance Council: Geneva.

IRGC, International Risk Governance Council (2007): *An Introduction to the IRGC Risk Governance Framework.* Policy Brief. International Risk Governance Council: Geneva

IRGC - International Risk Governance Center (2018): *IRGC Guidelines for the Governance of Systemic Risks.* EPFL: International Risk Governance Center: Lausanne

IRGC- International Risk Governance Council (2019): *Introduction to the IRGC Risk Governance Framework. Revised Edition.* EPFL: International Risk Governance Center: Lausanne

Jaeger, C.C.; Renn, O.; Rosa, E.A. & Webler, Th. (2001): *Risk, Uncertainty and Rational Action.* Earthscan: London.

Japp, K. (1996): Soziologische Risikotheorie. Juventa: Weinheim and Munich.

Kasperson, R.E.; Renn, O.; Slovic, P.; Brown, H.. S.; Emel, J.; Goble, R.; Kasperson, J.X. & Ratick, S.(1988): The Social Amplification of Risk. A Conceptual Framework, *Risk Analysis,* 8 (2): 177-187.

Lawrence, M.; Janzwood, S. & Homer-Dixon, T. (2022): *What Is a Global Polycrisis? And How Is It Different from a Systemic Risk?* Discussion Paper 2022-4. Version 2.0. Cascade Institute: Toronto.

Lawrence, M.; Homer-Dixon, T. Janzwood, S.; Rockstöm, J.; Renn, O. & Donges, J. F. (2024): The Causal Mechanisms of Crisis Entanglement. *Global Sustainability,* 1:1-36. doi: https://doi.org/10.1017/sus.2024.1

Klinke, A. & Renn, O. (2014): Expertise and Experience: A Deliberative System of a Functional Division of Labor for Post-normal Risk Governance. *Innovation: The European Journal of Social Science Research* 27(4): 442-465.

Klinke, A. & Renn, O. (2021): The Coming Age of Risk Governance. Risk Analysis, 41(3): 544-557, https://doi.org/10.1111/risa.13383

Lipsky, I. (2009): Through the Looking Glass: The Links between Financial Globalisation and Systemic Risks. In: D.D. Evanoff; D.S. Hoelscher & G.G. Kaufman (eds.): *Globalisation and Systemic Risks.* World Scientific Publishers: New Jersey, pp. 3-11.

Lucas, K. & Renn, O (2020): Systemic Risks: Intersections between Science and Society with Policy Implications for Sustainability. *Int J Environ Sci Nat Resources,* 23 (2): DOI: https://doi.org/10.19080/IJESNR.2020.23.556107

Lucas, K.; Renn, O. & Jaeger, C. (2018): Systemic Risks: Theory and Mathematical Modeling. *Advanced Theory and Simulations,* 4: https://doi.org/10.1002/adts.201800051.

Luhmann, N. (1986): *Ökologische Kommunikation. Kann die moderne Gesellschaft sich auf ökologische Gefährdungen einstellen?* Westdeutscher Verlag: Opladen.

Luhmann, N. (1990): Technology, Environment, and Social Risk: A Systems Perspective. *Industrial Crisis Quarterly,* 4 (3): 223-231.

Luhmann, N. (1997): Grenzwerte der ökologischen Politik. Eine Form von Risikomanagement. In: P. Hiller & G. Krücken (eds): *Risiko und Regulierung. Soziologische Beiträge zu Technikkontrolle und präventiver Umweltpolitik.* Suhrkamp: Frankfurt/Main, pp 195-221.

Mansbridge, J. (2012). A Systemic Approach to Deliberative Democracy. In: J. Parkinson & J. Mansbridge (eds): *Deliberative Systems.* Cambridge: Cambridge University Press.

National Research Council (2010): *Advancing the Science of Climate Change.* The National Academies Press: Washington, D.C.

OECD (2003): *Emerging Risks in the 21st Century. An Agenda for Action.* OECD: Paris.

OECD (2020): *Innovative Citizen Participation and New Democratic Institutions: Catching the Deliberative Wave.* OECD: Paris, https://doi.org/10.1787/339306da-en.

Perrow, C. (1990): *Normale Katastrophen. Die unvermeidbaren Risiken der Großtechnik.* Campus: Frankfurt/Main.

Perrow, C. (1992): Unfälle und Katastrophen – ihre Systembedingungen. *Journal für Sozialforschung,* 32 (1): 61- 75.

Pidgeon, N. (1998): Risk Assessment, Risk Values and the Social Science Program: Why Do We Need Risk Perception Research. *Reliability Engineering and System Safety,* 59 (1): 5-15, https://doi.org/10.1016/S0951-8320(97)00114-2.

Randers, J. (2012): *2052. Der neue Bericht an den Club of Rome. Eine globale Prognose für die nächsten 40 Jahre.* Oekom: Munich.

Rayner, S. (1990): *Risk in Cultural Perspective: Acting under Uncertainty.* Kluwer: Dordrecht and Norwell.

Renn, O. (2008): *Risk Governance: Coping with Uncertainty in a Complex World.* Earthscan: London.

Renn, O. (2014): New Challenges for Risk Analysis: Systemic Risks. *Japanese Journal of Risk Analysis* 24 (3): 157-160.

Renn, O. (2016): Systemic Risks: The New Kid on the Block. *Environment: Science and Policy for Sustainable Development,* 58 (2): 26-36, DOI: https://doi.org/10.1080/00139157.2016.1134019.

Renn, O. (2019): Gefühlte Wahrheiten. Orientierung in Zeiten postfaktischer Verunsicherung. Budrich: Baden- Baden

Renn O. (2021a): Transdisciplinarity: Synthesis towards a modular approach, *Futures, 130*: 102744, https://doi.org/10.1016/j.futures.2021.102744

Renn, O.(2021b): New challenges for risk analysts: systemic risks. *Journal of Risk Research,* 2, https://doi.org/10.1080/13669877.2020.1779787

Renn. O. (2024): Systemic Risks and Polycrisis: The need for an integrative approach. In: W. Haber, M. Grambow & P. Wilderer (eds.): *An interdisciplinary discourse on regulation. Biotic Self-Regulation: Model for Man-made Systems?* TUM-IAS: Munich, pp. 25-29, https://euro-acad.eu/files/WHITE%20BOOK_Regulation.pdf

Renn, O.; Burns, W.; Kasperson, R.E.; Kasperson, J.X. & Slovic, P. (1992): The Social Amplification of Risk: Theoretical Foundations and Empirical Application. *Social Issues,* 48, (4): 137-160.

Renn, O.; A. Klinke & M. van Asselt. (2011): Coping with Complexity, Uncertainty and Ambiguity in Risk Governance: A Synthesis. *AMBIO* 40 (2): 231-246.

Renn, O., Lucas, K.; Haas, A. &, Jaeger; C. (2019): Things Are Different Today: The Challenge of Global Systemic Risks. *Journal of Risk Research* 22 (4): 401–415, doi: https://doi.org/10.1080/13669877.2017.1409252.

Renn, O.; Laubichler, M.; Lucas, K.; Schanze, J.; Scholz, R. & Schweizer, P.-J. (2020). Systemic Risks from Different Perspectives. *Risk Analysis,* 42 (9): 1902-1920, https://doi.org/10.1111/risa.13657.

Renn, O. & Schweizer, P.-J. (2020). Inclusive Governance for Energy Policy Making: Conceptual Foundations, Applications, and Lessons Learned. In: O. Renn; F. Ulmer & A. Deckert (eds.): *The Role of Public Participation in Energy Transitions.* Elsevier Academic Press: London, pp. 39-79.

Reusswig, F. (2011): Klimawandel und globale Umweltveränderungen. In: M. Groß (ed.): Handbuch Umweltsoziologie. VS Verlag für Sozialwissenschaften: Wiesbaden, pp. 692-720.

Rosa, E.A. (1998): Metatheoretical Foundations for Post-Normal Risk. *Journal of Risk Research* 1(1): 15-44.

Rosa, E.A.; O. Renn & McCright, A. (2014): *The Risk Society Revisited. Social Theory and Governance.* Temple University Press: Philadelphia.

Rowe, G. & Frewer, L.J. (2000). Public Participation Methods: A Framework for Evaluation. *Science, Technology and Human Values* 25 (1), 3–29.

Science Advise for Policy by European Academies (SAPEA) (2019): *Making Sense of Science for Policy under Conditions of Uncertainty and Complexity.* Report to the EU Commission (2019), SAPEA: Berlin, https://doi.org/10.26356/MASOS.

Schweizer, P.-J. (2019/2021): Systemic Risks. Concepts and Challenges for Risk Governance. *Journal of Risk Research,* 24 (1): 78-93, doi: https://doi.org/10.1080/13669877.2019.1687574.

Schweizer, P.-J. & Renn, O. (2019): Governance of Systemic Risks for Disaster Prevention and Mitigation. *Disaster Prevention and Management,* 28 (6): 854-866, https://doi.org/10.1108/DPM-09-2019-0282.

Schweizer, P.-J.; Goble, R. & Renn, O. (2021): Social perception of systemic risks. *Risk Analysis,* 3, https://doi.org/10.1111/risa.13831

Siegrist, M. & Arvai, J. (2020): Risk perception: Reflections on 40 years of research. *Risk Analysis,* 40 (11): 2191-2206, https://doi.org/10.1111/risa.13599.

Sjöberg, L. (2000): Factors in Risk Perception. *Risk Analysis,* 220 (1): 1-11.

Slovic, P. (1992): Perception of Risk: Reflections on the Psychometric Paradigm. In: S. Krimsky & D. Golding (eds): *Social Theories of Risk.* Praeger: Westport, pp. 117-152.

Stirling, A. (1998): Risk at a Turning Point? *Journal of Risk Research,* 1 (2): 97–109.

Stirling, A. (2003): Risk, Uncertainty and Precaution: Some Instrumental Implications from the Social Sciences .In: F. Berkhout, M. Leach & I. Scoones (eds.): *Negotiating Change.* Elgar: London, pp. 33-76.

Susskind L. & Babbitt E. (1992): Overcoming the Obstacles to Effective Mediation of International Disputes. In: J. Bercovitch & J.Z. Rubin (eds.): Mediation in International Relations. Palgrave Macmillan: London, pp. 31-51, https://doi.org/10.1057/9780230374690_2

Trabucco, D. (2022): The European Union Beyond the Polycrisis? Integration and Politicization in an Age of Shifting Cleavages. *TECHNE-Journal of Technology for Architecture and Environment,* 283-284, file:///Users/ortwinrenn/Downloads/12922-Article%20Text-40022-1-10-20220518.pdf

WBGU, Wissenschaftlicher Beirat der Bundesregierung Globale Umweltveränderungen (2000): *World in Transition: Strategies for Managing Global Environmental Risks.* Annual Report 1998. Springer: Heidelberg and Berlin.

World Economic Forum (2022): Global Risks 2022. 12. Edition. WEC: Geneva.

WPKS-Wissenschaftsplattform Klimaschutz (2023): Resilienz und Klimaschutz: Herausforderungen für Wissenschaft und Politik. Hintergrundpapier der Wissenschaftsplattform Klimaschutz. WPKS: Berlin, https://www.wissenschaftsplattformklimaschutz.de/files/WPKS_Hintergrundpapier_Resilienz_Klimaschutz.pdf.-

References

Homer-Dixon, T., Johan Rockström, J.: What Happens When a Cascade of Crises Collide?' The New York Times (2022)

Lawrence, M., Homer-Dixon, T., Janzwood, S., Rockstöm, J., Renn, O., Donges, J.F.: Global polycrisis: the causal mechanisms of crisis entanglement. Global Sustain. **2024**, 1–36 (2024). https://doi.org/10.1017/sus.2024.1

IRGC - International Risk Governance Center: IRGC Guidelines for the Governance of Systemic Risks. Lausanne: EPFL: International Risk Governance Center (2018)

IRGC- International Risk Governance Council: Introduction to the IRGC Risk Governance Framework. Revised Edition. Lausanne, EPFL International Risk Governance Center (2019)

Kasperson, R.E., et al.: The social amplification of risk. A conceptual framework. Risk Anal. **8**(2), 177–187 (1988)

Lawrence, M., Janzwood, S., Homer-Dixon, T.: What Is a Global Polycrisis? And How Is It Different from a Systemic Risk? Discussion Paper 2022-4. Version 2.0. Cascade Institute (2022)

Renn, O.: New challenges for risk analysts: systemic risks. J. Risk Res. **2** (2021b). https://doi.org/10.1080/13669877.2020.1779787

Ortwin, R.: Transdisciplinarity: Synthesis towards a modular approach. Futures **130**, 102744 (pr) (2021a). https://doi.org/10.1016/j.futures.2021.102744

Renn, O.: Die drei Transformationen: Globalisierung, Digitalisierung, Sustainabilisierung. DDBW aktuell. Das Magazin für Arbeitsmedizin in Deutschland **9**(1), 4–5 (2020)

Renn, O., Laubichler, M., Lucas, K., Schanze, J., Scholz, R., Schweizer, P.-J.: Systemic risks from different perspectives. Risk Anal. **42**(9), 1902–1920 (2022)

Renn, O., Lucas, K., Haas, A., Jaeger, C.: Things are different today: the challenge of global systemic risks. J. Risk Res. **22**(4), 401–415 (2019). https://doi.org/10.1080/13669877.2017.1409252

SAPEA - Science Advice for Policy by European Academies: Strategic crisis management in the European Union. SAPEA, Berlin (2022). https://doi.org/10.26356/crisismanagement

Schweizer, P.-J.: Systemic risks – concepts and challenges for risk governance. J. Risk Res. **24**(1), 78–93 (2021). https://doi.org/10.1080/13669877.2019.1687574

Schweizer, P.-J., Goble, R., Renn, O.: Social perception of systemic risks. Risk Anal. **3** (2021). https://doi.org/10.1111/risa.13831

Trabucco, D.: The European Union Beyond the Polycrisis? Integration and Politicization in an Age of Shifting Cleavages. TECHNE-Journal of Technology for Architecture and Environment, pp. 283–284 (2022). file:///Users/ortwinrenn/Downloads/12922-Article%20Text-40022-1-10-20220518.pdf

WPKS-Wissenschaftsplattform Klimaschutz: Resilienz und Klimaschutz: Herausforderungen für Wissenschaft und Politik. Hintergrundpapier der Wissenschaftsplattform Klimaschutz. WPKS: Berlin (2023). https://www.wissenschaftsplattform-klimaschutz.de/files/WPKS_Hintergrundpapier_Resilienz_Klimaschutz.pdf

Disaster Science: Updating Theory to Guide Disaster Risk Reduction in the Future

David Alexander(✉)

University College London, London, UK
david.alexander@ucl.ac.uk

> "*A little sincerity is a dangerous thing, and a great deal of it is absolutely fatal.*"(Oscar Wilde)

1 Introduction

To begin with, systemic risk is not quite the right word: instead it is *systematic* risk. The difference is that the former is caused by a single entity which threatens to bring down the entire system whereas the latter occurs when vulnerability proliferates throughout the system, but not necessarily from a single source. However, systemic risk has been adopted as a working term in disaster risk reduction (Schweizer and Renn 2019) and thus it must remain. It is clear that the phenomenon will engender massive challenges in order to reduce disaster risks in the future.

The oft-lamented lack of connection between the various disciplines and endeavours has been described as "creating silos" (De Ruiter et al. 2021). It is not a good metaphor. I have never seen a silo that is not well connected: if it were not, it could not function as a silo. That, however, is a quibble, because, despite years of effort to join up the many elements of DRR into some kind of coherent whole, the process has never been concluded satisfactorily. There are at least 42 disciplines and professions involved and they are very varied in their outlook. What is required is a common language that contributes to a common culture. For those who work in disaster risk reduction it is as important to understand what other people do as it is to know what one is required to do oneself. The emergency response manager must create a dialogue with specialists in the contributory disciplines. He or she must speak enough of their language to be on the same wavelength and thus create mutual understanding.

Despite this imperative, there is often a somewhat naive belief that scientific information is automatically welcomed by those who might benefit from employing it. Science does enlighten those who believe in it, but there is little sign that merely providing scientific information to leading decision makers will improve their decisions (Gorman and Gorman 2021). Something is missing here. The purpose of this essay is to probe that deficiency.

H. Tatano et al. (Eds.): GSRIDRR 2023, DRRGADRIBS, pp. 94–99, 2025.
https://doi.org/10.1007/978-981-96-6609-6_8

2 Context: The World as It Is

Theory is a roadmap for the interpretation and management of disaster. If it works, it illuminates complex and apparently incomprehensible situations by connecting them to more general strands of relevant experience. One of the shortcomings of academic scholarship on disasters (perhaps "applied disasterology") is that theory tends to develop slower than change occurs, and at the moment the world is changing very fast indeed. The lag in theory induces us to treat disasters as if they had occurred 30 or 40 years ago, yet we live in the age of dynamic digital connection, which in previous eras simply did not exist. Hence, we must update theory much more quickly.

Concomitantly with the need to accelerate the development of theory, there is a pressing need to take more account of the *context* in which disaster risk is set. In the simplest terms, a person cannot be made resilient against, for example, floods, if he or she is not resilient against the slings and arrows of life in general, including homelessness and unemployment. What this means for disaster risk reduction is that *there are factors which affect our ability to cope with disaster that are in no way directly related to it.* The answer to disaster-related problems may thus lie in issues that do not directly impinge on DRR.

In the period 1970–1973 a process started which came to have the most profound implications for the world. This is the divergence of wealth between the rich and the poor (Taylor 2009). The gap has continued to widen ever since. During the first year of the Covid-19 pandemic, the world's 2,200 billionaires increased their net wealth by 24 per cent, while the world's poor lost an average of US$1,000 per family (Binns and Low 2021). Disaster acted as an *anti-catalyst*. There are 87 tax havens in the world and they contain a greater proportion of global wealth than the USA dedicates to its armed forces. At the same time, we know that expenditures which are small by comparison could ameliorate many of the world's most serious problems. The moral and ethical imperative has not been matched by even the smallest international tax on excess earnings, one which would preserve the wealth of the richest people but provide an income stream to solve major global problems (Adam and Miller 2021).

The Nobel Prize-winner Amartya Sen wrote that to be prosperous a country needs a good health system so that its workers can be productive (Sen 2008). After half a century of transfer of wealth from the poor to the rich, it is clear that the resources are not being made available for such systems to be created in the public domain.

In synthesis, in 1966 the American geographer William W. Bunge demonstrated that we are all suffering from fragility because we are all on the front line of potential hostilities, such is the range and power of modern weaponry (Bunge 1966). The same could be said now for threats from other forms of global change.

3 Power and Ideology

'Polarisation' is a word that is commonly used to describe the situation in the world today (e.g., Esmond and Atkins 2022). It stems from the adoption of rigid positions and the presence of hegemonies that create opposition as an imperative among disadvantaged people. Modern life has become a struggle against dearly held but highly destructive ideologies.

Why, for example, does the United Kingdom's National Resilience Framework, released in December 2022, contain no reference whatsoever to people with disabilities (UK Government 2022)? They constitute between 16 and 19 per cent of the population and are thus a very large minority.[1] Experience shows very clearly that failing to take them into account and work with them disadvantages them (Alexander 2021). Similar issues prevail in many parts of the world, not merely with the disabled, but also with gender and ethnic minorities.

In many cases, the root of the problem is a prevailing hostility to the concept of 'welfare'. This is a word that is seldom spelled out in operational terms, and yet it is not difficult to define. It is, perhaps, "The provision of care to a minimum acceptable standard to people who are unable adequately to look after themselves" (my definition). Politically, it is seldom if ever expedient to define it, or else decision makers might be held to honour their own decisions (Tselios and Tomkins 2019).

During the period of the main Covid-19 pandemic, leadership in some countries was arrogant and denialist ("it's not a problem"), ignorant (failing to consult experts). Indecisive or erratic (with a lack of attention or lack of a system), negligent (failing to act on imperatives) or exploitative (favouring vested interests, most of which were economic). Some 88 countries used the pandemic as a means of restricting human rights (Human Rights Watch 2021). These are necessary, not least because they endow citizens with access to information and the opportunity to act in their own best interests, and those of society as a whole.

Although the early days of the pandemic were characterised by confusion resulting from uncertainty about the characteristics and performance of the disease, it quickly became true that there was no shortage of hard, reliable scientific information. By November 2020, 75,000 papers and 1,300 books had been published in English in the mainstream scientific press (Else 2020). The number has since increased immeasurably. It was the greatest single endeavour of knowledge creation in the history of science. This stupendous development was in net contrast to the decision making stance that prevailed in those countries in which highly unscientific ideologies prevailed.

4 Emergency Planning and Management

In the United Kingdom, the Scientific Advisory Group for Emergencies (SAGE) advised the national government throughout the pandemic (McKee et al. 2022). This body consisted of 52 epidemiologists, virologists and medical experts, and four politicians. It contained no experts on emergency planning, management and response. As a result, decisions about communications, logistics and the management of the population were not taken by qualified experts, but by a combination of medical doctors and politicians. The results were predictably bad: failed test-and-trace initiatives, condemned personal protective equipment, delays, miscommunications, and 15 abrupt, confusing changes of policy in a single year (Calvert and Arbuthnot 2021).

Similar deficiencies have been revealed by recent analyses of disasters in the UK, Germany and Sweden (e.g., Fekete and Sandholz 2021). Moreover, the President of

[1] Research by UK Government sources gives the proportion of disabled people in the United Kingdom as 24% (House of Commons 2023), which is much higher than the global average.

Turkey frankly admitted that arrangements to manage the earthquakes of February 2023 were insufficient. His excuse, that the disaster was "too big to manage", is not acceptable (Aksoy and Çevik 2023).

What all this amounts to is a pervasive tendency to underestimate the importance of good preparation for emergencies, despite all that has been written and said about this imperative. We lack standards and professionalism in emergency planning and management (Alexander 2005). While no one would dispute the need to move away from a reliance on response towards disaster prevention and mitigation, this must not happen by abandoning and neglecting response. In the future we will need civil protection systems that are not only complete, integral and functional, but also *an order of magnitude bigger and more effective than the ones we have now*. Response to disaster is a form of mitigation, as in the future one of its major roles will be *damage limitation*.

5 Foresight

Emergency plans need to be based, as far as is possible, on suites of scenarios for likely events (CCRD 2020). The scenarios should not be predictions of future outcomes, but should be structured explorations of possible future impacts, together with estimates of the resources required adequately to respond to such events. We face a future in which we will all have to cope with (a) the proliferating failure of the critical infrastructure on which we all depend for life support, (b) intensifying meteorological phenomena associated with climate change, (c) unplanned mass migration on a large scale, and (d) the effect of conflict, especially all-out warfare.

We also have to contend with intersectionality. This word, borrowed from studies of race and gender (Collins and Bilge 2020), refers here to four kinds of coincidence or intersection: (a) between different sources of disaster causality, (b) between different kinds of disaster and crisis, (c) between disaster and its context, and (d) between disaster and social circumstances. One of the characteristics of this intersectionality is that people who are particularly disadvantaged are especially sensitive to the consequences of decisions about how to deal with disaster risk. As they lack buffering resources, they are profoundly affected by such decisions.

6 Conclusion: A Strategy

Regardless of international guidelines and initiatives, countries should seek to reduce disaster risk as a matter of course. It is a moral imperative (and often a legal requirement) that the collective state should as far as possible seek to guarantee the safety of its citizens (Waldron 2006). Anything else is negligence. However, there are so many instances in which this does not happen automatically. It needs the support of the citizens, who need *to demand security and safety from their governments*, as well as work towards these things in their own right.

The institutional approach to disaster risk reduction is bound to fail. It will be outflanked by power and ideologies which may have very different agendas. Many of these have been supercharged by the disaffection of citizens and misplaced faith in authoritarianism and demonstrably false viewpoints. In many analyses, the missing factor is human nature. For example, the history of antiseismic construction in Turkey is one of evading the laws and ignoring the imperatives of safe building (Cogurcu 2015). Is this merely a failure of regulation by the leadership of Turkey or is it a collective failure to demand safety—from oneself as much as from the government?

There is no doubt that in the last half-century the world *has* made progress in disaster risk reduction. For the most part, gains have not been spectacular, but they have been real. There is a danger, however, in creating a misplaced belief in progress. For example, the reduction in death tolls in disasters is entirely illusory. At the world scale, disaster deaths are "spiky", and hence there is always the potential for a major lethal event (Holzer and Savage 2013). By way of example, a congratulatory editorial published in the journal *Disaster Prevention and Management* in 2005 congratulated the world in getting the annual total of disaster deaths down to 59,000 for the previous five years (Wilson 2005). At almost the same time, the Indian Ocean Tsunami killed as many people as had died world-wide in the previous half-decade.

For many years we have known the answers to disaster-related problems, but have we known the questions? We should search for them amid the *context* of DRR. Here, much of modern life has become a struggle against dearly held but highly destructive ideologies. The logic and ethics of disaster risk reduction clash with these dogmas and creeds. As George Bernard Shaw wrote (unfortunately in the sexist language of his times, when 'man' stood for 'women and men'): "A reasonable man adapts himself to the world. An unreasonable man persists in trying to adapt the world to himself. Therefore, all progress depends upon the unreasonable man." Hence, it is time to launch an ideological counter-offensive in favour of public safety, security, peace and disaster reduction.

One final thought needs to be aired. Disaster risk reduction requires democracy as much as we all require oxygen to breathe (Platt 1999). Democracy allows citizens to access information and act upon it in order to improve human safety. It is therefore sobering to note that in less than two decades the number of full democracies in the world has dwindled from 28 to 23: they are now outnumbered two to one by fully authoritarian regimes (Repucci and Slipowitz 2021). History does not repeat itself, but it does offer lessons about the present and the future. Hence, are we for the second time facing the end of enlightenment (Whatmore 2023), and what does that mean for attempts to bring down the increasing burden of disasters?

References

Adam, S., Miller, H.: The economic arguments for and against a wealth tax. Fisc. Stud. **42**(3–4), 457–483 (2021)

Aksoy, H.A., Çevik, S.: Political and economic implications of the Turkish earthquakes: centralisation of power has eroded state capacity. SWP Comment 2023/C19, Stiftung Wissenschaft und Politik, Berlin, 6 pp.

Alexander, D.E.: Towards the development of a standard for emergency planning. Disaster Prev. Manag. **14**(2), 158–175 (2005)

Alexander, D.E.: Disabled persons in viral pandemics: the example of Covid-19. Working Paper, Council of Europe, Strasbourg (2021)

Binns, C., Low, W.Y.: The rich get richer and the poor get poorer: the inequality of Covid-19. Asia Pac. J. Public Health **33**(2–3), 185–187 (2021)
Bunge, W.W.: Theoretical Geography. Gleerup, Lund (1966)
Calvert, J., Arbuthnot, G.: Failures of State: The Inside Story of Britain's Battle with Coronavirus. Mudlark (Harper-Collins), London (2021)
CCRS 2020: Scenario Best Practices: Developing Scenarios for Disaster Risk Reduction. Cambridge Centre for Risk Studies, Lighthill Risk Network, 2020. Judge Business School University of Cambridge, Cambridge, UK
Cogurcu, M.T.: Construction and design defects in the residential buildings and observed earthquake damage types in Turkey. Nat. Hazard. **15**(4), 931–945 (2015)
Collins, P.H., Bilge, S.: Intersectionality, 2nd edn. Polity Press, Cambridge (2020)
De Ruiter, M.C., Couasnon, A., Ward, P.J.: Breaking the Silos: an online serious game for multi-risk disaster risk reduction (DRR) management. Geosci. Commun. **4**(3), 383–397 (2021)
Else, H.: How a torrent of Covid science changed research publishing - in seven charts. Nature **588**, 553 (2020)
Esmond, B., Atkins, L.: Education, Skills and Social Justice in a Polarising World: Between Technical Elites and Welfare Vocationalism. Routledge, Abingdon (2022)
Fekete, A., Sandholz, S.: Here comes the flood, but not failure? Lessons to learn after the heavy rain and pluvial floods in Germany 2021. Water **13**(21), 3016 (2021)
Gorman, S.E., Gorman, J.M.: Denying to the Grave: Why We Ignore the Science That Will Save Us. Oxford University Press, Oxford (2021)
Holzer, T.L., Savage, J.C.: Global earthquake fatalities and population. Earthq. Spectra **29**(1), 155–175 (2013)
House of Commons: UK Disability Statistics: Prevalence and Life Experiences. House of Commons Library, UK Parliament, London (2023). https://commonslibrary.parliament.uk/research-briefings/cbp-9602/. Accessed 06 Feb 2024
Human Rights Watch: World Report 2021: Events of 2020. Human Rights Watch, New York (2021)
McKee, M., et al.: Open science communication: the first year of the UK's Independent Scientific Advisory Group for Emergencies. Health Policy **126**(3), 234–244 (2022)
Platt, R.H.: Disasters and Democracy: The Politics of Extreme Natural Events, p. 320. Island Press, Washington, D.C. (1999)
Repucci, S., Slipowitz, A.: Democracy under Siege. Freedom in the World 2021. Freedom House, Washington, DC (2021). 12 pp.
Schweizer, P.-J., Renn, O.: Governance of systemic risks for disaster prevention and mitigation. Disaster Prev. Manag. **28**(6), 862–874 (2019)
Sen, A.: Why and how is health a human right? The Lancet **372**(9655), 2010 (2008)
Taylor, S.: Wealth, health and equity: convergence to divergence in late 20th century globalization. Br. Med. Bull. **91**(1), 29–48 (2009)
Tselios, V., Tomkins, E.L.: What causes nations to recover from disasters? An inquiry into the role of wealth, income inequality, and social welfare provisioning. Int. J. Disaster Risk Reduct. **33**, 162–180 (2019)
UK Government: The UK Government Resilience Framework. HM Government of the United Kingdom, London (2022)
Waldron, R.: Safety and security. Nebraska Law Rev. **85**(2), 454–507 (2006)
Whatmore, R.: The End of Enlightenment: Empire, Commerce, Crisis. Allen Lane, London (2023). 496 pp.
Wilson, H.C.: Editorial. Disaster Prev. Manag. **14**(1), 4 (2005)

Panel Group Discussion I: Towards GADRI Objectives of Achieving a Sustainable and Resilient Society Against Disasters

Report of the GADRI Committee on Networking: Networking to Promote Disaster Risk Reduction

Charles Scawthorn[1](✉) and Ritsuko Yamazaki-Honda[2]

[1] SPA Risk LLC and University of California, Berkeley, USA
cscawthorn@gmail.com

[2] National Research Institute for Earth Science and Disaster Resilience (NIED), Tsukuba, Japan

Abstract. A conference session on "Networking to promote disaster risk reduction" was organized by the GADRI Committee on Networking during the 6th Global Summit of GADRI held in March 2023 in Kyoto. The session sought to explore communication and exchanges to promote disaster research, resilience and risk reduction, specifically three key questions: (1) How can we foster exchange between individual GADRI researchers? (2) Which stakeholders should GADRI be networking with? And (3) How would GADRI approach networking with multitude of DRR consumers? These questions were addressed both broadly and with a focus on how networking might be facilitated within GADRI as well as between GADRI and the many other existing networks at regional and global scale working in the risk and disaster field. Four speakers from diverse backgrounds were invited to the panel discussion session which was held in a "world café" style – which involved an ad hoc organizing of participants into small groups and moving around in musical-chairs manner to deepen discussions on the presentations by the speakers. The speakers covered included how to enhance local innovation with other international and regional entities; the importance/tips of "story telling" to improve networking; the system of bringing together industry and academia employed in the US, citing an Industry-University Cooperative Research Center as a good practice; and the status of an ongoing database project to collect information on disasters, together with the project to link institutions and researchers, which will contribute to the research activities of the GADRI community. Discussion in both small groups and plenary followed the presentations, resulting in the following three recommendations for improving disaster-related networking: (1). GADRI should create a Global Disaster Researcher Directory, structured as a voluntary compilation of individuals engaged in disaster research, channeled through GADRI member institutes, perhaps building on ORCID IDs. (2) GADRI should develop a strong social media presence (Facebook, Instagram, Twitter, Linked-in, YouTube). Each platform requires different approach. YouTube allows the use of GADRI lectures and more detailed lengthier presentations appropriate to academic audiences and can be foundational for all 5 platforms. (3). GADRI should create GADRI Fellow and Young Researcher Programs.

Keywords: Networking · disasters · research · risk reduction · Global Alliance of Disaster Research Institutes · GADRI · communications

H. Tatano et al. (Eds.): GSRIDRR 2023, DRRGADRIBS, pp. 103–148, 2025.
https://doi.org/10.1007/978-981-96-6609-6_9

1 Introduction

The global disaster reduction community is both quite large, and extremely diverse. It ranges from local volunteers, NGOs and community organizers, through regional, national and indeed international professionals, academics, researchers and practitioners to internationally recognized heads of agencies and leaders. While not intimate members of the community, even heads of state, the UN Secretary General and CEOs of multinationals will periodically address this topic.

One of the leading groups within the global disaster reduction community is the disaster research community, consisting largely of academic disaster research institutes, but also including many government, industry-sponsored and private stand-alone institutes as well as even some individuals. Many of these disaster research institutes are comprised of dozens to hundreds of subject matter experts, so that the entire global disaster research community most likely numbers in the many tens of thousands. These researchers range across a very broad spectrum of the natural and social sciences, such as seismology, hydrology, meteorology, volcanology, sociology, psychology, economics, political science, anthropology, and so on. In fact, the breadth of expertise is so wide that, while researchers will be familiar with other researchers within their own relatively narrow field, they certainly will not be familiar with many researchers who in fact share mutual interests, whether those interests be data, analytical or experimental methods, or applications. Therefore, enormous opportunities for collaboration and synergy are being missed.

One way to improve this situation and foster disaster research is to improve communication among these researchers. The Global Alliance of Disaster Research Institutes (GADRI) was in fact established with this as one of its goals. GADRI membership now includes more than 200 disaster research institutes (DRIs) from Asia, Europe, North and South America, Australia and Oceana, and the GADRI websites provides quick linkages to any and all of these member DRIs.

However, communication is ultimately between people, not institutions, and the current methods for researcher-to-researcher communication are cumbersome, consisting of personal contacts, some limited peer-to-peer networking, some exploration of professional society rosters, and cold calling. To address this situation, a session on "Networking to improve disaster reduction" was organized by the GADRI Committee on Networking during the 6th Global Summit of GADRI held in March 2023 in Kyoto. Below, we summarize the session and its recommendations.

Ch10-1: Group Photo

2 Session

The session was chaired by Prof. Charles Scawthorn and Ms. Ritsuko Yamazaki-Honda, and was organized in the following manner:

- Introductions by co-chairs
- Presentations by 4 Speakers (40 min)
- "World Café" (15 min × 2 rounds)
- Summary from World Cafés
- Panel Discussion by the 4 speakers (15 min)

The "world café" style involved an ad hoc organizing of participants into small groups who discussed the questions and presentations, and the moving/mixing around in musical-chairs manner to repeat the process and deepen discussions. This "moving/mixing" proved popular and appeared to really improve discussion.

The key questions addressed during the session were:

1. **How can we foster exchange between individual GADRI researchers?**
   - Would a directory of GADRI institution researchers contribute to this goal?
   - Would monthly on-line discussion/seminars on specific topics be of interest?
2. **Which stakeholders should GADRI be networking with**? (e.g. international agencies, MDBs, NGOs, etc.)

- What would be the specific goals in doing so?

3. **How would GADRI approach networking with** multitude of DRR consumers? **How can GADRI support/assist its members in need?**

Four speakers from diverse backgrounds addressed these questions:

- Rajib Shaw, Professor, Keio University, Tokyo, who presented how to enhance local innovation with other international and regional entities, and highlighted specific programs to motivate youth and young professionals;
- Ronan Sato, Associate, Energy & Environment Investment, Tokyo, who demonstrated the importance/tips of "story telling" to improve networking;
- Jean-Paul Pinelli, Professor, Florida Institute of Technology, who elaborated on a system of bringing together industry and academia, citing an Industry-University Cooperative Research Center as a good practice;
- Hiroyuki Goto, Disaster Prevention Research Institute (DPRI), Kyoto University, who outlined the ongoing database project to collect information on disasters, together with the project to link institutions and researchers, which will contribute to the research activities of the GADRI community.

In More Detail:

Prof. Rajib Shaw shared information on other similar networks and alliances such as the U-Inspire Alliance; IRDR Young Scientists Program; and Resilience Innovation Knowledge Academy (RIKA India).

In summary, Prof. Shaw concluded that:

- Networking starts with 1) purposefulness, 2) collectiveness, 3) connectedness, 4) openness and 5) synergy
- Entrusting the regional entities (where available) to foster regional networks
- Need specific programs (training/action research/advocacy) to capitalize the networks
- Youth and young professionals play an important role in networking
- Enhance GADRI Young Scientist Network
- GADRI needs to foster incubation hub in different member institutions
- Promote GADRI DRR Innovation and Incubation Hub

Ch10-2: Presentation by Prof. Rajib Shaw

Ronan Sato, in his presentation "Better Networking through Storytelling" stated that today, everybody is part of the attention economy – from fellow academics to bureaucrats to the general population. He stated that stories are just sequences of events – "scientists using MRIs have discovered that **the parietal lobe** reduces in activity when listing to sequences of events". He continued to talk how to prevent attrition by replacing "ANDs" with "BUTs". He stated that "real life is a sequence of ANDs, but the more BUTs you have in a story, the more you will reduce attrition. If your story has plenty of BUTs, people (and their parietal lobes) won't be able to resist!". Finally, a good story needs FOCUS. "For example, the story formula could be "I am researching "X" and it is interesting because of "X". Example – I am doing a story on the USS John C Stennis and it is interesting because the ship is run by teenagers!" If you have to add further clarification to your (Y), it's probably not interesting enough.

CH10-3: Presentation by Dr. Ronan Sato

Prof. J.P. Pinelli described the organization of IUCRC as follows:

- Center can have any number of universities as independent SITES
- Each university site must bring in at least 4 industry members
- Universities can be from outside the USA, but will not receive NSF subsidies
- Industry members can be private companies, professional organizations, laboratories, government agencies
- Industry members can be from any country in the world
- Industry members drive the research agenda: they decide what to fund.

Its mission is "to pursue research that are of interest to industry to enhance the resiliency of buildings and infrastructure to extreme windstorms such as hurricanes and tornadoes".

IUCRC provides many benefits for its members:

- Opportunity to prioritize research projects
- Realize high return on research investment
- Access world class facilities at discounted price, faculty expertise and motivated students
- Connect with students and gain advantage in recruiting
- Gain royalty-free rights to intellectual property
- Join peers to innovate new risk models, structural and component systems
- Fulfill social responsibility by mitigating damage and anguish

In summary:

- An IUCRC like the WHIP-C brings together academia and industry for the benefit of society
- It is open to a variety of members from different areas and countries
- It facilitates the interaction between the members and fosters cooperation
- It is a conduit for the transfer of mitigation and resilience research into practice
- The framework could be replicated in other parts of the world.

Prof. Hiroyuki Goto, outlined the ongoing database project to collect information on disasters, together with the project to link institutions and researchers, which will contribute to the research activities of the GADRI community, making the following points:

- The database will be built in cooperation with the GADRI Committee on Data and Information Sharing that will contribute to the research activities of the GADRI Community and beyond.
- Once developed, the database allows users to search for websites, researchers, and documents related to disaster reduction research to take advantage of the wealth of data resources on disaster reduction research that exists around the world. It could automatically build a summary page for disaster response. It would allow, free keyword search which are automatically generated from the link information by use of an URL; it will provide an institution connection; and researcher connection with the help of ORCID ID. The project was initiated in 2022; and in 2023, it is in the stage of implementation of disaster collection data.

Their presentations appear below.

3 Discussion

Key points that emerged from the discussion of each question included:

1. **How can we foster exchange between individual GADRI researchers?**
    - **Young professionals/researchers** to collaborate and exchange in and across region. Online sessions or even better would face-to-face interactions
    - Use **social media** as a platform to connect internally and engage outside of GADRI and feed all information outcomes of events, research meetings, etc. (e.g. FB, Instagram, "GADRI YouTube Channel")
    - **Develop database** for external networking with institutions on DRR
    - **Storytelling** as researchers - a "People-centered GADRI" activity particularly aimed at vulnerable people for disaster mitigation measures. GADRI, as an umbrella, to facilitate storytelling.
2. **Which stakeholders should GADRI be networking with**?
    - **International or regional associations**
      e.g. Education … to reach out students of any school level;
    - At global level: UNESCO, online training

- At national level - education department at university level
- At decentralized/ local level e.g. School Safety Program in Thailand

3. **How would GADRI approach networking with multitude of DRR consumers? How can GADRI support/assist its members in need?**

- Creation of a **GADRI Fellow Program:**
  accomplished disaster researchers will be nominated as GADRI Fellows. E.g. GADRI will extend invitation for Fellows to visit/lecture DRIs on a case-by-case basis (at no financial burden to GADRI, to be explained). GADRI members support GADRI fellows in their area (e.g. hosting, sabbatical)
- **Young professional program in GADRI members** – inviting young professionals to be hosted in GADRI institutions to study and research for three – twelve-month period.
- Creation of a *Google Scholar-like of GADRI,* **Global Disaster Researcher Directory**: *voluntary compilation of individuals engaged in disaster research, channeled through GADRI member institutes*
- Existing common programs to involve different network/stakeholders

Ch10-1: Q&A Session

4 Recommendations

In summary, the Committee recommended the following next steps:

1. GADRI should **create a <u>Global Disaster Researcher Directory</u>**: voluntary compilation of individuals engaged in disaster research, channeled through GADRI member institutes, perhaps building on ORCID IDs.
2. GADRI should **develop a strong <u>social media</u> presence** (Facebook, Instagram, Twitter, Linked-in, YouTube). Each platform requires different approach. YouTube allows the use of GADRI lectures and more detailed lengthier presentations appropriate to academic audiences and can be foundational for all 5 platforms.
3. GADRI should **create <u>GADRI Fellow and Young Researcher Programs</u>**: This would be based on:
   (a) Nominations of accomplished senior disaster researchers, from which GADRI Fellows will be selected. GADRI will extend invitation for Fellows to visit/lecture DRIs on a case-by-case basis at no financial burden to GADRI.
   (b) A Young Researcher program to foster exchange of early career researchers.

These initiatives will be forwarded to the GADRI BOD for implementation and promotion.

5 Presentations by the Panelists

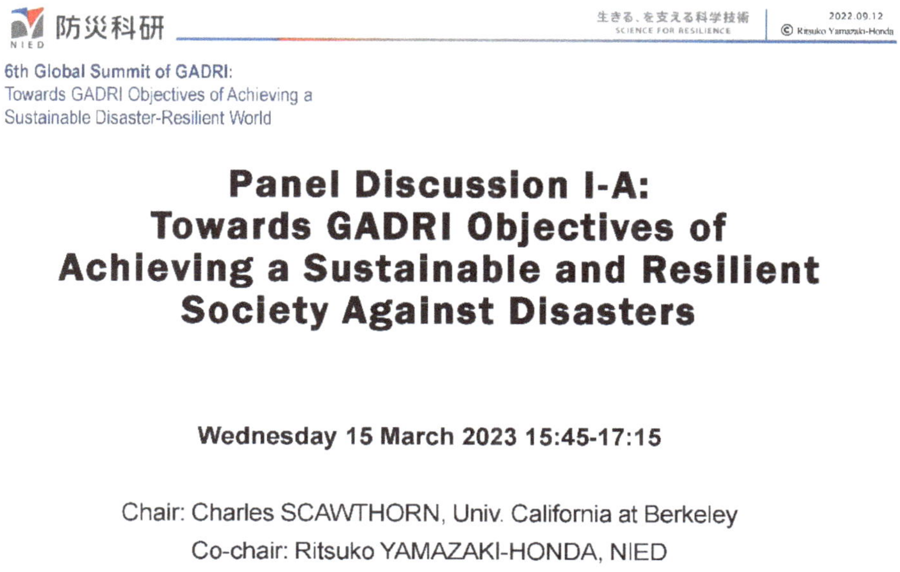

防災科研 NIED 生きる、を支える科学技術 SCIENCE FOR RESILIENCE 2022.09.12 © Ritsuko Yamazaki-Honda

Agenda

1. Introduction

2. Presentations by Speakers (40min)

- Rajib Shaw, Professor, Keio University.
- Ronan Sato, Associate, Energy & Environment Investment
- Jean-Paul Pinelli, Professor, Florida Institute of Technology
- Hiroyuki Goto, Associate Professor, DPRI, Kyoto University.

3. "World Café" (15min x 2 rounds)

4. Summary from World Café and Panel Discussion by the four speakers (15min)

防災科研 NIED 生きる、を支える科学技術 SCIENCE FOR RESILIENCE 2022.09.12 © Ritsuko Yamazaki-Honda

Introduction

GADRI Objectives

a. To establish a global research network that promotes and engages disaster research

The session contributes to networking;

◆**Presentations by 4 Speakers** (40min)

◆**"World Café"** (15min x 2 rounds)

● **Summary from World Café and Panel Discussion by the 4 speakers** (15min)

防災科研 NIED 生きる、を支える科学技術 SCIENCE FOR RESILIENCE 2022.09.12 © Ritsuko Yamazaki-Honda

Presentations by Speakers

(10 minutes each)

- **Rajib Shaw**

  Professor, Keio University.
- **Ronan Sato**

  Associate, Energy & Environment Investment
- **Jean-Paul Pinelli**

  Professor, Florida Institute of Technology
- **Hiroyuki Goto**

  Associate Professor, Disaster Prevention Research Institute (DPRI). Kyoto University.

防災科研
NIED
生きる、を支える科学技術
SCIENCE FOR RESILIENCE
2022.09.12
© Ritsuko Yamazaki-Honda
World Café with Speakers
1st Round (15 minutes)
Move to one group on one's choice
Rajib
Ronan
Jean-Paul
Hiroyuki
5

f

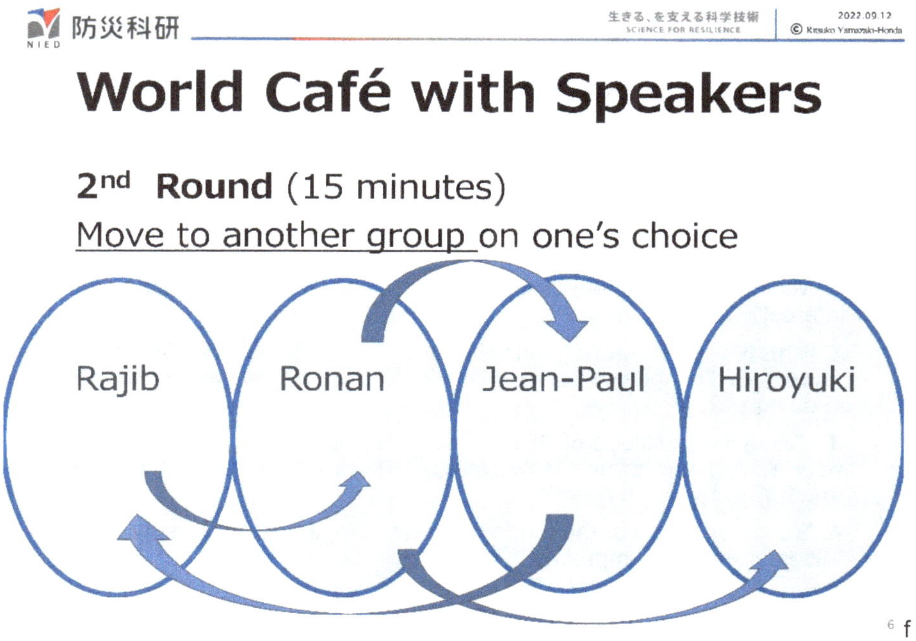
防災科研
NIED
生きる、を支える科学技術
SCIENCE FOR RESILIENCE
2022.09.12
© Ritsuko Yamazaki-Honda
World Café with Speakers
2nd Round (15 minutes)
Move to another group on one's choice
Rajib
Ronan
Jean-Paul
Hiroyuki
6

f

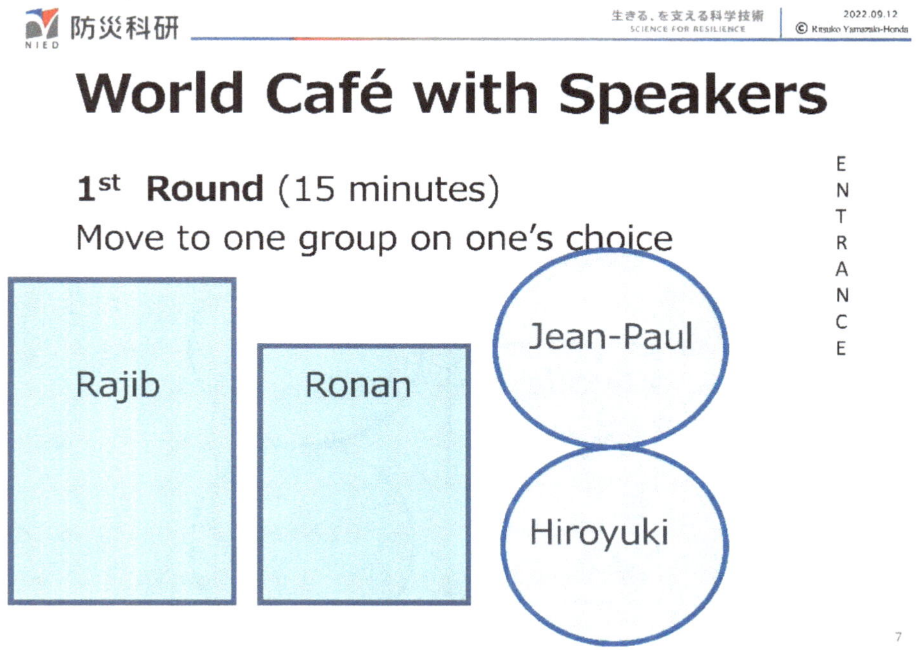
防災科研
NIED
生きる、を支える科学技術
SCIENCE FOR RESILIENCE
2022.09.12
© Ritsuko Yamazaki-Honda
World Café with Speakers
1st Round (15 minutes)
Move to one group on one's choice
ENTRANCE
Rajib
Ronan
Jean-Paul
Hiroyuki
7

防災科研 NIED 生きる、を支える科学技術 SCIENCE FOR RESILIENCE 2022.09.12 © Ritsuko Yamazaki-Honda

Key Questions

1. How can we foster exchange between individual GADRI researchers?

- Would a directory of GADRI institution researchers contribute to this goal?

- Would monthly on-line discussion/seminars on specific topics be of interest?

2. **Which stakeholders** (international agencies, MDBs, NGOs, etc.) **should GADRI be networking with**? What would be the specific goals in doing so?

3. Given the multitude of DRR consumers, **how would GADRI approach networking with them? How can GADRI support/assist its members in serving those in need?**

4. Are we agreed that the fundamental message in GADRI networking is "disaster research improves DRR"?

8

防災科研 NIED 生きる、を支える科学技術 SCIENCE FOR RESILIENCE 2022.09.12 © Ritsuko Yamazaki-Honda

World Café with Speakers

2nd Round (15 minutes)

Move to one group on one's choice

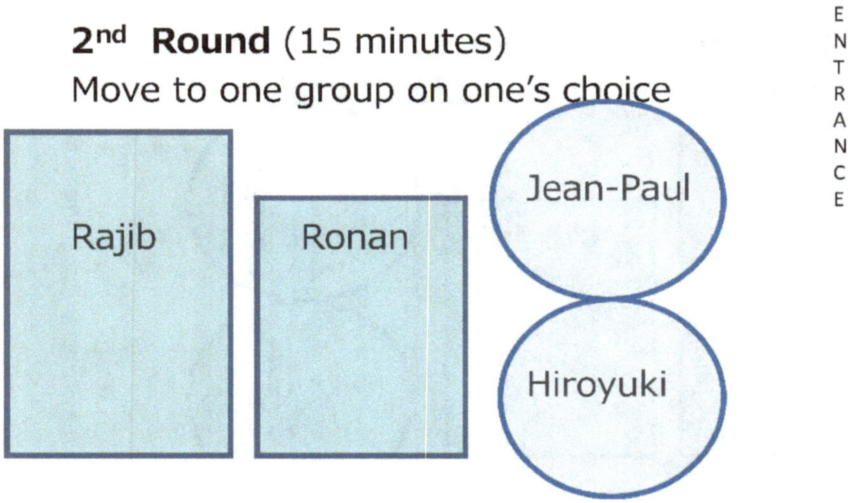

9

Summary from World Café & Panel Discussion

(15 minutes)

- **Rajib Shaw**

  Professor, Keio University.
- **Ronan Sato**

  Associate, Energy & Environment Investment
- **Jean-Paul Pinelli**

  Professor, Florida Institute of Technology
- **Hiroyuki Goto**

  Associate Professor, DPRI, Kyoto University.

10

What next?

1. How can we foster exchange between individual GADRI researchers?

- Creation of a Global Disaster Researcher Directory: *voluntary compilation of individuals engaged in disaster research, channeled through GADRI member institutes*
- *To be added from discussion*

防災科研 NIED 生きる、を支える科学技術 SCIENCE FOR RESILIENCE 2022.09.12 © Ritsuko Yamazaki-Honda

What next?

2. Which stakeholders should GADRI be networking with?

What would be the specific goals in doing so?

- *To be added from discussion*

防災科研 NIED 生きる、を支える科学技術 SCIENCE FOR RESILIENCE 2022.09.12 © Ritsuko Yamazaki-Honda

What next?

3. How would GADRI approach networking with them? How can GADRI support/assist its members in serving those in need?

- Creation of a GADRI Fellow Program: *accomplished senior disaster researchers will be nominated, from which GADRI Fellows will be selected. GADRI will extend invitation for Fellows to visit/lecture DRIs on a case-by-case basis (at no financial burden to GADRI, to be explained).*
- *To be added from discussion*

Fostering networking to enhance Global DRR

To enhance local innovation

Rajib Shaw

Professor, Keio University, Japan Distinguished Professor, Indian Institute of Technology (IIT), Guwahati, India Distinguished Professor, IDMR, Sichuan University, China Visiting Professor, Indian Institute of Science (IISc), Bengaluru, India	Academia
Co-Chair, United Nations Asia Pacific Science Technology Advisory Group (AP-STAG) Coordinating Lead Author (CLA), Asia Chapter, IPCC 6th Assessment Report	Science Policy interface
Co-Founder, Resilience Innovation Knowledge Academy (RIKA) https://rikaindia.com Co-Founder, RIKA Institute https://www.rikainstitute.org	Start-up and innovation
Chair of Board, SEEDS Asia https://www.seedsasia.org Chair of Board, CWS Japan https://www.cwsjapan.org	NPO / NGO

www.rajibshaw.org AND www.rikaindia.com

Five basic principles of networking

1. Win/Win Situations Produce Strong Connections
   - Based on both side's priority
2. Give More Than You Get
   - Need to show value of being involved
3. Activity Beats Inactivity
   - Being pro-active
4. Quality Over Quantity
   - Focusing on content is important
5. People Do Collaboration With People They Like
   - Person to person collaboration

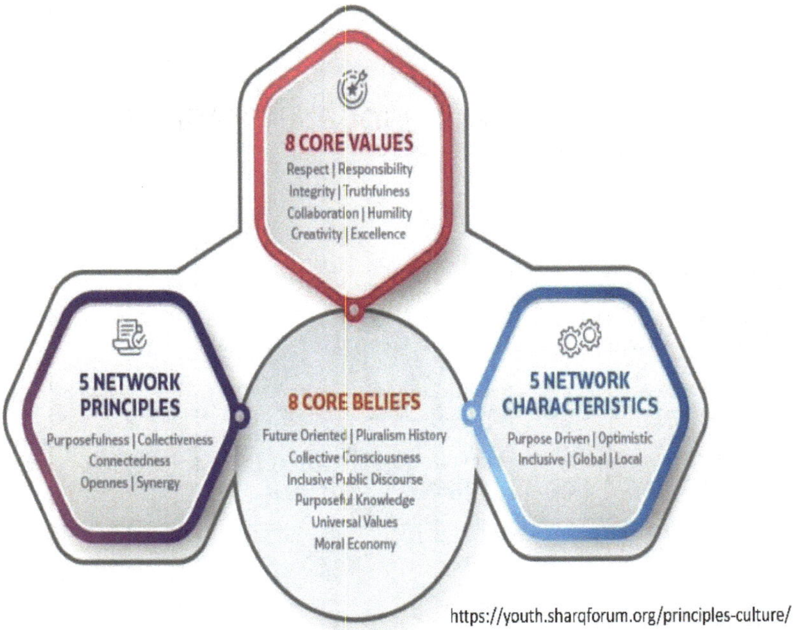
8 CORE VALUES
Respect | Responsibility
Integrity | Truthfulness
Collaboration | Humility
Creativity | Excellence
5 NETWORK PRINCIPLES
Purposefulness | Collectiveness
Connectedness
Opennes | Synergy
8 CORE BELIEFS
Future Oriented | Pluralism History
Collective Consciousness
Inclusive Public Discourse
Purposeful Knowledge
Universal Values
Moral Economy
5 NETWORK CHARACTERISTICS
Purpose Driven | Optimistic
Inclusive | Global | Local
https://youth.sharqforum.org/principles-culture/

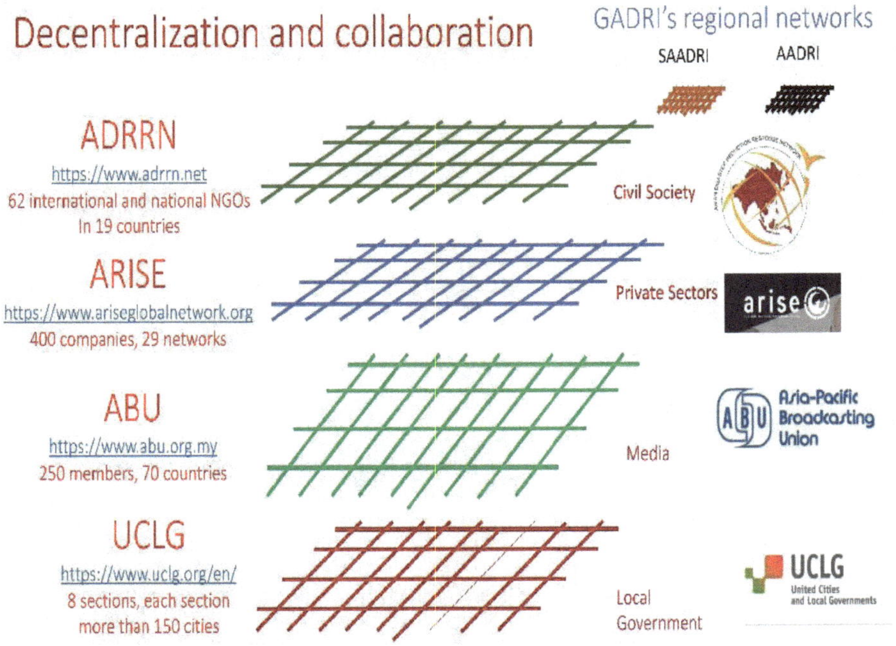
Decentralization and collaboration
GADRI's regional networks
SAADRI
AADRI
ADRRN
https://www.adrrn.net
62 international and national NGOs
In 19 countries
Civil Society
ARISE
https://www.ariseglobalnetwork.org
400 companies, 29 networks
Private Sectors
arise
ABU
https://www.abu.org.my
250 members, 70 countries
Media
Asia-Pacific Broadcasting Union
UCLG
https://www.uclg.org/en/
8 sections, each section
more than 150 cities
Local
Government
UCLG
United Cities and Local Governments

Youth and young professionals: U-Inspire Alliance

https://uinspirealliance.org

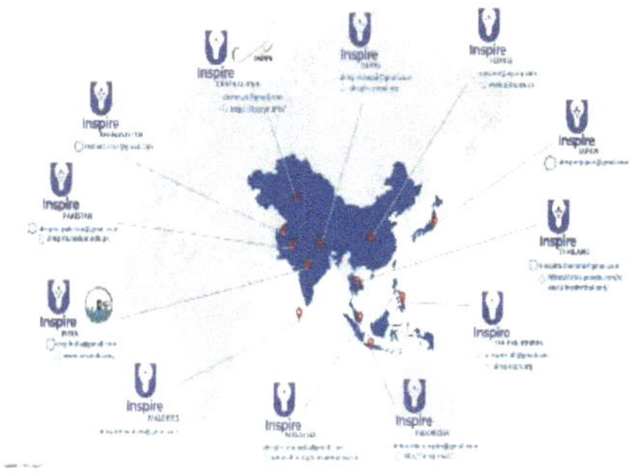

1. Connect Youth & Young Professional (in SETI) for DRR in Asia Pacific
2. Promote dialogue, capacity building and networking on DRR
3. Promote Innovation
4. Promote access to DRR information, data and technology
5. Monitoring and Evaluation

Disaster Risk Reduction: Learning from the past, present And link to Future

- Pro-active role in damage assessment in Palu Earthquake
- Developing professional network in some countries
- Developing BIPAD portal with Ministry of Home in Nepal

CRITICAL WRITING WORKSHOP

Co-organized by:

SAADRI, U-INSPIRE, BRIN & India Japan Lab- Keio University

Winter workshop: 13th December 2022

IRDR Young Scientist Program

Responsibility of IRDR Young Scientists

- Contribute to innovative research in the field of disaster risk reduction
- Be the ambassador of IRDR in different conferences and/or social media
- Develop and contribute to networking of young professionals
- Submit a 6-month report of activities in the prescribed format of IRDR
- Acknowledge the contribution of IRDR in the academic paper, thesis etc.

IRDR Young Scientists Programme

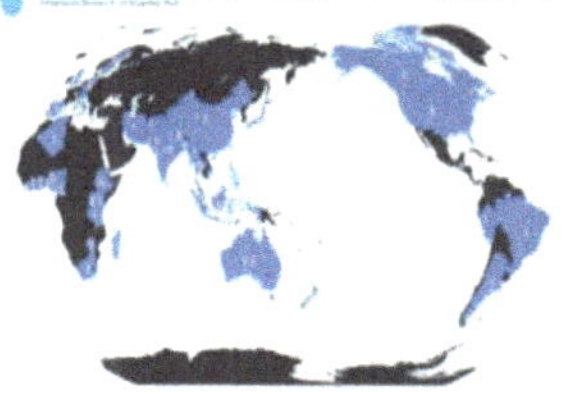

Country	Number	Country	Number	Country	Number
Nepal	16	Canada	3	Tanzania	1
Pakistan	14	Cameroon	3	Zimbabwe	1
China	13	Belgium	2	South Africa	1
India	12	Singapore	2	Madagascar	1
Australia	10	Kenya	2	Cote d'Ivoire	1
Nigeria	6	United Kingdom	2	Ghana	1
Bangladesh	6	Argentina	1	Liberia	1
Thailand	6	Brazil	1	Portugal	1
Indonesia	6	Switzerland	1	Spain	1
Ethiopia	6	Peru	1	Uganda	1
United States	5	Sri Lanka	1	Sweden	1
New Zealand	5	Iran	1	Algeria	1
Germany	4	Afghanistan	1	Togo	1
Japan	4	Myanmar	1	Italy	1
Philippines	4	Guatemala	1		
Malaysia	4	Vietnam	1		

162 IRDR young scientists from 46 countries

119 43

- Training
- Joint research
- Joint writing

Science-preunership: Science based entrepreunership

- Entrepreneur mindset and ecosystem : incubation hub
- Government – academia – enterprise linkage

How to bring Youth and Young Professionals to solve local problems and achieve the targets of SDGs?

Sciencepreuner (Scientist + Entrepreneur) bring research into the core of disaster management activities of the private sector and policy making

Resilience Innovation Knowledge Academy (RIKA)
www.rikaindia.com

Summary

- Networking starts with 1) purposefulness, 2) collectiveness, 3) connectedness, 4) openness and 5) synergy
- Entrusting the regional entities (where available) to foster regional networks
- Need specific programs (training / action research / advocacy) to capitalize the networks
- Youth and young professionals play an important role in networking
    - GADRI Young Scientist Network
- GADRI needs to foster incubation hub in different member institutions
    - GADRI DRR Innovation and Incubation Hub

JUNSEISHA
2023 March 15
GADRI Presentation - Better Networking through Storytelling
by Ronan Sato

Introduction
Venture capitalist from the private sector
Previously built power plants for Mitsui (Morocco, Brazil)
Today, an investor in clean energy and space startups
+ podcast producer!
CEC
SKYDRIVE
Astroscale
Synspective
Ronan Sato
Associate, Energy Environment Investment

Today I'll be talking about...

How GADRI members might improve improve networking by...

STORYTELLING!

Why storytelling?

Today, EVERYBODY is part of the attention economy.
(Fellow academics, bureaucrats, the general population...)

Agenda: the shortest storytelling course in the world

1. What stories are made of
2. The ABT model
3. The Story Formula

Stories are just sequences of events

Scientists using MRIs have discovered that **the parietal lobe** reduces in activity when listening to sequences of events.

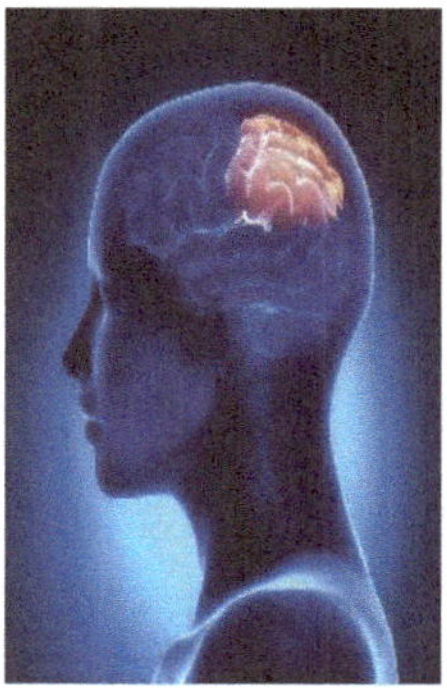

BUT!

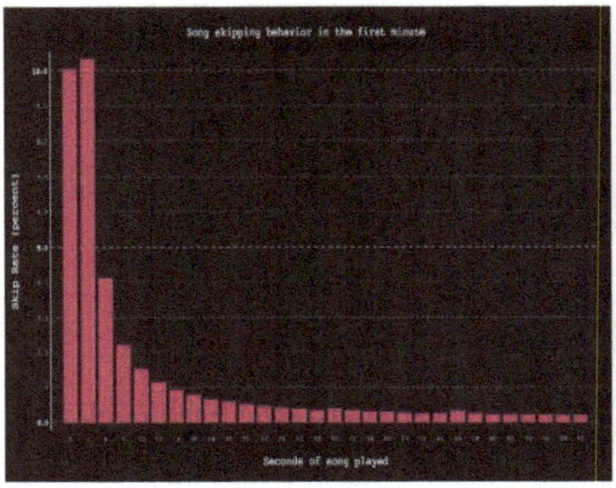

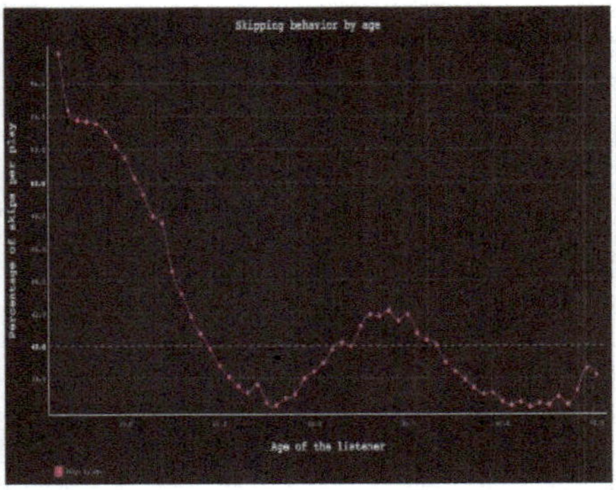

Attention attrition is vicious and GETTING WORSE..

Preveting attrition: replace ANDs with BUTs

Real life is a sequence of ANDs,

but the more BUTs you have in a story,

the more you will reduce attrition.

If your story has plent of BUTs,

people (and their parietal lobes)

won't be able to resist!

Finally, a good story needs FOCUS!

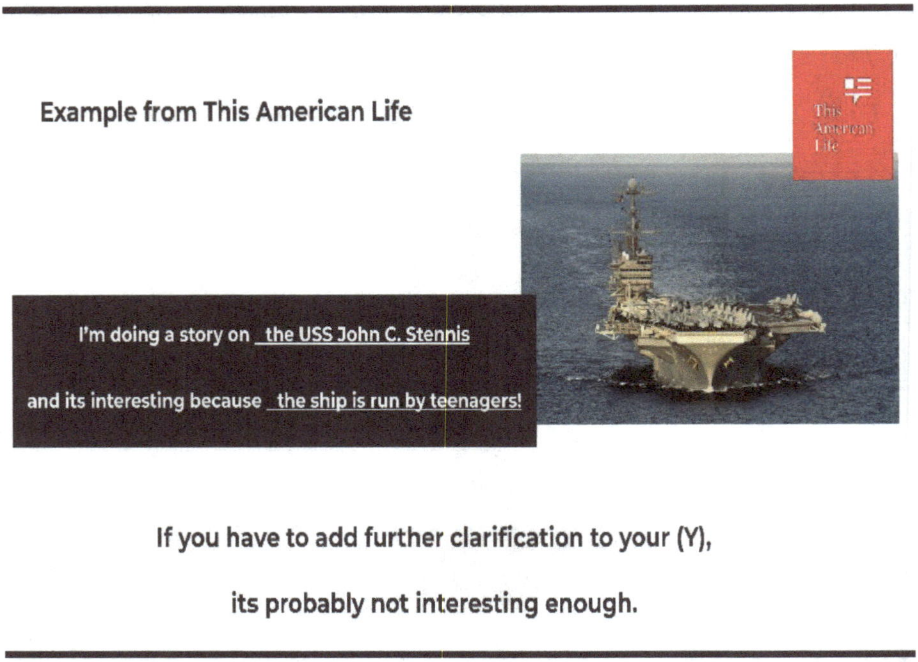
Example from This American Life
This American Life
I'm doing a story on the USS John C. Stennis
and its interesting because the ship is run by teenagers!
If you have to add further clarification to your (Y),
its probably not interesting enough.

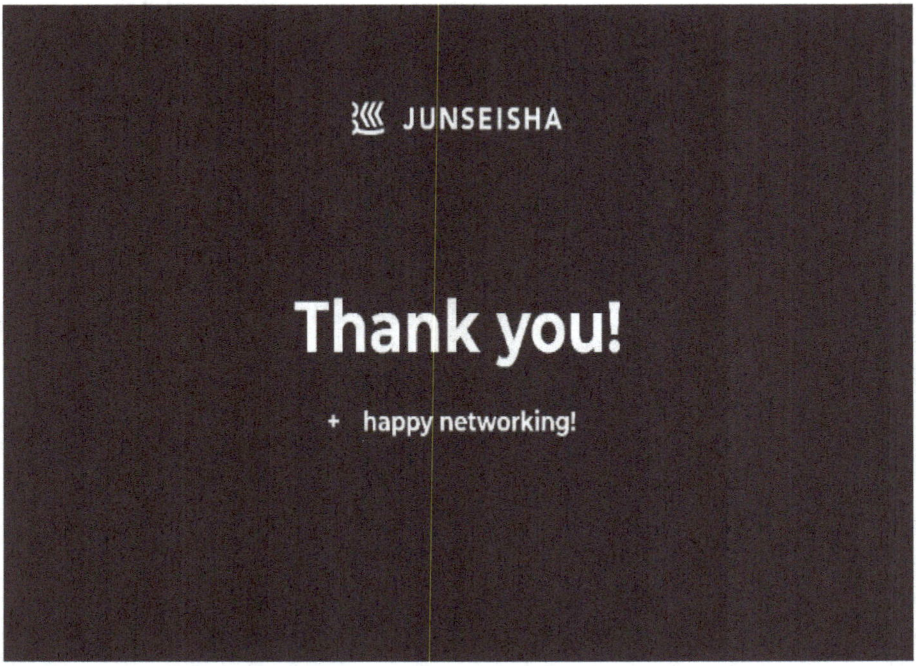
JUNSEISHA
Thank you!
+ happy networking!

Further reading

Randy Olson,

Marine biologist turned storyteller

WHIP-C

Wind Hazard and Infrastructure Performance

FIU

An Industry-University Cooperative Research Center

Bringing together industry and academia

a presentation to the

2023 GADRI SUMMIT

By

Dr. Jean-Paul Pinelli, P.E., Professor, Florida Institute of Technology

March 16, 2023

FLORIDA TECH

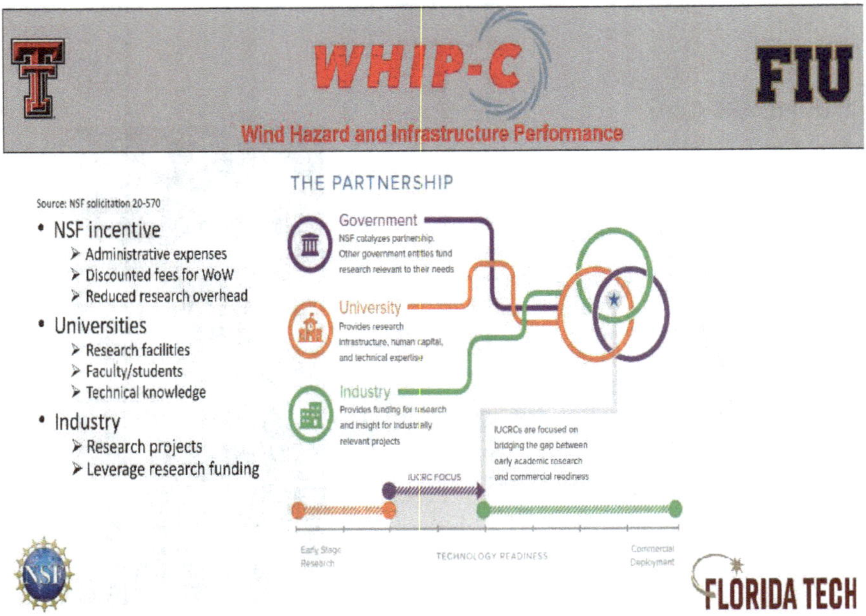
WHIP-C
Wind Hazard and Infrastructure Performance
FIU
THE PARTNERSHIP
Source: NSF solicitation 20-570
• NSF incentive
➢ Administrative expenses
➢ Discounted fees for WoW
➢ Reduced research overhead
• Universities
➢ Research facilities
➢ Faculty/students
➢ Technical knowledge
• Industry
➢ Research projects
➢ Leverage research funding
Government
NSF catalyzes partnership. Other government entities fund research relevant to their needs
University
Provides research infrastructure, human capital, and technical expertise
Industry
Provides funding for research and insight for industrially relevant projects
IUCRCs are focused on bridging the gap between early academic research and commercial readiness
IUCRC FOCUS
Early Stage Research
TECHNOLOGY READINESS
Commercial Deployment
FLORIDA TECH

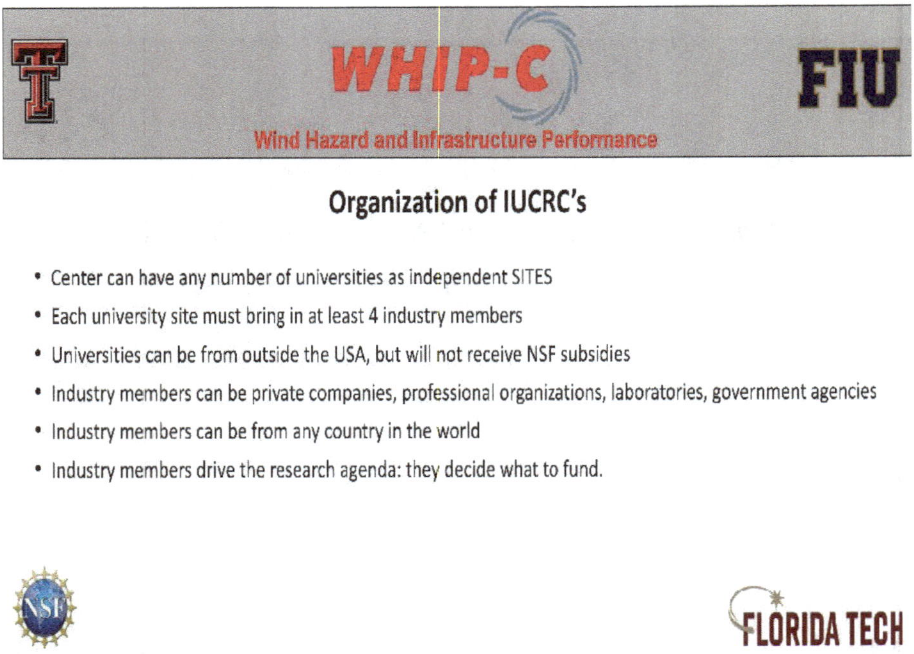
WHIP-C
Wind Hazard and Infrastructure Performance
FIU
Organization of IUCRC's
• Center can have any number of universities as independent SITES
• Each university site must bring in at least 4 industry members
• Universities can be from outside the USA, but will not receive NSF subsidies
• Industry members can be private companies, professional organizations, laboratories, government agencies
• Industry members can be from any country in the world
• Industry members drive the research agenda: they decide what to fund.
FLORIDA TECH

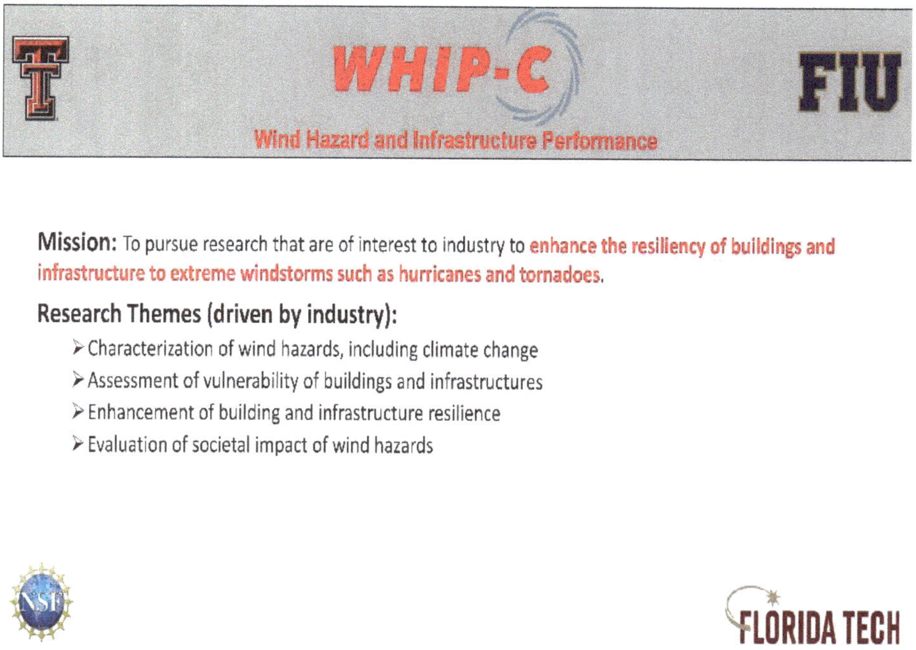
WHIP-C
FIU
Wind Hazard and Infrastructure Performance
Mission: To pursue research that are of interest to industry to enhance the resiliency of buildings and infrastructure to extreme windstorms such as hurricanes and tornadoes.
Research Themes (driven by industry):
➢ Characterization of wind hazards, including climate change
➢ Assessment of vulnerability of buildings and infrastructures
➢ Enhancement of building and infrastructure resilience
➢ Evaluation of societal impact of wind hazards
NSF
FLORIDA TECH

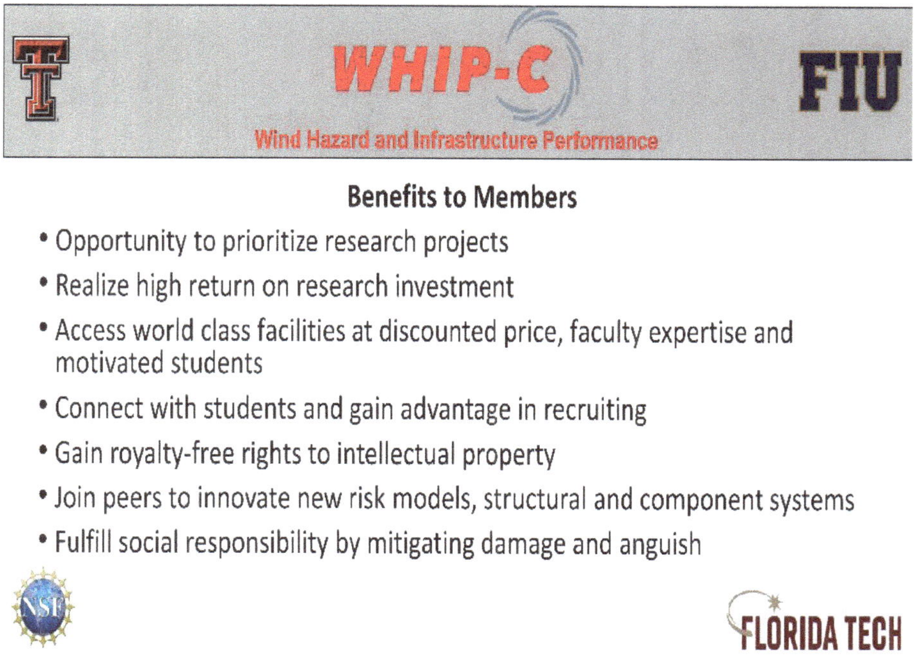
WHIP-C
FIU
Wind Hazard and Infrastructure Performance
Benefits to Members
• Opportunity to prioritize research projects
• Realize high return on research investment
• Access world class facilities at discounted price, faculty expertise and motivated students
• Connect with students and gain advantage in recruiting
• Gain royalty-free rights to intellectual property
• Join peers to innovate new risk models, structural and component systems
• Fulfill social responsibility by mitigating damage and anguish
NSF
FLORIDA TECH

WHIP-C
FIU
Wind Hazard and Infrastructure Performance
Current Industry Members
Risk Modeler
• VERISK (previously AIR Worldwide). Boston, MA
AIR
Insurance industry
• Berkshire Hathaway - Specialty Insurance. Boston, MA
Berkshire Hathaway Specialty Insurance.
• State Farm - Insurance. Bloomington, IL
StateFarm
• USAA. San Antonio, TX
USAA
Manufacturers
• ESDEC Solar Group
ESDEC SOLAR GROUP
• GAF Materials – Roofing. Parsippany, NJ
GAF
Government Agency
• National Institute of Standards and Technology
NIST
NSF
FLORIDA TECH

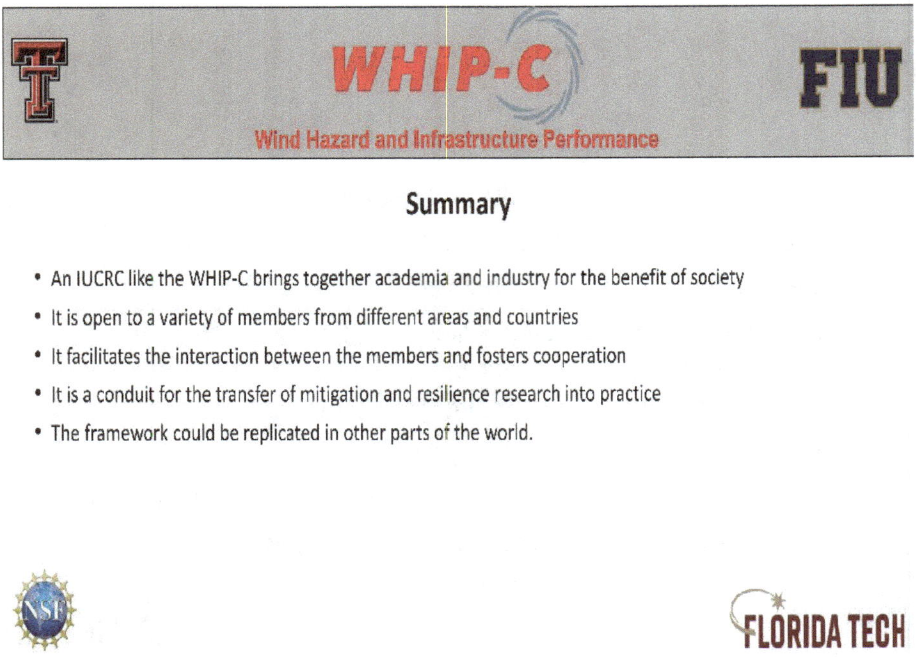
WHIP-C
FIU
Wind Hazard and Infrastructure Performance
Summary
• An IUCRC like the WHIP-C brings together academia and industry for the benefit of society
• It is open to a variety of members from different areas and countries
• It facilitates the interaction between the members and fosters cooperation
• It is a conduit for the transfer of mitigation and resilience research into practice
• The framework could be replicated in other parts of the world.
NSF
FLORIDA TECH

WHIP-C

Wind Hazard and Infrastructure Performance

Thank you!

www.whipc.org

Contacts:
Jean-Paul Pinelli, Ph.D., P.E.
E-mail: pinelli@fit.edu
Kishor Mehta, Ph.D., NAE, Dist. M. ASCE Director, WHIP Center
E-mail: kishor.mehta@ttu.edu;
Phone: 806-834-5679

Possible Contribution from GADRI Disaster Collection Database

Hiroyuki Goto
DPRI, Kyoto University

Disaster Collection Database

The database allows users to search for Websites, Researchers, and Documents related to disaster reduction research to take advantage of the wealth of data resources on disaster reduction research that exists around the world.

Automatically build a summary page for disaster response

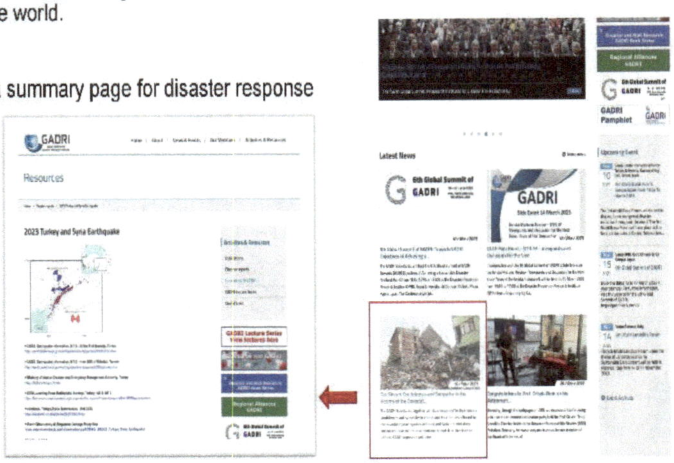

Disaster Collection Database

✓Free keyword search

✓Keywords are automatically generated from the link information

Only the URL is required!

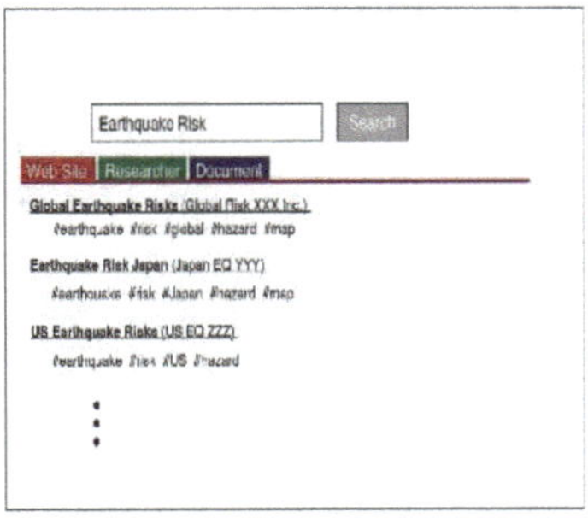

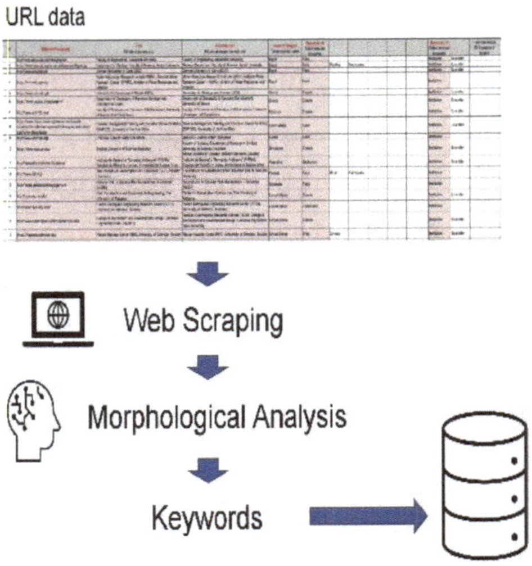

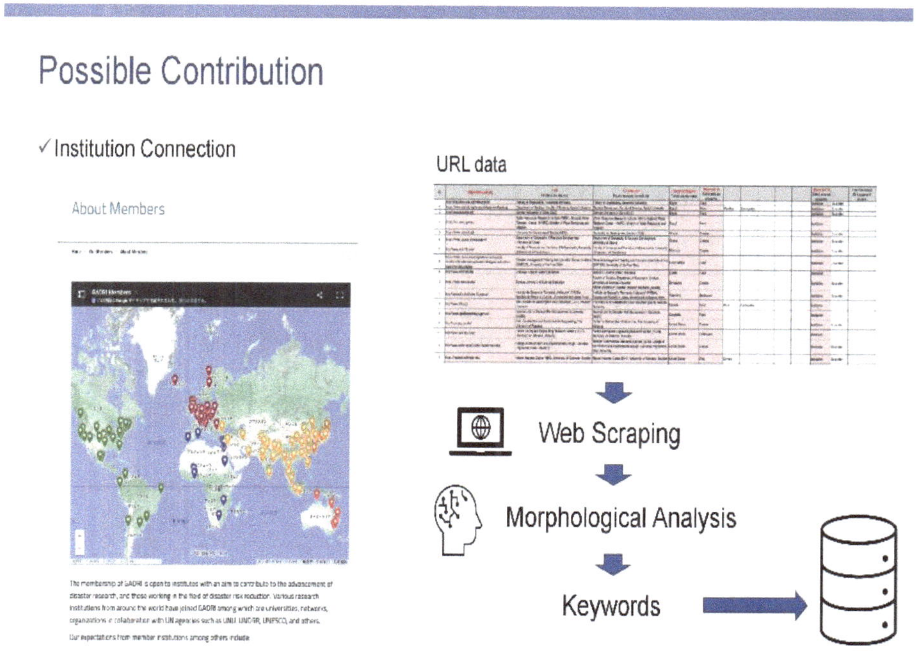
Possible Contribution
✓Institution Connection
About Members
URL data
Web Scraping
Morphological Analysis
Keywords

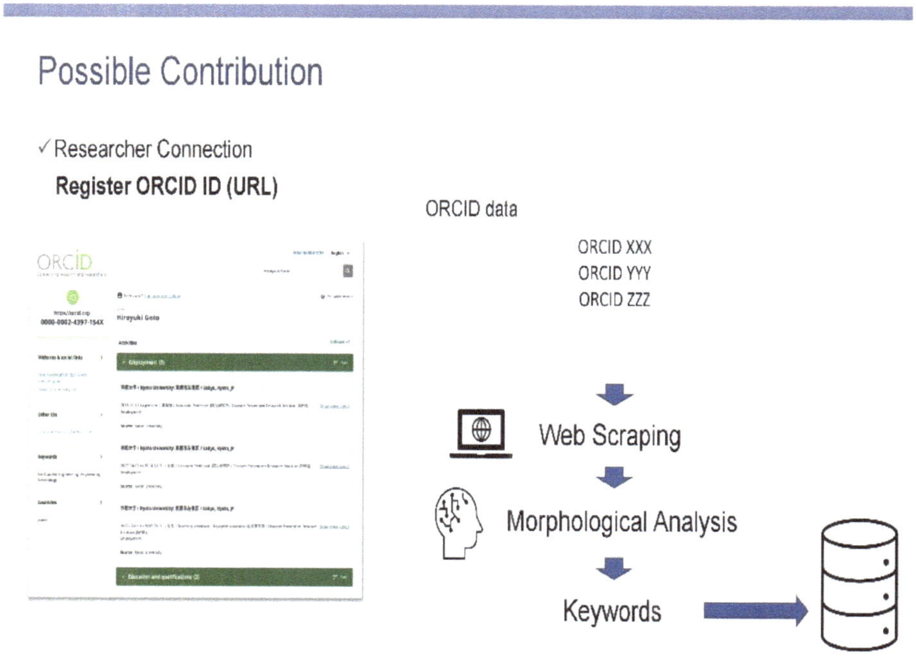
Possible Contribution
✓Researcher Connection
Register ORCID ID (URL)
ORCID data
ORCID XXX
ORCID YYY
ORCID ZZZ
Web Scraping
Morphological Analysis
Keywords

Schedule

FY2022

System design

Current

FY2023

Implementation of Disaster Collection Database

Issues

How do we collect researcher information?

DB system

The backbone of the databases is common.

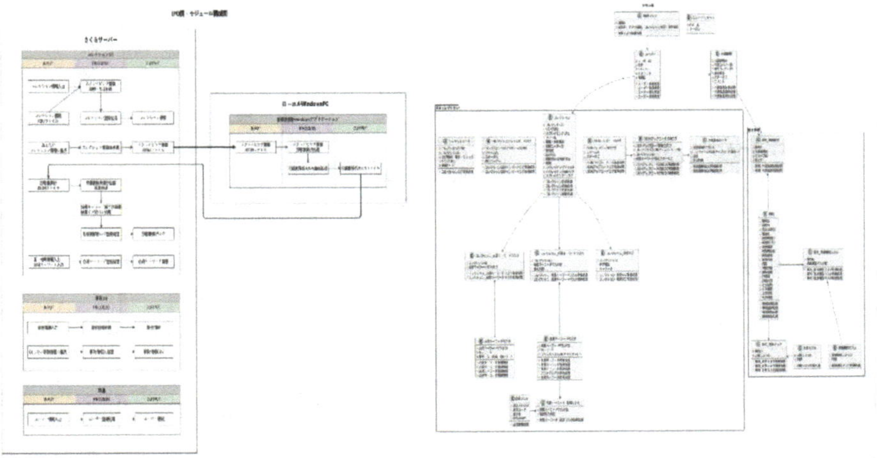

6th Global Summit of GADRI
15 - 17 March 2023
DPRI, Kyoto University, Uji Campus, Japan

Networking Contribution to Sendai Framework Midterm Review

Charles Scawthorn

Professor Emeritus, Kyoto University
Visiting Scholar, University of California, Berkeley

Sendai Framework – *in a nutshell*

- **Goal:** Over 15 years, substantially reduce losses
- **Priorities:**
  - 1: Understanding disaster risk.
  - 2: Strengthening disaster risk governance.
  - 3: Investing in disaster risk reduction
  - 4: Enhancing preparedness, "Build Back Better" and improve EWSs
- **Guiding Principles:**
- ...through international, regional...cooperation.
- ...open exchange and dissemination of data...science-based information
- ...

Midterm review – *in a nutshell*

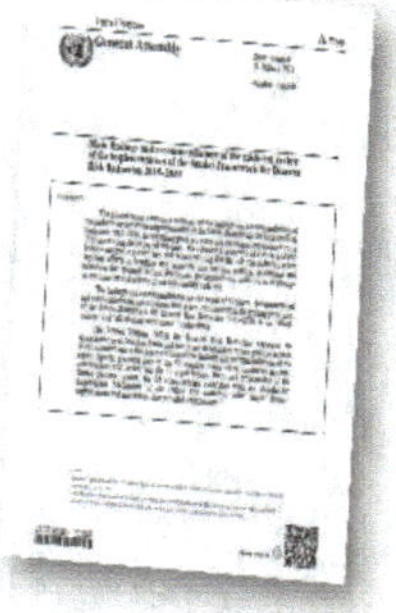

- **I – Retrospective:**
  - Shift from response to managing risk
  - Growing commitment
  - Access to data/knowledge remains inadequate
  - Substantially reduce losses: *mixed results*
- **II – Prospective**
  - States must improve the standard of official risk data ... risk assessments... data-sharing infrastructure...field data collection...databases...risk mapping ...statistics...taxonomies...methods...tools
  - Multi-hazard, vulnerability and exposure analyses must be used to inform high-level, multi-year socioeconomic planning

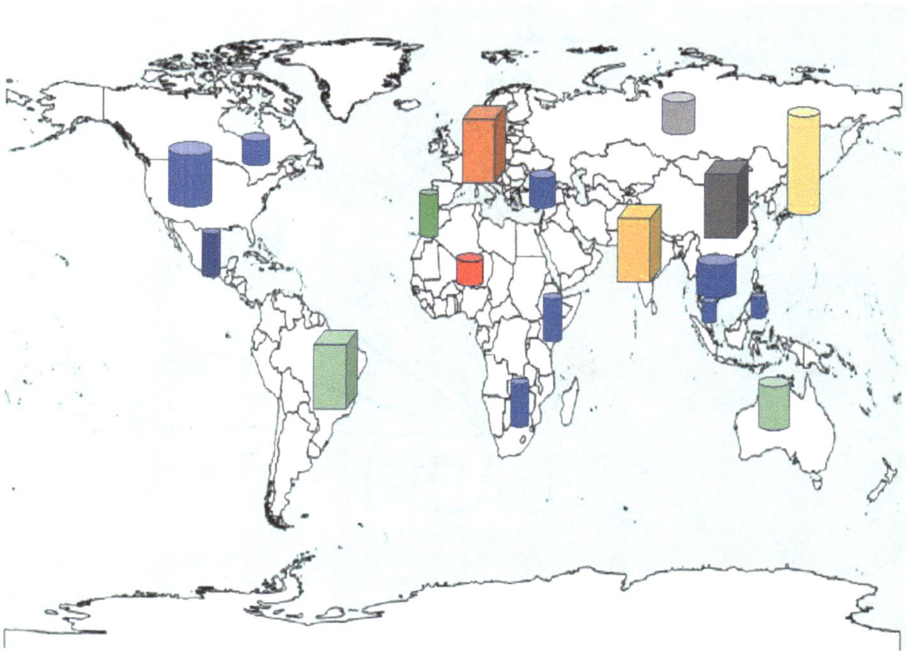

International cooperation / lack of silos

- Scientific data: geophysical (EQ, volcano...) / hydromet data sharing
- UN: DRR, GAR, UNDP...Red Cross...GFDRR...

Still Fractured / fragmented / siloed

- Exposure data (demographic, building, infrastructure, economic...)
- Standards and taxonomies for this data
- Risk modeling, despite GEM and other efforts
- Templates for disseminating / using / applying risk results

Need for all these
that GADRI can help fill – indeed, lead the way

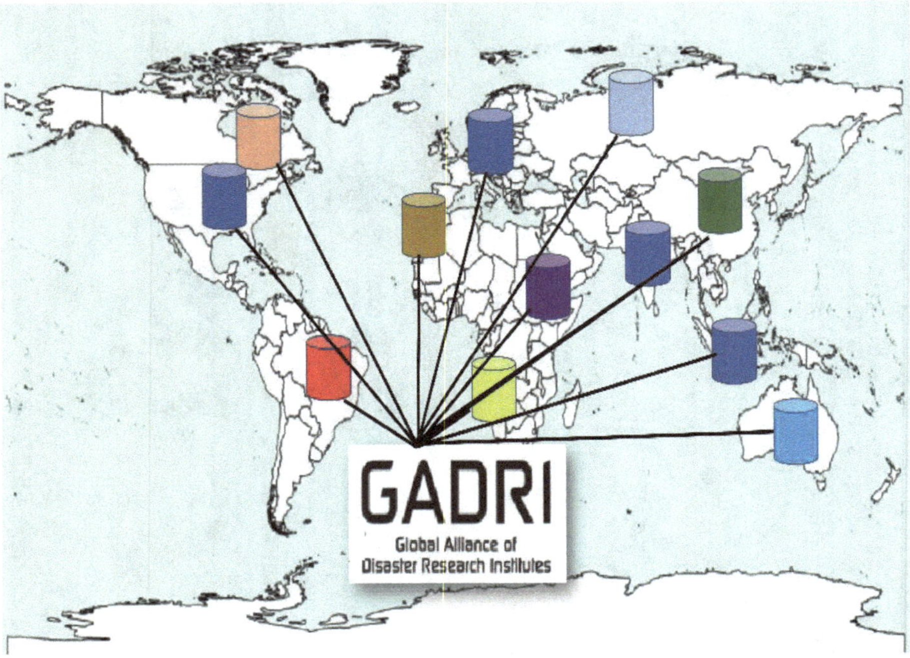

GADRI / Networking Contributions to the next 7 years of the SF/ MTR

GADRI needs to develop an Action Plan, consisting of:

- Support each GADRI member in interacting with their key national agencies
  - →develop criteria for and then designate a member as the "GADRI representative" to the country's disaster risk leadership
- Formulate and promulgate a minimum GDP % for DRR
- Identify key international DRR bodies to promote how GADRI can contribute to global DRR
  - →develop criteria for and then designate "GADRI representative" to these international DRR bodies
- Enhance GADRI internal development
  - →develop criteria for and then appoint "GADRI Fellows" to do visiting lectures at GADRI member DRIs

GADRI / Networking Contributions to the next 7 years of the SF/ MTR (cont.)

- Support member DRI outreach / service to DRR "consumers"

  →develop multimedia materials and network of outreach programs
- Support global data and standards re demographic, building, infrastructure, economic exposure and risk (e.g., build on the ECLAC PDNA model)

  → Organize international workshops leading to global data standards, taxonomies, methods and tools
- Support global multi-hazard risk models

  → Organize international workshops to formulate an open-source generic risk analysis tool (e.g., built on top of QGIS) that can be applied to any combination of linear and area source models with distance attenuating effects

Thank you / discussion

cscawthorn@gmail.com

6th Global Summit of GADRI

Panel Discussion I-A

Fostering networking to enhance Global DRR

Wednesday 15 March 2023 15:45-17:15

Ritsuo Yamazaki-Honda (NIED) and Charles Scawthorn (UC Berkeley)
Co-chairs

Panelists

- Rajib Shaw, Professor, Keio University, Tokyo.
- Ronan Sato, Associate, Energy & Environment Investment, Tokyo
- Jean-Paul Pinelli, Professor, Florida Institute of Technology
- Hiroyuki Goto, (DPRI). Kyoto University.

Agenda

1. Introduction and instruction (Charles & Ritsuko) 5min
2. Presentation by the four speakers (Rajib, Ronan, Jean-Paul, and Hiroyuki) 40min
3. World Café discussion by four groups, each moderated by one of the four speakers: Two round 30min
4. Summary from World Café and panel discussion by the four speakers 15min

Focus: how to foster networking, both within the DRR community, and by the DRR community with others including environment and information technology.

Purpose: to identify methods, channels and venues for DRR networking, with the objective of improving DRR research by the research community and enhancing utilization and impact of DRR globally.

Topics to be discussed:

- **Communication within the GADRI community:** what is the current state, is it satisfactory, and how can it be improved.
- **GADRI's communication with those that influence DRR (e.g., media, governments, MDBs, NGOs):** what is the current state, is it satisfactory, and how can it be improved.
- **GADRI member's communication with those in need of DRR (e.g., NGOs, local groups, vulnerable populations, private sector):** what is the current state, is it satisfactory, and how can it be improved.
- Of all the channels for networking, ranging from face-face and virtual meetings to websites and blogs, to broadcast and social media, which are the most effective for the various GADRI's objectives? Are they currently being employed, and what has the experience been?

Key questions to be addressed:

1. How can we foster exchange between individual GADRI researchers? Would a directory of GADRI institution researchers contribute to this goal? Would monthly on-line discussion/seminars on specific topics be of interest?

2. Which international agencies, MDBs, governments, NGO's and others should GADRI be networking with? What would be the specific goals in doing so?

3. Given the multitude of DRR consumers, how would GADRI approach networking with this audience? Is it feasible for GADRI to do this, or is this best left to individual GADRI member institutions? How can GADRI support/assist its members in serving those in need? Are there templates, canned DRR products or other things that GADRI could help produce, that could be used or adapted for many and different cultures.

4. Are we agreed that the fundamental message in GADRI networking is "disaster research improves disaster risk reduction"?

Results and recommendations

1. Creation of a Global Disaster Researcher Directory: voluntary compilation of individuals engaged in disaster research, channeled through GADRI member institutes
2. Creation of a GADRI FellowProgram: Nominations of accomplished senior disaster researchers will be nominated, from which GADRI Fellows will be selected. GADRI will extend invitation for Fellows to visit/lecture DRIs on a case-by-case basis (at no financial burden to GADRI, to be explained).

防災科研 NIED 生きる、を支える科学技術 SCIENCE FOR RESILIENCE © 2023.3.16

6th Global Summit of GADRI:
Towards GADRI Objectives of Achieving a
Sustainable Disaster-Resilient World

Summary Report from Panel Discussion I-A: Towards GADRI Objectives of Achieving a Sustainable and Resilient Society Against Disasters

Wednesday 15 March 2023 15:30-17:00

Chair: Charles SCAWTHORN, Univ. California at Berkeley
Co-chair: Ritsuko YAMAZAKI-HONDA, NIED

防災科研 NIED 生きる、を支える科学技術 SCIENCE FOR RESILIENCE © 2023.3.16

Agenda

1. Introduction

2. Presentations by Speakers (40+min)

- Rajib Shaw, Professor, Keio University.
- Ronan Sato, Associate, Energy & Environment Investment
- Jean-Paul Pinelli, Professor, Florida Institute of Technology
- Hiroyuki Goto, Associate Professor, DPRI, Kyoto University.

3. "World Café" (15+min x 2 rounds)

4. Summary from World Café and FLOOR Discussion (15min)

2

防災科研
NIED
生きる、を支える科学技術
SCIENCE FOR RESILIENCE
©
2023.3.16
GADRI Objective
a. To establish a global research network that promotes and engages disaster research
Integrating 200 disaster research institutions all over the world as one voice…
But also recognizing that we haven't maximized enough to leverage each member of GADRI.
3

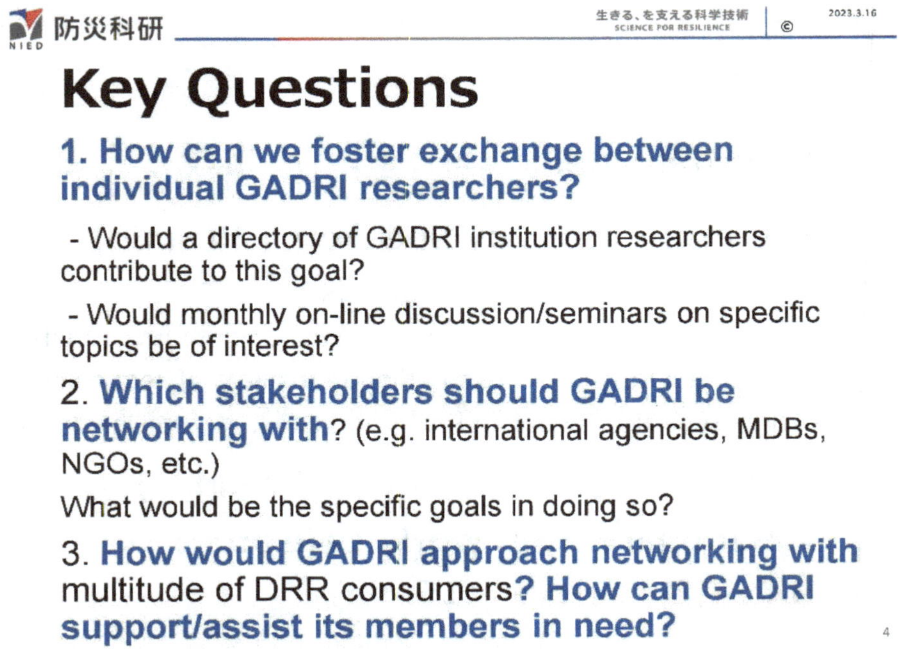
防災科研
NIED
生きる、を支える科学技術
SCIENCE FOR RESILIENCE
©
2023.3.16
Key Questions
1. How can we foster exchange between individual GADRI researchers?
- Would a directory of GADRI institution researchers contribute to this goal?
- Would monthly on-line discussion/seminars on specific topics be of interest?
2. Which stakeholders should GADRI be networking with? (e.g. international agencies, MDBs, NGOs, etc.)
What would be the specific goals in doing so?
3. How would GADRI approach networking with multitude of DRR consumers? How can GADRI support/assist its members in need?
4

防災科研 NIED 生きる、を支える科学技術 SCIENCE FOR RESILIENCE © 2023.3.16

1. How can we foster exchange between individual GADRI researchers?

- Young professionals/researchers to collaborate and exchange in and across region. online but face-to-face is better
- Use Social media or database *as a platform to connect internally and engage (FB, Instagram) all info such as events research into "GADRI Youtube Channel"
- Storytelling as researchers "People-centered GADRI" particularly vulnerable people for disaster mitigation measures. GADRI, as an umbrella, to facilitate storytelling
- Internal association or regional association

* External networking with institutions on DRR

6

防災科研 NIED 生きる、を支える科学技術 SCIENCE FOR RESILIENCE © 2023.3.16

2. Which stakeholders should GADRI be networking with?

■**<u>Education</u>** … students of any school level

To reach out for instance;

- At global level: UNESCO for online training
- At National level'

- Education department at University. which should have connections

- Disaster preparedness training for teachers

- At Decentralized, local level

e.g. (Thai) School safety program

7

防災科研 NIED 生きる、を支える科学技術 SCIENCE FOR RESILIENCE © 2023.3.16

3. How would GADRI approach networking with multitude of DRR consumers? How can GADRI support/assist its members in need?

- Creation of a <u>GADRI Fellow Program</u>:

accomplished disaster researchers will be nominated as GADRI Fellows. E.g. GADRI will extend invitation for Fellows to visit/lecture DRIs on a case-by-case basis (at no financial burden to GADRI, to be explained). GADRI members support GADRI fellows in their area (e.g. hosting, sabbatical)

- <u>Young professional program in GADRI members</u> for months
- Creation of a *Google Scholar-like of GADRI* , <u>Global Disaster Researcher Directory</u>: *voluntary compilation of individuals engaged in disaster research, channeled through GADRI member institutes*
- Existing common programs to involve different network/ stakeholders

8

Report of the Committee on Science and Technology (S&T) Roadmap

6th Global Summit of GADRI March 2023, Kyoto

Hirokazu Tatano[1], Selim Gunay[2], and Maria Camila Suarez Paba[1,2](✉)

[1] Disaster Prevention Research Institute, Kyoto University, Kyoto, Japan
maria.suarez@gestiondelriesgo.gov.co

[2] Pacific Earthquake Engineering Research (PEER) Center, University of California, Berkeley, CA, USA

Abstract. During the 6th Global Summit of GADRI held in March 2023 in Kyoto, the Committee on Science and Technology (S&T) Roadmap organized a session on March 15th. This session was chaired by professor Hirokazu Tatano on behalf of Prof. Khalid Mosalam and Co-chaired by Toshio Koike, Selim Gunay and Yuki Matsushi. The session focused on the S&T Roadmap's possible contributions to increasing the understanding of hazards, as well as technologies that help mitigate their negative consequences. Two main questions were addressed including 1) How to enhance the Science and Technology roadmap? and 2) How should we proceed with the Science and Technology Roadmap? Six panelists were invited to address these questions from diverse perspectives. Highlighting the importance of the relationship among different stakeholders, data sharing and understanding risks, with the role of a facilitator, to enhance sustainable development and resilience. Yet, community involvement and behavioral change are essential for strengthening DRR. Experts also mentioned that the Science and Technology roadmap should consider research innovation, policy, and planning, pilot projects, capacity building, and collaboration by touching upon ecosystem-based resilience. Furthermore, the S&T roadmap should also consider pathways to transfer science and engineering into climate actions as well as innovative resilience projects that determine the investment in key areas of research. Another perspective features the extreme events research and reconnaissance as crucial for DRR and the need for an interdisciplinary approach. The discussion focal point stated that the GADRI Science and Technology Roadmap will aim to promote sustainable and disaster-resistant society through shared vision, partnerships among academic disciplines, and data collection, analysis, and prescription. It will emphasize the importance of collaboration for effective solutions and sustainable development.

Keywords: Science and Technology · Data sharing · multi-stakeholder interaction · multi-hazard approach · science-policy interface · GADRI

H. Tatano et al. (Eds.): GSRIDRR 2023, DRRGADRIBS, pp. 149–159, 2025.
https://doi.org/10.1007/978-981-96-6609-6_10

1 Introduction

The Global Alliance of Disaster Research Institutes, or GADRI, seeks to provide a voice for the global disaster risk reduction community by fostering collaboration amongst institutes and exchanging insights on various viewpoints. GADRI seeks to address this area and make a contribution in order to fully support the objectives of the 2016 Science and Technology Roadmap and the implementation of the Sendai Framework for Disaster Risk Reduction 2015–2030, which mandates States to take targeted sector-specific action at the local, national, regional, and international levels. By combining knowledge and technologies from around the globe, the Science and Technology Committee plays a crucial role in GADRI's efforts to build disaster resilience by improving our understanding of disasters and identifying workable solutions.

Therefore, the goals of the Science and Technology (S&T) roadmap are to investigate risks holistically, create technologies utilizing data relevant to hazards, and enhance understanding of natural hazard events. The committee aims to apply lessons learned from COVID-19 to social, economic, and humanitarian preparedness for major disasters. It intends to combine cutting-edge engineering technologies with affordable solutions for disaster risk reduction, with an emphasis on vulnerable populations. It also envisions promoting the use of the newest technologies for catastrophe monitoring, reconnaissance, and recovery. The S&T Committee wants to collaborate with other GADRI committees to adopt and execute the S&T Roadmap in an efficient manner.

Based on the aforementioned, the S&T Committee has a number of obstacles in creating a plan that is both feasible and responsive to the real demands. Due to the significant divide between the scientific and technological communities and society at large, stakeholders in science and technology should be encouraged to improve our understanding of disaster risk by reorienting the focus from hazards to community vulnerability and exposure. Additionally, as disasters impact social systems in addition to the built and natural ecosystems, it is imperative to promote risk-informed practices.

Thus, the question of how humanity could withstand such risks and coexist with them in a sustainable manner emerges. In order to accomplish this, a number of factors are pertinent, such as the scientific community and a knowledge integrator (a facilitator) who can help close the gap between communities and scientists. In order to move towards a risk-informed system for DRR and support the implementation of science to reduce vulnerability, it is necessary to establish a linkage between the stakeholders in science and technology and the decision makers. This can be done through the use of interdisciplinary and transdisciplinary approaches, open science and open data, and an all-hazard approach that takes into account compound, cascading, and systemic hazards.

All these aspects were discussed during the S&T Committee session. Presentations, points of view and recommendations will be addressed in the following sections.

2 Science and Technology (S&T) Roadmap Session

The Science and Technology (S&T) Roadmap is an important component of GADRI for achieving disaster resilience. With this important direction and goal in mind, this session focused on the potential contributions of the S&T Roadmap in advancing the science of hazards, along with technologies that can minimize their harmful effects (Fig. 1).

Fig. 1. (L) Prof. Hirokazu Tatano and Prof. Toshio Koike, Chairs of the Session

The session invited the following experts to the panel discussion:

1) Prof. Toshio Koike, International Centre for Water Hazard and Risk Management under the auspices of UNESCO (ICHARM), Japan
2) Prof. Quinli Han, Integrated Research on Disaster Risk (IRDR), China
3) Prof. Yuicho Ono, International Research Institute of Disaster Science (IRIDeS), Tohoku University, Japan
4) Prof. Mahua Mukherjee, Indian Institute of Technology (IIT) Roorkee, India
5) Dr. Guirong Grace Yan, Missouri University of Science and Technology, USA
6) Dr. Selim Günay, Pacific Earthquake Engineering Research Center (PEER), University of California, Berkeley, USA

The session was chaired by Prof. Hirokazu Tatano, Disaster Prevention Research Institute, Kyoto University on behalf of Prof. Khalid Mosalam, Taisei Professor of Civil Engineering and Director of the Pacific Earthquake Engineering Research (PEER) Center; and Dr. Selim Gunay, Pacific Earthquake Engineering Research (PEER) Center, University of California, Berkeley, USA.

Questions addressed during the session included:

- How to enhance S&T Roadmap?
- How should we proceed with the S and T roadmap?

However, other key questions raised during the session, including the following:

1. How can we structure the S&T Roadmap to enhance the broader impacts of GADRI activities?
2. How can we maximize the effects of GADRI in shaping disaster-related policy changes in the world through the S&T Roadmap?
3. What are the tools and methodologies needed for developing holistic and consistent solutions to multiple hazards (e.g., earthquakes and climate change related hazards)?
4. What is the most effective way of establishing supportive communications between engineering, earth science, social sciences, and economic disciplines for disaster mitigation?
5. How can the GADRI S&T Roadmap realistically prevent hazards from becoming disasters?
6. How can the various advanced engineering technologies be merged with low-cost solutions to serve the most vulnerable populations to hazards around the world.

Each invited expert made a presentation with a different focus, bringing to the discussion different perspectives to address the key questions raised[1].

a) Panelist Prof. Toshio Koike

The relationship between disaster resilience and Disaster Risk Reduction (DRR) is critical, as effective DRR measures, including evacuation and rebuilding strategies, enhance resilience and contribute to quicker recovery post-disaster. Supporting stakeholders—national and local governments, organizations, and communities—necessitates better information sharing systems while balancing confidentiality concerns. Establishing an integrated knowledge-sharing framework, facilitated by skilled individuals who can communicate scientific data to diverse audiences, is essential for capacity building. This accumulated information can inform various aspects of planning, management, and investment, ultimately fostering sustainable development and strengthening resilience against future disasters.

In more detail Prof Toshio Koike's ideas included:

- Relationship between disaster resilience and DRR (Disaster Risk Reduction)
- Evacuation, rescue, build back better are factors that can add to resilience, and can also facilitate a faster rehabilitation and recovery.
- How to support stakeholders (national and local governments, companies, organizations, residents, communities). We need a system for better information sharing with these stakeholders, however confidentiality can be an issue in this process.
- It is needed to develop an integrated knowledge-open science, from which a quality-oriented society could benefit.
- Capacity building for stakeholders needs a facilitator to implement the proposed framework.

  • A facilitator should have the skills to translate, interpret, and be a mobilizer of data and scientific information to stakeholders.

[1] These opinions shared by each panelist were separately submitted by Dr. Selim Günay and Dr. Maria Camila Suarez Paba.

- The target audience for the facilitator to give the message should be defined.
- A mechanism for data sharing is also needed to use it effectively for capacity building and decision-making.

- Accumulated information can be used for planning and training, land use planning and management, infrastructure investment, and disaster prevention.
- As a consequence, sustainable development and resilience can be enhanced or strengthened.

b) Panelist Prof. Qunli Han

The analysis of the Sendai Framework (SF) roadmap from 2018 highlights several critical points, including the ambiguity surrounding its impact on national implementation and the necessity for effective progress measurement. There's a pressing need to integrate the understanding of systematic, cascading, and compound risks within the GADRI network, reflecting insights gained from the COVID-19 crisis. Furthermore, the role of facilitators in disaster risk reduction (DRR) is essential in fostering stakeholder collaboration and comprehension of risks. Overall, approaching DRR as a development issue, enhancing community involvement, simplifying scientific knowledge, and promoting open science and data sharing are vital for achieving sustainable development goals.

In more detail Prof Qunli Han's ideas included:

- Key points of 2018 to analyze the Sendai Framework (SF) roadmap:

  - Problem: we don't know how the roadmap has been added to the SF implementation, is it really helping countries?
  - We need to think about how to measure the progress of the SF.
  - It is needed to discuss how the GADRI Science and Technology Roamap can help to accelerate SF implementation or progress.
  - Systematic and cascading risks were not understood well even in 2019, but they became evident after Covid-19.
  - Urban areas under different risks are not necessarily highlighted in the SF agenda.

- We need to include systematic, cascading, and compound risk in GADRI's network.
- Facilitator's role in DRR, mentioned by Prof. Koike, is crucial for interaction among stakeholders and understanding of risks.
- We need to consider DRR as a development issue, otherwise, sustainable development will not be achieved.
- The involvement of communities and behavioral change are essential for strengthening DRR.
- Science knowledge/understanding needs to be simplified for this purpose.
- GADRI science-policy interface can be organized for enhanced society engagement.
- Open science, open data, data sharing and accessibility are needed, and GADRI needs to support these.

c) Panelist Prof. Mahua Mukherjee

An ecosystem-based resilience approach should be integrated into the Science and Technology roadmap to address the evolving urban risks exacerbated by climate change and anthropogenic actions. This involves aligning ecosystem restoration initiatives with Sustainable Development Goals while prioritizing benefit/cost analyses for nature-based solutions over mid to long-term time frames. Given the ambiguity surrounding the effectiveness of these solutions, a robust evaluation framework using scenario analysis is essential. Collecting and analyzing data resources related to ecosystem-based urban risk resilience is crucial for informing research, fostering innovation, bridging policy gaps in densely populated countries like India, and empowering collaborative pilot projects focused on mitigation and resilience strategies.

In more detail Prof Mahua Mukherjee's ideas included:

- Ecosystem-based resilience potentially incorporated to the Science and Tehnology roadmap.
- Ecosystem restoration, related to Sustainable Development Goals (SDGs) and initiatives by the global community.
- Urban risks are constantly changing, because of anthropogenic actions, and climate change issues.
- Urban risk: benefit/cost analysis for nature-based solutions. Should consider a time frame in the mid or long-term.
- The effectiveness of nature-based solutions is not clearly understood. So, an evaluation or validation approach is needed (e.g., analysis through scenarios can be a good approach).
- Data for ecosystem-based urban risk resilience is a resource that should be carefully analyzed, and determine its appropriate use.
- Collecting data is not the end of the task, but it is the starting point for researchers to develop nature-based solutions.
- In countries with mega-populations, e.g., India, policies have a gap because they are not properly addressing the current environmental issues.
- Science and Technology roadmap should consider research innovation, policy, and planning, pilot projects, capacity building, and collaboration. Therefore, it is essential to understand the mitigation, resilience, and their interaction.

d) Panelist Dr. Grace Yan

To effectively translate science and engineering advancements into actionable climate responses, a synergistic approach is essential, emphasizing interdisciplinary collaboration among researchers, industry leaders, government officials, and educators through both top-down and bottom-up strategies. This involves enhancing climate modeling resolution, fostering high-impact research through robust funding mechanisms, and employing multi-hazard risk assessments that consider social science impacts and engineering solutions. Education initiatives must address gaps in resilience science, with a curriculum encompassing natural hazards, community resilience strategies, and overcoming barriers to implementation. Furthermore, innovative resilience projects should be prioritized for investment, particularly encouraging insurance companies to focus on community-based resilience bonds and adapting FEMA's cost-benefit analyses to promote informed decision-making about safer living environments.

In more detail Prof Grace Yan's ideas included:

- Pathways to transfer science and engineering into climate actions
- Focus on collaboration with research (research integration-advance science and engineering), industry, government, education (bottom-up approach)
- Interdisciplinary research, downscaling climate change modeling (resolution is too low)
- Mechanisms for researchers to collaborate to do high-impact research (higher grants/funding)
- Multi-hazard risk assessment approaches for the characterization of hazards
- Impacts of hazards on people (social science), solutions to reduce the impacts of hazards (engineering and social science)
- Collaboration among stakeholders is needed among countries, and governments, and no country should be left behind. What kind of approach is needed for collaborating with governments? Top-down or bottom-up?
- Resilience related points

  - Education gap: resilience science and engineering
  - The facilitator should take a resilience science and engineering class (proposed) and a broader perspective of DRR-related courses.

    o Natural hazards
    o Hazard impacts on built environment and society
    o Hazard mitigation and adaptation
    o Overcoming barriers to resilience implementing

  - Innovative resilience projects are needed (where to invest the money)
  - Improvement of resilience should be a target of the investment (insurance companies are a profitable business).
  - Insurance companies are changing their focus to look more at a community level (resilience bond)
  - FEMA is likely to change their cost-benefit analyses
  - We need to encourage people to have a wider scope to choose a place to live different from that of a cheaper place even if they are located in a risky area

e) Panelist Dr. Selim Günay

Learning from disasters through interdisciplinary reconnaissance efforts is essential for advancing disaster risk reduction (DRR) and enhancing community resilience. Initiatives like the Structural Extreme Events Reconnaissance Network (StEER), alongside organizations such as PEER and GEER, emphasize collaborative post-disaster research that informs effective rebuilding strategies. Through platforms like the RAPID NHERI and DesignSafe-CI, stakeholders can access vital data and tools needed for informed decision-making. A recent report on the Turkiye earthquake highlighted urgent needs for retrofitting structures with known vulnerabilities and called for enhanced seismic instrumentation, real-world testing of code-designed buildings, and the development of cost-effective protective systems globally, to address severe earthquakes that surpass expected return periods.

In more detail Prof Selim Günay's ideas included:

- Learning from natural disasters through reconnaissance provides opportunities for rebuilding and building back better
- Extreme events research and reconnaissance are crucial for DRR and need an interdisciplinary approach for hazard (e.g., Pacific Earthquake Engineering Research Center, PEER), geotechnical (e.g., Geotechnical Earthquake Engineering Reconnaissance Network, GEER), building and lifelines (e.g., Structural Extreme Events Reconnaissance Network, StEER), and socioeconomic (e.g., Social Science Extreme Events Research, SSEER)
- This approach is completed by enable agencies in the US, e.g., RAPID NHERI-Natural Hazards Reconnaissance Facility (open access), and DesignSafe-CI (open access)- Web interface of the NHERI that provides the data and software tools
- Under the NHERI umbrella---StEER: Structural extreme events reconnaissance (www.steer.network). Builds societal resilience by generating new knowledge on the performance of the built environment through impactful post-disaster reconnaissance disseminated to affected communities. This is a network that aims at collaboration among research, practice, and policy
- StEER aims at finding solutions for our challenges as a society in mitigating natural hazards by focusing on 1) affected communities, 2) field observations, 3) research and development, 4) regulatory mechanisms
- StEER facilities learning from earthquakes and enhance resilience through reconnaissance apps
- In a recent application, a joint report is prepared for the Turkiye earthquake together with EERI

  - Several buildings with well-known structural weaknesses collapsed
  - Immediate action plans are needed for retrofitting this kind of building not only in Turkiye but worldwide
  - Earthquake severity of 2475 years return period (Maximum Considered Earthquake, MCE) was exceeded
  - The instrumentation of structures should be promoted and increased
  - The seismic response of code-designed structures should be explored when subjected to such extreme shaking with realistic shaking table tests or other testing methods that are closer to reality
  - Increased usage of protective systems should be promoted all around the world. The feasibility of using cost-effective protective systems should be explored for the reconstruction of the largely impacted cities

3 Discussion

The discussion touched upon creating context networks for natural hazards analyses which requires the integration of diverse tools and techniques, emphasizing resilience independently of specific hazards, while adopting a roadmap tailored to individual hazards or aligned with Sustainable Development Goals (SDGs). Solutions can include data-driven frameworks that facilitate disaster risk reduction (DRR) strategies, supported by institutions like GADRI, which can bridge global knowledge and local expertise to enhance disaster prevention and mitigation. Effective strategies must promote international collaboration, acknowledging the importance of local context to inform and improve global policies, ensuring that both global initiatives and local insights converge to foster resilient communities (Figs. 2 and 3).

Fig. 2. Panel discussion session

Fig. 3. Panel discussion session

Further ideas discussed included:

- Ideas on creating context networks for natural hazards analyses
- Tools, techniques, and items for research. Which solutions can we achieve with these?
- If focused on hazard, the roadmap will be different for each of the hazards
- If not focused on the hazards, but on the SDGs, it can be a more holistic approach
- Resilience is independent of the natural hazard and one approach is to focus directly on resilience
- Question: What is needed for achieving this solutions to provide strategies focused on DRR? Answer: GADRI can be a facilitator to integrate knowledge and data in terms of DRR efforts (strategy)
- A global perspective is needed to share experiences and enhance disaster prevention and mitigation policies
- We need international collaboration to address hazards that have impacts worldwide (a global approach). Global aspects make an impact at the local level as well.
- Local context and knowledge are also very important and should be included in the global perspective of GADRI. Therefore, the global-local interrelation is helpful.

Furthermore, there were some more specific topics discussed during the session, as mentioned below.

1. Advancing the science and understanding of various natural hazard phenomena, including earthquakes, wildfires, volcanic eruptions, droughts, hurricanes, tornadoes, sea-level rise, floods, and their relationships to climate change, using theoretical, numerical, and data-driven approaches.
2. Developing or adopting technologies to use vast amounts of hazard-related data available in recent years due to reconnaissance activities.
3. Developing tools, technologies, and methods that explore all hazards from a holistic perspective that considers commonalities and interactions between them. In addition, reducing the consequences of each hazard individually and preventing them turning into disasters.
4. Using the experiences from the COVID-19 pandemic, formulating, and preparing for the world-wide consequences of large-scale disasters from social science perspective (e.g., potential changes on the social fabric of communities), economic (e.g., worldwide supply-chain issues), and humanitarian (e.g., physical, and mental health) perspectives.
5. Using the same experiences of COVID-19 (e.g., developing vaccines in unprecedently short time frames), formulating methods to bring available science and advanced engineering methods together for finding timely solutions for disaster risk reduction.
6. Promoting and advocating for the development and use of latest technologies for disaster monitoring, reconnaissance activities (view from the ground), short-term and long-term recovery, and related city- and regional-scale simulations, e.g., deep learning, satellite imagery (view from the sky), drone technologies, computer vision, natural language processing, high performance computing, and low cost and accurate sensors.
7. Developing methods to merge advanced engineering technologies with low-cost solutions that can facilitate the world-wide usage of modern technologies of the fourth (digital) industrial revolution, both in the developed world and underdeveloped and developing countries, with a special emphasis on the most vulnerable populations, for disaster risk reduction.
8. Working across-board with other committees (Institutional Capacity Building, Networking, Data Information Sharing, and Advocacy) for effective implementation and adoption of the S&T Roadmap.

4 Final Conclusions of the Session

The GADRI S&T Roadmap will aim to establish a collaborative framework that outlines future science and technology directions, fostering disaster resilience and sustainable development through cross-disciplinary partnerships. It will emphasize the necessity of re-evaluating and restructuring research priorities to achieve effective disaster risk reduction (DRR) policies. Additionally, the roadmap will highlight the importance of systematic data collection and analysis from field activities that can inform policy improvements, ultimately encouraging a transformative impact on societal resilience.

Finally, more specifically, to enhance the Science and Technology roadmap and to proceed with the Science and Technology Roadmap, GADRI S&T Committee will provide:

- o future vision of directions of S&T activities to be shared among members of GADRI and hopefully influence policy and systems transformation towards disaster resilient and sustainable society.
- o Importance of collaborations among different academic discipline to provide effective, affordable and sense-making solutions and implementation strategies for DRR policies and sustainable development. There is a need to re-allocate and reshape the S&T Roadmap. This will be the starting point for the S&T Roadmap.
- o Data collection, analysis and prescription based on the field activities should be organized to share knowledge reflected in policy improvement in the real world.

Report: GADRI Committee on Institutional Capacity Building for Disaster Risk Reduction

Wei-Sen Li[1](✉), Gretchen Kalonji[2], and Desmond Manatsa[3]

[1] NCDR, New Taipei City, Taiwan
li.weisen@ncdr.nat.gov.tw
[2] IMDR, Sichuan University, Chengdu, China
[3] Bindura University of Science Education, Bindura, Zimbabwe

Abstract. The Committee on Institutional Capacity Building of the Global Alliance of Disaster Research Institutes (GADRI) is focused on leveraging networking capabilities and resources among all partners under GADRI to build capacity to cope with existing and future challenges. The committee aims to share knowledge management opportunities, best practices, and training opportunities on enhancing disaster resilience research and risk reduction. This report summarizes the discussions and key points from the recent session chaired by Prof. Wei-Sen Li alongside co-chairs Prof. Gretchen Kalonji and Prof. Desmond Manatsa.

Keywords: Capacity building · institutional capacity building · DRR · institutional capacity development gaps

1 Introduction

Institutional capacity building is not just critical, but it is the backbone of disaster risk reduction (DRR). It strengthens organizations' abilities to perform effectively, efficiently, and sustainably. The GADRI Committee on Institutional Capacity Building, recognizing the invaluable contributions of all its members, has been working towards fostering collaborations and creating opportunities for exchange and learning. This report, a testament to your work, outlines the discussions during the recent session, highlighting key challenges, successful strategies, and your crucial role in shaping the future directions for institutional capacity building in DRR (Figs. 1 and 2).

H. Tatano et al. (Eds.): GSRIDRR 2023, DRRGADRIBS, pp. 160–169, 2025.
https://doi.org/10.1007/978-981-96-6609-6_11

Fig. 1. Chairs of the Session (L) Prof. Desmond Manatsa; and Prof. Gretchen Kalonji

Fig. 2. Chairs of the Session – (L) Prof. Wei-Sen Li, Prof. Manatsa; and Prof. Kalonji

2 Keynote Address

2.1 Prof. Reini Wirahadikusumah, Rector of the Bandung Institute of Technology (ITB), Indonesia

Prof. Reini Wirahadikusumah delivered a keynote address: a speech and a call to real action. She emphasized the importance of a strategic focus on research and education in disaster science, sharing her experiences in building ITB's capacity to address practical improvements in education and research in disaster risks. Her words set the tone for the session and inspired us all, infusing us with renewed energy and highlighting the need for robust institutional frameworks and strategic planning in DRR (Fig. 3).

Fig. 3. Online Keynote Address by Prof. Reini Wirahadikusumah, Rector of the Bandung Institute of Technology (ITB), Indonesia

3 Panelist Contributions by Five Panelists

3.1 Mr. Bill Ho, Director of the Asia Disaster Preparedness Center (ADPC), Thailand

During the discussion, Mr. Bill Ho highlighted the challenges and successful strategies in ADPC's training programs in Southeast Asia. He pointed out several challenges in in-service capacity development, such as the lack of a clear plan, ad-hoc training approaches, and inadequate support. ADPC addressed these challenges by conducting comprehensive assessments to customize training programs to specific organizational needs. Additionally, ADPC successfully established specialized DRR (Disaster Risk Reduction) training facilities, which provided a consistent and high-quality training environment. Collaborating with universities also played a vital role in developing a skilled trainer pool, ensuring the sustainability and quality of training programs (Fig. 4).

Fig. 4. Online Presentation by Mr. Bill Ho, Director of the Asia Disaster Preparedness Center (ADPC), Thailand

3.2 Prof. Geoffrey Mukwada, University of the Free State, South Africa

Prof. Mukwada first emphasized the critical role of data infrastructure and highlighted the data infrastructure needs for DRR in Africa. A robust data system enables accurate risk assessments and informed decision-making. He emphasized the importance of capacity building in data analysis and interpretation, suggesting that universities should incorporate these skills into their curricula. This approach prepares future professionals and ensures that current practitioners have access to the necessary skills and tools.

3.3 Prof. Dimeter Velev, University of National and World Economy, Bulgaria

With his insightful observations, Prof. Velev focused on developing capacity in artificial intelligence (AI) for DRR. He discussed various AI technologies, such as machine learning algorithms, deep learning algorithms, and natural language processing, that can be applied in DRR. However, he also acknowledged the limitations, particularly the challenges related to data quality and privacy concerns. Prof. Velev advocated for developing ethical guidelines and robust data governance frameworks to address these issues and ensure that AI technologies are used responsibly and effectively.

3.4 Prof. Ailsa Holloway, Auckland University of Technology, New Zealand

Prof. Holloway shared her experiences with the Peri-Peri University, a network of 12 African universities focusing on institutional capacity development for DRR. By connecting universities across Africa, the network facilitated knowledge exchange and best practices, significantly enhancing institutional capacity for DRR. She stressed the importance of collaboration and shared learning among universities to build effective DRR programs and that similar networks could be established in other regions, fostering a global community of practice dedicated to DRR.

3.5 Prof. Krishna Pribadi, Bandung Institute of Technology (ITB), Indonesia

Prof. Pribadi discussed the efforts to bring together 100 universities in Indonesia to highlight the importance of educational networks in meeting DRR challenges. By developing study programs and training on DRR, these universities can play a pivotal role in building national capacity. Prof. Pribadi also emphasized the importance of global networking, suggesting that international collaborations can provide valuable insights and resources to support local DRR efforts (Fig. 5).

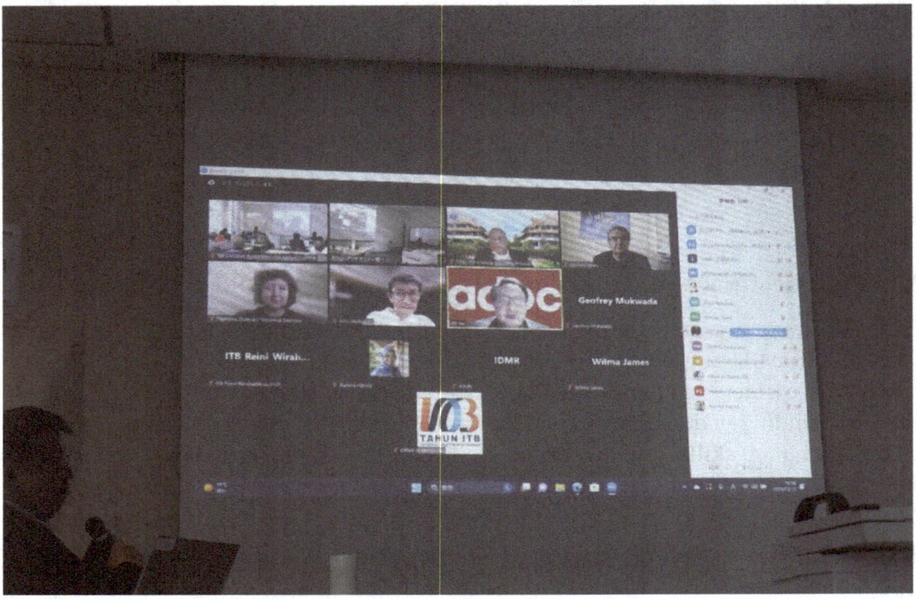

Fig. 5. Group of Panelists

4 Key Questions Addressed During the Discussion

1. **Engagement of GADRI Member Institutes**: How to engage member institutes to share their expertise and demands of disaster risk management and resilience.
2. **Defining Scope of Institutional Capacity Building**: The scope includes intra-GADRI and external entities ranging from international agencies to local communities.
3. **Creating a Roadmap for Institutional Capacity Building**: Develop a roadmap encouraging physical and virtual exchanges involving various stakeholders.
4. **Identifying Channels for Capacity Building**: Exploring diverse channels for creating opportunities for exchange and learning.
5. **Institutional Capacity Building at the GADRI Member Institution Level**: This may include cooperative research programs, joint research exchanges, webinars, and other training initiatives.
6. **Facilitating Dialogues and Identifying Gaps**: Facilitating dialogues at the societal level to exchange ideas and identify gaps related to disaster resilience.

5 Summary of Discussion Session

The session brought together participants from diverse backgrounds, leading to a comprehensive discussion on enhancing institutional capacity for DRR. The primary objective was to examine how research institutes can better contribute to research, education, and societal service in DRR.

5.1 Challenges Identified Based on the Discussion

1. Lack of a clear mandate and commitment within institutions. The following points could help identify the insufficiency

   - Develop Institutional Policies: Institutions must establish clear policies and frameworks that make DRR a central focus. This involves integrating DRR into strategic plans and making it a key performance indicator for relevant departments.
   - Leadership Engagement: Get institutional leaders engaged in promoting DRR initiatives. This can be accomplished by organizing workshops and seminars to emphasize the significance of DRR, showcasing successful case studies of DRR implementation, and illustrating the long-term benefits of prioritizing DRR.
   - Incentives and Recognition: Establish incentive structures to motivate departments and researchers to prioritize DRR. This could involve creating awards, recognition programs, and funding opportunities for innovative DRR projects.

2. Limited research and collaboration among institutions. Possible solutions could be:

   - Establishing Research Networks: Work on creating networks of universities and research institutions to support joint research on DRR. These networks can combine resources, exchange data, and co-author research papers to maximize their collective impact.

- Joint Research Grants: Advocate for funding agencies to allocate grants specifically for collaborative DRR research. These grants should prioritize projects involving multiple institutions and interdisciplinary teams.
- Collaborative Platforms: Create online platforms for researchers to collaborate, share resources, and discuss ongoing DRR projects. These platforms can also host virtual conferences and workshops to facilitate knowledge exchange.

3. Insufficient investment in DRR education and research in developing countries. The potential solutions could be:

   - Advocate for Increased Funding: Engage with governments, international organizations, and private sector stakeholders to advocate for increased funding for disaster risk reduction (DRR) education and research in developing countries. Emphasize the economic and social benefits of investing in DRR.
   - Partnerships with International Organizations: Partner with international organizations focusing on DRR, such as the United Nations Office for Disaster Risk Reduction (UNDRR). These partnerships can provide financial support, expertise, and technical assistance.
   - Capacity Building Programs: Implement capacity-building programs to provide training and resources to educators and researchers in developing countries. This may involve providing fellowships, scholarships, and exchange programs with institutions in more developed regions.

4. Higher educational networks are needed to support disaster management, as outlined in the Sendai Framework for DRR. The possible approaches are:

   - Regional and Global Networks: Establish regional and global networks of higher education institutions focused on DRR. These networks should facilitate sharing curricula, best practices, and research findings.
   - Curriculum Development: Develop standardized DRR curricula that universities can adopt worldwide. This curriculum should cover various topics, including risk assessment, disaster response, recovery planning, and resilience building.
   - Accreditation Programs: Create accreditation programs for DRR education and training programs to ensure high standards and consistency across institutions. Accredited programs can help in building trust and recognition among stakeholders.

5. Limited numbers of universities providing degree programs and activities on DRR.

   - Expand Degree Programs: Encourage universities to develop and offer programs specifically focused on DRR. This can include undergraduate, graduate, and doctoral programs that cover various aspects of disaster risk management.
   - Interdisciplinary Courses: Promote the development of multidisciplinary courses that integrate DRR with other fields such as environmental science, engineering, public health, and social sciences. This approach can attract a broader range of students and professionals.
   - Online Education and MOOCs: Utilize online education platforms and Massive Open Online Courses (MOOCs) to provide accessible DRR education to a global

audience. These courses can be particularly beneficial for students in regions where DRR programs are not readily available.

5.2 Other Successful Strategies to Meet the Challenges

1. Establishing clear plans linking local and national DRR-related human resource needs to academic programs.
2. Creating new interdisciplinary programs that break down silos between disciplines.
3. Strengthen connections with local communities and link research directly to them.
4. Including social perspectives in DRR solutions, not just engineering approaches.
5. Building strategic alliances at national and international levels.

5.3 Future Directions and Recommendations

1. **Develop Clear Institutional Mandates**: Establish clear mandates within institutions to prioritize DRR research and education.
2. **Enhance Collaboration**: Foster collaboration among research institutions, disaster management agencies, and other stakeholders.
3. **Invest in DRR Education and Research**: Increase investment in DRR educational programs and research funding, particularly in developing countries.
4. **Create Higher Educational Networks**: Develop networks to support disaster management as expected in the Sendai Framework for DRR.
5. **Expand Global Networking**: Engage in global networking to share knowledge and best practices related to DRR.
6. **Leverage Technology**: Utilize AI, big data, and other technologies to enhance DRR capabilities.
7. **Focus on Social Perspectives**: Ensure that DRR solutions incorporate social perspectives and address the needs of vulnerable communities.
8. **Strengthen Policy Frameworks**: Advocate for robust policy frameworks that support institutional capacity building for DRR.

By following these recommendations, institutions can build their capacity to effectively address disaster risks and contribute to creating resilient societies.

6 Conclusions

This extended report provides a comprehensive overview of the discussions and key points from the session on institutional capacity building for disaster risk reduction. It highlights the challenges, successful strategies, and future directions for enhancing the capacity of research institutes to contribute to DRR efforts globally. The key takeaways are:

6.1 Engaging GADRI Member Institutes

The engagement of GADRI member institutes is crucial for effectively implementing DRR strategies. By leveraging the expertise and resources of various member institutes, GADRI can enhance its capacity to address disaster risks. Surveys and feedback mechanisms can be utilized to understand member institutes' specific needs and expertise, allowing for tailored capacity-building initiatives.

6.2 Defining the Scope of Institutional Capacity Building

Institutional capacity building within GADRI should encompass both internal and external collaborations. Internally, it involves enhancing the capabilities of GADRI member institutes through knowledge sharing, joint research, and training programs. Externally, it includes partnerships with international agencies, NGOs, national governments, media, and the public. This holistic approach ensures a comprehensive capacity-building framework that addresses various aspects of disaster risk management.

6.3 Creating a Roadmap for Institutional Capacity Building

Developing a roadmap for institutional capacity building involves identifying key milestones and activities that contribute to enhancing DRR capabilities. This roadmap should encourage interdisciplinary and transdisciplinary exchanges involving stakeholders from research, practical implementation, education, and policy-making. Regular reviews and updates to the roadmap can ensure that it remains relevant and effective in addressing emerging challenges.

6.4 Identifying Channels for Capacity Building

Diverse channels should be explored to create opportunities for exchange and learning. These channels include webinars, workshops, conferences, online platforms, and collaborative research projects. Using multiple channels, GADRI can reach a wider audience and facilitate more inclusive and effective capacity-building initiatives.

6.5 Institutional Capacity Building at the GADRI Member Institution Level

At the member institution level, capacity building can take various forms. Cooperative research programs allow institutes to share resources and expertise on specific DRR topics. Joint research exchanges enable researchers to gain new insights and develop innovative solutions. Webinars and training sessions provide opportunities for continuous learning and professional development. By implementing these initiatives, GADRI member institutes can enhance their DRR capabilities and contribute to the overall mission of GADRI.

6.6 Facilitating Dialogues and Identifying Gaps

Facilitating dialogues at the societal level involves engaging various stakeholders, including local communities, policymakers, researchers, and practitioners. Through these dialogues, GADRI can identify gaps in current DRR strategies and develop targeted initiatives to address them. This collaborative approach ensures that DRR efforts are comprehensive and inclusive, addressing the needs of all stakeholders.

The session on institutional capacity building for disaster risk reduction provided valuable insights into the challenges, successful strategies, and future directions for enhancing DRR capabilities. By leveraging the expertise and resources of GADRI member institutes, fostering collaborations, and utilizing diverse channels for capacity building, GADRI can create a comprehensive framework for institutional capacity building. The recommendations outlined in this report provide a roadmap for institutions to enhance their DRR capabilities and contribute to the overall mission of GADRI.

Institutional capacity building is a continuous process that requires commitment, collaboration, and innovation. By addressing the challenges and leveraging the opportunities identified in this report, GADRI member institutes can play a crucial role in building resilient societies and reducing disaster risks.

Outcomes from the Committee on Data and Information Sharing

Andrew E. Collins[1(✉)] and Tom De Groeve[2]

[1] DDN, Northumbria University, Newcastle Upon Tyne, UK
andrew.collins@northumbria.ac.uk
[2] European Commission Joint Research Centre, Ispra, Italy

Abstract. The GADRI Committee on Data and Information Sharing convened a session of the Global Summit that brought together some of its component interest areas and the challenges there are in making data work better for the Disaster Risk Reduction sector. Underpinning dilemmas represented by relatively well-known FAIR (Findable, Accessible, Interpretable, Reusable) principles, successes and challenges as experienced by a small sample of member case studies via their work with data in key areas of disaster research and the recent concerns of ongoing gaps highlighted by the Sendai Mid Term Review provided focus. Common themes were identified, and these contributed to identifying some intervening priorities for GADRI, conclusions from the discussions and a statement that formed part of a response to the Mid Term Review as further elaborated in this section of the proceedings.

Keywords: Data Sharing · Active Principles · Gaps and Priorities · Disaster Reduction

1 Introduction and Background

Whilst the production of knowledge from disaster data is more available, the demand for initiatives that bring this information forward for action orientated impact in increasingly at-risk settings has intensified. GADRI set up this Committee to be available to help broker partnerships for data sharing and impact across its membership and beyond. This has been informed through multiple experiences of working in disaster science, practice and policy facing initiatives including through data that is active within an all of society approach.

Based on the Committee concept note, the GADRI Summit session on data and information sharing aimed to advance the agenda by sharing understanding about the implications of promoting a greater emphasis on *data for action orientated knowledge impact*. Examining this would require varied data providers, managers and users at all levels to commit to reflecting on what data works for bringing about change in disaster risk and resilience. It had been acknowledged by this committee from the outset that there is much already in place that can be built upon to this end. In particular, the principles for greater usability of data are well reflected in the FAIR (Findable,

H. Tatano et al. (Eds.): GSRIDRR 2023, DRRGADRIBS, pp. 170–178, 2025.
https://doi.org/10.1007/978-981-96-6609-6_12

Fig. 1. (L) Chairs Dr. Tom De Groeve and Prof. Andrew Collins

Accessible, Interpretable, Reusable) data principles, that could be implemented more widely. The principles would need to underpin active data management in the interests of moving forward disaster prevention and response research activity more effectively and equitably. The session was also set up to consider how this is especially the case when considering particularly marginalized situations of some demographics, such as children, old people and women and girls living in the poorest environments of low-income countries where disaster impact is most prevalent.

From a multi-level DRR governance perspective questions emerge. For example, both from the top-down perspective as well as from the bottom-up view, especially in the context of citizen/community-generated risk data. In the former, these specifically relate to the tension between accessibility and openness. From the community and local authority point of view, openness is related to questions of data ownership, understandability, and local meaning as well as control over the further uses of local-level and citizen-generated data for DRR.

The following members formed the panel discussion and gave presentations (Figs. 1, 2 and 3):

- o Prof. Andrew Collins (Chair) (University of Northumbria, UK) – *Why have a Data and Information Sharing Committee?*
- o Dr. Tom de Groeve (Co-Chair)(Acting Head of Unit at European Commission Joint Research Centre, EC) – *EC leadership and contributions in data and information sharing.*

Fig. 2. Panel Members (L) Dr. Hiroyuki Goto, Dr. Becky Richardson, Prof. Jean-Paul Pinelli, Prof. Norio Maki

- o Dr. Hiroyuki Goto (DPRI, Kyoto University) – *GADRI Database Project – Disaster Collection Database and Case Study Database.*
- o Prof. Norio Maki (DPRI, Kyoto University) – *Good practices in DRR database development.*
- o Prof. Jean-Paul Pinelli (University of Florida, USA) – *Enabling Data Analytics and High-Performance Computing in Natural Hazard Engineering.*
- o Becky Richardson (Northumbria University, UK) - *Listening to children's perspectives on risk to understand the narrative of those living in extreme environments.*

The session included an introduction to the Committee rationale, the series of presentations and an open discussion amongst all participants in attendance, some of whom expressed an interest in joining up going forward.

Highlights extracted from a summary report presented by Dr. Tom De Groeve as feedback from the session to the main Summit on behalf of the committee included the following:

Committee objective will need to further emphasise the promotion of exchange and sharing of data and information for scientific research across the globe. Through the formulation of this focus GADRI has included the promotion of "action data", "bridging gaps to knowledge sharing", and developing an "active database". Further, the Mid Term Review of the Sendai Framework for Disaster Risk Reduction confirms a comprehensive array of gaps in progress on data under its priority one (understanding risk). The GADRI

Fig. 3. Presentation by Prof. Hiroyuki Goto

Committee would need to continue, in many ways, the work of the Global STAG that had now been disbanded by UNDRR. Referring to **go-fair.org** Dr. De Groeve emphasised how a strategic view on data in general, and particularly for DRR, is that it should be findable, accessible, interoperable and reusable. It is not sufficient to have data put on a website and to hope that it will be used. However, it takes significant effort to make this work. The presenters in the session in their various ways confirmed the need for progress across multiple fronts but some common themes are that:

- FAIR principles require a common taxonomy and ontology for hazard definitions
- Recommended data governance policies could be provided for GADRI research organisations. It is noted that some public organisations do not have an open data policy to protect their interests, to sell their data, or for security reasons. Therefore, the data is not open. From the research point of view, it would be good to have a common policy to publish data so that it can be used.
- Data policy and consequent ethical consideration across the GADRI membership needs to recognise that there is a lack of open data policy due to commercial, security and other constraints on institutions in their everyday operation. However, it would therefore be better to also establish the data principles in times of crises and especially for during the crisis. During a crisis changing circumstances can make it difficult to access data. One clear recent example was where the Turkish government implemented change by opening an archive of aerial imagery immediately after the recent earthquakes. This was very useful, as the science community had hitherto minimum

access to the area and the data released was able to be used in a timely way. When data is released at some later points after the event it is often too late to be used to the best effect. In relation to this issue, GADRI could play a role in thinking about how fair principles and policies can help its membership and the UN to take this further.
- The technology is important and there is plenty of it evolving fast; as such sustainable systems are needed to enable buying this via the Information Communication Technology (ICT) scientists.
- Accessibility is a major ongoing issue since in particular members from developing countries can encounter serious hurdles in accessing data where it is considered linked to a commercial interest. Some data is sold at cheaper prices but even at lower rates it can be too expensive for researchers in developing countries due to scarce research funds available there. GADRI could explore how it might be able to help in this respect.

2 Further Background, Elaboration and Forward-Looking from This Session

The Committee Chair and Co-chair elaborated on the orientation of the earlier concept note that had identified the need for "data informed and action orientated knowledge partnerships that steer progress in disaster reduction worldwide". This includes the ambition to:

i) Promote exchange and sharing of data and information for scientific research globally
ii) Promote the flow and application of active data for disaster reduction impact

Source: GADRI (2022) Data and Information Sharing Committee Concept Note

Influential discussions on the orientation of the Committee had notably included via the former UNDRR STAG Data Working Group (2020) that from its period of existence concluded specifically on the need for more active data. The previous GADRI Summit of 2021 on bridging gaps to knowledge sharing, the DPRI active database approach, and most recently the Mid Term Review of the Sendai Framework (Jan 2023) all pick up on this theme in various ways. The Mid Term Review array of significant gaps in progress on disaster risk data under the Sendai Framework Priority One: Understanding Risk including through the following statements:

- Global access to disaster data and applicable risk knowledge, including multi hazard early warning systems, remains inadequate. (para 4)
- Data ecosystems, including for disaggregated data, need to be strengthened, including through enhanced interoperability across systems, as well as the inclusion of local, traditional and Indigenous knowledge, feedback and expert opinion. (para 15)
- Improvements are needed in strengthening and mainstreaming monitoring, evaluation and learning processes and underlying knowledge management platforms. (para 18)
- Challenges remain regarding participation, including in data collection, with significant data gaps on women, the elderly, persons with disabilities and children. Without such data, "problems remain invisible and thus are not solved within the policy framework". (para 19)

- The scarcity of quality, interoperable or accessible data remains a roadblock to effective disaster risk reduction. Even when data is available and tools such as weather station networks exist, lack of capacity to interpret data and develop risk information impedes risk-informed decision-making and policy uptake. (Para 20)

Source: United Nations (2023) MTR of Implementation of the Sendai Framework for Disaster Risk Reduction

A focus of the Committee for the coming years would therefore need to include working to bring about "GADRI Data Sharing, Action Data and Active Data Base" together with further items from the outcomes of this session at the GADRI 6th Summit being taken into the GADRI Outcomes and Resolutions provided with these proceedings. The outcomes include this Committee's input in relation to GADRI's suggested revisions to the UNDRR Mid Term Review that were to be sent by GADRI to UNDRR.

The Chair's orientations in the opening included reference to an example of the systemic nature of some core issues represented by the committee (Fig. 4). This is being used to frame ongoing discussions within and more widely beyond the committee regards action data in the context of DRR. Inevitably there are other further systemic representations; this one was already available in various formulations. The point here is to determine in what way data processes work in terms of making a disaster reduction impact and to also emphasis that all of these involve data sharing.

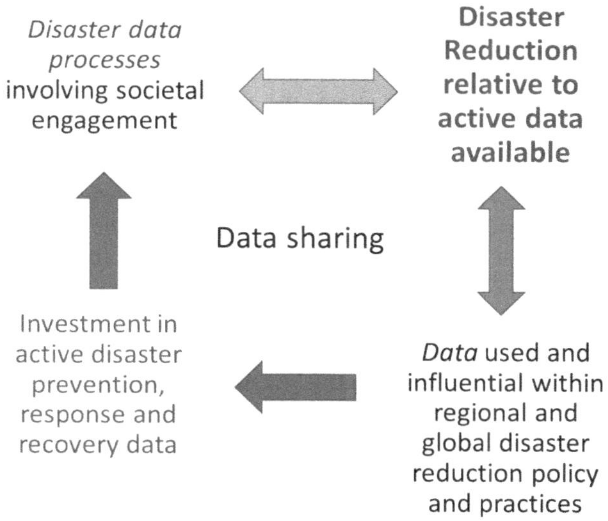

Fig. 4. Making data work for disaster reduction (Collins, A.E.)

Tom De Groeve's presentation on the European Commission Leadership and contributions in data and information sharing demonstrated that comprehensive commitment to data and information sharing was being promoted with the Commission and that GADRI would benefit from maintaining open dialogue with this process. A range of key detailed examples were provided with further information on progress in this area being released via the Disaster Risk Management Unit of the Joint Research Council.

Presenting the GADRI Database Project, Hiroyuki Goto explained how at DPRI they aimed to build a database that will contribute to the research activities in the GADRI community. Two databases are prepared: These are the Disaster Collection Database and the Case Study Database. The Disaster Collection Database allows users to search for websites, researchers, and documents related to disaster reduction research to take advantage of the wealth of data resources that exist around the world. The Case Study Database is for cases related to disaster reduction research. Data items to be stored in this database need to be entered in a fixed format, to provide a consistent approach to implementation, dates and locations so that all items are searchable.

Presenting in the context of his leadership of the engineering team of the Florida Public Hurricane Loss Model (FPHLM), funded by the Florida Office of Insurance Regulation, Jean-Paul Pinelli provided significant details on how enabling data analytics and high-performance computing was an active pursuit with positive outcomes in Natural Hazard Engineering. The DesignSafe cyberinfrastructure has been developed as part of the Natural Hazards Engineering Research Infrastructure (NHERI) and allows researchers to more effectively share, find, mine, analyze, and publish data; perform numerical simulations and utilize high-performance computing (HPC); and integrate diverse datasets. He described how various DesignSafe functionalities facilitate research and data sharing and re-use. Cyber-infrastructures like DesignSafe could be a game-changer in hazard engineering research, with the potential for additional DesignSafe functionalities that could benefit hazard engineering research.

The societal engagement theme within the Committees was in this instance exemplified by the work of Becky Richardson who presented her research on listening to children's perspectives on risk to understand the narratives of those living in the extreme environments of an informal settlement in Nairobi, Kenya. To achieve the demand for an all-of-society approach to Disaster Risk Reduction (DRR), all the different groups in society inclusive of those living in marginalized settings need to be reached and heard. Collecting data with groups in marginalized settings can be a challenge due to limited access and ethical considerations. The talk presented an ongoing case study of action research with one of the most central groups to an all-of-society approach, children. The research applied a child-centred approach to understanding risks children are exposed to in school and their neighborhoods which impact their health and wellbeing. It explored how these risks can be effectively communicated through listening to children's experiences and perspectives, then linked to local level solutions and policies to achieve impactful action data. The methods applied at field level, in this case an informal settlement in Nairobi Kenya, were participatory qualitative approaches using creative and arts-based tools creating enhanced engagement, communication and trust between the researcher and participants. Methods included deep mapping, walking interviews, model-making, storytelling, and drawing. These approaches demonstrated possibilities of retrieving data out of extreme environments such as this case study, but also giving a voice through that data to those marginalized groups who are traditionally unheard.

In terms of priorities the committee agreed the following emphasis for GADRI:

- o Sustainable implementation of FAIR principles by all members including in the development of databases.
- o Identification of platforms or databases that can help make GADRI data FAIR, able to combine quantitative and qualitative data, being supported by data scientists and domain experts.
- o Collation of data that can assist in forming narratives about vulnerable people, being inclusive in terms of accessibility and suitability for member related data policy and governance purposes. This is to enable appropriate recommendations to all DRR actors considering data principles during times of crises.
- o The emphasis on reusable data is to encourage Members to monitor ongoing data application to and engagement with the purposes of DRR.

3 Conclusions

The GADRI session contributed to elaborating a series of conclusions as formulated for the GADRI Outcomes and Resolutions from the overall summit. These are listed here and then reiterated in that section of this volume alongside the other Outcomes and Resolutions:

- o GADRI is committed to promoting "action data", "bridging gaps to knowledge sharing" and developing an "active database".
- o The GADRI Committee is acutely aware of data issues flagged in the Sendai Framework MTR under priority one (understanding risk). The GADRI Committee is committed to continuing in many ways the work of the Global STAG working group.
- o The Committee emphasises the importance of the various elements of FAIR data: Findable, Accessible, Interoperable and Reusable.
- o Inputs at the Summit brought influential examples of; databases and their sustainability and the use of big data and high performance computing (DesignSafe) supporting engineering research and publishing workflows; the GADRI database project that intends to regularly scrape URLs from members and analyse the pages with keywords, so the data becomes searchable; local level data from narratives on risk from local, marginalized communities in particular children, which responds to the Sendai objective to understand the risk for the most vulnerable people; new action-oriented collection methods for data through drawings, art, stories, which works well for health-related risks and impact of wellbeing.

The following statement was provided to UNDRR for its consideration in the context of its Mid Term Review of the Sendai Framework:

"Sound and reliable evidence based on data collection and analysis, including risk assessment, must be used to further enhance the connection between risk information and DRR decisions and actions and this should be supported by scientific and academic partnerships. Whilst good data is a basis for improved decision making towards achieving Global Targets for Disaster Risk Reduction and Sustainability, there is a need to analyse how it is contributing at all levels. Understanding what data works requires guidance for data providers and users regards its authenticity and impact in diverse contexts so that it is active as a voice and enabler of disaster risk reduction, particularly in marginalised at-risk locations." GADRI (2023) Summary Comments on the Sendai Framework Midterm Review.

Further opportunities to develop work contributing to the focal interest areas of the committee are considered ongoing by committee members and will be returned from time to time when they become more fully elaborated.

References

GADRI: Data and information sharing committee concept note (2022)

UNDRR: STAG data working group: report on concept, activity and implication. United Nations Disaster Risk Reduction (2020)

United Nations: MTR of implementation of the Sendai framework for disaster risk reduction (2023)

GADRI: Summary Comments on the Sendai Framework Midterm Review. DPRI, Kyoto University, GADRI Secretariate (2023)

Forging Alliances: Collaborative Advocacy Efforts in Advancing Science and Technology for Disaster Risk Reduction

Irasema Alcántara-Ayala[1](✉), Rajib Shaw[2], and Genta Nakano[3]

[1] Institute of Geography, National Autonomous University of Mexico (UNAM), Av. Universidad 3000, Coyoacán, 04510 Mexico City, Mexico
ialcantara@geografia.unam.mx

[2] Graduate School of Media and Governance, Keio University, Endo 5322, Fukisawa 252-0882, Japan
shaw@sfc.keio.ac.jp

[3] Disaster Prevention Research Institute, Kyoto University, Gokasho, Uji 611-0011, Kyoto, Japan
nakano.genta.8n@kyoto-u.ac.jp

Abstract. The Sendai Framework for Disaster Risk Reduction (SFDRR) is a testament to the transformative power of science and technology in disaster risk reduction through evidence-based decision-making. The Science and Technology Roadmap was designed to support the implementation of the Sendai Framework for Disaster Risk Reduction 2015–2030. Several bodies globally focus on science-based decision-making, such as G-STAG (Global Science Technology Advisory Group), Regional STAG, and IRDR (Integrated Research on Disaster Risk). The DRR can draw lessons from the IPCC (Intergovernmental Panel on Climate Change) style science-policy interface, as exists in the climate change regime. There are many good practices at the national and local levels of science-based decision-making in some countries and with some NGOs (non-government organizations). Mass media can also play a significant role in science advocacy and enhancing preparedness before a disaster. This chapter offers insights into the GADRI´s commitment to leading science-based advocacy in collaboration with different stakeholders.

Keywords: Advocacy · Disaster Risk Reduction · Alliances · Science and Technology · Global Alliance of Disaster Research Institutes · GADRI

Institute for Catastrophic Loss Reduction (ICLR), Canada; Prof. Irasema Alcántara-Ayala, Institute of Geography, UNAM, Mexico; and Prof. Rajib Shaw, Keio University, Japan

H. Tatano et al. (Eds.): GSRIDRR 2023, DRRGADRIBS, pp. 179–188, 2025.
https://doi.org/10.1007/978-981-96-6609-6_13

Fig. 1. Session Chairs: (L) Prof. Paul Kovacs, Chair, Board of Directors of GADRI; and Executive Director,

1 Introduction

1.1 The Sendai Framework for Disaster Risk Reduction 2015–2030: On the Road

The Sendai Framework for Disaster Risk Reduction aims to reduce disaster risk and ensure global resilience. It underlines the significance of understanding disaster risk, reinforcing disaster risk governance, allocating resources to DRR and resilience, and bettering disaster preparedness for efficient response and recovery. To meet its ambitions, the framework places high importance on collaborating with various entities, including governments, civil society, academia, the private sector, and other stakeholders. It endeavours to instil a culture of resilience and propagate sustainable development practices that decrease disaster risk and enhance resilience across all levels [1].

The Science and Technology Roadmap was designed to support the implementation of the Sendai Framework for Disaster Risk Reduction 2015–2030 [2]. This framework aims to reduce the risk of disasters by fostering cooperation among science communities and other stakeholders to integrate tools and instruments relevant to DRR. The roadmap emphasizes four primary outcomes and 58 actions that align with the four Priorities for Action of the Sendai Framework. The regional reports for Europe and Central Asia [3], and the Americas and the Caribbean [4] highlight the need for scientific-based advocacy to achieve the desired results. Therefore, the science and technology communities must collaborate with stakeholders, from policymakers to grassroots communities, to achieve optimal outcomes. Overall, the Science and Technology Roadmap is a comprehensive guide for all stakeholders involved in DRR to work together and achieve a safer and more resilient future.

1.2 Beyond Borders: The Global Alliance of Disaster Research Institutes (GADRI)

The Global Alliance of Disaster Research Institutes (GADRI) is a collaborative platform that shares knowledge and fosters partnerships to reduce disaster risks and enhance resilience. It is driven by organizational values committed to DRR, and its activities are directed towards supporting the objectives of the 2016 Science and Technology Roadmap for implementing the Sendai Framework for Disaster Risk Reduction 2015–2030 [5].

Working closely with organizations worldwide, GADRI strives to enhance DRR and resilience by facilitating the exchange of information, knowledge, experiences, and ideas on disaster research. To achieve its goals, GADRI has established a global research network, provided a research roadmap, and organized disaster research groups. It also aims to promote the capacity development of disaster research institutes, facilitate researcher and student exchange, and encourage the exchange and sharing of data and information for scientific research worldwide. Additionally, GADRI is an advocacy organization that speaks with one voice to influence decision-making processes [5].

GADRI has established diverse Committees: Committee on Networking, Committee on Institutional Capacity Building, Committee on Science and Technology Roadmap, Committee on Data and Information Sharing and Committee on Advocacy.

The Committee on Advocacy seeks to serve as an advocacy organization presenting evidence-based approaches that influence decision-making processes. Against this backdrop, during the *6th Global Summit of GADRI: Towards GADRI Objectives of Achieving a Sustainable Disaster-Resilient World*, key challenges and potentials of science-based advocacy processes were identified through reflections, and different stakeholders' roles were determined (Fig. 1).

1.3 Towards GADRI Objectives of Achieving a Sustainable Disaster-Resilient World: The 6th Global Summit of GADRI

During the Advocacy Committee session, special attention was given to identifying the key needs of critical advocacy from stakeholder perspectives, learning from science based advocacy in the climate change sector, and identifying good practices of national and/or local science-based advocacy in DRR (Fig. 2).

The discussion began by exploring the essential criteria for effective advocacy, considering various viewpoints, such as those of national and local governments and grassroots organizations. Then, dialogues delved into the insights gained from IPCC's science-based advocacy and how they can be leveraged in developing DRR strategies. Furthermore, the role of media in advancing science-based advocacy was examined.

The following questions were addressed during the discussion: What are the key barriers to science-based advocacy? What is the need for "critical advocacy" at different levels, from national to local governments to grassroots? What lessons can be drawn from climate sectors? How can the media play a critical role in science-based advocacy?

Fig. 2 Session organized by the Committee on Advocacy during the 6th Global Summit of GADRI: Towards GADRI Objectives of Achieving a Sustainable Disaster-Resilient World (Source: photo courtesy of Wilma James).

2 Collective Advocacy Endeavors

2.1 Need for Critical Advocacy and Identification of Obstacles: National Government Perspectives

In today's rapidly changing world, it is essential to promote critical thinking to ensure informed decision-making at both national and local levels. National governments face many intricate challenges, such as climate change, biodiversity conservation, and unforeseen risks like oil spills. Governments depend on scientific advice based on thorough research and practical implications to address these complex issues. This guidance is indispensable in policymaking, providing a solid foundation for decision-making.

Climate change represents a pressing issue with far-reaching consequences for ecosystems, economies, and human well-being at a global scale. Governments seek scientific insights tailored to their unique geographical contexts, which can help them navigate climate change's complex and evolving dynamics while identifying effective mitigation and adaptation strategies. Moreover, preserving biodiversity requires a nuanced approach that accounts for ecosystem resilience, species conservation, and sustainable resource management. Achieving this goal calls for interdisciplinary expertise that brings together biologists, ecologists, policy analysts, and sociologists. They can develop holistic solutions that transcend traditional disciplinary boundaries by working collaboratively.

In addition, new risks, such as oil spills, highlight the importance of the government taking anticipatory measures and responding quickly, guided by scientific knowledge. The containment and mitigation of such environmental disasters require technological innovations and well-informed policy frameworks based on scientific evidence. At the local level, it is necessary to adapt national policies to meet the diverse needs of different communities. This emphasizes the need for interdisciplinary collaboration, where scientists, policymakers, and local stakeholders work together to create solutions that are relevant to the context and socially just. Effective advocacy is essential for governments to navigate the complexities of modern challenges, promoting resilience, sustainability, and social well-being.

2.2 IPCC: Science Policy Advocacy

Intergovernmental organizations, like the Intergovernmental Panel on Climate Change (IPCC), have a crucial role in connecting scientific evidence with policy development. Effective advocacy for science policy depends on having a clear purpose and building constructive partnerships between scientists and policymakers.

The IPCC is responsible for evaluating the scientific foundation of climate change, its effects, and ways to mitigate them. It represents the vital intersection of science and policy. By promoting collaborative partnerships among scientists, policymakers, and stakeholders, the IPCC cultivates an atmosphere that encourages informed decision-making on a global level.

Institutionalizing science policy advocacy within organizations like the IPCC is pivotal for sustained impact and relevance. When advocacy initiatives are incorporated into an organization's structure and processes, scientific research is expertly conveyed and translated into practical policy recommendations. The IPCC assessment reports, serving as a universal interface, play a pivotal role in disseminating advocacy messages to policymakers and the general public. These reports are meticulously peer-reviewed and condensed by renowned experts, lending legitimacy to advocacy efforts and amplifying their authority in shaping policy agendas.

Successful advocacy for science policy entails much more than merely technical reports. It requires engaging in high-level conversations, organizing outreach events, and communicating through various media channels. Through effective engagement with policymakers and the initiation of advocacy campaigns, we can encourage political leaders to adopt policies grounded in evidence. Outreach events and media campaigns can further amplify advocacy messages, increasing public awareness and support for science-informed policy action. The IPCC plays a pivotal role in convening stakeholders, facilitating dialogue, and disseminating scientific knowledge to diverse audiences globally.

Fostering the development of the next generation of scientists is a critical aspect of science policy advocacy. By identifying and nurturing young researchers, we can equip them with the skills to become effective champions for evidence-based policy. Organizations like the IPCC offer valuable opportunities for mentorship, professional growth, and capacity-building, enabling these promising scientists to reach their full potential.

Recognizing their contributions through awards, fellowships, and leadership roles can inspire excellence, encourage innovation, and promote inclusivity within the scientific community. Ultimately, science policy advocacy within intergovernmental organizations such as the IPCC is essential for driving collective action towards addressing the most pressing challenges of our time, such as climate change and sustainable development.

2.3 On the Grass-Roots Perspectives for Disaster Risk Reduction Advocacy

The involvement of grassroots perspectives is integral in advocating for DRR as it helps to connect scientific knowledge with the needs of local communities. Civil society plays a crucial role in DRR by providing a link between policymakers, scientists, and grassroots communities. Civil society organizations utilize grassroots initiatives to directly engage with communities, promote awareness, enhance capacity, and encourage individuals to mitigate and respond to disasters effectively.

Customising strategies to suit communities' unique needs and contexts emphasizes the significance of disaster management from a localized perspective. Strong advocacy ensures alignment between disaster risk reduction policies and programs and this localized approach. Facilitating communication between scientists and grassroots communities is crucial for effective advocacy, as it fosters the exchange of knowledge and expertise. By merging scientific insights with local knowledge and experiences, advocacy campaigns at the grassroots level can develop more precise and effective solutions to tackle disaster risks.

A community-centric innovation ecosystem can powerfully drive science-based advocacy in DRR. Such an ecosystem encourages collaboration between various stakeholders, including scientists, community leaders, government agencies, and NGOs. By leveraging local knowledge and resources, community-centric innovation empowers communities to develop and execute DRR initiatives tailored to their specific circumstances. Moreover, by promoting bottom-up approaches, this ecosystem fosters community ownership of DRR efforts, enhancing their resilience and sustainability in the face of disasters.

Successful grassroots advocacy entails generating awareness about essential measures to mitigate the risks associated with disasters and cultivating a profound comprehension of the factors that lead to their occurrence and the development of risk. By initiating conversations with communities on the underlying causes of disasters, grassroots advocates can inspire significant transformation. Such advocacy endeavours can address fundamental issues like poverty, inequality, and environmental degradation, supporting the creation of more resilient and equitable societies. By responding to local needs and priorities, grassroots advocacy is significant in promoting sustainable development at the community level and advancing DRR agendas.

2.4 Advocating for Disaster Risk Reduction Through Communication and Mass Media

Efficient communication through the mass media is critical in promoting DRR and encouraging public involvement in disaster prevention initiatives. Clear and concise scientific communication is essential in raising awareness of potential hazards and encouraging proactive risk reduction measures. By presenting intricate scientific concepts in an easy-to-understand format, media channels can empower individuals and communities to make informed decisions and implement suitable measures to mitigate disaster risks.

Promoting a culture of resilience necessitates effective communication and collaboration among scientists, media professionals, and the general public. Media platforms are critical conduits for sharing insights, exchanging best practices, and co-creating innovative solutions to tackle emerging threats. Promoting media literacy is equally important to ensure accurate comprehension and dissemination of information about disaster risk reduction. Scientists are instrumental in partnering with media professionals to provide dependable data and insights that counter misinformation and encourage evidence-based decision-making.

Japan serves as an excellent model for advocating disaster risk reduction. In this country, media professionals are educated about DRR principles from the outset. Rather than simply reacting once disasters occur, proactive measures are taken to educate media personnel about the critical importance of disaster prevention and preparedness. This approach guarantees that media coverage of disasters is informative and highlights the value of risk reduction measures and community resilience. By incorporating DRR principles into media training programs and fostering continuous collaboration among scientists, policymakers, and media outlets, nations can raise public awareness and develop a more robust society capable of successfully mitigating the effects of disasters.

3 GADRI: Advancing Science and Technology for Disaster Risk Reduction

The role of media is paramount in promoting science-based DRR by effectively communicating risk information to diverse audiences, including decision-makers and communities. Using various media platforms, such as social media, television, radio, and community-based channels, scientists can convey complex scientific concepts in simple language, increasing awareness and informed decision-making. This, in turn, ensures that crucial information reaches vulnerable communities, empowering them to reduce disaster risks and enhance resilience proactively. Additionally, media engagement can influence policymakers by highlighting the importance of investing in science-based DRRstrategies, shaping policy agendas, and fostering greater political commitment to risk reduction efforts (Fig. 3).

Tailoring science advocacy messages to specific audiences and contexts is essential for engaging key change agents who help bridge the gap between researchers and communities. Advocacy efforts should be customized to meet stakeholders' unique needs and concerns, from local community leaders to national policymakers. Community-based media platforms are especially valuable for directly engaging at-risk communities, building trust, and facilitating dialogue around local priorities and solutions. By empowering

Fig. 3. Dr. Irie Sayaka, Matsumoto University (ex NHK); and Mr. Takeshi Komino, ADRRN, and CWS Japan.

these key change agents, science advocacy can inspire grassroots action, drive positive change, and strengthen community resilience.

Achieving impactful science advocacy for DRR requires collaboration between researchers, civil society, stakeholders, and policymakers (Fig. 4). This joint effort allows advocates to develop comprehensive solutions that address the root causes of risk and promote sustainable development. The GADRI questionnaire survey conducted in 2021 revealed that 20 GADRI's institutes contribute to the national DRR platforms, which provide the scientific research outcomes of disaster modelling and monitoring and capacity building to support the evidence-based DRR with multi-stakeholder participation. In addition, 16 GADRI institutes belong to the local DRR platforms. Adopting a collaborative approach makes advocacy efforts inclusive, relevant to local contexts, and seamlessly integrated into policy and practice. Engaging stakeholders at every level ensures greater support, fosters collective ownership, and maximizes the impact of DRR interventions.

The points raised in [6] and [7] concern establishing an academically sound, science-based assessment process for disaster risk reduction to provide rigorous knowledge, assist governments worldwide in formulating policies and goals, and identify research gaps. Likewise, a global group of disaster risk reduction specialists encompassing various industries and civil organizations should be created. The objective of this body is to promote awareness from grassroots communities to the highest levels of government. It is our shared belief that sustainability governance should integrate interconnected issues,

Fig. 4. Scientific conversation between Mexican and Japanese scientists and local municipal officers for evidence-based DRR in Zihuatanejo de Azueta, Guerrero, Mexico (Source: photo courtesy of Genta Nakano).

not just disaster risk. An International Panel on Disaster Risk, similar to the IPCC, could periodically evaluate the science of disaster risk and generate policy-relevant assessments [7].

Based on the discussions held with all the participants, the following key takeaways were identified:

- Media is a vital tool for science advocacy, enabling effective communication of risk information to diverse audiences, including communities and decision-makers.
- Customized science advocacy through various media channels, from traditional to social and community-based platforms, engages key change agents to bridge the gap between researchers and communities.
- Cross-cutting science advocacy, spanning grassroots to policy levels, necessitates collaborative efforts in co-designing solutions with researchers, civil society, stakeholders, and policymakers.
- Strengthening science-government-community interactions through media platforms enhances knowledge exchange, collaboration, and the impact of advocacy efforts in disaster risk reduction.
- Communication and advocacy training are essential for scientists to convey research findings in accessible language to policymakers and society, fostering greater understanding and action.

- Commitment to producing Gender and Disaster Risk Reduction Initiative and Policy briefs at various levels, in collaboration with initiatives like IRDR, enhances the accessibility and relevance of scientific knowledge for policy-making.
- Exploring the creation of a Periodic Disaster Risk Assessment, distinct from UNDRR's Global Assessment Report (GAR), can provide timely, context-specific data to inform policy decisions and enhance disaster risk reduction initiatives.
- Embracing a culture of advocacy strengthens scientists' and stakeholders' ability to contribute collectively to building a more resilient and equitable society.

4 Concluding Remarks

Science advocacy is essential for better risk reduction and informed decision-making. Several avenues of science advocacy can be done through formal channels in the government, especially with the Science Technology Advisory Groups at the national level. This practice exists in many countries; two great examples are from the Philippines and Malaysia. Media also plays an important role in science advocacy, and Japan's example of NHK playing a critical role in risk communication can be considered a very good example. Also, grass-roots science-based decision-making practices exist in many countries, especially working closely with local non-government organizations (NGOs). GADRI, as a global network, has the potential to be a strong advocate for science-based decision-making. For this, collaboration with other stakeholders and synergies of efforts are required.

Acknowledgements. The authors would like to extend their sincere gratitude to Ms Antonia Loyzaga, Secretary, DENR, Philippines; Prof. Joy Pereira, UKM, Malaysia; Mr. Takeshi Komino, ADRRN, and CWS Japan; and Dr. Irie Sayaka, Matsumoto University (ex NHK), Japan for their valuable contributions during the discussion session. Their active participation and candid responses were instrumental in shaping the outcomes presented in this chapter.

References

UNISDR (United Nations International Strategy for Disaster Reduction). Sendai framework for disaster risk reduction 2015–2030 (2015). http://www.wcdrr.org/uploads/Sendai_Framework_for_Disaster_Risk_Reduction_2015-2030.pdf. Accessed 17 Apr 2024

Aitsi-Selmi, A., et al.: Reflections on a science and technology agenda for 21st century disaster risk reduction: based on the scientific content of the 2016 UNISDR science and technology conference on the implementation of the Sendai framework for disaster risk reduction 2015–2030. Int. J. Disaster Risk Sci. **7**, 1–29 (2016)

Amaratunga, D., et al.: United Nations Office for Disaster Risk Reduction Regional Assessment Report on Disaster Risk Reduction 2023: Europe and Central Asia (2023)

UNDRR, Regional Assessment Report on Disaster Risk in Latin America and the Caribbean, United Nations Office for Disaster Risk Reduction (UNDRR) (2021)

GADRI Homepage. https://www.gadri.net/. Accessed 14 Apr 2024

Cutter, S.L., et al: Global risks: pool knowledge to stem losses from disasters. Nature **522**(7556), 277–279 (2015)

Alcántara-Ayala, I., Burton, I., Lavell, A., Oliver-Smith, A., Brenes, A., Dickinson, T. Forensic investigations of disasters: past achievements and new directions. Jàmbá: J. Disaster Risk Stud. **15**(1), 1–11 (2023)

Panel Group Discussion II: New Challenges for Action by GADRI

Big Science for DRR: Large-Scale Experiment

Nobuhito Mori(✉) and Shinichi Matsushima

DPRI, Kyoto University, Kyoto, Japan
mori.nobuhito.8a@kyoto-u.ac.jp

Abstract. GADRI aims to serve as an advocacy organization to speak with one voice in an effort to influence decision-making processes. The purpose of this session is to discuss next step of large-scale research facilities supported by research communities. Large-scale experiments, along with numerical simulations, are major tools for clarifying phenomena in disaster prevention research. For example, in earthquake research, the usefulness of large-scale shaking tables has been demonstrated. In this session, large scale experimental apparatuses, instruments, observational facilities, and computational environment to promote research and development on disaster prevention is discussed. Discussion about what kind of experimental instruments and facilities are necessary for future disaster prevention research and science, and how to utilize them. Furthermore, the utilization of these large-scale facilities, such as joint use, is discussed from the perspective of the disaster research community.

Keywords: Disaster risk reduction · Big science · Large-scale experiment · Data sharing · Digital twin

H. Tatano et al. (Eds.): GSRIDRR 2023, DRRGADRIBS, pp. 191–200, 2025.
https://doi.org/10.1007/978-981-96-6609-6_14

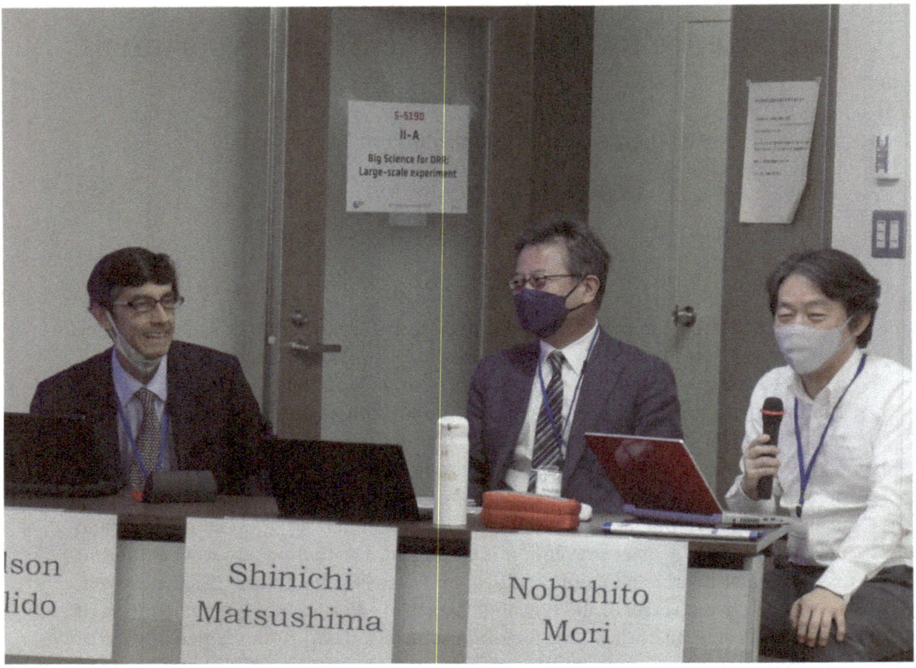

Ch15–1: Session Chairs (R) Prof. Nobuhito Mori, and Prof. Shinichi Matsushima, DPRI, Kyoto University; with Dr. Nelson Pulido, National Research Institute for Earth Science and Disaster Resilience (NIED), Tsukuba, Japan

1 Introduction

Large-scale experimental apparatuses, instruments, observational facilities, and computational environments are important to promote research and development on disaster prevention. It is also important to determine what experimental instruments and facilities are necessary for future disaster risk reduction research and science and their effective utilization. For example, the first step of understanding unknown phenomena is an observation, and an experiment is a powerful approach, generally. Furthermore, the scaling law for physical phenomena can allow us to change the size of experiments from a natural to an experimental size. However, different physical phenomena have different scaling laws and sometimes conflict. Therefore, large-scale experimental facilities are important to cover multi-physical phenomena in the controlling experimental conditions. For these cases, the large-scale experimental facilities are important to accelerate the understanding, assessment, and demonstration of a big science for Disaster Risk Reduction (DRR).

Although large-scale experimental facilities or observational facilities are important, these big science facilities are very expensive. Therefore, sharing facilities is one choice to increase efficiency and economic performance (Timofejevs and Avotins, 2018). The collaborative and open science, a shift from new knowledge to finding new solutions, new ways of governance of science, and the formation of a new generation of top scientific talent. Although the impact of increased utilization is long-term and difficult to measure in the short term, it can be concluded that there is a significant positive impact at both the national and local levels. The evaluation of output and outcome are generally strictly checked. The absurdity of consistently using simple publication counts to indicate the productivity and quality of science and concludes that such measures are irrelevant to an organization-level analysis of science and should be replaced by a qualitative assessment of the content of the science produced (Hallonsten, 2014).

We must develop new technologies, including large-scale experimental or observational facilities for the DRR research development. Here, we discussed the next steps for large-scale research facilities supported by research communities.

2 Review of the Current Status of Large-Scale Research Facilities in Different Hazards

There are several large-scale research facilities for natural hazards for different targets. Here are typical examples of hazards.

- Earthquakes
- Tsunamis

The National Research Institute for Earth Science and Disaster Resilience (NIED) operates an Observation Network for earthquakes, tsunamis, and volcanoes in Japan called MOWLAS (Monitoring Of Waves on Land And Seafloor) (Aoi et al., 2022) (Fig. 1). "Moura" means comprehension in Japanese. NIED maintains Achievement and future research perspectives on earthquake-disaster risk reduction with E-Defense shake table (Fig. 2). Port and Airport Research Institute operates super large experiments on coastal structures using large hydrogeo fluids. DPRI, Kyoto University operates Large coastal wave flumes for tsunami research (Fig. 3). DPRI, Kyoto University also operates a Large-scale shaking table (Fig. 4), Geotechnical centrifuge (Fig. 5), Boundary layer wind tunnel (Fig. 6) for collaborative use.

Furthermore, large-scale test facilities for earthquake engineering research are being developed worldwide (e.g., Elgamal. 2027). At the same time, advances in information technology (IT) are creating unprecedented opportunities for (i) remote access and presence during long-term remote experiments, (ii) hybrid simulation of entire structural systems with multi-site experiments and computational whole models, (iii) near real-time data archiving, processing, and sharing. Current facilities are recognized as being used, shared, and improved following research and community support.

Challenges of large-scale experimental facilities for DRR:

1. Sharing facilities is important – large-scale experimental facilities are very expensive. The sharing of facilities to conduct experiments was widely discussed and recommended. The US-based NEES, founded by the National Science Foundation and Japan-based NIED supported by MEXT, supports such facilities. DPRI also supports the sharing of experimental facilities.

2. Sharing dataset with systematical format-- each facility has historical experimental datasets. It is not easy to use this outside of the facilities. / Digital foundation of the experimental dataset is also important to share.
3. Development of new instruments and sensors – use different types of sensors from economical to sophisticated /Digital twins using large-scale experiments.
4. Benchmark tests for numerical models, including structure modeling or tsunami modeling – there are many different numerical models but lack knowledge of the performance of the individual numerical models. Through large facilities, it is possible to test the performance of numerical models using the benchmark-test type of research.
5. Hybrid physical-numerical modeling – this part has been developed in structural engineering. Expanding to hybrid, physical, and numerical experiments to other fields such as structure, soil or tsunami soil interaction would be good.
6. Disaster education – this is beyond this session.

Observations and recommendations submitted by the Chairs to the Board of Directors of GADRI:

- Seismic and tsunami monitoring to improve accuracy and speed for Early Warning System
- Detection of earthquakes in ocean regions by seafloor observation
- Use of new technologies such as Distributed Acoustic Sensing (DAS) to lower the cost of observation for continuous extension.
- Utilize ocean optic fiber lines to observe seismic motion and water pressure in the sea.
- Extend from full-scale single-structure experiments to experiments of a system with multiple structures, including structure-structure or soil-structure interaction.
- Provide existing experimental data together with metadata so that the users can fully understand the conditions of the data.
- Large-scale experiments are expensive, and the number of experiments conducted will be limited. Therefore, establishing a method to combine numerical simulation and large-scale experiments will be essential.
- Collaboration between institutes with large-scale experimental facilities, such as PARI, and institutes capable of numerical simulations is necessary.
- Data obtained by experiments at DPRI is not collected, but if it can be available to the community, it would benefit the community, although an environment to share the data needs to be fostered.
- It would be ideal if the observed data of buildings and infrastructure were shared among the community to understand the safety of existing structures.
- A new facility to consider the coupling effect for compound hazards (multi-hazard) needs to be developed, but it would be very expensive, and it will be difficult to consider many combinations of hazards. Hence, the proposal is that environments such as Digital Twin should be utilized to consider multi-hazard effects.
- A less expensive protective system, such as a base-isolation system, should be developed to be utilized in earthquake-prone countries worldwide.
- The effect of soil-structure interaction should be considered more to understand the response at sites where soil-structure interaction can have a large impact.
- Cost-effective instruments can be used for system monitoring.

- Blind prediction exercises of large-scale experiments will benefit the community by helping them understand and update the capability of numerical simulation codes.
- The use of large-scale experiments should not be limited to researchers but would also be effective in educating students and the general public. The senses can have a greater impact than virtual reality because of the real sound, pressure, and smell.

Ch15–2: Panellists (L) Dr. Kentaro Tabata, National Research Institute for Earth Science and Disaster Resilience (NIED), Tsukuba, Japan; and Dr. Kojiro Suzuki, Director, Dept. Coastal Hydraulic Engineering, National Institute of Maritime, Port and Aviation Technology, Port and Airport Research Institute, Japan.

3 Summary

Large-scale experimental facilities for disaster risk reduction (DRR) face several challenges due to the complexity of managing such facilities and the diverse nature of the disasters they seek to address. Some common challenges are listed below:

Cost: The construction and maintenance of large experimental facilities is enormously expensive. These facilities often require specialized equipment, infrastructure, and personnel, all contributing to high operating costs. Securing funding for construction and ongoing operations can be a significant challenge.

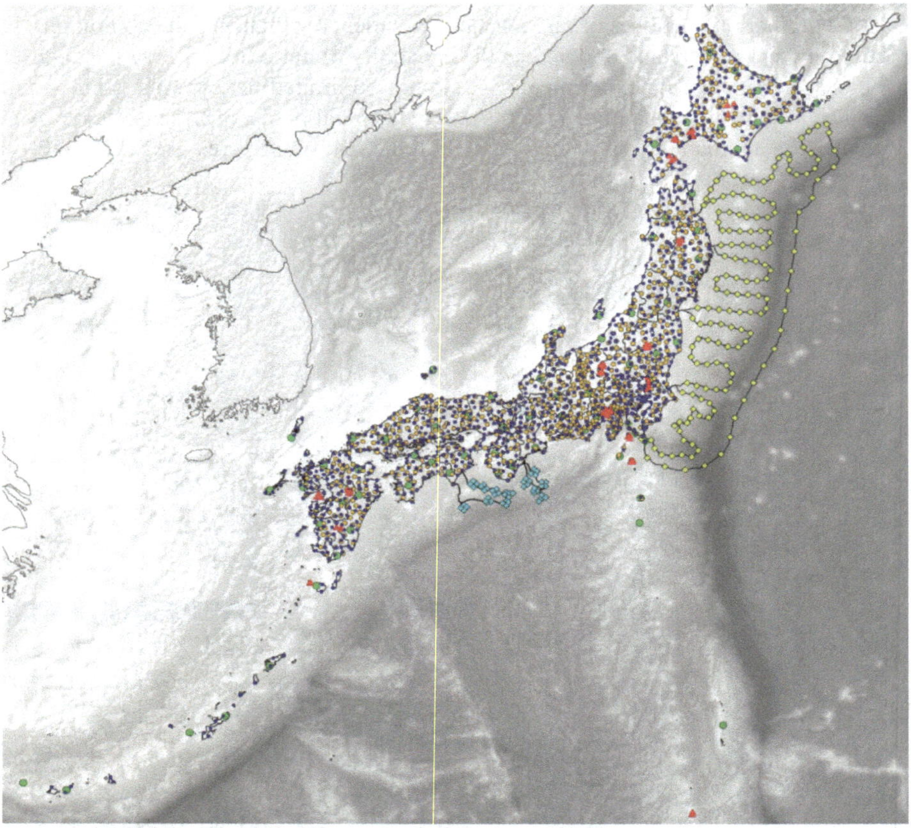

Fig. 1. Observation Network for Earthquake, Tsunami and Volcano operated by NIED, MOWLAS.

Technical Complexity: Designing and operating a large-scale laboratory facility for disaster risk reduction requires expertise across multiple disciplines, including engineering, environmental science, and disaster management. Coordinating these diverse technical aspects is complex and may require the cooperation of various stakeholders.

Safety and Ethics: Large-scale disaster experiments involve inherent risks for researchers and the surrounding community. The safety of participants, researchers, and the surrounding community is paramount. In addition, ethical considerations must be addressed, particularly concerning potential harm to the environment and community involved in the experiment.

Scale and practicality: Achieving a realistic scale in an experimental facility is difficult. Many disasters, such as earthquakes and floods, involve large-scale phenomena that are difficult to reproduce accurately in a controlled environment. Striking a balance between practicality, utility, and safety is a constant challenge for facility operators.

Interdisciplinary Collaboration: Effective disaster risk reduction often requires collaboration across multiple disciplines, including engineering, social sciences, public

Fig. 2. E-Defence 3-D shake table and surrounding facilities operated by NIED in Miki, Hyogo, Japan. Inset figure is the shake table with a size of 50 m × 50 m.

health, and policy. Large-scale laboratories must facilitate interdisciplinary collaboration to address the complex nature of disaster risk.

Data Collection and Analysis: Collecting high-quality data from large-scale experiments is critical to inform disaster preparedness and response measures. However, collecting, analyzing, and interpreting data from complex experiments can be challenging. Facilities must invest in a robust data collection and analysis infrastructure to derive meaningful insights from experiments.

Fig. 3. Tsunami reproduction device in Ujigawa Open Laboratory of DPRI.

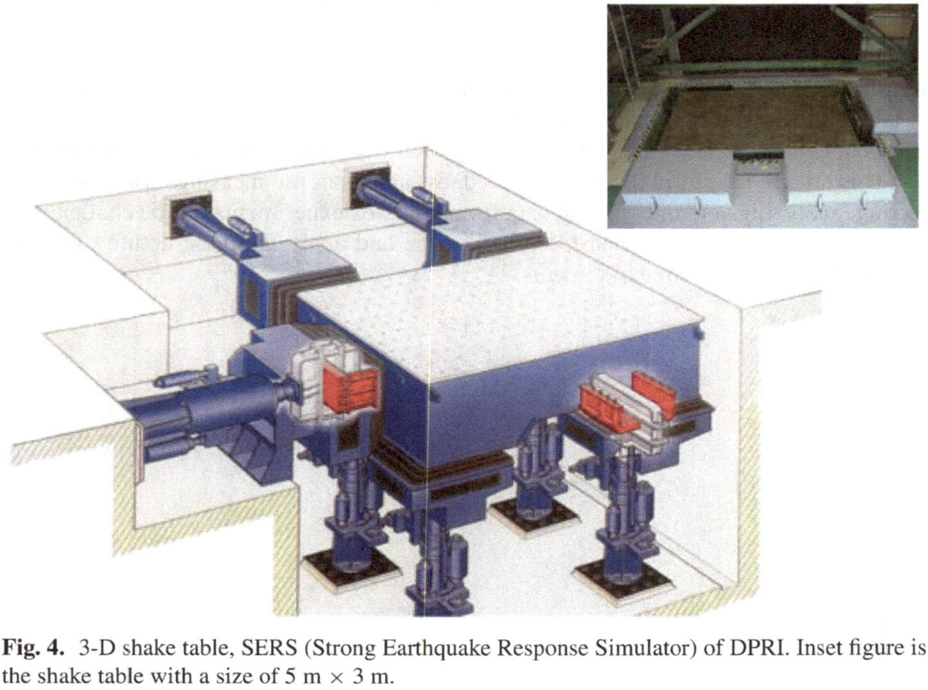

Fig. 4. 3-D shake table, SERS (Strong Earthquake Response Simulator) of DPRI. Inset figure is the shake table with a size of 5 m × 3 m.

Fig. 5. Geotechnical centrifuge with an effective radius of 2.5m of DPRI. Inset figure is the shake table device connected to the end of the arm.

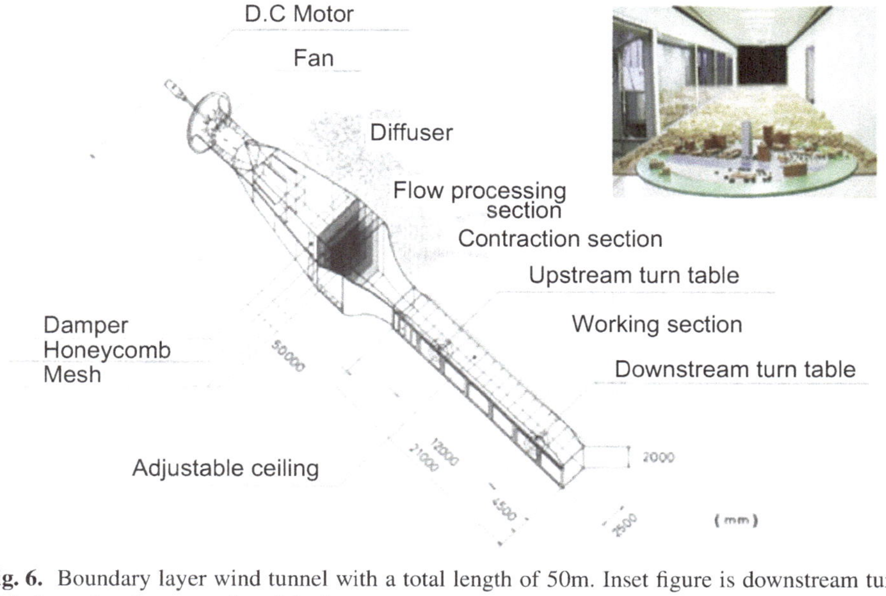

Fig. 6. Boundary layer wind tunnel with a total length of 50m. Inset figure is downstream turn table from the view opposite of the fan.

Acknowledgement. The session chairs thank and appreciate valuable and construction inputs from the panelists; Dr. Nelson Pulido, National Research Institute for Earth Science and Disaster Resilience (NIED), Tsukuba, Japan; Dr. Kojiro Suzuki, Director, Dept. Coastal Hydraulic Engineering, National Institute of Maritime, Port and Aviation Technology, Port and Airport Research Institute, Japan; and Dr. Kentaro Tabata, National Research Institute for Earth Science and Disaster Resilience (NIED), Tsukuba, Japan.

References

Aoi, S., et al.: MOWLAS: NIED observation network for earthquake, tsunami and volcano. Earth Planets Space **72**(126) (2020). https://doi.org/10.1186/s40623-020-01250-x

Elgamal, A., et al.: A review of large-scale testing facilities in geotechnical earthquake engineering. In: Earthquake Geotechnical Engineering: 4th International Conference on Earthquake Geotechnical Engineering-Invited Lectures, pp. 93–129. Springer, Netherlands (2007)

Hallonsten, O.: How expensive is Big Science? Consequences of using simple publication counts in performance assessment of large scientific facilities. Scientometrics **100**(2), 483–496 (2014). https://doi.org/10.1007/s11192-014-1249-z

Timofejevs, A., Avotins, V.: The large-scale research facilities–the impact of increased utilization rate. Space Rev. **5**, 66–97 (2018)

Report of the Discussion Session on Sustainable Disaster Risk Reduction: Integrating Climate Action, SDGs and Field DRR and Data (Experience) Sharing - New Challenges for Actions by GADRI

Tetsuya Takemi[1(✉)] and John van de Lindt[2]

[1] Disaster Prevention Research Institute (DPRI), Kyoto University, Kyoto, Japan
takemi@storm.dpri.kyoto-u.ac.jp

[2] Center for Risk-Based Community Resilience Planning, Colorado State University, Fort Collins, USA

Abstract. Water-related disasters from extreme weather phenomena are anticipated to be more destructive and disastrous under future global warming. To mitigate and prevent water and weather-related disasters we need to assess the impacts of global warming on extreme weather. This session discussed how the GADRI community influence or direct policy makers for achieving SDGs and climate change adaption. Field measurements for a sustained period of time is critically important to monitor environmental and climate changes. Data and/or experience sharing from field measurements were also discussed.

Keywords: Sustainable Disaster Risk Reduction · climate change · weather related disasters · global warming · Hydro-Meteorological Disasters

1 Purpose

Water-related disasters such as floodings, inundation, and landslides result from the impacts from extreme weather such as tropical cyclones, long-lasting precipitating systems, and rapidly developing thunderstorms. Under global warming in a future climate, such extreme weather phenomena are anticipated to intensify and hence to become more destructive and disastrous. Therefore, to mitigate and prevent water and weather-related disasters we need not only to understand the impacts of extreme weather under current meteorological conditions but also to quantitatively assess the impacts of changing climate on extreme weather (Mori and Takemi 2016; Takemi et al. 2016; Nakakita et al. 2018; Takemi et al. 2020; Mori et al. 2021; Nayak and Takemi 2022, 2023; Takemi 2023).

H. Tatano et al. (Eds.): GSRIDRR 2023, DRRGADRIBS, pp. 201–211, 2025.
https://doi.org/10.1007/978-981-96-6609-6_15

In the past GADRI meetings, we set up a session group discussing climate change impacts and adaptation in order to understand and share the experiences and knowledge on the ideas and methods from the studies on the scientific, engineering, and implementation aspects. The present discussion session group focuses on discussing sustainable disaster risk reduction through exchanges the experiences and knowledge on integrated climate action, SDGs, and field observations aimed at disaster risk reduction. In order to respond to or to contribute to the assessment reports published by IPCC, the GADRI institutes have been making significant efforts in assessing the impacts of climate change on natural hazards and in developing new concepts and ideas on the adaptation strategy against climate change.

The purpose of this group discussion session is to share the current status on the scientific and engineering studies on climate change impact assessment and climate change adaptation among the GADRI institutes. Climate change is a global concern for both the natural environment and human health/lives. In addition to conduct such studies, this discussion session shares experiences on the interaction and cooperation among academic sectors, practitioners, and policy makers, and the practitioners. Through sharing those issues, this session discusses ways of how the academic community can influence policy makers, national/local government officials, news media, etc. This session also discusses engagement mechanisms and research linkages required to explore such ways. Among the SDGs, it is important to recognize that climate change adaptation is listed as "Goal 13: Take urgent action to combat climate change and its impacts". It is also important to recognize that climate change adaptation requires accurate assessments of climate change on natural hazards, water resources, and eco-systems. Therefore, this session further discusses the adaptation strategy for climate change. Experience on field measurements among the participating institutes will be also shared to discuss the significance of data sharing.

2 Panel Presentations

Photo IIB_1: Chairs of Session: Prof. Tetsuya Takemi and Prof. John van de Lindt.

The following speakers were invited as panellists:

- Prof. Weiqiang Ma, Professor, Institute of Tibetan Plateau Research, Chinese Academy of Sciences, China,
- Dr. Fatima Akter, Associate Professor, Department of Meteorology, University of Dhaka, Bangladesh
- Prof. Mahua Mukherjee, Professor, Department of Architecture and Planning, and Joint Faculty and Ex-Head of Centre of Excellence in Disaster Mitigation and Management (CoEDMM), Indian Institute of Technology Roorkee, India
- Mr. Novvria Sagita, Badan Meteorologi, Klimatologi, dan Geofisika (BMKG), Indonesia and PhD Program, Graduate School of Science/Disaster Prevention Research Institute, Kyoto University, Japan

The abstracts of the presentations provided by these panelists are listed here.

Prof. Weiqiang Ma (Institute of Tibetan Plateau Research, Chinese Academy of Sciences)

"Introduction to the Observational System of Land-Air Interaction on the Tibetan Plateau and Disaster Related Work"

The environmental change of the Tibetan Plateau is closely related to the westerly monsoon process. The thermal and dynamic effects of the Tibetan Plateau and its atmospheric water cycle process have an important impact on the weather and climate of China and Asia, as well as the global atmospheric circulation. Based on this, our group has conducted field observations and experiments, satellite remote sensing and numerical model research on the Plateau atmospheric boundary layer process on various underlying surfaces of the plateau in the past 20 years. Our research achievements include: 1) Aiming at the complexity of the Plateau atmospheric boundary layer, we mainly participated in the design and establishment of the Plateau atmospheric boundary layer comprehensive observation network on different underlying surfaces; 2) It has broken through the bottleneck of the previous satellite remote sensing and simulation of the plateau surface heat flux initial field lack and verification difficulties, and its work has promoted the development of China's observation of the Plateau atmospheric boundary layer energy and water exchange process. The above data accumulation and analysis play a very important role in local disaster research (Ma et al. 2023a; Ma et al. 2023b; Ma et al. 2023c; Hu et al. 2023).

Photo 2: Presentation by Prof. Weiqiang Ma

Prof. Fatima Akter (Department of Meteorology, University of Dhaka)

"Frequency and Intensity of Extreme Weather Events in the Changing Climate Scenarios of Bangladesh"

Bangladesh, as a geographically sensitive country, is subjected to extreme weather events such as flash floods, monsoon floods, droughts, thunderstorms and tropical cyclone. The influence of extreme weather events on agriculture, environment, infrastructure, economics, and also life and livelihood, is the focus of this research. Climate change plays an important role in the frequency of extreme weather events and sometimes these are exaggerating disaster risk. Climate change impact cannot be prevented in a short duration; rather, it is a long process that can be improved by solutions developed at the local level by farmers and experts and policy makers. This research contributes new knowledge about the spatial distribution of different extreme weather events that can be very significant in terms of identifying the changing climatic condition throughout Bangladesh. The research can contribute to identify the connection between extreme climate events and agricultural production.

Photo 3: Presentation by Prof. Fatima Akter

Prof. Mahua Mukherjee (Department of Architecture and Planning, Indian Institute of Technology Roorkee)

"GeoSM-NatE-based Integrated Adaptation and Mitigation Strategies for Future Climate Scenarios"

Integrated approach to reach consultative alignments on SFDRR, SDG and Paris Climate Agreement demand innovative approaches like the Geo-Spatial Mapping of Nature-based Ecosystem (GeosM-NatE) tool. This uses the earth-observation data on GIS platform, to identify hotspots and provides opportunity to simulate scenario for adaptation and mitigation-based DRR. The flexibility of the tool allows adoption of future climate projections (RCP & SSP) to be integrated for assessment of urban risks like urban heat island, urban flood, GWD, etc.

Mr. Novvria Sagita (Badan Meteorologi, Klimatologi, dan Geofisika (BMKG), Indonesia and PhD Program, Graduate School of Science/Disaster Prevention Research Institute, Kyoto University)

"How Indonesia Response Hydro-Meteorological Disasters"

Indonesia is an island country in the tropic region that is threatened by many hydro-meteorological disasters. Hydro-Meteorological Disasters cause economic loss, infrastructure damage and death in Indonesia. There have been around 24,000 hydro-meteorological disasters from 2014 to 2022 (BNPB, 2023). Thunderstorm activity is one indicator of severe weather because this phenomenon is followed by heavy rainfall, strong wind, and hailstorm. The average thunderstorm event in Indonesia is around 100–200 days a year (Mori et al., 2014). This study will simulate the response of thunderstorms to variability phenomena such as El Nino Southern Oscillation (ENSO), Dipole Mode, MJO and Inter Tropical Convergence Zone (ITCZ). This condition challenges the Indonesian Agency for Meteorology, Climatology and Geophysics (BMKG) to decrease impact and mitigate the condition. BMKG have already used National Digital Forecast (NDF). It is a website system to integrate weather prediction in Indonesia. NDF help weather forecaster in Indonesia to make weather information product and disseminate it to society. Another support of BMKG is to build a community that is aware of mitigation and adaptation through field works and school programs for society.

3 Outcomes

Topics to be discussed are listed as follows.

- Share the recent research activities on climate changes assessment and adaptation from the GADRI institutes.
- Discuss what engagement mechanisms and research linkages are needed to influence research directions among various sectors.
- Experiences on the linkages among the research community, the policy makers, and the practitioners will be shared.
- Discuss adaptation and implementation strategies for climate change.
- Discuss the significance of data sharing from field measurements.
- Discuss how the GADRI community influence or direct policy makers for achieving SDGs and climate change adaption.

Key questions that are intended to be addressed in the session are as follows.

- How the GADRI community influence policy makers for achieving SDGs and climate change adaption?
- How can we make best use of existing data from field measurements and monitoring for climate change assessment?

Outcomes of the session shared by the Chair, Prof. Tetsuya Takemi:

The session shared current activities on studies related to climate change and their impact assessments from the GADRI member institutes.

- Highlight experiences on the linkages among the research community, the policy makers, and the practitioners.
- Try to understand of the importance of data and/or experiences from field measurements for a sustained period of time in order to monitor environmental and long-term climate change effects.

With the objective we shared current understanding on climate change assessment from scientific and engineering viewpoints. Further, promoted discussion adaptation strategy for climate change; as well as experiences on field measurements among the participating institutes. The session also discussed the significance of data sharing.

Some specific points shared by the panelists are summarized as follows.

Prof. Weiqiang Ma (Chinese Academy of Sciences)

- Shared experiences of field observations in the Tibetan Plateau region for long periods of time
- Emphasized the importance of observations especially, in data-sparse regions in order to monitor long-term climate change and the impacts of extreme weather
- Emphasized that generating long-term datasets requires continuous effort.

Prof. Fatima Akter (University of Dhaka)

- Shared the research progress on cyclone disasters in Bangladesh
- Demonstrated the impact of SSTs on cyclones, intensity and tracks, and how those cyclones are felt in the local scale hazards
- On Disaster management in Bangladesh
    - Should change our mind moving from reactive humanitarian relief to proactive disaster risk reduction
    - Emphasized the need for establishing early warning systems
    - Number of Institutional commitments: technical monitoring, capacity building, preparedness & responses
- Shared efforts by Bangladesh government on strategy on DRR
    - Monitoring/observation, forecasting, warning, information dissemination
    - As an example, she demonstrated the establishment of multi-purpose cyclone shelters, coastal greenbelt; and
    - Mass awareness and educational programs

Prof. Mahua Mukherjee (Indian Institute of Technology Roorkee)

- Emphasized interconnectedness of COP21 (Paris Agreement), Sendai FW, SDGs
- Demonstrated the Understanding of urban risk: urban heat island, air pollution
    - Demonstrated the importance of geospatial mapping of natural ecosystem
        - Identify hotspots at risk at local scales
        - Use of GIS platforms to identify such urban risks
- Emphasis on how to incorporate SFDRR at local and urban scales.
    - Understanding risk
    - Strengthening disaster risk governance
    - Investing in disaster risk reduction for resilience
    - Enhancing disaster preparedness for effective response

Mr. Novvria Sagita (BMKG/Disaster Prevention Research Institute)

- Demonstrated the weather forecasting in Indonesia on hydro-meteorological hazards such as thunderstorm, landslides, and floods
- BMKG early warning services provides:
    - multi-hazard early warning systems
    - high-resolution weather prediction
    - impact-based weather forecasting
    - special event weather forecasts for tourism, sports event, and other important functions
    - Such early warning information is sent no later than 30 min before a severe weather event
- Emphasized the importance of impact-based forecast – not only forecasting weather phenomena themselves, and how those extreme weather impacts the society
- Introduced the activity in building community awareness
    - Climate field schools, fishing weather schools, aviation weather schools
- Emphasized the importance of analysis of climate trend of risk score in order to identifying risk zones

In addition, we welcomed a presentation by an attendee of this session. The summary is here.

Prof. Vibha Vaishnav (Sardar Patel Univ)

- Introduced activities at Sardar Patel University on issues related to green and sustainable development
- Green practice at Sardar Patel University
    - Education, multidisciplinary approach, community awareness, sustainability initiative, eco-system restoration
- Shard experiences of Sardar Patel University establishing community science center

4 Outcomes and Recommendations Submitted by the Chairs of the Session to the Board of Directors of GADRI

After these presentations by the panellists and the attendee, we discussed how the GADRI community share our knowledges and experiences and collaborate with each other. The final outcomes and recommendations of this session are summarized as follows.

- The GADRI community should recognize the importance of long-term in-situ field observations and data accumulation for climate change monitoring in a changing climate especially for vulnerable regions. Sharing such observational data will be useful not only for climate change research purposes but also for regional collaborations and cooperation.
- The GADRI community should emphasize that, and should encourage practitioners and regular citizens to understand that, impact-based forecasting and future projection are important to build community to be aware of disaster mitigation and climate change adaptation.
- The GADRI community should emphasize to practitioners, policy makers, and regular citizens that future projections for hazard, vulnerability and exposure will help societies for sustainable risk financing.
- The GADRI community should recognize that alignments of targets of different frameworks (SDG, SFDRR, climate change) will help to address real world problems with same strategic interventions.
- The GADRI community should recognize the importance of multi-disciplinary approach; connection between environmental analysis or climate research and solid-earth analysis is an example for such a multi-disciplinary research topic.

Photo 4: General Q&A Session

Photo 5: Participants at the end of the session

References

BNPB: Data and Information on Disasters in Indonesia (DIBI) [Online]. National Disaster Management Agency (2019). Retrieved on March 01, 2023. https://dibi.bnpb.go.id/home/index2

Hu, W., Ma, W., Yang, Z., Ma, Y., Xie, Z.: Sensitivity analysis of the Noah-MP land surface model for soil hydrothermal simulations over the Tibetan Plateau. J. Adv. Modeling Earth Syst. **15**, e2022MS003136 (2023). https://doi.org/10.1029/2022MS003136

Ma, Y., et al.: Comprehensive study of energy and water exchange over the Tibetan Plateau: a review and perspective: from GAME/Tibet and CAMP/Tibet to TORP, TPEORP, and TPEITORP. Earth Sci. Rev. **237**, 104312 (2023). https://doi.org/10.1016/j.earscirev.2023.104312

Ma, Y., et al.: QOMS: a comprehensive observation station for climate change research on the top of the earth. Bull. Am. Meteor. Soc. **104**(3), E563–E584 (2023). https://doi.org/10.1175/BAMS-D-22-0084.1

Ma, Y., et al.: Earth summit mission 2022: Scientific expedition and research on Mt Qomolangma helps reveal the synergy between westerly winds and monsoon and the resulting climatic and environmental effects. Adv. Atmos. Sci. **40**(2), 187–193 (2023). https://doi.org/10.1007/s00376-022-2166-3

Mori, N., Takemi, T.: Impact assessment of coastal hazards due to future changes of tropical cyclones in the North Pacific Ocean. Weather Clim. Extremes **11**, 53–69 (2016). https://doi.org/10.1016/j.wace.2015.09.002

Mori, N., et al.: Recent nationwide climate change impact assessments of natural hazards in Japan and East Asia. Weather Clim. Extremes **32**, 100309 (2021). https://doi.org/10.1016/j.wace.2021.100309

Mori, S., Arbain, A., Syamsudin, F., Yamashita, K., Takahashi, Y., Yamanaka, M.: Lightning climatology around Jakarta, Indonesia, based on 13-years SYNOP observation and GSMaP rainfall data. Japan Geoscience Union (2014)

Nakakita, E., Tachikawa, Y., Takemi, T., Mori, N., Tanaka, K.: Future changes of extreme weather and natural disasters due to climate change in Japan and Southeast Asia. In: Bridging Science and Policy Implication for Managing Climate Extremes, Hong-Sang Jung and Bin Wang, Ed., World Scientific Publishing Co. Pte Ltd., pp. 101–117 (2018)

Nayak, S., Takemi, T.: Assessing the impact of climate change on temperature and precipitation over India. In: Wadi Flash Floods - Challenges and Advanced Approaches for Disaster Risk Reduction and Water Harvesting in Arid and Semi-Arid Regions, Tetsuya Sumi, Sameh A. Kantoush and Mohamed Saber, Ed., Natural Disaster Science and Mitigation Engineering: DPRI reports, pp. 121–142. Springer, Singapore (2022). https://doi.org/10.1007/978-981-16-2904-4_4

Nayak, S., Takemi, T.: Chapter 9 -Climate change scenario over Japan: trends and impacts. In: Khare, N. (ed) The Role of Tropics in Climate Change, Elsevier, pp. 161–185, 526 (2023). ISBN 978–0–323–99519–1. https://doi.org/10.1016/B978-0-323-99519-1.00005-3

Takemi, T.: Analysis and predictability of mesoscale precipitating systems in moist environments, pp. 331–364. In: Park, S.K. (eds) Numerical Weather Prediction: East Asian Perspectives. Springer Atmospheric Sciences, p. 581. Springer, Cham (2023). https://doi.org/10.1007/978-3-031-40567-9_14, Hardcover ISBN 978–3–031–40566–2, eBook ISBN 978–3–031–40567–9

Takemi, T., Okada, Y., Ito, R., Ishikawa, H., Nakakita, E.: Assessing the impacts of global warming on meteorological hazards and risks in Japan: Philosophy and achievements of the SOUSEI program. Hydrol. Res. Lett. **10**, 119–125 (2016). https://doi.org/10.3178/hrl.10.119

Takemi, T., Yoshida, T., Horiguchi, M., Vanderbauwhede, W.: Large-eddy-simulation analysis of airflows and strong wind hazards in urban areas. Urban Clim. **32**, 100625 (2020). https://doi.org/10.1016/j.uclim.2020.100625

Gender and Inclusivity in DRR Policy and Practice

Punam Yadav[1], Ana Maria Cruz[2], Shigeo Tatsuki[3], Miwako Kitamura[4], and Peter Sammonds[1](✉)

[1] UCL Institute for Risk and Disaster Reduction, London, UK
{p.yadav,p.sammonds}@ucl.ac.uk
[2] Disaster Prevention Research Institute (DPRI), Kyoto University, Kyoto, Japan
cruznaranjo.anamaria.2u@kyoto-u.ac.jp
[3] Department of Sociology Department, Doshisha University, Kyoto, Japan
statsuki@mail.doshisha.ac.jp
[4] IRIDeS, Tohoku University, Sendai, Japan
miwako.kitamura.b8@tohoku.ac.jp

Abstract. Disasters are not natural. They are socially constructed. In disasters, women and girls are impacted disproportionately due to discriminatory gender norms and power relations. In disaster risk reduction (DRR) policy and practice, the LGBTQ+ community have been much overlooked. The presence of these pre-existing inequalities in society, evident in daily life even before disaster events, necessitates a tailored approach to support the shift from a traditional disaster risk reduction (DRR) perspective that prioritized equality—treating everyone the same—towards recognizing and addressing the pre-existing societal inequalities. This paper discusses disaster risk reduction in terms of gender, disability, gender and sexual minorities, and culture, taking an intersectional approach. Recommendations are made for GADRI resolutions to take forward to the UNDRR Sendai Framework Mid-term Review.

Keywords: Gender · intersectionality · LGBTQ+ · disability · inclusivity · DRR policy and practice

1 Introduction

The 2022 UN DRR (United Nations Disaster Risk Reduction) Global Platform for Disaster Risk Reduction in Bali was criticized for not providing a space where pervasive issues of gender inequality were engaged with seriously and addressed by all parties and participants (Bradshaw et al. (2022). Gender inequality and discrimination against marginalised groups needs to be addressed with urgency if DRR policy and practice is to meet the needs of all. But how should this be done? Which voices should be heard? Whose interests should be considered? In a session, which aimed to put gender back

H. Tatano et al. (Eds.): GSRIDRR 2023, DRRGADRIBS, pp. 212–226, 2025.
https://doi.org/10.1007/978-981-96-6609-6_16

onto the agenda, an expert panel of Punam Yadav, Ana Maria Cruz, Shigeo Tatsuki and Miwako Kitamura, chaired by Peter Sammonds, discussed and put forward policy changes on Gender and Inclusivity in DRR Policy and Practice. Ideas from this panel were taken forward in plenary at the 2023 Global Summit of GADRI (Global Alliance for Disaster Research Institutes). These formed the basis for recommendations to UN DRR for the mid-term review of the Sendai Framework for Disaster Risk Reduction (2015–2030).

Ch17-1: Session Chairs, (L) Prof. Punam Yadav and Prof. Peter Sammonds, UCL Institute for Risk and Disaster Reduction, London UK

Ch17-2: Panelists: Dr. Miwako Kitamura, Prof. Ana Maria Cruz, and Prof. Shigeo Tatsuki

GADRI aims to serve to serve as an advocacy organization to speak with one voice in an effort to influence decision-making processes. The purpose of this session was to provide a space where these difficult issues of tackling inequalities in DRR policy and practice could be discussed. As Reliefweb notes (reliefweb.int/report/world/gender-and-disasters-part-i) disasters do not discriminate, but their impact does. Disproportionate impacts occur due to gender inequality. While women and girls are constructed a vulnerable group, the power inequalities embedded in this categorization are rarely addressed in DRR. It is also important to understand that gender does not mean just women. Gender is a social structure that impacts men, women and gender minorities differently (Connell, 1985). People of all genders, classes, ages, ethnicities, abilities and sexual orientation have several identities (Blanchard et al, 2023). Intersectionality is a framework for understanding how people's identities combine to create discrimination and privilege (Crenshaw, 1991). Adopting an intersectional approach ensures that any DRR policy and initiative is mindful of whose voices are heard, whose needs, experiences and interests are addressed, why and how' (www.grripp.net/post/gender-in-drr-mainstreamed-into-invisibility). Tackling structural inequalities (e.g., social, institutional, legal and cultural inequalities) should be a priority for implementing the Sendai Framework.[1]

Key questions to be addressed by the panel included: 1) Gender in DRR – why does it matter? What is the real problem? 2) What should be measured, and which data collected to support an intersectional approach? 3) What are the key opportunities for change?

[1] https://www.undrr.org/implementing-sendai-framework/what-sendai-framework.

Following the session, participants were urged to engage their home disaster research institutes on the need to listen, research and lead on the implementation of the Sendai Framework for all.

2 Gender and Disaster by Punam Yadav

Disaster is not natural. It is socially constructed. This has now been well established in disaster studies. Behind every disaster, there are humans. While this may sound like a common sense, I wanted to bring the attention once again to the 'human element' to examine gender and inclusivity in the current DRR policies and practices.

Not all humans are equal: we experience disaster differently due to structural inequalities and discriminatory cultural practices. Women and girls are impacted disproportionately in disasters due to discriminatory gender norms and power relations (Yadav et al., 2021). Men are also affected due to gendered social norms and cultural practices. Certain forms of masculinity, enforced through the social structure, create certain gendered expectations that are exacerbated at times of disasters. However, gendered impacts of disaster on men are often overlooked, leading to limited research on the subject. Furthermore, women are not a homogeneous category. Despite often being grouped together, their experiences are diverse, influenced by factors such as age, class, caste, ethnicity, education, and economic background. Therefore, their vulnerability in disasters is dependent on their personal and social circumstances.

It is also interesting to observe how women are conceptualized within disaster policy frameworks. They are often viewed as vulnerable groups and are lumped together with children and the elderly, despite their diverse experiences and unique needs.

Another important point to consider while studying disasters or working on disaster is to examine the impacts of disasters as relational. We are related to one another through certain power relations, and these power relations are gendered. Hence, even the impacts of disasters are relational. This is different from the cascading effects of disaster. When we say the impacts are relational, we are not just looking at how one impact leads to another, but we go beyond that and look into how pre-disaster conditions impact people's experiences in disaster and post-disaster contexts. For instance, if a woman has been a victim of gender-based violence, how has her experience been in the post-disaster period? Likewise, if government support for women pre-disaster was poor, then how has that impacted her post-disaster experience? Additionally, if a woman was a victim of sexual violence post-disaster in a male-dominated patriarchal society where family honor carries more weight than the life of a woman, what happens to them afterwards? These relationalities of impacts and the continuity of their experiences are important to consider. In this short presentation, we will not be able to discuss in detail the relationality of impacts, but we will try to question all the assumptions that we work with in the disaster context and see what that means when we put gender at the heart of our analysis. Hence, let's look at some questions.

The first set of questions is around the concepts and definitions that we use in DRR policies and practices. When we say inclusive DRR, what do we mean by that? Do we mean including women in DRR policies and practices, or do we mean going beyond the gender binary and including gender minorities in our conversations? When thinking about vulnerability, what kind of image do we have in our minds? Who do we think of when we think of vulnerability? We generally consider women as a vulnerable category, but when it comes to providing support, we narrow it down to certain categories of women, such as pregnant, disabled, elderly women, and we do not include other women in the conversations, even though they could also be vulnerable in disasters due to various personal and social conditions. Vulnerability is seen as static/fixed in this conceptualization of women as vulnerable groups. People's experiences are different due to the multiple identities each one carries. Our experiences are dynamic and change over time - they're not static. Likewise, vulnerability is not static. I can be very resilient now, but in ten seconds or ten minutes, I might be vulnerable to some form of disaster. Therefore, we cannot level one category over others as vulnerable and non-vulnerable. We need to think of vulnerability as a dynamic process. The same assumptions can be found in the conceptualization of other terminologies in disasters, such as resilience. What do we mean by resilience? Who are resilient, and can one always be resilient while the other vulnerable? Can this change? Can their experience change? Similarly, building back better - what do we mean by building back better? Do we want to go back to the exact same situation pre-disaster, and what do we want to build back better - social structure or physical infrastructure? Another keyword is wellbeing. What do we mean by wellbeing? Are we securitizing wellbeing? In what sense has this been defined within disaster discourse? Likewise, what do we mean by risk? And what do we mean when we say capacity building or capacity to rebuild? And finally, what does a resilient society look like? How do we imagine the future of a resilient society? These are some of the questions that we need to ask even at the conceptual level.

The other set of questions is related to the institutional structure. We have discussed quite a lot in the last two days at GADRI about the siloed (e.g. disciplinary boundaries) approach. We are moving beyond a single hazard approach to a multi-hazard approach and trying to address the siloed approach. However, do we include conflict in it? Even though there are links between conflict and disaster, often in our conversations, conflicts are left outside of DRR policies and frameworks not only in academic work, but also in policy frameworks. If you look at the government, there are separate ministries for conflict and disaster. And if you look at the funding stream, there are separate streams of funding for conflict and disaster. Likewise, there are disciplinary boundaries between disaster and conflict. We don't cross the boundaries.

The final set of questions is around methodology. How do we study disaster risk reduction? Do we use qualitative methods? Do we use quantitative methods? What's our dominant methodological approach, positivist, post-positivist approach? Also, are we putting efforts into decolonizing knowledge? Do we really question where the knowledge is coming from and where do we situate ourselves in our own work? We talk about gender-disaggregated data, but what kind of data are we disaggregating? Although we call it gender-disaggregated data, it's often sex-disaggregated data, not gender-disaggregated data. Finally, disasters are often viewed as consensus events, which emphasize community cohesion and mutual aid. Yet, the impacts of gender relations are ignored in DRR studies, policies, and interventions. Although gender mainstreaming seems to be considered for DRR interventions, gender is often reduced to binary categories, excluding gender minority groups and intersectional aspects of human experiences. In conclusion, disasters are often viewed as consensus events, with much emphasis on community cohesion and mutual aid. Yet, the impacts of gender relations are ignored during times of crisis (Yadav & Fordham, 2022). Gender mainstreaming is an important part of DRR interventions, yet gender is reduced to binary categories, recognizing only the special needs of special categories of women (disabled, pregnant, and elderly), leaving the vast majority of women with no support e.g., during the first COVID-19 lockdown, although working women faced a huge burden of work in both the home and the workplace, there was no support available to them. Disasters are often seen as amenable to a technical fix. During COVID-19, we saw rules coming with immediate effect such as the 'stay home, stay safe' policy. The question for us is, was home safe for everyone? Also, women are seen as largely victims in disaster discourse. Even when their agency is recognized, it is looked at through a victim lens. For instance, the discourse on women's agency in disaster has not got much further than "they need to be included, 'promoted', 'empowered' by someone else". Lastly, despite ongoing efforts, disaster as a field remains very much male-dominated.

3 Pre-existing Inequality, Inclusivity and Equity by Shigeo Tatsuki

Pre-existing inequality, inclusivity and equity within society are three interconnected themes.

My analysis is grounded in a repeated-ANOVA study of the Great East Japan Earthquake (GEJE) Natori longitudinal study conducted in the city of Natori, located south of Sendai City. A key metric I have utilised since 1995 is the sense of life recovery score, which serves as the vertical axis in my analysis. The findings indicate that individuals experiencing stagnation in their recovery process were those with severe mental or physical health issues, unemployment immediately following the disaster, and those living in single elderly households (Fig. 1). While seemingly intuitive, this evidence underscores that the most significant struggles were faced by those already encountering challenges in their pre-disaster daily lives.

Secondly, I explored housing recovery rates using objective data on the length of stay in temporary housing. Contrary to the subjective sense of life recovery, the analysis of temporary housing duration inversely correlates with housing recovery speed, highlighting disparities based on household size (Fig. 2). Families with more members tended to recover more swiftly, whereas single-person households experienced more stagnation, particularly in Natori city.

This pattern was also observed in Sendai City, with data available through our open data source.[2] Moreover, when examining recovery by the gender of the household head in both Natori and Sendai City, it was evident that households headed by women experienced slower recovery. This finding aligns with broader research indicating that pre-existing inequalities—such as age, family size, health status, disability, socioeconomic class, gender, poverty level, and immigration status—are significant determinants of recovery speed post-disaster. These variables can predict which groups are likely to recover the slowest.

[2] https://www.city.sendai.jp/saiken-kikaku/shise/daishinsai/fukko/sekatsu/kasetsu_opendata.html.

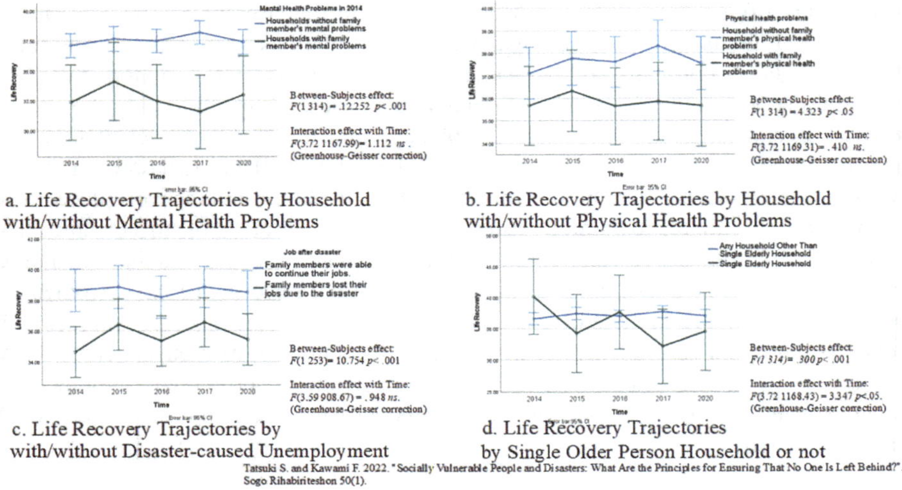

Fig. 1. Data showing recovery trajectories for different vulnerable groups

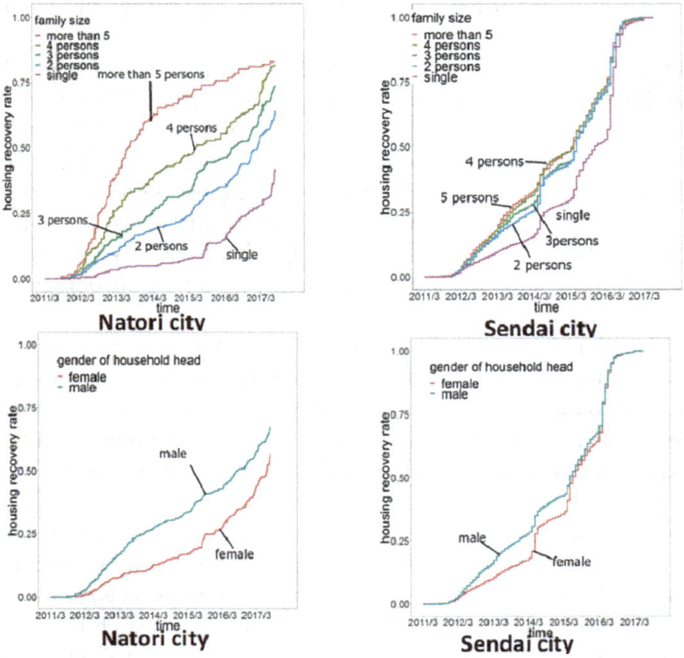

Fig. 2. Housing recovery rate for two cities looking at family size and gender-headed households.

In addressing equity, we observe distinct variations in recovery rates within Natori, particularly concerning household size and the gender of the household head. Notably, the disparity is much less pronounced in Sendai City. Figure 2 shows Natori exhibits a significant gender gap, whereas Sendai records a narrower one. The reason behind this difference is Sendai City's implementation of equitable measures, specifically disaster case management. This analysis incorporates data from 14 municipalities, enabling an estimation of the hypothetical scenario had Sendai not adopted disaster case management. The actual data show that the speed of transitioning to temporary housing in Sendai was considerably faster from April 2015 onwards, coinciding precisely with the city's introduction of a systematic approach aimed at supporting the most vulnerable populations, including households headed by women and single, elderly families. Each of these identified vulnerable clients was allocated a disaster case manager, who played a crucial role in facilitating access to necessary infrastructure, policies, and resources. By offering such reasonable accommodations, the process of housing recovery in Sendai, specifically the move to permanent housing, was significantly expedited. This culminates in the final point: the shift from a traditional disaster risk reduction (DRR) perspective that prioritized equality—treating everyone the same—towards recognizing and addressing the pre-existing societal inequalities.

The Sendai experience underscores the importance of providing reasonable accommodations to achieve true equity. It is through such measures that equity becomes a means to attain equality, showcasing the pivotal role it plays in disaster recovery efforts. The implementation of reasonable accommodations significantly accelerated the housing recovery process in Sendai, demonstrating a more efficient transition to temporary housing. This approach represents a departure from traditional disaster risk reduction (DRR) perspectives, which have historically emphasized equality, advocating for uniform treatment of all individuals. However, the presence of pre-existing inequalities in society, evident in daily life even before disaster events, necessitates a tailored approach to support. Sendai's experience highlights the importance of equity, illustrating that equitable measures are essential for addressing the varied needs of affected populations.

4 Culture, Inequality and Disasters by Ana Maria Cruz

I'm not a gender specialist, but I am a woman. And I am a woman working in Japan since 2014. At the beginning of 2023, I was the only tenured female professor in DPRI at Kyoto University. But things are changing. And so, I was thinking about my role as a woman in work, but also about the research that I have done with one of my Ph.D. students, a woman from Afghanistan. Why does it matter? I think first, because if you do not address gender bias, then sometimes you do not even know that gender bias exists. That in many ways is a problem in many countries, including Japan, which has the lowest proportion (16.6%) of female researchers in science, technology, engineering, and mathematics (STEM), among all OECD[3] countries (Kakuchi 2023). So, we need to ensure that policies for disaster risk reduction are specifically addressing potential gender biases.

Let's take a look at this research example of my student in Afghanistan. She wanted to understand the vulnerability of women to disasters. I thought it must be part of culture

[3] Organization for Economic Cooperation and Development.

and religion but she was convinced that it was not. She conducted focus groups in remote areas of Afghanistan, where it is not common for women to travel alone. With the help of a non-profit organization, and a kind taxi driver that waited for her, she carried out a series of focus groups with women (only), men (only), NGOs and government officials at disaster risk management organizations. The study identified some of the problems including for example the fact that at the time of the study, disaster education programs were being provided only to men (Hamidzada et al. 2019). This was in part due to the fact that there were no female staff available to deliver them; also, the lack of female staff among aid workers resulted in a lack of medical services for women. This is because male doctors cannot see women due to cultural imperatives and the interpretation of religion in some rural areas.

During the study, some interventions by NGOs provided educational programs that specifically had female staff for women in small villages in Afghanistan. We found that the disaster education programs helped improve disaster preparedness and positively impacted other areas related to women's daily lives. For example, following the educational program, women initiated their own early warning system to communicate via cell phones between different villages to warn of possible flooding. Furthermore, the study showed that the educational program also empowered women to carry out joint activities such as the preparation of food or other things to sell, and so on, or to support women-only households (Hamidzada *et al.* 2019; Hamidazada and Cruz 2023). The above study points to ways in which addressing gender inequities can lead to overall improvements in other areas of people's lives despite religious and cultural constraints.

The Ph.D. student herself is a good example of how women academics can rise to the imposed challenges and limitations, even in countries that are considered so closed off and difficult to investigate. Coming back to the case of Japan, where gender biases are deeply embedded in culture, I believe that explicit policies targeting gender biases are needed. As it is very challenging to change these stereotypes and introduce changes.

5 Gender and Sexual Minorities and Disasters by Miwako Kitamura

I am going to bring attention to an area that appears to be much overlooked: the LGBTQ+ sexual minority community and using Japan as an example. This group has received scant attention, mainly due to the extreme fear and discrimination they face. Within the LGBTQ+ community, particularly transgender individuals and those living with HIV, there are unique medical needs. This discussion aims to shed light on these needs. Even before disasters strike, these groups are placed in exceedingly difficult situations. However, LGBTQ+ individuals encounter various complex barriers when accessing medical care, including issues related to stigma, economic resources, and connected mental health problems. Mental health among the LGBTQ+ population tends to be worse than that of non-LGBTQ+ individuals in Japan. According to a random consumption survey conducted in Osaka City in 2019, 6.9% of transgender people reported experiencing significant mental health issues, the highest among the groups surveyed. This study demonstrates the suffering of LGBTQ+ individuals even before disasters. Thus, it raises the question: How are LGBTQ+ individuals affected by disasters? Although numerous studies have been conducted globally, research in Japan remains exceedingly limited. In this discussion, I aim to explore three main research questions: How do LGBTQ+ individuals perceive their disaster experiences differently, and what unique needs do they have? To what extent does the government understand the needs of LGBTQ+ individuals in disasters, and what policies are in place to meet these needs? Lastly, what unique capacities and knowledge do LGBTQ+ individuals and their communities possess?

We aim to underscore the disparity between the needs of LGBTQ+ individuals and the government's understanding of these needs. Our methodology involved conducting interviews, ranging from 15 min to nearly 2 h, with participants aged between their twenties and sixties. The interviews covered various topics, including sexual and gender identity, community disaster experiences, personal preparedness, and medical support requests to the government. A significant finding was that LGBTQ+ individuals often feel unable to seek help due to uncertainty about whether their identity will be accepted, for fear of being outed. The concern can pose life-threatening risks, particularly as some LGBTQ+ individuals opt not to use evacuation centers to avoid the risk of being exposed. This preliminary conclusion indicates that LGBTQ+ individuals encounter specific and unique challenges at all stages of the disaster cycle, including preparation, response, and recovery. Special consideration for LGBTQ+ people is crucial, yet their needs are often perceived as secondary or selfish. Some participants reported uncomfortable, negative, and mixed experiences with researchers and support workers. However, respondents who felt more comfortable advocating for their needs experienced less discrimination in their everyday lives. Groups that are openly expressive about their gender identity are more inclined to vocalise their specific needs during disasters. Conversely, groups defined by sexual orientation, who regularly face discrimination, find it challenging to disclose their privations. This predicament is particularly prevalent in rural areas perceived as small and conservative. This discrepancy highlights a reported lack of community support among LGBTQ+ individuals in Japan.

This research is ongoing, and there remain many unanswered questions. Our initial findings underscore the urgent need for further investigation. Our research has identified a significant gap between the needs of LGBTQ+ individuals and the level of understanding by governments. While local governments are willing to enact policies that support LGBTQ+ communities, there is notable uncertainty regarding the most effective actions to take. Communication between departments is needed to ensure the development and implementation of these policies, suggesting a systemic issue within governmental structures. This indicates a potential foundation on which more targeted and effective policies could be built, provided there is greater awareness and intentionality. The barriers to action most frequently cited include a scarcity of time and resources, perceived cultural and demographic challenges such as the 'backwardness' of rural areas, an older population, and narratives framing LGBTQ+ needs as 'selfish'. These factors contribute to the complexity of addressing LGBTQ+ needs in disaster preparedness and response efforts.

6 Discussion Points

During the active discussion session, several questions were raised by the discussants, and some of the responses are presented below.

Ch17 -3: Discussion Session

Discussant 1:
How can we tie the insights on how women are fighting in disasters into the Sendai Framework?

Response by Punam Yadav:
After two decades of gender mainstreaming, we know that the gender mainstreaming approach has not been very helpful in addressing gender-specific needs. Hence, we are proposing that gender has to be the starting point for any policy or intervention. Gender is relational. We are all related to one another through gender relations. In the Sendai Framework, women are portrayed as victims, not as agents of change. That has consequences for how we design an intervention. Likewise, women are seen as a homogeneous category. There is diversity among women. There are women leaders out there. We cannot put them in one basket and call them victims or vulnerable. Therefore, we need to look at the portrayal of women within the policy frameworks and then propose something that captures the diversity among women.

Discussant 2:
We should take care of our vulnerable people because they are the most impacted people. They may have low income; they may live at low elevation more often. So we do want to take care of them. But how can we make them un-vulnerable?

Response by Shigeo Tatsuki:

A primary factor exacerbating the plight of those who suffer post-disaster is the issue of siloization. Typically, these individuals rely on social services during normal times. However, once a disaster occurs, responsibility shifts to a different department focused on recovery and assistance. This division between standard operations and disaster response creates a fundamental barrier to effective support for people with disabilities. A stark reality is that those who endure the most hardship post-disaster are often those already facing challenges, such as illness or poverty. Recognizing this, professionals like social workers and case managers, who provide regular social services, should incorporate post-disaster recovery planning into their case management strategies, ensuring continuity of assistance across all phases.

Discussant 3:
How about we educate the general public to relook at LGBTQ+ people to bias?

Response by Miwako Kitamura:
Unlike disabled individuals who can openly declare their status and receive governmental assistance, sexual minority individuals may be reluctant due to the fear of facing extreme discrimination. This reluctance is especially pronounced in disaster scenarios. For example, it's improbable to expect sexual minority individuals to identify themselves in social care settings or evacuation centres for fear of discrimination. This presents a significant challenge in locating participants for interviews, necessitating the establishment of a trustful relationship with organizations that can provide a secure environment for these interviews to take place. This process is inevitably time-consuming. However, initiating change through educational interventions from an early age can foster a society that values equality, where every individual has the opportunity to express themselves freely. Although it's a long-term strategy, we are committed to making gradual progress. We envisage a society where equality prevails to such an extent that the challenges currently faced by sexual minorities in disaster situations become non-existent. This is the future we hope to achieve.

Discussant 4:
For disaster risk reduction as a field of study, there are disparities particularly in the domination of research from the Global North. Seems like it would really sort of color that knowledge generation from our field. Are there methodologies that promote a course correction for that knowledge generation process?

Response by Punam Yadav:
Many young scholars have engaged with the question of decolonisation, and they would like to do something about it. But how do we do it? Nobody talks about it in a holistic manner. Hence, I would suggest, for now, being aware of your own positionality, stating who you are and how you position yourself within the context of your research, not claiming to be the global leader, but presenting what you found within the context of your research, highlighting your limitations, and expanding your reading list to include scholarship from the global south.

Discussant 5:
How do you measure intersectionality, because you cannot add oppression, you cannot multiply oppression? The second aspect is more like the LGBTQ+ research. So I'm also doing research, also doing looking at gender, sexual minority inclusion within disaster.

And some of the things that emerged from that research was actually a bit contrary to the common narrative that we see from gender and sexual minorities. A lot of the things they were talking about were like, well, 'my identity is not only my sexuality, so I use my other identities to be part of those governance processes'. So it was more like a comment. But that was quite interesting because that it shows the importance having that localized contextual knowledge to see the discrepancies within different geographies.

Response by Punam Yadav:
Some may find intersectionality a very useful approach for certain contexts, while others may not find it as useful. We don't have to use it all the time; we only use it when it is important and useful to us. For instance, in the 2015 Nepal Earthquake, more women died than men. My question then would be, who were they? What kind of women were they? Were they working women, housewives, pregnant women, disabled women, poor women, rich women? What religion did they belong to, and so on? That is where the concept of intersectionality becomes useful to understand who actually were the victims of the 2015 Gorkha earthquake.

Response by Shigeo Tatsuki:
I draw upon my insights from studying intersectionality. My learning journey has led me to a profound realization regarding societal roles in discrimination. Initially, I categorized people into three groups: those who discriminate, those who do not discriminate, and a hypothetical neutral group. However, my studies have illuminated that by not actively combating discrimination—simply claiming non-discrimination—we inadvertently support the discriminators. This understanding underscores that the concepts of intersectionality, disability, and inequality are intrinsically linked; they represent two sides of the same coin. Therefore, passive non-discrimination is not a viable stance. This revelation has been a driving force behind my engagement in disability and inequality studies, advocating for active measures against discrimination.

Discussant 6:
Culture is very fluid. Culture changes extensively, especially based on age. Many young Japanese people are actively against these old structures, especially, the Japan salary men culture, this idea of working over hours, the loyalty to the workplace, and that is from both men and women. The newer generation which has been exposed to international influences are more open to gender, sexuality, exploration of minorities to discuss gender roles. The generational and the gender discourse are closely connected and they are related to the intersectionality.

Response by Miwako Kitamura:
Perceptions of the United States as more inclusive of sexual minorities may not fully align with reality. The experiences of individuals seeking hormone therapy or other gender-affirming treatments are strikingly similar in both the United States and Japan, with many facing the fear of discrimination. This highlights a global need for a deeper understanding of gender medical care, suggesting that this field requires further study. Understanding the nuances of these experiences worldwide could inform strategies to mitigate discrimination, both in everyday life and during disasters.

7 GADRI Draft Resolutions

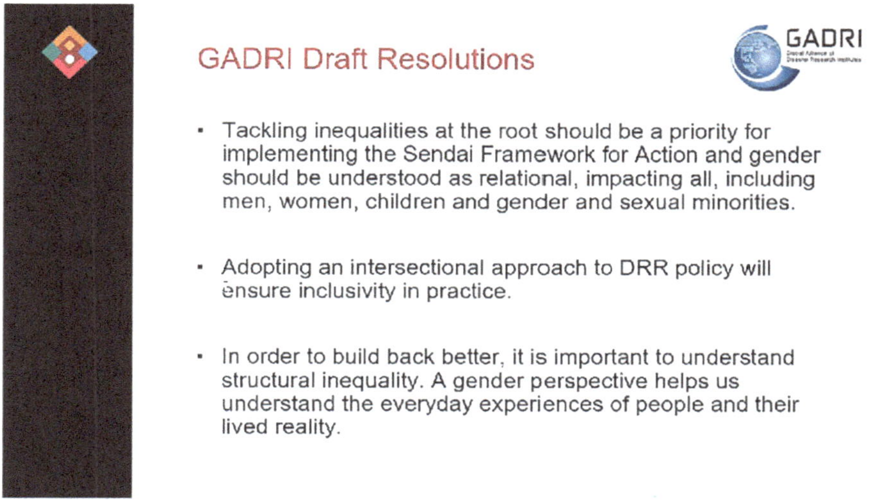

Fig. 3. Original GADRI resolutions to be discussed.

Comment from Peter Sammonds:
I think this is probably the only space we have had a talk about gender at the GADRI Summit. GADRI does not have any policy on its gender and inclusivity, which we would like to include these in the resolutions from the GADRI Summit. At the plenary session we will discuss these resolutions. But also, we wish to feed into the GADRI response to the UNDRR Midterm Review of the Sendai Framework (Fig. 3).

Comment from Punam Yadav:
I would like to emphasize that although gender disaggregation has been included, it is actually sex disaggregated data that the draft Mid-Term Review talks about. Disaggregation of data by age and disability a step forward. However, within the disability data, gender disaggregation is not looked at. Likewise, conflict has been included without being specific on the nature and types of conflict. Women have again been portrayed within the framework of victims, which defeats everything that we discuss today.

Comment from Shigeo Tatsuki:
I would like to include 'adopting an intersection approach to their policy will ensure inclusivity and equity in practice'.

Acknowledgements. Dr Zahra Khan, UCL IRDR, assisted with the organization of the panel discussion and editing the transcript of presentations. The UKRI Collective Fund award 'Gender Responsive Resilience and Intersectionality in Policy and Practice (GRRIPP) - Networking Plus Partnering for Resilience' was funded by the Global Challenges Research Fund (AH/R005370/1).

The manuscript was much improved by reviews of Irene Petraroli and Ilan Kelman, The authors gratefully acknowledge this support.

References

Blanchard, K., Chuck, E., Fordham, M., Khan, Z., Roberts, J., Walmsley, O.: Gender and Sexual Minorities in Disaster Risk Reduction: A Reference Guide. GRRIPP. Programme. IRDR Centre for Gender and Disaster, University College London, London (2023)

Bradshaw, S., et al.: Gender in DRR - Mainstreamed into Invisibility (2022). www.preventionweb.net/news/gender-drr-mainstreamed-invisibility

Gender and Disasters (Part I): reliefweb.int/report/world/gender-and-disasters-part-i (2020)

Connell, R.W.: Theorising gender. Sociology **19**(2), 260–272 (1985). https://doi.org/10.1177/0038038585019002008

Crenshaw, K.: Mapping the margins: intersectionality, identity politics, and violence against women of color. Stanford Law Rev. **43**(6), 1241–1299 (1991). https://doi.org/10.2307/1229039

Hamidazada, M., Cruz, A.M., Yokomatsu, M.: Vulnerability factors of afghan rural women to disasters. Int. J. Dis. Risk Sci. **10**(4), 573–590 (2019). https://doi.org/10.1007/s13753-019-00227-z

Hamidazada, M., Cruz, A.M.: Disaster Vulnerability, displacement and resilient livelihoods among the afghan urban women. In: Rezaul Islam, M. (ed.) Disaster, Displacement and Resilient Livelihoods: Perspectives from South Asia, pp. 55–80. Emerald Publishing Limited (2023). https://doi.org/10.1108/978-1-80455-448-720231004

Kakuchi, S.: Quotas for women in universities: Once step towards equity. University World News, Japan 17 June 2023. https://www.universityworldnews.com/post.php?story=20230615151132720. Accessed 29 Mar 2024

Kelmanm, I.: Do Disasters Kill More Women Than Men? Disaster death rates may result from societal roles (2020). www.psychologytoday.com/gb/blog/disaster-choice/202012/do-disasters-kill-more-women-men

Yadav, P., Fordham, M.: Why aren't we talking to each other?: Thinking gender, conflict and disaster as a continuum. In: Smith, S., Yoshida, K. (eds.) Feminist Conversations on Peace, pp. 168–180. Bristol University Press (2022). https://doi.org/10.2307/j.ctv2rr3hcn.16

Yadav, P., Saville, N., Arjyal, A., Baral, S., Kostkova, P., Fordham, M.: A feminist vision for transformative change to disaster risk reduction policies and practices. Int. J. Dis. Risk Reduct. **54**, 102026 (2021). https://doi.org/10.1016/j.ijdrr.2020.102026

Putting Health into Disaster Risk Reduction (DRR) and Recovery

Virginia Murray[1](✉), Andrew Collins[2], and Ryoma Kayano[3]

[1] Global Disaster Risk Reduction, UK Health Security Agency, Bristol, UK
virginia.murray@ukhsa.gov.uk
[2] DDN, Northumbria University, Newcastle-Upon-Tyne, UK
[3] WHO Kobe Office, Kobe, Japan

Abstract. The Sendai Framework for Disaster Risk Reduction marked a step change in the way health is recognized as core to managing disasters effectively. The recent COVID 19 global emergency further sensitized governments, societies and the research community of the need to build prevention into the recovery process. The World Health Organization (WHO) published the Health and Disaster Risk Management Framework in 2019. This is part of an approach that includes multiple aspects of the now established health and disasters agenda.

From this the WHO Health Emergency and Disaster Risk Management Research Network, launched in 2018, has developed opportunities to drive progress in its increasingly people-centered and impact orientated research agenda.

Keywords: Health into disaster risk reduction; preparedness · response and recovery for future health and well-being

The session explored action on preparedness, response and recovery for future health and well-being, cognizant that health and wellbeing within disaster research is pivotal to multiple, if not all, of the objectives aimed at achieving a sustainable disaster-resilient world (Fig. 1).

This session aimed to advance the agenda of putting health into disaster risk reduction and recovery through short interventions and a facilitated discussion.

Speakers of the panel session were:

- Dr. Ryoma Kayano, WHO Kobe
- Dr. Genta Nakano, DPRI, Kyoto University
- Prof. Virginia Murray, UKHSA

1 Presentation of the Outcomes of the Session by Prof. Virginia Murray

Prof. Virginia Murray expressed her delight to see a session on Health included in the programme of the 6th Global Summit of GADRI. Referring to the pre-conference survey undertaken by GADRI, Prof. Murray stated that there were a very few GADRI institutions who are working on health disaster related activities. Having attended quite

H. Tatano et al. (Eds.): GSRIDRR 2023, DRRGADRIBS, pp. 227–231, 2025.
https://doi.org/10.1007/978-981-96-6609-6_17

Fig. 1. Session Chairs: (R) Prof. Virginia Murray, Head, Global Disaster Risk Reduction, UK Health Security Agency, UK; and Co-chairs Prof. Andrew Collins, DDN, Northumbria University, UK; and Dr. Ryoma Kayano, WHO Kobe Office, Japan

a few GADRI meetings, she stated that it was timely that health was further incorporated into the GADRI conference agenda.

Objectives of the Session

The session discussed putting health into disaster risk reduction and recovery. This is within the context of COVID-19 and some of the impacts were reflected on.

- Discussions centred around the Sendai Framework itself; and how to engage with WHO.
- WHO has been committed to trying to supporting UNDRR and the health domains of the Sendai Framework by delivering the WHO Health and Disaster Risk Management Framework in 2019; and trying to bring that approach across a wider domain of all hazards.
- The value of research networks such as GADRI are critical and young scientists are future leaders on addressing and learning about health impacts from health emergencies and disasters.
- The session reflected on how preparedness, response and recovery for future health and well-being is vital and the need to find sustainable disaster resilience in a world that is so complex.
- The session was about putting health into disaster risk reduction and recovery (Fig. 2).

Brief summary of panel presentations

Fig. 2. Presentation by Dr. Ryoma Kayano, WHO Kobe Office, Japan

- Dr. Ryoma Kayano, WHO Kobe
- Dr. Ryoma Kayano shared an introduction to the WHO Kobe Centre with its WHO Health Emergencies and Disaster Risk Management team's organization of a global research network, WHO Thematic Platform for Health Emergency and Disaster Risk Management Research Network. He talked about the function of the research network that facilitates international research collaboration and how all people can be participants of it. He also shared some of the outputs that have come out which are very important.
- He presented the WHO Guidance on Research Methods for Health Emergency and Disaster Risk Management, revised 2022 which was initially published in 2021, and a supplement on COVID epidemic added in 2022, and it is planned that it will be updated in 2025.
- Dr. Genta Nakano Disaster Prevention Research Institute, Kyoto University (Fig. 3)

o Dr. Genta Nakano presented on Community-based disaster risk reduction in an ageing society – shifting from social welfare to health maintenance. He provided the example of how the elderly can walk in the event of a tsunami or other event and move to an

Fig. 3. Presenter Dr. Genta Nakano, DPRI, Kyoto University, Japan

evacuation center and shared some of the principles; and how to manage the evacuation centers.

o It was requested to Dr. Nakano, that he share with the group whatever he publishes on the important work he is conducting.

- Prof. Virginia Murray, UKHSA

  - Referencing various sources, she presented on the standardizing of hazard definitions. This follows the UN Office for Disaster Risk Reduction (UNDRR) and the International Science Council (ISC) establishing of a Technical Working Group to this end in 2019. A key aim of this work has been to identify more comprehensive identification of hazards implicated by the Sendai Framework for Disaster Risk Reduction. This will enable countries and other actors to interpret and apply risk reduction policy and practice. The Hazard Information Profiles (HIPs) were published 2021 and can be accessed at UNDRR/ISC Hazard Information Profiles.
  - A multi-sectoral range of experts from the United Nations, research community, state, interstate and private sector across the world are reviewing the HIPs

(review). This is taking place in advance of the 2025 Global Platform for DRR. An updated output will then continue to support a wide range of professionals linking into the Sendia Framework global targets, such as A on mortality and B on people affected (paragraph 18).

Young Scientists Session on Youth and DRR - The Role of Youth and Young Professionals in Data and Knowledge Sharing in Disaster Risk Management

Gretchen Kalonji[1(✉)], Wei-Sen Li[2], and Mizan Bisri[3,4]

[1] IDMR, Sichuan University, Chengdu, China
gretchen.kalonji@gmail.com
[2] NCDR, Xindian, Chinese Taipei
[3] Kobe University, Kobe, Japan
[4] U-INSPIRE, Jakarta, Indonesia

Abstract. The prolonged worldwide pandemic, intensifying climate change impacts, and cascading risks have taught us that conventional approaches to disaster risk planning and management are insufficient for developing sustainable and resilient communities. Substantial improvement in our effectiveness requires not only more systematic identification of processes, techniques, evidence, challenges and opportunities, but also that we unleash to a far greater degree the latent capacity of diverse sets actors for contribution and collaboration. Among these actors, youth and young professionals (YYP's), and their organizations, have profoundly important roles to play in addressing the complex challenges related to disaster risk and climate change, the ongoing digital transformation, and intergenerational communication and collaboration. Numerous efforts worldwide have already illustrated the impact that such leadership roles of YYP's can have. It is important that the broader disaster risk management community learn from examples of successful youth leadership and accelerate efforts to support YYP's and their organizations to yet further expand their transformative capacity. This session highlighted global examples of three roles of YYPs in data and knowledge sharing for multi-hazards disaster risk management, specifically as "producer", "broker", and "synthesizer". In the first role, as "**data and knowledge producers**", YYPs devote themselves primarily to capturing natural and social phenomena essential for understanding the landscape of disaster risk. In the second role, as "**brokers**", YYPs provide essential services to ensure stakeholders can access the most relevant disaster-related data and knowledge based on their needs. In the third role, as "**synthesizers**", YYPs access and craft the available disaster-related data and knowledge - consolidating, combining, and providing meaning, such as to inspire or trigger societal changes, including for policy-making, implementation of research results, program ideation, and capacity-building. The session showcased examples from multiple countries and regions that illustrate each of these roles and explored ongoing challenges and opportunities for collaboration for YYPs from diverse professional backgrounds, including, but not limited to, earth scientists, social scientists, engineers, policy-makers, and emergency management professionals.

H. Tatano et al. (Eds.): GSRIDRR 2023, DRRGADRIBS, pp. 232–238, 2025.
https://doi.org/10.1007/978-981-96-6609-6_18

Fig. 1. Chairs of the Session-(R) Prof. Gretchen Kalonji and Prof. Wei-Sen Li

In addition, the session included a specific focus on water-related disaster management. Water is fundamental for humankind and the earth's survivability, especially in the face of increasing climate impacts. When water resources are not managed properly, various crises occur. There are cases in which water hazards, such as floods, droughts, and accidental water pollution due to natural hazards triggering technological disasters in international watercourses, lead to tension among countries. At the same time, there are multiple cases in which water-related hazards become opportunities to foster peace among riparian countries, including by sharing critical information and by transnational support amongst affected countries. As such, the session also discussed the experience of YYPs engagement in water-related disaster management, through the work of the U-INSPIRE Alliance (Youth and young professionals Network), in connection with the High-Level Panel on Water and Disasters (HELP), and its Alliance of Alliances on Research and Education on Water and Disasters (the AoA), as well of through the work of the Asia Water Council Young Professionals over recent years. Conclusions and recommendations on how to elevate the roles of YYPs more substantially was fed into the 10th World Water Forum in Bali in May 2024 (Figs. 1 and 2).

Discussion Session

The session conducted an interactive talk show and panel discussion for YYPs to share experiences in data & knowledge sharing for disaster-risk management. Participants included:

- Dr. Mizan B. F. Bisri, PhD (Kobe University / U-INSPIRE Indonesia)
- Ryo Tsuchida / Ph.D. Student Kyoto University / U-INSPIRE Japan
- Dr. Nuraini Rahma Hanifa, Researcher, National Research and Innovation Agency (BRIN); and Representative, UINSPIRE Alliance, Indonesia

Fig. 2. Group Photo of Session

- Dr. Ranit Chatterjee (Resilience Innovation Knowledge Academy (RIKA), U-INSPIRE India)
- Dr. Maria Camilla Suarez Pablo, UNGRD, Colombia
- Dr. Chipo Mudavanhu, Bindura University of Science and Education, Zimbabwe (Fig. 3)

The session was facilitated by Dr. Mizan Bisri, who also presented the session outcome. The following is a summary of the discussion session

Dr. Misri began by noting that half of the panelists were members of the Global Alliance of Disaster Research Institutes (GADRI) with exposure to the U-INSPIRE movement and were attending in-person, and half were not. Dr. Bisri also acknowledged Dr. Mark Ashley Perry from UK who brought in the European perspective to the session.

Dr. Bisri stated that the proposition, that youth and young professionals can play roles as data information and knowledge producers, brokers or synthesizers, through their activities in education and research, applies not only to natural sciences, but also to social sciences. Through their research activities, YYPs generate information and knowledge leading to concrete outcomes, with relevance for resilient development.

He also described his own personal involvement with the UN Major Group for Children and Youth (UNMGCY) and also for the introduction of SFDRR article 36.2A - to include the phrase – "children and youth as Agents of Change".

Inclusion of "youth and children" as agents of change is not sufficient, though. There is a need to push the young professionals, young scientists and also young politicians and young entrepreneurs, to secure their involvement in disaster risk management.

Fig. 3. Panelists Dr. Chipo Mudavanhu, Dr. Maria Camila Suarez-Paba, and Mr. Ryo Tsuchida

The discussion focused on what would be the best formulation of the roles of Youth and Young professionals. Dr. Bisri shared the gist of the panelists' statements as follows:

- Ryo Tsuchida showcased local and national collaboration and the various roles of Youth and Young Professionals within U-INSPIRE Japan. To highlight the essence of his presentation – the Japanese youth clearly can shift at the time of disaster. They play roles as volunteers. But once disasters are over and it is time for the recovery phase, and future risk reduction, they play roles within the labour market and that is the key to their involvement.
- Nuraini Rahma Hanifa, who is also a United Nations Office for Disaster Risk Reduction (UNDRR) Rising Star Award recipient and served as Secretary General of the U-INSPIRE Alliance, showed the history, the model, the movement, of U-INSPIRE. After the launch of the Sendai Framework, it quickly emerged in 14 Asia Pacific countries with over 1,000 young professionals. She highlighted that it is not only scientists, but also politicians, innovators, and also entrepreneurs.
- Ranit Chatterjee highlighted the importance of ecosystem creation for entrepreneurship. Not only in volunteer works, not only for capacity building for students and for early career researchers, but genuinely for job creation. He showed the U-INSPIRE India model, RIKA, which is has also been applied in the context of Indonesia and Nepal.

- Maria Camila Suarez Paba highlighted the experiences from Colombia; prospects and needs to create a critical mass of engagement by Youth and Young professionals. Not only to become future leaders but also to adopt current leadership roles.
- Chipo Mudavanhu brought in the experiences of Zimbabwe and stressed why youth is important. He also brought up an important question - how to mitigate brain drain, and ensure that the biggest contributions should be made at country and local levels.

U-INSPIRE and GADRI contribute to ecosystem development for promotion of leadership of YYP's in disk risk reduction and management. Three U-INSPIRE models that have worked well over the years, and to which GADRI could further contribute, include:

- In India, RIKA is a social venture and social entity which has a close engagement with U-INSPIRE India and at regional and global levels through the U-INSPIRE Alliance mechanism.
- In Indonesia, including among the Indonesian diaspora abroad, demonstrating the engagement between volunteers, young businesses, and young scientists through U-INSPIRE Indonesia mechanism and also through the U-INSPIRE Alliance. For example, in Indonesia, there are many young entrepreneurs with drone companies and their services are being widely used to conduct real-time ariel observations for plantation, for mining, etc. Once a disaster happens, they become volunteers. They send their drones to collect emergency related observations that the government really needs. There are also AI companies, such as CARI, that works on disaster knowledge in Indonesia. At country level, they are competing but at the same time cooperating inshowing the potential of Youth and Young Professionals
- In Nepal, they have developed Youth Innovation Labs and other social ventures interacting with U-INSPIRE network in Nepal as the umbrella at country level and leveraging from the multinational network of U-INSPIRE Alliance.

Where does GADRI fit in all this?

It is expected to have closer ties and more productive collaboration with GADRI Secretariat and GADRI members highlight and connect science and research with job creation, entrepreneurship and also policy advocacy by Youth and Young Professionals.

It is about pushing for early adopters in science and technology and also to diffuse Asia-Pacific experience and models for wider outreach.

Similar demands and trends are also followed in Africa, America and also in Europe, to continuously prepare and produce present and future science, political, business young leaders for a sustainable disaster resilient world.

Dr. Bisri, in particular, highlighted the following three points:

- Youth and Young Professionals in DRM should shift beyond volunteering. There should be more opportunities for providing decent jobs and they should be integrated in political decision making. It was also noted that there is a changing landscape within institutions before and after COVID pandemic, to provide better opportunity for youth and young professional.
- What is Young? Discussion on this led to being young at heart for advancing DRM which is essentially a multidisciplinary and interdisciplinary field.

- **Outcomes of the session and recommendations for the Board of Directors of GADRI Secretariat:**
- Youth and Young Professionals' role is validated as not limited to timely knowledge sharing. It is extended to data information producers but also to provide opportunities to be improved and become efficient collectors of data information. For example – the models in Asia Pacific countries can be used - through which the engagement and deployment of youth having businesses in drone and ariel observation, crowdsource application and numerator, AI based scrolling engine, was as were highlighted in the session.
- Knowledge Brokers and Facilitators are also catalytic beings among Youth and Young Professional as demonstrated by U-INSPIRE which also showed agility in advocating for political decision making at country level, as well as within regional and international domains
- Youth and Young Porfessional also serve as knowledge base synthesizers and implementers. These activities fall within the GADRI Committees on Institutional Capacity Building, Data and Information Sharing, and Advocacy.

Fig. 4. Discussion Session

Through the discussion it was shown that the models for youth and young professional engagements are in the field of DRM have proven their usefulness; and the ability of youth to take action to mitigate disaster and disaster risk management. Now it is time to put it in to action beyond the Asia Pacific region.

References

1. Global Alliance of Disaster Research Institutes (GADRI) - https://gadri.net/members/members/
2. U-INSPIRE Alliance (Youth and young professionals Network- Innovation in Science, Engineering, and Technology for Disaster Resilience at National and Global Levels)
3. https://www.preventionweb.net/organization/u-inspire-indonesia
4. United Nations Office for Disaster Risk Reduction (UNDRR) - https://www.undrr.org/
5. UNDRR Sendai Framework Agenda - https://www.undrr.org/implementing-sendai-framework/what-sendai-framework
6. Resilience Innovation Knowledge Academy (RIKA) - https://rikaindia.com/
7. UN Major Group for Children and Youth (UNMGCY) - https://www.unmgcy.org/

Regional Alliances

Progressing the United Kingdom Alliance for Disaster Research

Andrew E. Collins[(✉)] and Mark Ashley Parry

Northumbria University, Newcastle-Upon-Tyne, United Kingdom
andrew.collins@northumbria.ac.uk

Abstract. The United Kingdom Alliance for Disaster Research (UKADR) has remained active since around the time of the Sendai Framework for Disaster Risk Reduction in 2015. Its work is focused on the allied processes of disaster risk reduction and its contexts, through being a scientific, practice and policy informed research alliance. Its institutional partners and activities have expanded and become more embedded, but also outward looking, with several notable additional steps achieved since it last reported to a GADRI summit. The alliance is reviewing its research priorities on a continual basis and has also established an Early Career Researchers and Practitioners [ECRP] Group.

1 Introduction

The following text is provided by the UKADR as its own description of the alliance's function. It is reproduced here by way of increasing its exposure within the GADRI community:

> "The primary motivation for the UK Alliance for Disaster Research (UKADR) is to bring together the UK's rich and diverse disaster research community. This is to facilitate collaboration and partnership to aid representation of the research community at government level in the UK, and, where appropriate, help with the implementation of the Sendai Framework for Disaster Risk Reduction. The Alliance is independent and managed by voluntary contributions from the UK research community. Membership of UKADR is open to staff and PhD students of any research-and-capacity- building institute based in the UK that is active in researching disaster risk and its management. This includes universities and other educational institutions, charitable trusts, think tanks and research departments within government agencies, private businesses or civil society organisations. Disaster risk reduction, prevention and preparedness, disaster response, reconstruction and recovery are all of areas of interest". (UKADR website http://www.ukadr.org/home.html)

The alliance has since the Sendai Framework, and as championed by some of its members well over a decade earlier, often diversified its focus to reflect on interconnected natures of disaster risk reduction, emphasising for example, dual disaster and

H. Tatano et al. (Eds.): GSRIDRR 2023, DRRGADRIBS, pp. 241–249, 2025.
https://doi.org/10.1007/978-981-96-6609-6_19

development synergies and inseparable issues of climate adaptation and resilience. Members bring intersecting expertise from physical, social and health sciences and the legal profession, amongst others.

2 Progression and Orientations

Despite the quite broad reaching agenda recognisable in this alliance, the overriding focus is the common understanding of risk management for adverse risk reduction as involving the following as within the realm of this institution:

> "All contexts and systems of disaster risk recognising of the UK role in this field across an interconnected world. UKADR understands science on disaster risk and its management to be interdisciplinary. There is an emphasis on integrated and applied dimensions of knowledge production rooted in fundamental science. This includes work from the natural sciences, engineering and technology, medical and health sciences, agricultural science, social sciences, and humanities and includes information technology, economics, and behavioural sciences." (UKADR website http://www.ukadr.org/home.html)

Since its formation the UKADR has demonstrated support of excellence in science and has enhanced the opportunities for impact on policy and practice. It has done this by facilitating partnership within the UK institutions, advocating UK disasters related science to the UK Government, being a contact point for when the Government is looking for the best science in this area available and as a mechanism for international efforts in this field. As such its international connectivity, including via GADRI is all important to this agenda.

2.1 Core Institutional Partners

The UKADR has twenty-seven core partners, located in 21 towns and cities across three of the four nations that make up the United Kingdom. These are listed below and located in Fig. 1.

1. Anglia Ruskin University
2. Aston Crisis Management Centre, Aston University
3. British Geological Survey
4. Cabot Institute for the Environment, Bristol University
5. Cardiff University / Prifysgol Caerdydd
6. Centre for the Observation and Modelling of Earthquakes, Volcanoes and Tectonics (COMET), University of Leeds
7. Coventry University
8. Geology for Global Development
9. Heriot-Watt University
10. Institute of Hazard, Risk and Resilience, Durham University
11. King's College London
12. National Centre for Resilience, University of Glasgow

13. Northumbria University
14. Overseas Development Institute (ODI)
15. Oxford Brookes University
16. THINKLab, University of Salford
17. UCL Department of Risk and Disaster Reduction (RDR)[1]
18. UK Centre for Ecology & Hydrology (UKCEH)
19. UK Health Security Agency[2]
20. University of Cambridge
21. University of East Anglia
22. University of Edinburgh
23. University of Glasgow
24. University of Huddersfield
25. University of Leicester
26. University of Plymouth
27. University of Reading

2.2 Recent Activities

Some recent examples of actions from the UKADR prior to the 2023 GADRI summit included facilitating a substantive discussion around the potential contributions by the UKADR to the wider UK funding landscape, focusing on structure, mechanisms, and values of funding. It also provided formal partnership support to UK Government Research and Innovation (UKRI) in the context of the COP26 Adaptation and Resilience Event Series for which a contribution to identifying priority research themes proposed through this process. It's partnership building initiatives over the pre-GADRI summit period included with the Adaptation Research Alliance (ARA), GADRI and the South Asia Alliance of Disaster Institutes. The UKADR has also been identified as a regional link for the International Science Council co-sponsored programmes of IRDR.

The UKADR runs an annual conference series which at the time of going to press had taken place as follows and located in Fig. 2:

1. UKADR Annual Conference 2017 – Integrated Research that Makes a Difference [King's College London; 9-10th January 2017]
2. UKADR Annual Conference 2018 – Complex Hazards / Complex Vulnerabilities [University of Bristol; 27th-28th March 2018]
3. DwD-GCRF-UKADR-DRG-UKCDR International Conference, 2019 – New points of departure in transitioning disaster reduction and sustainability challenges [Northumbria University; 17-19th July 2019]
4. UKADR 2022 Conference – Disaster research and innovation at a time of global uncertainty [Edinburgh Climate Change Institute, University of Edinburgh, British Geological Survey, Heriot Watt University; 7th-8th December 2022]
5. UKADR Annual Conference 2023 – 2030 and Beyond: Risk-informed decision making, investment and behaviour [University of Huddersfield, 18th-19th December 2023]

---

[1] - Until July 2024, it was known as 'UCL Institute for Risk and Disaster Reduction (IRDR)'

[2] - Until March 2021, this government agency was known as 'Public Health England'

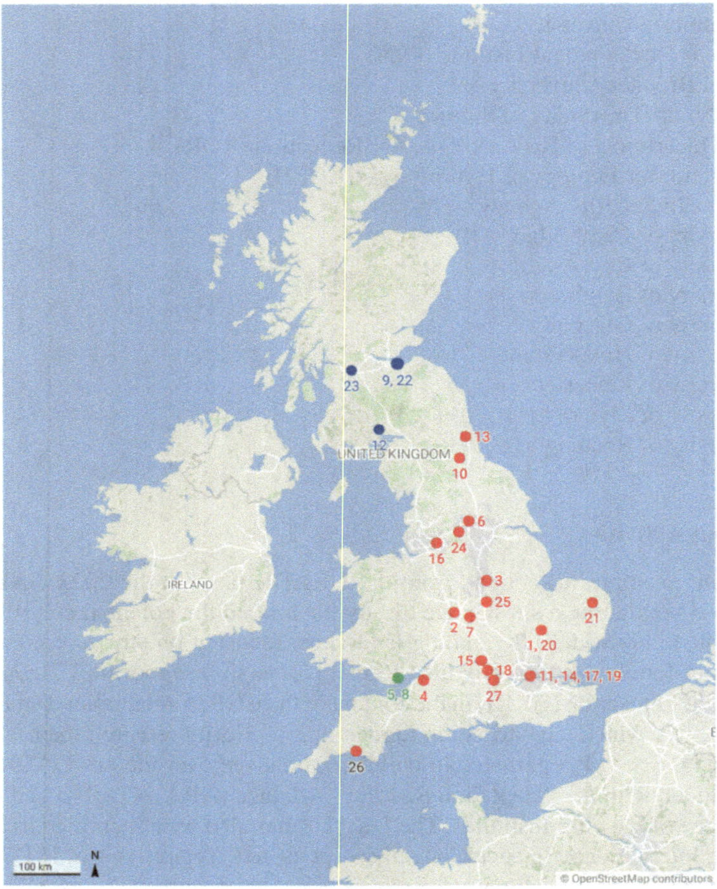

Created with Datawrapper

Fig. 1. Location of the UKADR Core Partners (as of 31st August 2024). [Source: Authors] Key: Blue Markers = Scottish Based Partners; Green Markers = Welsh Based Partners; Red = English Based Partners

6. UKADR Annual Conference 2024 – Collaborative Research to Make a Difference [UCL Institute for Risk and Disaster Reduction, 26th-27th July 2024]

2.3 Membership Culture

The alliance uses principles of open access and rotational representation of its members in the Steering Group subject to nominations and election on a three yearly basis. The Co-Chairs have also been rotated on this basis. At the time of writing the first author of this entry for the GADRI proceedings had served as a founding Chair and has now been rotated off the Committee. The existence of an advisory group allows for the currently elected Steering Committee, if it wishes in in interests of the alliance, to identify senior

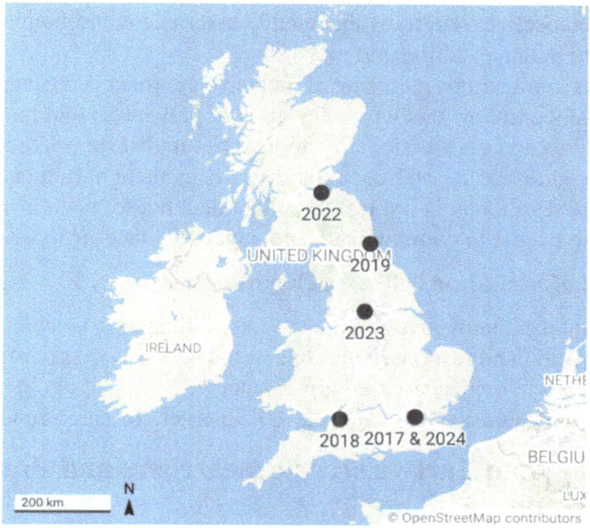

Fig. 2. List of UKADR Annual Conferences Between 2017 and 2024 (Source: Authors)

advisors, including former chairs of UKADR, to continue to assist in the work of the alliance through this group. Significant recent developments have also included the establishment of an early career researcher network and this is presented in more detail under Sect. 3 of this proceedings section.

2.4 UKADR Research Priorities

As expressed by its membership in the period preceding the GADRI summit through a consultation and discussion on research priorities within the UKADR communities the following are considered ongoing priorities. These will be added to indefinitely being reviewed periodically via the Steering Group and annual conference events.

i) Understanding Risk

- social, economic and environmental conditions that perpetuate disaster risk reduction and disaster risk creation across all societies
- extensive and intensive disaster risk
- cascading disasters, particularly in relation to climate change
- cumulative impacts of recurrent hazards on recovery and adaptive capacity
- health security and well-being for disaster resilience and climate adaptation
- intensified displacement through climate and other environmental changes
- ecosystems based disaster risk reduction and nature-based solutions
- intergenerational equity and justice
- disaster risk management education

ii) Disaster science methodologies

- critical perspectives and tools that enable assessment and monitoring, including of loss and damage, action and inaction
- active data applications requiring of inclusive approaches to data acquisition and management and how this works for disaster prevention and management
- social groups and intersectionality approaches to disaster risk reduction
- co-produced research, interdisciplinarity, engagement and impact mechanisms
- disasters research that is not just to collect data but to try to bring about change supporting people to communicate and engage in disaster risk reduction

iii) Mechanisms of investment and research governance

- New, long-term, innovative and multi-sectoral financing mechanisms for disaster risk prevention and response solutions based on likely disaster trajectories
- Effectiveness in evidence-based governance
- Structure, mechanisms and values of UK disasters research funding

3 United Kingdom Early Career Researchers and Practitioners [ECRP] Group

The United Kingdom Early Career Researchers and Practitioners Group conceptually existed since the first UKADR annual conference in January 2017. However, it was not until the advent of the COVID-19 pandemic that the concept was implemented. This was, in part, driven by a desire within UKADR and its associated networks to increase support levels for students and those who are starting out in this field, especially during very challenging times.

Until February 2024, the group was known as the 'United Kingdom Early Career Researchers' (UKADR ECR) Group. However, the change to the current name was prompted by the growing awareness that the existing name implied it was only open to young researchers. This change, therefore, recognizes the importance of bridging researchers and non-age specific practitioners in the disasters field of study and enterprise.

Since then, the group has developed its own sub-committee, which provides feedback to the main steering group committee of UKADR, as demonstrated in Fig. 3. This indicates how there are three representatives who are members of the steering group committee and who also serve as leads of the ECRP sub-committee. However, one of the changes since its formation is that the ECRP representatives ideally should include at least one representative from each of the academic and practitioner worlds. This is to better represent the diverse field of disaster research and promote both interdisciplinarity and transdisciplinary connections between members, especially since there are representatives at the lead of the ECRP group.

The ECRP group has eight direct objectives as follows.

i. **Networking and Collaboration:** This was one of the original objectives of the ECRP group. It serves as a platform facilitating the connection of different early career researchers from various institutions across the United Kingdom. This platform aims to foster connections, share research ideas, and potentially initiate collaborate on research projects among participants. The objective has been to create networking opportunities that could lead to partnerships between researchers

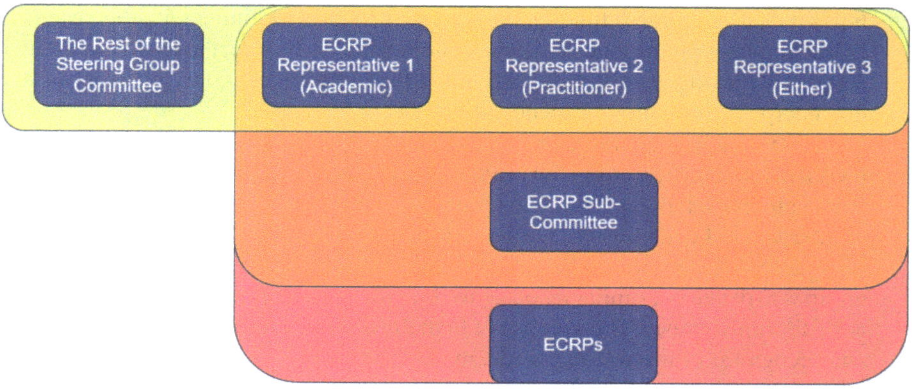

Fig. 3. The Structure of the ECRP Sub-Committee (Source: Author)

and institutions, the opportunities of co-authorship in academic papers, and foster interdisciplinary collaborations – a critical aspect in our modern interconnect world, especially in disaster research. All of these elements are vital for professional development and are facilitated through various means, including bi-monthly informal meetings, a LinkedIn Group, and an ECRP mailing list.

ii. **Mentorship and Support:** As the majority of our ECRPs are either PhD students or have recently completed their PhDs, they are still in the process of establishing themselves in their academic careers. The advice from senior researchers, can significantly guide and assist ECRPs, given the challenges of working with academia, especially considering that the UKADR comprises many senior researchers who are a key part of the support for the next generation of researchers. This willingness was evident in the 2-hour pre-conference webinar held in 2023, which was designed around grant funding and how to develop a research profile. Additionally, the 'UKADR in conversation…' series covers diverse topics, including advisory subjects such as getting published, as well as more general disaster-related topics.

iii. **Professional Development:** The ECRP group hosting of a monthly 'In conversation with…' series is open to all members of UKADR and the wider academic community, albeit is primarily aimed at the ECRP group. This series focuses on specific advisory subjects and provides more general disaster-related topics, which might spark interdisciplinary research among the ECRPs. Furthermore, as the ECRP network has expanded, it has led to the establishment of an ECRP subcommittee. This development means that that these events and meetings are now organised and planned by members of the ECRP subcommittee, offering vital skills that ECRPs might require in their future work.

iv. **Advocacy and Representation:** At the heart of the ECRP group it is recognised that the UK research community, as often the case elsewhere too, can be very challenging, and consequently, there are specific concerns that directly relate to ECRPs within the academic community. The ECRP group serves as a platform to address these concerns, enabling the ECRP to the wider UKADR committee. Additionally, these representatives explore ways to best serve the ECRP community.

For instance, as research funding has become increasingly limited, it often results in ECRs missing opportunities to attend conferences. Therefore, the ECRP committee is actively seeking ways to gaining funding for conferences and trying to encourage the incorporation virtual elements into these conferences.

v. **Visibility and Recognition:** One of the benefits of participation in the ECRP group is enhanced visibility within the academic community. The group offers opportunities to present internal research to fellow peers before presenting at either national or international conferences. Additionally, the ECPR group has developed a repository of PhD theses undertaken at British institutions. The next stage of the ECRP group is to develop a similar resource tool for master's dissertations also undertaken at British universities. The purpose of both these repositories is to provide visibility and recognition for students' work. In the case of PhD theses, they are typically left in universities' online libraries, while in the case of master's dissertations, they are often not published to the public domain, a practice common in many countries globally but not in the United Kingdom.

vi. **Information Sharing and Resources:** The ECRP group provides a bi-monthly newsletter that is available to all members of UKADR. The newsletter offers information about future conferences, job opportunities, research grants, and resources such as new books and journal articles that members might find beneficial. It can be argued that this information is beneficial to all members, but particularly to those early in their careers, as it can be crucial for career growth. In addition, the ECRP group has started to maintain a Jisc mailing list, enabling members to share information more instantly if needed about various elements that might be useful to other career researchers and practitioners.

vii. **Fostering a Supportive Community:** It is well recognized that academic research can be isolating at times, especially for early career academics. This, in turn, has resulted in higher-than-normal levels of mental health ailments among them (Woolston, 2021; Friedrich et al., 2023). Therefore, it has long been hoped that the ECRP group will provide a community of peers who understand the situation, offering each other emotional support while fostering a sense of belonging and friendship among the members.

viii. **Diversity and Inclusivity:** It has been important for us to ensure the promotion of diversity and inclusivity within the ECRP group, regardless of gender, sexuality, race, age, disability, or financial situation. This is to ensure the creation of a more equitable research community. As highlighted earlier, we have further opened the community to include practitioners who may not see themselves as researchers.

4 Conclusions

The trajectory of the UKADR to date indicates that the appetite for alliancing regionally has significant potential to bring together core research bodies dedicated to disaster and allied research topics. The success of the initiative is relative to the commitment of the component institutional partners remaining active in attempting to progress the intended outcomes of the Sendai Framework for Disaster Risk Reduction and associated parts of the 2015-2030 agenda of the United Nations. As a regional alliance, this activity needs to continue to seek inflection points for progress nationally and locally across prioritised research topics and a through a wide demographic of researchers working under varied professional and intermediate working environments.

References

Friedrich, J., et al.: "How is your thesis going?" – Ph.D. students' perspectives on mental health and stress in academia. PLoS ONE **18**(7), e0288103 (2023). https://doi.org/10.1371/journal.pone.0288103

Woolston, C.: Depression and anxiety 'the norm' for UK PhD students'. Nature, 14th December (2021). https://www.nature.com/articles/d41586-021-03761-3. Accessed 7 Mar 2024. https://doi.org/10.1038/d41586-021-03761-3

Selected Presentation Papers (Arranged By Country Order)

Building Smallholders' Resilience to Floods in Northern Ghana. Do Integration of Indigenous and Scientific Adaptation Strategies Matter?

Hamdiyah Alhassan[1,2](✉), Gabriel Ayum Teye[3], Mensah Tawiah Cobbinah[4], and James Amowine[4]

[1] Kazuhiko Takeuchi Centre for Sustainability and Resilience, University for Development Studies, Tamale, Ghana
ahamdiyah@uds.edu.gh

[2] Department of Economics, School of Economics, University for Development Studies, Tamale, Ghana

[3] Department of Animal Science, Faculty of Agriculture, Food and Consumer Sciences, University for Development Studies, Tamale, Ghana

[4] Department of Agribusiness, Faculty of Agriculture, Food and Consumer Sciences, University for Development Studies, Tamale, Ghana

Abstract. This paper investigates farmers' indigenous and scientific flood adaptation strategies for maximizing crop production in floodplains. A total of two focus group discussions were held in Pwalugu and Kpasenkpe, both in northern Ghana, to collect qualitative data for the study. Express Scribe Transcription Software was used to transcribe the interviews. The transcripts were coded and analyzed using the qualitative data analysis software MaxQDA. Farmers produce food crops in floodplains by taking advantage of flood benefits, such as replenishing soil fertility and retaining groundwater. Floods, on the other hand, shorten the period between cultivation and harvest, requiring farmers to implement indigenous and scientific flood adaptation strategies that enhance production of food crops. Scientific flood adaptation strategies adopted by farmers include supplementary irrigation, early maturity varieties, and water-saving/water-harvesting technologies, and scientific weather information received from organizations. Also, indigenous practices such as farming on multiple plots, diversifying crops, changing crops and planting dates, and reading weather based on beliefs are used by farmers. Policies that integrate indigenous customs with scientific adaptation strategies are needed to help farmers meet floodplain production deadlines. Further, easy access to weather forecasting information, energy-efficient irrigation technologies, and early maturity cultivars should be provided to farmers. Esoko and Ghana Meteorological Agency should be resourced to provide weather information to farmers in flood-prone communities.

Keywords: Recurring floods · indigenous strategies · focus group discussion · northern Ghana

H. Tatano et al. (Eds.): GSRIDRR 2023, DRRGADRIBS, pp. 253–266, 2025.
https://doi.org/10.1007/978-981-96-6609-6_20

1 Introduction

Floods occur due to heavy or prolonged downpours, storms or surges, dam or levee rupture or spillage (Tellman et al. 2021; Schneider et al. 2007). Floods strike abruptly and fast in a violent manner without forewarning (Zickgraf, 2021). Yet, they can cause deaths or massive damage to lives and property, as well as instill psychosocial trauma in victims (Alderman et al. 2012; Milojevic et al. 2011; Ahern et al. 2005). Apart from being catastrophic and costly, floods were the world's most frequently occurring natural hazards in 2021 (CRED 2022). Worldwide, more than 800 million people are at risk of floods (Kundzewicz et al. 2014). Most flood victims and those at risk of floods live in Asia and Sub-Saharan Africa (Tellman et al. 2021).

Agriculture is no exception to the effects of floods (Wang et al. 2022; Ulubaşoğlu et al. 2019; Posthumus et al. 2009). In Ghana, agriculture accounts for 37.1% of the active population and 65.2% of the active rural population and contributed 19.7% of Ghana's current GDP and about 30% of export earnings in 2021 (Cornille et al. 2022). Yet, the sector is vulnerable to climate change and related hazards, of which floods are among the most catastrophic (Ntim-Amo et al. 2022; Yiran and Stringler 2016; Lolig et al. 2014). Ghana is most prone to floods (Cornille et al. 2022; Amoateng et al. 2018) and experiences some of the worst floods in West Africa (Karley 2009). Among the most severe was the Accra flood event of 2015, which occurred after a lit cigarette fell into flood water with petrol on the surface. The fire ran back into the leaking petrol tank in a state-owned gasoline station and caused an explosion. Both explosion and fire killed 152 people and destroyed properties worth millions of dollars (Quarshie et al. 2018; Yankson et al. 2017). Between 1990 and 2021, 26 serious flood events occurred in Ghana, which affected 5,016,412 people, 522 of whom died (Ntim-Amo et al. 2022). The economic damage from these floods is more than US$600 million (Ntim-Amo et al. 2022). The intensity of floods is expected to increase in Ghana due to climate change (IPCC 2012; World Bank 2011). The savannah ecological zone of Ghana is highly vulnerable to climate change (Armah et al. 2010) and carries a higher proportion of the burden of climatic hazards such as floods (Alhassan, 2020).

In northern Ghana, floods occur primarily due to heavy or prolonged downpours and or the spilled water from the Burkinabe Bagre Dam, which raises the water levels of the Black and White Volta Rivers and subsequently seeps into downstream communities (Fiasorgbor et al. 2018; Yiran and Stringler 2016). Since the opening of the Bagre Dam, many communities in northern Ghana have been exposed to recurring floods. However, the most devastating ones happened in 2007, 2010, 2012, and 2017–2020 after water released from the Bagre Dam coincided with heavy downpours (Cornille et al. 2022; Fiasorgbor et al. 2018; Yiran and Stringler 2016). Between 2007 and 2020, floods caused by high-intensity rainfall combined with water releases from the Bagre Dam affected 582,503 people and destroyed 102,459.6 hectares of farmland (Cornille et al., 2022). The floods caused damage to livestock and food storage facilities, generated sudden price changes, and blocked or submerged roads (Kassim et al. 2021; Fiasorgbor et al. 2018; Yiran and Stringler 2016; Armah et al. 2010). The resultant effects of floods have been food shortages and livelihoods destroyed (Fiasorgbor et al. 2018; Yiran and Stringler 2016).

Adaptation to floods can be effective if farmers blend indigenous practices with modern adaptation strategies (Guodaar et al. 2021). However, little is known about this integrated approach to flood adaptation. Conversely, several studies have focused on examining the choice of modern adaptation strategies to climate hazards in Ghana (Yiran et al. 2022; Asante et al. 2021; Kassim et al. 2021; Alhassan 2020, Tangonyire and Akuriba 2020; Williams et al., 2019; Afriyie et al. 2018; Fagariba et al. 2018; Ndamani and Watanabe 2016; Al-Hassan et al. 2013; Kuwornu et al. 2013). This study derives intuition by addressing the following question: what indigenous and modern adaptation strategies are adopted by farmers to maximize crop production in floodplains? Policy-wise, this study will lay a foundation for decision-makers in developing an integrated adaptation approach for effective flood adaptation in Ghana.

2 Research Methodology

2.1 Recruitment of Participants

Sixty participants, made up of 33 men and 27 women, were selected for the focus group discussions (FGDs). The FGDs were organized in Pwalugu in the Talensi District of the Upper East Region and Kpasenkpe in the West Mamprusi Municipality of the North East Region, with the help of two National Disaster Management Organization officers. Pwalugu and Kpasenkpe lie by the White Volta River, making them highly prone to floods any time water is released from the Bagre Dam in Burkina Faso (Cornille et al., 2022). The Pwalugu-FGD included 18 men and 12 women, while the Kpasenkpe-FGD included 15 men and 15 women. In Pwalugu, the FGD was conducted in the village square on 23rd April 2022, while in Kpasenkpe, it was conducted in the chief's palace on 24th April 2022. Snowball sampling was used to select the participants. This method was adopted because it enabled us to locate participants who owned farms in the flood plains. Hence, they shared similar and relevant characteristics to the study (Powell and Single, 1996). However, the participants were selected voluntarily for the focus groups. The participants gave informed consent. They were also informed of the purpose of the study and the confidentiality of their information.

2.2 Focus Group Procedure and Data Collection

The FGDs were moderated by two authors using a checklist with 15 semi-structured questions. The qualitative data were collected by a research assistant using digital audio and video recorder. The discussions proceeded along the following questioning path. We started the discussions by introducing the research team members and the study participants. In the second part, the participants were asked to provide a history of water bodies in their communities. In the third part, the participants explained the benefits of floods to agriculture. The fourth section asked about indigenous and modern flood adaptation strategies. Each FGD lasted for about two hours.

2.3 Data Analysis

Two tape recordings were accessed in Windows Media Player and written verbatim in Microsoft Word using the Express Scribe Transcription Software. Express Scribe is professional audio playback control software designed to assist in the transcription of audio recordings. The authors and the research assistant transcribed the data to produce a full narrative of the two discussions. The transcripts were analyzed using the qualitative data analysis software MaxQDA. Transcripts were subjected to an open coding system by two independent authors for the extraction of thematic constructs. The authors then grouped similar codes, which were combined into subcategories and further categorized into main categories. Disagreements were discussed with all authors until a consensus was reached. The data were analyzed through a mixed content analysis approach that included both inductive and deductive approaches. Deductive content analysis is chosen when there is sufficient prior research on the subject matter (thus, modern flood adaptation strategies) to select and define the categories (Goel et al., 2019). It is used to test a previous theory in a different situation or to compare categories in different periods (Elo and Kyngäs, 2008). Inductive content analysis, on the other hand, is chosen when there are very few studies examining indigenous flood adaptation practices (Elo and Kyngäs, 2008). Detailed information on the preparation, organization, and resulting phases of the content analysis process can be found in Elo and Kyngäs (2008).

3 Results and Discussions

This section presents and discusses the findings on the sources of recurring floods, benefits of flooding and the adaptation strategies adopted by the smallholder farmers in Pwalugu and Kpasenpke, Ghana.

3.1 Source of Recurring Floods in Pwalugu and Kpasenpke

Before delving into the benefits of floods, the participants were asked to give a narrative of the White Volta River and why it overflows its banks to cause floods. The White Volta River has a long and storied history, according to the locals. The river was an important route for transporting slaves and goods during the slave trade. The river provides several social and economic benefits for locals and neighboring communities in that "we get enough water for irrigation from the river, allowing us to produce twice as much in a season and harvest more. We can grow high-value crops such as chilies, onions, cabbage, and calabash thanks to the river. Farmers use petrol or diesel pumps to extract water from the river for irrigation" (P03, Male, 67 years). "The White Volta River is to us just like the Nile is to the Egyptians. The White Volta River is more or less like a company or factory, which employs labourers from neighbouring communities. Therefore, it is creating employment for the youth in terms of fishing and irrigation farming" (P14, male, 44 years). "The river is a source of fishing for us. It is also creating employment for the youth and reducing migration. The monies we get from irrigation farming, we use them for household upkeep, payment of school fees, and hospital bills" (P23, Female, 56 years). A snapshot of the river is shown in Fig. 1.

Fig. 1. The White Volta River at Kpasenkpe, North East Region of Ghana **Source: Interviews, 2022**

The river flood in Pwalugu is a well-known problem in Ghana. "The biggest flooding occurred in 1994. We only have three months to produce. "Yes, it is true that the floods always submerge our crops and destroy them and also displace people from their homes. Yet, we derive a lot of benefits from floods" (P14, male, 44 years). "In the olden days, we use to experience flood once in three years. But, this is not the same nowadays. We now experience every year since 1994 and this is usually caused by heavy rains combined with the opening of the Bagre Dam in Burkina Faso"(P03, Male, 67 years).

3.2 Benefits of Recurring Floods to Farming Societies

In all the FGDs, the participants agreed that floods bring substantial advantages to crop farmers. The benefits of floods were categorized into two themes and three subthemes. The main themes were crop production and fishing. The subthemes under crop production show that floods replenish soil fertility, retain groundwater for crop growth, and allow for multiple cropping. On the other hand, floods create natural ponds, which are used for fishing. The subthemes are presented in the Table 1 below.

Table 1. Benefits of recurring floods to crop production and livelihoods

Theme	Subthemes
Crop production	Replenish soil fertility Retain groundwater for plant growth Allow for multiple cropping
Fishing	Create natural ponds

Source: Interviews, 2022

(1) *Replenish soil fertility*

The participants remarked that their farmlands become fertile year after year because of the flooding. "*We do not use fertilizer on the land near the river*" (P07, male, 32 years). This farmer has a three-acre pepper farm (see Fig. 2) at Pwalugu near the White Volta River that is doing well without fertilizer application. According to the participant, the pepper farm in the flood plain performs better than farms on uplands, even when fertilizer is applied. Another participant remarked on this same issue, "*We reap numerous benefits from recurring floods. We don't use fertilizer on our farms along the river since the soils around the river are fertile. Second, we plant twice along the river's side; we normally plant maize and then, shortly after harvesting, we plant beans*" (P39, female, 43 years).

Fig. 2. Pepper farm with no fertilizer at Pwalugu, Upper East Region of Ghana **Source: Interviews, 2022**.

(2) *Retain groundwater for crop growth*

Participants were unanimous in their answers that floods deposit water on the ground, which supports plant growth. "*Due to the flooding, I produce in the dry season without irrigation. The land becomes moist after flooding, allowing us to produce crops in the dry season*" (P17, female, 41 years). "*The flooding deposits water on the ground, which makes the soil moist. We have enough groundwater to produce maize and vegetables for three (3) months without needing water for irrigation*" (P51, Male, 49 years).

(3) *Allow for multiple cropping*

During the FGDs, many of the participants agreed floods enable farmers to engage in multiple cropping due to the all-year-round availability of water. "*Floods supply water for irrigation, enabling farmers to engage in multiple cropping. In the rainy season, we get flood water for irrigation when the rains are delayed. In the dry season, we normally use machines to pump the flood water to irrigate our vegetables such as onion, pepper, kenaf, cabbage, tomatoes, and even fresh maize*" (P04, female, 31years).

(4) *Create natural ponds*

During the FGD, the participants indicated that the floods leave the water with fish in valleys, gutters, and dams, which serve as a source of livelihood for them. Examples of natural ponds are displayed in Fig. 3. This revelation was shared by all of the participants. But for brevity, we presented a few responses from the FGDs: *"For us, the flooding also creates a natural pond. Because the pond is always full of large fish, our women go fishing in the natural pond after the flood. In fact, for most households, it is a major source of income and food"* (P05, male, 42 years).

"We the women who do not have canoes, fishing nets, or hook and line to harvest fish depend on the natural ponds to earn a living. It is safe to do fishing in the natural ponds because they are quite shallow and we do not require any of the harvesting equipment to do so" (P14, female, 40years).

Fig. 3. Natural pond created by flood at Pwalugu (A) and Kpasenkpe (B) **Source: Interviews, 2022.**

3.3 Flood Adaptation Strategies

The majority of the farmers interviewed have continuously farmed in the floodplains using both indigenous and modern (scientific) flood adaptation strategies. Table 2 shows the flood adaptation strategies implemented by farmers for maximizing crop production. Flood adaptation measures learned from experts and organizations were categorized as modern or scientific. Indigenous flood adaptation strategies, on the other hand, were established on one's initiative and passed down from farmer to farmer.

(1) Supplementary irrigation

Most farmers engaged in supplementary irrigation to meet the rain and avoid delays in planting. "*There is no means of preventing or controlling flooding. Our production season has been shortened due to flooding. However, because the riverbank is so fertile, we have no choice but to devise methods of producing crops before flooding begins. I do supplementary irrigation with my small holes (see* Fig. 4 *) before the rains begin. We have been following some of these practices for a long time, and they have not let us down"* (P23, male, 37 years).

(2) Early maturity varieties

Farmers plant early maturity varieties to harvest their crops in three months because *"they help us to harvest earlier. The varieties have a 75–90 day growth cycle. With these*

Table 2. Flood adaptation strategies

Theme	Subthemes
Modern flood adaptation strategies	Supplementary irrigation Early maturity varieties Water-saving and water-harvesting technologies Scientific weather information
Indigenous flood adaptation strategies	Farming on multiple plots Crop diversification Changing crops and planting dates Read weather based on beliefs

Source: Interviews, 2022

Fig. 4. Microirrigation practices by some farmers in Pwalugu **Source: Interviews, 2022.**

varieties, I can start cultivating my farm in May and harvest in July. There is nothing to harvest on the farm by the time the flood hit in August or September. The only challenge is that early maturity crops are not easily available for buying" (P05, female, 42 years).

(3) Use of water-saving and water harvesting technologies

Most farmers use mulching and bunding to control nutrient leaching in floodplains. "*We leave weeds on the farm to keep the soil moist and prevent erosion. We also built a bund to keep water in place and avoid nutrient leaching"* (P11, male, 58 years). *"Flood water contains a lot of nutrients. Bunds help to retain some of the nutrients on the farm. If there are no bunds or weeds on the surface of the soil, flood water will move existing nutrients from the farm to other regions, causing you to lose nutrients"* (P05, female, 42 years).

(4) Accessing scientific weather information

Few farmers reported using Esoko and the media to get weather forecasting information. "*I frequently sit in front of the television or radio to listen to daily weather information, which helps me organize my production. When I receive constant news of forthcoming rains for one week, I begin preparing my land"* (P35, male, 52 years). *"On my phone, I have weather forecasting information, which helps me know when it will rain. If I know the rains will be late, I either plant early maturity crops or wait until after the flooding to plant more"* (P35, male, 52 years).

(5) Changing crops and planting dates

During the FGDs, farmers placed importance on changing crops and planting dates owing to the unpredictable weather. *"When we plant early in the valley this year, we shift the planting date a month or two ahead the following year to avoid flooding. We do this because the timing of the flooding is not consistent. Some flooding occurs in August, while others occur in September. Also, we often plant early maturing crops when we know the flood will be late"* (P30, male, 46 years).

(6) Farming on multiple plots

Most farmers agreed that it is not good to put all eggs in one basket. *"That is suicidal for a farmer who decides to invest all his resources to producing crops in the flood. I am working on two different plots. One is located on the upland, while the other is located in the lowland. I always make sure I have a small farm on the upland that I can fall back on if my lowland farm is destroyed by flooding. I cultivate the lowland plot first, followed by the upland plot"* (P44, male, 38 years).

(7) Crop diversification

Crop diversification is used by the majority of farmers because *"it allows us to get some rewards from our investment." On the same area, we cultivate green vegetables and maize. The leafy vegetables are harvested before the river flood arrived to submerge the maize."* (P35, male, 52 years).

(8) Reading the weather based on beliefs

"I always check the weather to figure out what crops to plant and when to plant them. To forecast the weather, I rely on my faith. For example, if rain starts in the middle of the night, I know there will be plenty of rain. I can tell whether the rains will be good or bad if I hear a specific frog cry. If the wind also blows in a windward direction toward the mountain areas, it means there will be plenty of rain. Although my predictions are not always correct, I have benefited greatly from them, and other farmers always seek my permission before producing, and they have done very well as a result" (P31, male, 61 years).

Discussions

Farmers depend on floods to produce crops and engage in fishing. Farmers go fishing in the flood water deposited in valleys. The natural ponds were a source of income for women who do not have canoes and the physical strength to do river fishing. In crop production, floods also improve productivity by replenishing soil fertility, retaining groundwater for crop growth, and allowing for multiple cropping. Floods make the farms fertile and thus farmers did not need chemical fertilizer application to maintain the fertility of the soil. This finding is consistent with Fujita and Shaw (2019), who established that floods increase land production by replenishing soil fertility. In arid regions of North Eastern Africa, farmers use seasonal flash floods to sustain agricultural production (Fadul et al. 2019). This benefit is because flood water draws up fertile soils or nitrogen-rich sediments from upstream or distant mountains and deposits them bit by bit on farmlands as it travels (Tran et al. 2021; Motsholapheko et al. 2011; Montgomery 2007). This free fertilizer sustains agriculture by increasing the land productivity of farmers in floodplains (Cunfer 2021). In Pwalugu and Kpasenkpe, farmers engage in multiple cropping due to water availability, which is induced by floods. After floods, enough groundwater is retained, which helps farmers resort to multiple cropping.

Floods, on the other hand, shorten the growing season and destroy farms if farmers fail to plant on time. The farmers interviewed implement flood adaptation strategies that accelerate the production process before flooding. With flood adaptation strategies, farmers harvest good yields twice in a year, as early as February/March-April/May (first season) and June/July-August/September (second season) before the flooding. These strategies were classified into (1) modern adaptation practices: supplementary irrigation, early maturity varieties, water-saving/water-harvesting technologies, and scientific weather information received from organizations) and (2) indigenous adaptation practices: farming on multiple plots, diversifying crops, changing crops and planting dates, and reading weather based on beliefs. Generally, farmers adopt various on-farm practices or adjust their farming activities to reduce the adverse effects of flood events on crop production (Alhassan 2020; Tambo, 2016; Ndamani and Watanabe, 2015). Similar adaptation strategies, such as improved seed varieties, irrigation, and soil improvement techniques, including mulching were also reported by Asare-Nuamah et al. (2022) and Ndamani and Watanabe (2016).

Farmers engaged in supplementary irrigation with micro-irrigation technologies early before the rains begin. This finding is consistent with Naz et al. (2018), who found that farmers are not able to produce crops before floods due to dry spells, forcing them to adopt irrigation to deal with drought risks. Irrigation protects against drought and ensures the availability of food reserves (Agyei-Antwi et al. 2018). Several studies in Ghana have also revealed irrigation as one of the key flood adaptation strategies (Agyei-Antwi et al. 2018; Limantol et al. 2016; Ndamani and Watanabe 2015). Farmers also cultivated early maturity varieties or flood-tolerant varieties to harvest before the flooding. This is consistent with Tambo (2016), who found that farmers adopt early maturing crop varieties to reduce the adverse effects of flood events on crop production. Early maturity crop varieties enhance the smallholder farmers' resilience to floods (Kuwornu et al. 2013). Furthermore, farmers interviewed diversified crops on the farm. Farmers grew a mix of crops: some crops were less tolerant to droughts, while others were tolerant to floods. This finding is consistent with Ndamani and Watanabe (2015), who found crop diversification as one of the important flood adaptation strategies used by farmers.

The study also found that most farmers cultivated multiple plots. When floods destroyed farms in floodplains or lowlands, there was usually one plot in the uplands to act as security. Mulching and bunding were the main water-saving and water-harvesting technologies adopted by the majority of farmers interviewed. This finding is congruent with Kuwornu et al. (2013), who identified bunds as flood adaptation strategies by farmers. Farmers in the Savannah agroecology suffer severely from unpredictable weather, making it critical for them to shift crops and planting dates. This finding is consistent with Naz et al. (2018) who found that farmers adopt crop substitutes to address the situation caused by sudden floods and reduce crop damages. To adapt to dry spells in floodplains, farmers adopted deep-water crops that could absorb water for certain periods and reduce crop damage (Naz, 2018). A similar finding is reported in northern Ghana by Ndamani and Watanabe (2015). Switching to crops that are less sensitive to climate stress was also reported by Tambo (2016). Farmers have various beliefs that certain movements in the atmosphere and their environment foreshadow severe rains and flooding. This finding is consistent with Dube and Munsaka (2018), who revealed that communities understand weather patterns and could predict imminent flooding after studying trees and clouds, and the behaviours of certain animal species.

3.4 Conclusion and Recommendations

This paper took an in-depth look at flood adaptation strategies used by farmers in floodplains of northern Ghana. Qualitative data collected through focus group discussion was analysed using content analysis. Based on the findings, floods improve land productivity by replenishing soil fertility and retaining groundwater for crop growth. Most farms in floodplains do not require fertilizers since the soils become fertile and moist after the flood. This benefit entices farmers in floodplains to implement flood adaptation strategies that accelerate the production process before flooding. Farmers meet production deadlines and produce more crops per area by combining indigenous practices with scientific adaptation measures. Modern flood adaptation strategies such as supplementary irrigation, early maturity varieties, water-saving/water-harvesting technologies, and scientific weather information received from organizations were employed. Others used indigenous practices such as farming on multiple plots, diversifying crops, changing crops and planting dates, and reading weather based on beliefs. Policies and programs that integrate indigenous customs with scientific adaptation strategies are needed to enhance the farmers resilient to flood. Farmers should have easy access to weather forecasting information, energy-efficient irrigation technologies, and early maturity cultivars. Researchers and seed producers should work together to increase the production and supply of early maturity varieties so that they can plant and harvest on time. Esoko and Ghana Meteorological Agency (GMA) should be resourced to provide weather information to farmers in flood-prone communities. Farmer-to-farmer learning should be encouraged to aid in the spread of flood adaptation measures in flood-prone communities.

References

Afriyie, K., Ganle, J.K., Santos, E.: 'The floods came and we lost everything': weather extremes and households' asset vulnerability and adaptation in rural Ghana. Climate Dev. **10**(3), 259–274 (2018)

Ahern, M., Kovats, R.S., Wilkinson, P., Few, R., Matthies, F.: Global health impacts of floods: epidemiologic evidence. Epidemiol. Rev. **27**(1), 36–46 (2005)

Alderman, K., Turner, L.R., Tong, S.: Floods and human health: a systematic review. Environ. Int. **47**, 37–47 (2012)

Alhassan, H.: Farm households' flood adaptation practices, resilience and food security in the Upper East region. Ghana. Heliyon **6**(6), e04167 (2020)

Al-Hassan, R.M., Kuwornu, J.K., Etwire, P.M., Osei-Owusu, Y.: Determinants of choice of indigenous climate related strategies by smallholder farmers in Northern Ghana. British J. Environ. Climate Change **3**(2), 172–187 (2013)

Amoateng, P., Finlayson, C.M., Howard, J., Wilson, B.: A multi-faceted analysis of annual flood incidences in Kumasi, Ghana. Int. J. Disaster Risk Reduction **27**, 105–117 (2018)

Antwi-Agyei, P., Dougill, A.J., Stringer, L.C., Codjoe, S.N.A.: Adaptation opportunities and maladaptive outcomes in climate vulnerability hotspots of northern Ghana. Clim. Risk Manag. **19**, 83–93 (2018)
Armah, F.A., Yawson, D.O., Yengoh, G.T., Odoi, J.O., Afrifa, E.K.: Impact of floods on livelihoods and vulnerability of natural resource dependent communities in Northern Ghana. Water **2**(2), 120–139 (2010)
Asante, F., Guodaar, L., Arimiyaw, S.: Climate change and variability awareness and livelihood adaptive strategies among smallholder farmers in semi-arid northern Ghana. Environ. Dev. **39**(2), 100629 (2021)
Asare-Nuamah, P., Antwi-Agyei, P., Dick-Sagoe, C.: Mitigating the risks of climate variability and change on mango seedlings in Ghana: Evidence from mango seedlings producers in the Yilo Krobo Municipality. Environmental Challenges, p.100594 (2022)
Banerjee, L.: Effects of flood on agricultural productivity in Bangladesh. Oxf. Dev. Stud. **38**(3), 339–356 (2010)
Cornille, O., Bocquet, J.M., Wiafe, K.: Pwalugu Multipurpose Dam Project: story of a major collaborative project. In: E3S Web of Conferences, vol. 346, p. 03008. EDP Sciences (2022)
CRED, 2022. '2021 Disasters in numbers', Brussels: https://cred.be/sites/default/files/2021_EMDAT_report.pdf
Cunfer, G.: Soil fertility on an agricultural frontier: the US Great Plains, 1880–2000. Soc. Sci. Hist. **45**(4), 733–762 (2021)
Dar, M.H., et al.: Transforming rice cultivation in flood prone coastal Odisha to ensure food and economic security. Food Secur. **9**(4), 711–722 (2017)
Elo, S., Kyngäs, H.: The qualitative content analysis process. J. Adv. Nurs. **62**(1), 107–115 (2008)
Fadul, E., Masih, I., De Fraiture, C.: Adaptation strategies to cope with low, high and untimely floods: Lessons from the Gash spate irrigation system, Sudan. Agric. Water Manag. **217**, 212–225 (2019)
Fiasorgbor, D., Wiafe, E., Tettey, C., Abasiyam, M.: Assessment of the coping strategies of flood victims in the Builsa District. Environ. Sustainability **2**(1), 17–25 (2018)
Fujita, K., Shaw, R.: Preparing International Joint Project: use of Japanese flood hazard map in Bangladesh. Int. J. Disaster Risk Manage. **1**(1), 62–80 (2019)
Goel, A., Ganesh, L.S., Kaur, A.: Deductive content analysis of research on sustainable construction in India: current progress and future directions. J. Clean. Prod. **226**, 142–158 (2019)
Guodaar, L., Bardsley, D.K., Suh, J.: Indigenous adaptation to climate change risks in northern Ghana. Clim. Change **166**(1), 1–20 (2021)
IPCC: Managing the risks of extreme events and disasters to advance climate change adaptation. The World Bank, Washington, DC (2012). https://doi.org/10.1596/978-0-8213-8845-7
John, K.M., Etwire, P.M., Osei-Owusu, Y.: Adaptation strategies of smallholder farmers to climate change and variability: evidence from Northern Ghana. Inf. Manage. Bus. Rev. **5**(5), 233–239 (2013)
Karley, N.K.: Flooding and physical planning in urban areas in West Africa: Situational analysis of Accra, Ghana. Theoretical and Empirical Researches in Urban Manage. **4**(4 (13), 25–41 (2009)
Kassim, Z., Alhassan, H., Appiah-Adjei, C.: Farm households' choice of strategies in response to floods in the Builsa-North district. Ghana. Climate Dev. **13**(6), 484–492 (2021)
Kundzewicz, Z.W., et al.: Flood risk and climate change: global and regional perspectives. Hydrol. Sci. J. **59**(1), 1–28 (2014)
Lolig, V., et al.: Households' coping strategies in drought-and flood-prone communities in Northern Ghana. J. Disaster Res. **9**(4), 542–553 (2014)
Milojevic, A., et al.: Long-term effects of flooding on mortality in England and Wales, 1994–2005: controlled interrupted time-series analysis. Environ. Health **10**(1), 1–9 (2011)

Montgomery, R.R., et al.: Functional assay of VWF measured by its interaction with platelet GPIb without the use of ristocetin. Blood **110**(11), 3158 (2007)

Motsholapheko, M.R., Kgathi, D.L., Vanderpost, C.: Rural livelihoods and household adaptation to extreme flooding in the Okavango Delta, Botswana. Physics and Chemistry of the Earth, Parts A/B/C **36**(14–15), 984–995 (2011)

Munsaka, E., Dube, E.: The contribution of indigenous knowledge to disaster risk reduction activities in Zimbabwe: A big call to practitioners. Jàmbá: Journal of Disaster Risk Studies **10**(1),1–8 (2018)

Namara, R.E., et al.: Typology of irrigation systems in Ghana, vol. 142. IWMI (2011)

Naz, F., Doneys, P., Saqib, S.E.: Adaptation strategies to floods: a gender-based analysis of the farming-dependent char community in the Padma floodplain, Bangladesh. Int. J. Disaster Risk Reduct. **28**, 519–530 (2018)

Ndamani, F., Watanabe, T.: Farmers' perceptions about adaptation practices to climate change and barriers to adaptation: a micro-level study in Ghana. Water **7**(9), 4593–4604 (2015)

Ndamani, F., Watanabe, T.: Determinants of farmers' adaptation to climate change: a micro level analysis in Ghana. Scientia Agricola **73**, 201–208 (2016)

Ntim-Amo, G., et al.: Farm households' flood risk perception and adoption of flood disaster adaptation strategies in northern Ghana. Int. J. Disaster Risk Reduct. **80**, 103223 (2022)

Posthumus, H., Morris, J., Hess, T.M., Neville, D., Phillips, E., Baylis, A.: Impacts of the summer 2007 floods on agriculture in England. J. Flood Risk Manage. **2**(3), 182–189 (2009)

Quarshie, A.M., Leuschner, R.: Cross-sector social interactions and systemic change in disaster response: a qualitative study. J. Bus. Ethics **150**(2), 357–384 (2018)

Schneider, S.H., et al.: Assessing key vulnerabilities and the risk from climate change. Climate Change 2007: Impacts, Adaptation and Vulnerability. Contribution of Working Group II to the Fourth Assessment Report of the Intergovernmental Panel on Climate Change, M.L. Parry, O.F. Canziani, J.P. Palutikof, P.J. van der Linden and C.E. Hanson, Eds., Cambridge University Press, Cambridge, UK, pp. 779–810 (2007)

Svetlana, D., Radovan, D., Ján, D.: The economic impact of floods and their importance in different regions of the world with emphasis on Europe. Procedia Econ. Financ. **34**, 649–655 (2015)

Tangonyire, D.F., Akuriba, G.A.: Socioeconomic factors influencing farmers' specific adaptive strategies to climate change in Talensi district of the Upper East Region of Ghana. Ecofeminism Climate Change **2**(2), 50–68 (2020)

Tellman, B., et al.: Satellite imaging reveals increased proportion of population exposed to floods. Nature **596**(7870), 80–86 (2021)

Tran, D.D., Huu, L.H., Hoang, L.P., Pham, T.D., Nguyen, A.H.: Sustainability of rice-based livelihoods in the upper floodplains of Vietnamese Mekong Delta: Prospects and challenges. Agric. Water Manag. **243**, 106495 (2021)

Ulubaşoğlu, M.A., Rahman, M.H., Önder, Y.K., Chen, Y., Rajabifard, A.: Floods, bushfires and sectoral economic output in Australia, 1978–2014. Econ. Rec. **95**(308), 58–80 (2019)

Wang, X., Liu, Z., Chen, H.: Investigating flood impact on crop production under a comprehensive and spatially explicit risk evaluation framework. Agriculture **12**(4), 484 (2022)

Williams, P.A., Crespo, O., Abu, M.: Adapting to changing climate through improving adaptive capacity at the local level–The case of smallholder horticultural producers in Ghana. Clim. Risk Manag. **23**, 124–135 (2019)

World Bank: Disaster risk management programs for priority countries. The World Bank, Washington, DC (2011)

Yankson, P.W.K., Owusu, A.B., Owusu, G., Boakye-Danquah, J., Tetteh, J.D.: Assessment of coastal communities' vulnerability to floods using indicator-based approach: a case study of Greater Accra Metropolitan Area. Ghana. Natural Hazards **89**(2), 661–689 (2017)

Yiran, G.A., Stringer, L.C.: Spatio-temporal analyses of impacts of multiple climatic hazards in a savannah ecosystem of Ghana. Clim. Risk Manag. **14**, 11–26 (2016)

Yiran, G.A.B., Atubiga, J.A., Kusimi, J.M., Kwang, C., Owusu, A.B.: Adaptation to perennial flooding and food insecurity in the Sudan savannah agroecological zone of Ghana. Environ. Res. **214**(3), 114037 (2022)

Zickgraf, C.: Climate change, slow onset events and human mobility: reviewing the evidence. Current Opinion Environ. Sustainability **50**, 21–30 (2021)

What Do We Need to Learn from Three Recent Major Disasters Caused by Extreme Hydrological Events in Brazil?

Masato Kobiyama(✉), João Gabriel Fontes Maciel, Erika Gabriella Ruoso, Alessandro Gustavo Franck, and Marina Refatti Fagundes

Graduate Program of Water Resources and Environmental Sanitation, Federal University of Rio Grande do Sul, Av. Bento Gonçalves, Porto Alegre, RS 9500, Brazil
masato.kobiyama@ufrgs.br

Abstract. Considering only the human fatalities, the hydrological disasters that occurred in Petrópolis city (Rio de Janeiro state) in February 2022, in the Metropolitan Region of Recife in May and June 2022, and on the north coast of São Paulo state in February 2023, could be considered the largest disasters in Brazil in 2022 and 2023. Therefore, this study aims to highlight the main characteristics of these three hydrological disasters and to propose important lessons that Brazilians and their society should learn from them. The current problems in the integrated disaster risk management in Brazil, both in terms of local populations and public institutions, are fundamentally due to a deficient system of Disaster Risk Reduction Education. For developing this type of education, it is essential to establish school catchments and utilize them through citizen science. Furthermore, given the changing types of disasters, it will be increasingly necessary to pay special attention to disasters occurring in mountain regions where landslides, debris flows and flash floods are frequent.

Keyword: Disaster risk reduction education · School catchment · Mountain environment · Citizen science

1 Introduction

COVID-19 threatened the global population starting from the end of 2019. Soon later, the World Health Organization (WHO) characterized this event as a pandemic on March 11th, 2020. After an intensive global battle spanning over 3 years, the WHO declared the end of the pandemic on May 5th, 2023, informing that the disease no longer constituted a Public Health Emergency of International Concern (Lenharo 2023). While COVID-19 continues to affect all the society, it is surely the climate change that most-profoundly impacts socio-economic and environmental conditions today and will continue to do so in the future. This global phenomenon must be urgently addressed, because human interventions' failures against it can be fatal to the world society (WEF 2023, Wynes et al. 2022). Therefore, all the world population must learn the necessary actions. Since correct and urgent measures against climate change represent one of humanity's greatest

H. Tatano et al. (Eds.): GSRIDRR 2023, DRRGADRIBS, pp. 267–291, 2025.
https://doi.org/10.1007/978-981-96-6609-6_21

current challenges, this issue is dealt as the 13th Sustainable Development Goal in the 2030 Agenda (UN 2015).

Climate change affects several aspects in the world, among which natural disasters stand out. The relationships between climate change and natural disasters have been discussed for many decades (Rädler 2022, Thomas and López 2015, Van Aalst 2006). Although there are several discussions about terminologies related to "disasters" and "natural disasters", the present study follows the definition of Chaudhary and Piracha (2021), in which natural disasters are caused by natural hazards such as floods, landslides, debris flows, and droughts, when they interact with social factors like settlements and infrastructure.

There are always reasons (functional and structural factors) for disasters to occur. The places where disasters occur are precisely those where society exists and/or interacts. Human involvement in disaster occurrence is very clear. In other words, human presence is absolutely essential for disasters to occur. In this sense, we can say that there is no reason why we need to call social disasters. Indeed, disaster consists in two main factors: (i) intrinsic (or structural) factors associated to vulnerability and exposure; and (ii) extrinsic (or functional) factors characterizing natural or technological hazards (Fig. 1).

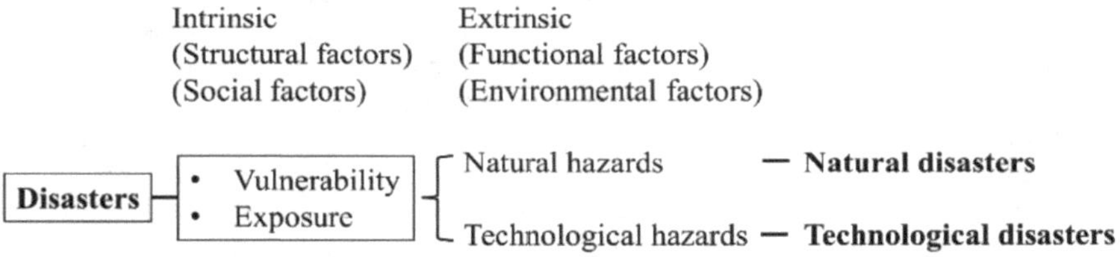

Fig. 1. Disasters and natural disasters.

According to UNISDR (2009), natural hazards are defined as a natural phenomenon that can cause loss of life, injury, or other health impacts, damage to property, loss of livelihoods and services, social and economic disruption, or environmental damage. Therefore, floods, landslides, debris flows, droughts, and so forth are initially just natural phenomena, which occur due to the local or regional characteristics such as vegetation, climate, topography, soil, and rocks, and can be considered natural hazards. When they occur in areas where humans live, they can cause damage and are then treated as natural disasters (Kobiyama et al. 2020). These natural hazards in Brazil are mainly associated to the climate.

Climate is characterized by various physical parameters contributing to the climate change debate. Among these parameters, rainfall and temperature are considered keys (Espinosa et al. 2022, Petroliagkis and Alessandrini 2021). Regarding rainfall, extreme hydrological events have been affecting many countries worldwide, including Brazil, in recent years. In Brazil, until early 2022, the national and official record for rainfall in 24 h was 407.6 mm, recorded in Maceió, Alagoas state, on April 28th, 1979. An approximately-equal intense rainfall in 24 h (404.8 mm) was recorded in Florianópolis, Santa Catarina state, on November 15th, 1991. According to Saldanha et al (2012), a more intense rainfall in 24 h recorded scientifically, but not officially, in Pelotas, Rio Grande do Sul state, was 418 mm, 464 mm and 479 mm with three different rainfall gauges at the same locality (research institute) on January 28th, 2009. However, the rainfall recorded in Petrópolis, Rio de Janeiro state, on March 20th, 2022, broke these national records, reaching 534.4 mm in 24 h. The record was broken again on February 18th and 19th, 2023, with rainfall reaching 682 mm in 24 h in Bertioga, São Paulo state. Hence, rainfall records across the country have been frequently changing, which creates anxiety, concern, and fear for disaster risk reduction (DRR) managers and the general public.

In the early 21st century, two severe disasters occurred in Brazil: one in Itajaí valley in 2008 (Frank and Sevegnani 2009), and the other in the Fluminense mountain region in 2011 (Marques and Baesso 2021). Based on the classification of the Centre for Research on the Epidemiology of Disasters – CRED (Below et al. 2009), both disasters are classified as hydrological. These two tragedies led the Brazilian government to implement a significant political reaction, namely, the approval of Law No. 12,608/12, which established the National Policy on Civil Protection and Defense – PNPDEC (Brasil 2012).

Although covering prevention, mitigation, preparation, response, and recovery actions, this law places special emphasis on the prevention stage through integration with education and environmental policies, among others. According to Silva et al. (2021b), this resulted from seeking harmony with the international scenario. Thus, there is a global and national movement to consider education as a crucial factor in DRR. UNESCO-UNICEF (2011) emphasized that education is a central topic in building resilient cities against natural hazards. The importance of education in DRR is also strongly highlighted in the Sendai Framework (UNISDR 2015).

Following this trend, some practices of education in DRR have been reported and analyzed in various countries, for examples, India (Petal 2011), Brazil (Matsuo and Silva 2021, Silva et al. 2021a), Nepal (Nakano and Yamori 2021), and Indonesia (Amri et al. 2022). Furthermore, to understand how DRR is being incorporated into school curricula, Selby and Kagawa (2012) conducted case studies in 30 countries. All these countries recognize the education importance in DRR and show their interest in investing in it. However, the above-mentioned reports present various problems such as the lack of adequate training for teachers, the lack of after-care for already-trained teachers, and the absence of continuity in transmitting DRR-related knowledge to students.

In addition to these issues in the teaching-learning system, teaching materials are lacking, especially in Brazil. Due to this situation, Zanandrea et al. (2023) published a set of records of the 18 main disasters that occurred in Brazil during the period 1941–2021. This type of record is crucial in DRR because it can be used for various purposes such as construction planning of dams, dikes, artificial canals, and also mapping of hazard and risk areas. Michel et al. (2021) highlighted the importance of using records for calibration and validation of computational models adopted in predicting natural hazards such as debris flows, landslides, and flash floods.

The record, i.e., historical description, is essential not only for scientists and managers of natural disasters and risks, but also for local inhabitants. According to Garnier and Lahournat (2022), in Japan, the local community that suffered from disasters has the habit of placing a memorial or monument such as stone monuments. This record provides a better understanding of the occurrence, and the community prepares for upcoming disasters. On the other hand, in the case of instructional and bibliographic materials, the monument about a disaster occurrence can be a detailed description of that event, along with its analysis of how and why it happened. Additionally, a proposal for preparation for the future can be included. Some good examples as detailed description in Brazil are: the tsunami and earthquake in the Northeast in 1755 (Veloso 2015); the flood in Porto Alegre, Rio Grande do Sul state, in 1941 (Guimaraens 2009); the flood in Tubarão, Santa Catarina state, in 1974 (Machado 2005); and the historical occurrences of floods in Joinville, Santa Catarina state, from 1851–2008 (Silveira et al. 2009).

In Brazil, due to its immense territory (8,510,345 km^2) and a large population (> 200 million inhabitants), there are many diversities such as geodiversity (different lithological and geomorphological units), biodiversity (different biomes), climatological diversity, sociodiversity (social inequality), and religious diversity. This naturally leads to a large diversity of natural disasters in the country, which strongly requires society to record each disaster occurrence to prepare for future incidents (Fig. 2).

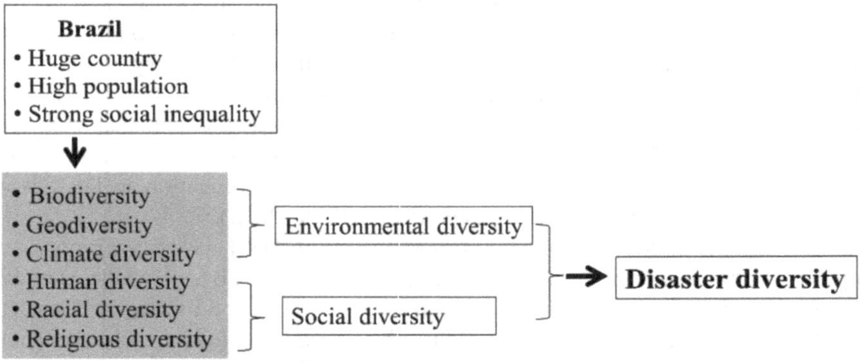

Fig. 2. Reasons for Brazil to have a high disaster diversity.

It is worth noting that the eastern part of the country has several mountainous regions, such as the Serra do Mar and Serra Geral. This part is very near to the ocean and the mountain environment as well as to their interaction. According to Groisman et al. (2005) and Marengo et al. (2009), a significant increase in total rainfall and rainfall intensity has been observed in this region under a current climate change. This surely increases concerns about the occurrence of large-scale debris flows which may become more frequent in mountain regions where urbanization and economic development occur without a scientific basis, increasing social vulnerability (Kobiyama et al. 2019). From the DRR perspective, municipal, state, and federal governments need to pay more attention to reducing social vulnerability, especially in the eastern part of the country.

In this context, Brazil suffered three catastrophic hydrological disasters in 2022–2023, precisely in the above-mentioned region. Therefore, the objective of the present study was to describe, for the educational purposes, the main characteristics of these three disasters and to discuss some lessons learned from them.

2 Description of Three Devastating Hydrological Disasters in Brazil Occurring in 2022 and 2023

Very large hydrological disasters, specifically sediment-related disasters, occurred in three locations in Brazil in the years 2022 and 2023 (Fig. 3 and Table 1).

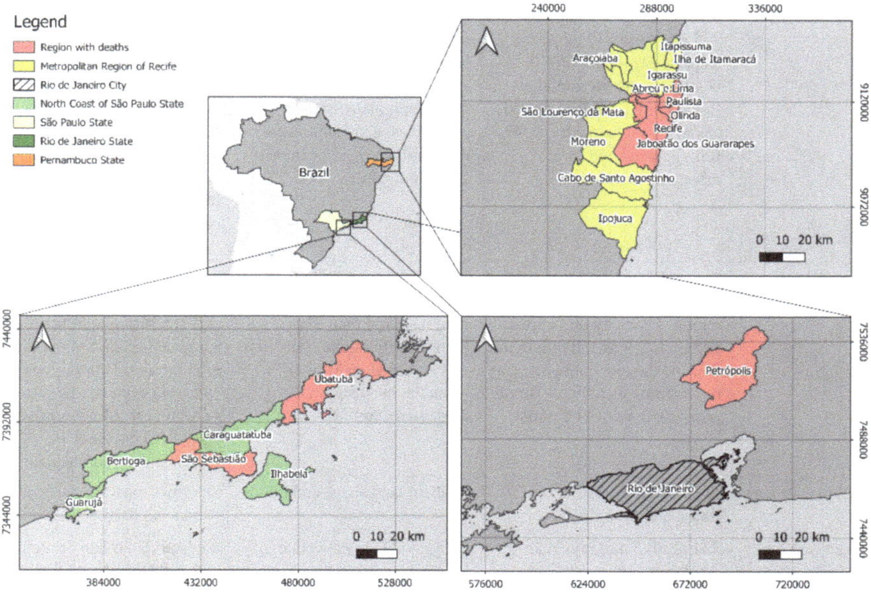

Fig. 3. Locations of three major hydrological disasters that occurred in Brazil in 2022 and 2023. Note: The Metropolitan Region of Recife consists of 14 municipalities. The five municipalities marked in pink are the ones that reported loss of human lives due to the disaster. The northern coast of São Paulo is composed of five municipalities, with the two marked in pink reporting deaths due to the disaster.

2.1 Petrópolis (Rio de Janeiro) in February 2022

Table 2 shows the history of all the hydrological disasters with the number of victims registered by the Civil Defense of Petrópolis from 1966 to 2022. According to Kobiyama et al. (2019), Petrópolis is historically the municipality with the highest number of debris flow disasters and fatalities in Brazil. These statistics indicate that this municipality is frequently subject to water and sediment-related disasters. Therefore, it is easy to imagine that there are many internal and external factors causing these disasters in this place.

In 2012, it was estimated that the municipality of Petrópolis had 500 areas considered at landslides risk. In 2017, the Municipal Risk Reduction Plan reported that over 28,000 households were within high or very high-risk areas, representing one-third of the total area of the municipality at that time (Xavier 2022).

Table 1. Comparison of three major hydrological disasters that occurred in Brazil in 2022 and 2023.

Location	City of Petrópolis	Metropolitan Region of Recife	North Coast of São Paulo
Geographical aspects	The population, area, and average altitude above sea level of the municipality of Petrópolis, State of Rio de Janeiro, are 278,881 (in 2022), 795 km^2, and 838 m, respectively. Petrópolis belongs to the Metropolitan Region of Rio de Janeiro and has many tourism attractions. It is characterized by a mountain environment	The population, area, and average altitude of Recife, the capital of Pernambuco state, are 1,488,920 (in 2022), 218 km^2, and 4 m, respectively. This city is located in an alluvial plain along the Atlantic Ocean coast, with many islands, peninsulas, and mangroves. Its metropolitan region comprises 14 municipalities	The north coast of São Paulo state consists of 5 municipalities (Bertioga, São Sebastião, Caraguatatuba, Ilhabela, and Ubatuba). The total population and total area of this region are 408,517 (in 2022) and 2,431 km^2, respectively. The region is characterized by the presence of the Atlantic Forest and is located in the Serra do Mar, featuring a mountainous and coastal environment with few flat areas
Dates and rainfall characteristics	The intensity of the rainfall that occurred in a very small region of the city on February 15th, 2022, was 230 mm in just 3 h, and the maximum intensity recorded by the Alto da Serra rain gauge was 199.2 mm/h	The event lasted 22 days (May 22nd to June 12th, 2022), totaling 926.4 mm. The highest daily rainfall (190 mm) was recorded on May 28th, when the most significant damages occurred	The torrential rainfall event occurred from February 18th, 2023 (Saturday during the Carnival holiday) in the early evening until February 19th (Sunday) evening. The amount of rainfall in 24 h in Bertioga was 682 mm, which is currently the national record
Natural phenomena	Shallow landslides, debris flows, and flash floods	Shallow landslides, and floods	Shallow landslides, debris flows, and flash floods
Damages and losses	234 dead, 3 missing, 994 displaced, and an economic loss of about US$ 123 million (considering only direct damages to businesses)	130 dead (121 deaths due to landslides), 9,134 displaced, and 119,523 temporarily relocated in the Metropolitan Region of Recife	65 dead (64 in São Sebastião and 1 in Ubatuba), 1,815 displaced, and 2,251 temporarily relocated in the north coast of São Paulo state

Table 2. Number of human deaths due to hydrological disasters in Petrópolis from 1966 to 2022.

Year	Deaths	Year	Deaths	Year	Deaths	Year	Deaths
1966	80	2001	51	2010	1	2018	4
1977	11	2003	17	2011	73	2021	1
1979	87	2007	3	2013	34	2022	234
1988	171	2008	9	2016	2	-	-
1997	6	2009	6	2017	1	-	-

An inappropriate (and often illegal) land-occupation by the vulnerable population and inefficient territorial management by public institutions have increased social vulnerability and population exposure. Without authorization of the municipal government, residents have been building their houses in these areas on their own. Thus, risk areas have been occupied by a poor population that has no other choice. On the other hand, as this municipality offers various types of tourism, there is a strong tendency to reduce the width of the natural river channel for road and building construction, which certainly reduces the river transport capacity. Vegetation areas have also decreased.

According to Pinto and Iory (2022), the Petrópolis City Hall invested about US$ 25,714 with its Civil Defense in 2020. This amount represents only 0.009% of the municipal revenue of US$ 286 million in 2020. In 2021, the City Hall further cut expenses with Civil Defense, investing only US$ 17,170, equivalent to only 0.007% of the revenue of US$ 245 million in 2021. Thus, it is observed that the actions of Petrópolis' whole society including residents and public agencies served to increase vulnerability in the municipality.

In the 2022 February disaster, Petrópolis Civil Defense reported 325 occurrences in 24 h, including 269 landslides (METSUL 2022). Figure 4a shows one of the shallow landslide occurrences in the municipality. Most of the locations where disasters occurred are located at the foot of steep hillslopes characterized by precarious land occupations. In cases where the amount of rock and soil material was larger and/or the steep hillslope was more extensive, debris flows occurred (Fig. 4b).

Fig. 4. Sediment-related disasters in Petrópolis municipality in February 2022: (a) shallow landslide (Source: METSUL (2022)); (b) debris flow (Source: Modesto et al. (2022)).

The National Center for Monitoring and Alerts of Natural Disasters (CEMADEN) notified that there would be heavy rain in Petrópolis by sending a "very intense climatic event" alert to Petrópolis municipality two days before the tragedy, but without providing information about the likely intensity of the rainfall. According to one meteorologist of CEMADEN, the clouds practically did not move, causing the rainfall to concentrate mainly in Petrópolis until the clouds dissipated (Candal and Tadeu 2022). This phenomenon might be similar to the phenomenon called "training". In this case, the equipment and monitoring system at that moment were not efficient. Thus, there were no models, methodologies, and tools that could predict this very high amount of rainfall at a very small area for a very short time. Understanding the mechanism of occurrence of this phenomenon and seeking measures against it are a major challenge for current meteorology worldwide (Min et al. 2021, Schumacher and Johnson 2008, Unuma and Takemi 2016).

2.2 Metropolitan Region of Recife, Pernambuco State, in May and June 2022

The current situation of Recife, the capital of Pernambuco state, which plays an important cultural and economic role, has stimulated the influx of people into the city from within the state, outside the state, and even abroad. The population of Recife and its Metropolitan Region in 2022 were 1,488,920 and 3.7 million, respectively. As Recife is known the 'Brazilian Venice', its geomorphic features are highly prone to coastal and river floods as well as tidal surges.

In the early stage of the city's implementation, poorer individuals began to live predominantly in vulnerable areas, especially in coastal zones, wetlands, and mangroves. These areas are naturally susceptible to coastal floods. Due to this history of population settlement, the residents of Recife often suffered from extensive and severe floods. The two largest recorded disasters with floods occurred in 1966 and 1975 (Table 3). Alves (2022) reported that in the past, a significant portion of the poor population living in wetlands or floodplains was evacuated and transported to hillslopes and cliffs. Conducting interviews with various experts in Recife, this author highlighted the housing deficit and the urgent establishment of appropriate housing policies. In summary, it can be said that the increase in population density on the slopes and cliffs of the city contributed to social problems, namely sediment-related disasters. According to IPCC (2007), Recife is considered one of the most vulnerable cities in Brazil, precisely due to all the geographical and social issues: very low altitude, very high population density, and socio-economic inequality.

Poor residents of the coastal region used to suffer from disasters related to river and coastal floods. As mentioned above, they were forcibly removed from these floodplains to the upper or lower parts of steep slopes. According to Lins et al. (2022), today, approximately 600 thousand people (about 40% of the population of Recife) live in precarious houses built at the top and base of cliffs and on steep slopes, i.e., at dangerous locations. Most of these areas are very prone to shallow landslides. In fact, there is a growing trend of landslides occurring in Recife. These authors investigated the telephone calls from residents reporting landslides, recorded by the Civil Defense of Recife during the period from 2014 to May 2022, and reported a total of 7,340 recorded calls (Table 4).

In terms of inequality, the Metropolitan Region of Recife is the worst among Brazilian metropolises. In this city, 13% of the region's population lives in extreme poverty (with income below US$ 32/month), and 39.7% live in poverty (US$ 93/month) (Salata and Ribeiro 2022). On the other hand, Recife has been playing an important role in climate change issues and greenhouse gas emissions, showcasing its leadership nationally and internationally (Leão et al. 2021). While these actions are undoubtedly helpful in improving the safety of residents and should be positively evaluated, the results of these actions have not yet reached people in socioeconomically vulnerable situations.

Table 3. Three main disasters that occurred in Recife (Metropolitan Region).

Year	Locality	Date	No. of deaths	Characteristics
1966	Recife	May 30th	175	Predominantly floods. The Capibaribe River overflowed. In many places, the water was over 2 m deep. There were no issues with cliffs
1975	Recife	July 17th-18th	107	Predominantly floods. The Capibaribe River overflowed. 80% of the city was flooded. 350,000 people were temporarily displaced. At one moment, over 50% of the city's houses were without electricity. There were no issues with cliffs
2022	Metropolitan Region of Recife	May 22nd to June 12th (May 28th, main)	130	Predominantly shallow landslides. Over 100 people were buried alive in soils and debris on the slopes and died

Table 4. Number of calls related to landslides in Recife from 2014 to May 2022.

Year	2014	2015	2016	2017	2018	2019	2020	2021	2022
Calls	486	1043	1124	710	639	1064	253	1145	876

Note: The number of calls in 2022 is only until May

Figure 5 shows a typical scene of a shallow landslide at the border between the municipalities of Recife and Jaboatão dos Guararapes. This border region, characterized by hills and cliffs, was severely affected in May 2022, with many people being buried alive. Table 5 presents the number of human deaths per municipality and also the mean monthly rainfall and the recorded rainfall for the months of May and June 2022. In other municipalities that had a similar amount of rainfall to Recife and Jaboatão dos Guararapes, a high number of deaths was not recorded. The main cause for these two municipalities having the highest number of deaths may be this geomorphic feature. In other words, the main determining factor was whether or not there are inhabitants in areas susceptible to shallow landslides.

Fig. 5. Typical scene of a shallow landslide occurrence on the slope at the border between the municipalities of Recife and Jaboatão dos Guararapes in May 2022. (Source: Alves (2022)).

Table 5. Number of human deaths and monthly rainfall for the months of May and June recorded in each municipality.

Municipality	No. of deaths	Rainfall in May (mm/month)		Rainfall in June (mm/month)	
		2022	Average	2022	Average
Camaragibe	7	684.5	294.6	393.5	339.1
Jaboatão dos Guararapes	64	675.0	310.1	436.0	352.5
Olinda	6	496.1	325.8	508.9	384.7
Paulita	1	653.5	310.1	447.7	363.0
Recife	50	686.4	328.9	449.2	389.6
Bom Conselho*	1	114.0	86.0	240.8	95.6
Limoeiro*	1	316.7	135.6	190.7	186.5

* The municipalities of Bom Conselho and Limoeiro do not belong to the Metropolitan Region of Recife, being located at 233.41 km and 64.75 km, respectively, from Recife.

Based on the information presented by Aguiar (2022) examining public data of various government agencies, Table 6 succinctly demonstrates the fatality data by category. It is clear that the disaster occurrence in the metropolitan region of Recife in 2022 was predominantly characterized by shallow landslides causing 121 deaths.

As demonstrated in Table 5, the extremely high amount of rainfall was decisive for the disaster occurrence in the Metropolitan Region of Recife in 2022. Figure 6 presents the daily rainfall recorded in Recife from May 20th to June 14th, 2022. This event lasted 22 days (May 22nd to June 12th), with a total amount of 926.4 mm and a mean daily rainfall of 42.1 mm during this period. The highest amount of rainfall (190 mm) occurred

Table 6. Characteristics by categories of fatalities caused by the disaster in the Metropolitan Region of Recife (Pernambuco) in May and June 2022.

Category	Description	Fatalities
Gender	Male Female	58 72
Type of hazard	Translational landslide Flood	121 9
Age	0–12 years old 13–17 years 18–30 years 31–65 years > 65 years	25 8 24 63 10

on May 28th and caused the most significant damage. It is important to note that the mean annual rainfall in Recife is about 2,155 mm.

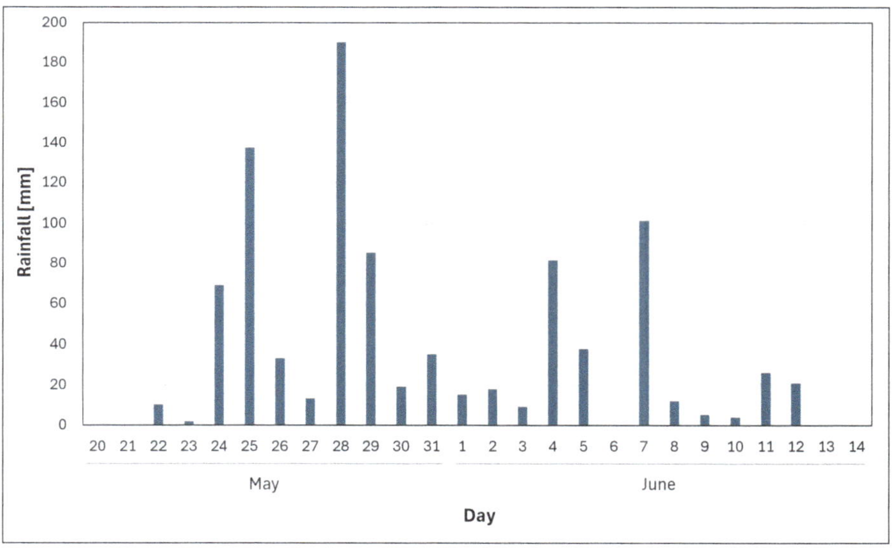

Fig. 6. Daily rainfall record in Recife from May 20 to June 14, 2022.

The CEMADEN issued a geo-hydrological bulletin on May 25th, warning of the "high risk" of heavy rainfall and landslides in the Metropolitan Region of Recife. Even with this alert in effect, it was only on May 27th (Friday), when the State Agency of Waters and Climate of Pernambuco (APAC) issued another statement informing about the forecast of heavy rainfalls for the weekend, that the Recife City Hall finally activated the contingency plan. After this alert from APAC, the Civil Defense of Recife acted immediately on May 27th in the morning, calling more than 3,000 municipal employees to be on duty from the same day. Furthermore, APAC sent the alert to more than 32 thousand families living in risk areas through the Short Message Service (SMS). However, some community-leaders commented that local residents were not informed that they would need to leave the area. Moreover, the residents did not receive assistance such as vehicles, to move them to safer shelters (Madeiro 2022).

2.3 North Coast Region of São Paulo State in February 2023, with Focus on the Municipality of São Sebastião

Bordering the state of Rio de Janeiro, the northern coast region of São Paulo state has the best beaches in the state, characterized by the Atlantic Forest and the Serra do Mar. Souza et al. (2018) demonstrated that the spatial distribution of slopes in the north coast is 20.33% (< 0.02 m/m), 3.29% (0.02–0.05 m/m), 19.14% (0.05–0.15 m/m), 42.57% (0.15–0.30 m/m), and 14.68% (> 0.30 m/m). This data indicates that there are few flat (safer) areas between the sea and steep hillslopes. In these flatter or safer areas, luxurious hotels and condominiums for wealthy peoples have been constructed. Workers serving these high-class categories live on or near the slopes, i.e., in areas prone to mass movements and/or flash floods.

According to Schmidt (2023), the City Hall of São Sebastião has received at least four alerts about landslide and flood risks since 2013, in which the State University of Campinas investigated landslide risk areas and provided their map with a warning to the City Hall. In 2018, the Technological Research Institute (IPT) pointed out 52 areas at landslides risk in São Sebastião. Their report showed that there were 161 houses in high-risk areas and 2,043 houses in medium or low-risk areas in São Sebastião. Upon identifying landslide risk areas in Sahy Village in São Sebastião in 2020, the State Public Prosecutor's Office of São Paulo filed a lawsuit against the City Hall to demand action, and concluded that Sahy Village would experience a "true tragedy". Finally, in 2022, the Housing and Urban Development Company of the State of São Paulo (CDHU) suggested that families occupying Sahy Village should be relocated to a neighboring area which is outside these landslide risk areas.

UOL (2023) reported a comment of the current mayor of São Sebastião: "People do not leave these areas, and continue to live there. Talking is easy. When we started the demolition program, the public outcry is enormous; the pressure was intense". This mayor's comment makes it clear that society as a whole, along with the strong determination of each resident, has been increasing its vulnerability. This strength often stems from the fact that residents have nowhere else to move.

In this social situation, a tragedy occurred during the Carnival of 2023. Here, based on the report by Casemiro and Matos (2023) and Xavier (2023), the main actions of various organizations are presented chronologically. On February 16th, 2023 (Thursday), with 48 h in advance, CEMADEN sent a disaster forecast warning about 'very intense rainfalls with the potential to cause disasters' to the federal government and the State Civil Defense of São Paulo (DC-SP), including possible disaster locations. On February 17th (Friday), still in the morning, CEMADEN held a meeting with the National Center for Risk and Disaster Management (CENAD) and DC-SP. Therefore, the warning of a possible incident was well in advance and was notified to the competent authorities and local authorities. On February 18th (Saturday), CEMADEN issued over 60 alerts to various municipalities such as Ubatuba, Ilhabela, Caraguatatuba, and São Sebastião. At noon, DC-SP sent an SMS to the population warning about heavy rainfalls in the region. However, these messages sent to part of the population had no effect. On Saturday, around 11 pm, the storm was already extremely intense, and the municipal civil defense of São Sebastião tried to reach Sahy Village, but there was no more access due to landslides blocking the roads. On February 19th (Sunday) at 6 am, a new SMS from DC-SP warned about the occurrence of rainfall, wind, and lightning and asked people to leave the area in case of "tilting and cracks in the walls" observation. However, these messages still did not mention anything about landslides. At that moment, there had been already deaths, missing persons, and people buried.

Figure 7 shows a typical occurrence of sediment-related disasters. Mainly in the northern coastal region of São Paulo and especially in the case of São Sebastião, the Serra do Mar practically collapses onto the seashore, making the coastal zone very narrow. This geomorphic feature might be decisive for this tragedy. There is nowhere to run in an evacuation, forming a true "death corridor". The lower region received sediments abruptly and violently descending from steep hillslopes (Fig. 7a). The poor people, whether they lived in old local communities or came to work on construction sites, had to occupy the foothills of the mountains, areas formed geologically by the accumulation of soil and rocks caused by the landslide's deposition. Therefore, such areas can be affected again by the similar geomorphic event (Fig. 7b).

Fig. 7. Sediment-related disasters in the municipality of São Sebastião in February 2023: (a) Mud and debris flow covering an area of houses of the low-income population (Source: São Sebastião City Hall); (b) Shallow landslide in a typical location of precarious residences on the hillslopes (Source: Amanda Perobelli/Reuters).

Table 7 presents the accumulated rainfall in 24 h in various municipalities of the north coast of São Paulo. Considering that the mean monthly rainfall for February is 225 mm in this region, it can be said that the amount of rainfall during this event was extremely high. The rainfall forecast was about 200 mm for between February 18th (Saturday) and 19th (Sunday). Although 200 mm of rainfall in 24 h was even very large, the actual event turned out to be much larger.

On February 18th, the most intense cold front in the last 54 years arrived from the south of the continent. At the same time, there was a transport of heat and humidity from the Amazon region. These two meteorological dynamics met in a low-pressure zone. This low pressure caused an increase in wind from the sea. Thus, this event dragged

Table 7. Rainfall in 24 h in the northern coast of São Paulo.

Municipality	Rain in 24 h (mm)	Number of Deaths
Bertioga	682	0
São Sebastião	626	64
Guarujá	388	0
Ilha Bela	337	0
Ubatuba	335	1
Caraguatatuba	234	0

moisture and raised the sea level. This rise in sea level hindered the surface water runoff to the sea. All these factors led to the heaviest rainfall in 24 h ever recorded in Brazil, i.e., 682 mm.

The distribution of 24-h rainfall is shown in Fig. 8. The extreme values were highly concentrated in the municipalities of Bertioga and São Sebastião. It is worth remembering that in Caraguatatuba, a heavy rain in March 1967 (584.8 mm in 48 h) caused flash floods and debris flows, killing 487 people.

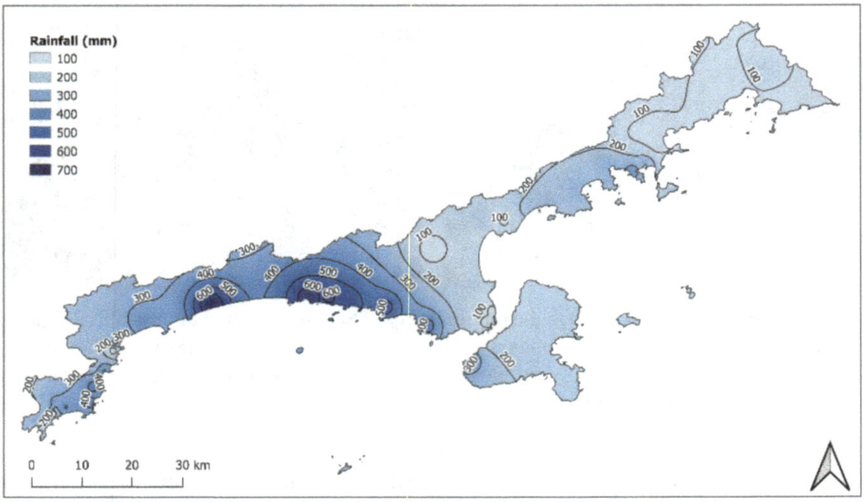

Fig. 8. Spatial distribution of rainfall in 24 h on the north coast of São Paulo state from February 18th to 19th, 2023.

Figure 9 presents the hyetograph obtained at two rainfall gauge stations: (i) Praia de Guaratuba (municipality of Bertioga); and (ii) Barra do Una (municipality of São Sebastião), both maintained by CEMADEN. It was observed that in Bertioga there was an extremely intense rain event with 273 mm in 2 h and 25 mm in 10 min on February 18th at night. On the same day, during the night, Bertioga received more than 400 mm of rainfall over 6 h. This high intensity and total amount of rainfalls were certainly the main triggering factor of this tragedy.

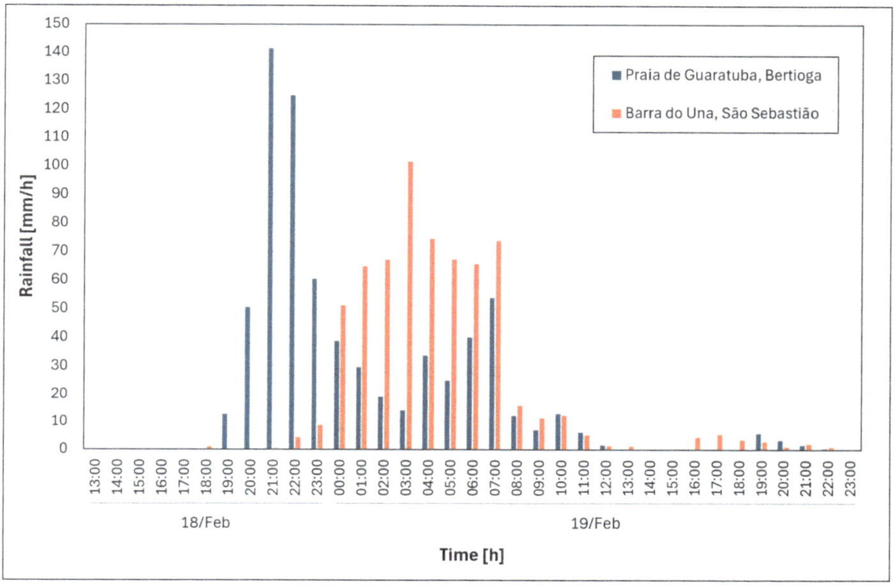

Fig. 9. Hyetograph obtained from two rain gauge stations: Praia de Guaratuba (municipality of Bertioga) and Barra do Una (municipality of São Sebastião) from February 18 at 1:00 p.m. to February 19 at 11:00 p.m.

If only this rainfall factor were considered, the current disaster could be simply regarded as a "purely natural disaster", or just "natural disaster". However, other factors within the local society need to be considered, which significantly contributed to the tragedy, i.e., intrinsic (or social or structural) factors.

It is noteworthy that this tragedy occurred during the Carnival period, one of the busiest moments for the Brazilian tourism. The lack of communication of the alerts, mentioned above, may have happened by chance, or the announcement may have been ignored thinking about the possibility of driving tourists away from the city during Carnival. In any case, this failure to transmit information caused the damage to increase. According to a statement from São Sebastião City Hall, 500,000 people were expected for the four days of Carnival (Casemiro and Matos 2023).

The fact is that the heaviest rainfall in Brazil's history hit the north coast of São Paulo during the Carnival weekend. Therefore, many tourists who were unfamiliar with the geography and local reality perished in this tragedy. If this historic rainfall had occurred outside the Carnival season, the disaster could have been less severe. In this sense, it can be perceived that there are problems not only on the part of the public administration but also on the part of the residents. In other words, the fact that education for DRR was not effectively disseminated on both sides led to the severe disaster on the north coast of São Paulo.

3 Lessons

3.1 General Aspects

Considering the information on these three disasters, we suggest the following actions related to weather/climate, which are necessary for DRR:

- Improvement and strengthening of the monitoring system: the number of rainfall gauge stations should be increased, especially in mountain regions, and they should provide data with temporal resolution suitable for the analysis of rainfall events.
- Conduction of researches on mechanism of the "training" (a type of thunderstorm) occurrence and its prediction.
- Establishment and strengthening of the communication and information transmission system among weather forecasting agencies, local government, and inhabitants.

The excessive rainfall cannot be considered the sole cause of the three mentioned tragedies. Social factors for reducing vulnerability should be always considered as a key point in DRR, for which the following actions are suggested:

- Awareness campaigns for the local population residing in mountain environments, about the concept of landslides, debris flows, and flash floods, and how they occur.
- Evacuation training for emergency situations, especially during events like Carnival and other holidays. Since many tourists may be unfamiliar with the area during a disaster, touristic cities need to seek strategies to train tourists and also to make clear the routes to safe locations.
- Improvement of hydrological and geomorphic hazard mapping, and effective use of these maps for land management (population settlement and planned urbanization). Based on the "build back better" concept, it is necessary to prohibit the construction of houses in high-risk areas after a disaster.
- Elimination of housing deficits in safe zones and urgent establishment of appropriate housing policies.
- Construction of appropriate infrastructure for DRR in high-risk areas.
- Ruralization in urban areas, where ecosystem services of various natural resources such as vegetation, water, and soil should be utilized (Kobiyama et al. 2021).
- Reduction of social inequality. Social inequality can be a major structural and intrinsic factor in disasters, because poor people generated and/or affected by social inequality often live at high-risk areas without preventive infrastructure. They are often unaware of the risks and receive little training from public institutions. This situation results from a lack of general education for all citizens.
- Thorough teaching of the DRR principle "protect your own life by yourself" to residents.

3.2 Hydrological Disasters Evolution: From Water-Related to Sediment-Related

Kobiyama et al. (2010) explained the relationship between the evolution of hydrological disasters and urban land occupation along a city formation (Fig. 10). An example of this evolution can be observed in Recife, where the poor people started living in mangroves or flat, low, and humid areas. Then, inhabitants of these areas often suffered from significant floods. After facing the two severe flood disasters in 1966 and 1975, resulting in 175 and 107 deaths, respectively, the drastic increase in population density on the slopes and cliffs of the city contributed to the occurrence of sediment-related disasters with 130 deaths in 2022 (Table 3). It is noteworthy that, in the case of the north coast of São Paulo where there are not many flat and safe areas, the construction of several communities has already begun on steep hillslopes.

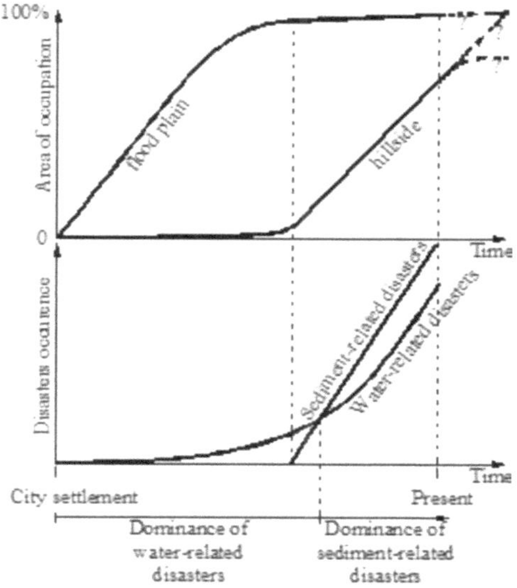

Fig. 10. Hydrological disasters evolution due to changes in land use and covers (Source: Kobiyama et al. (2010)).

Although disasters caused by floods and mass movements are grouped together into the classification of hydrological disasters, they have different characteristics (Kobiyama et al. 2018). The possibility of fatalities occurring with mass movements is much higher than with floods. Therefore, the Brazilian society must pay much more attention to sediment-related disasters from now on.

3.3 Integrated Management in Mountainous Environments

Natural phenomena such as landslides, debris flows, and flash floods are very common and inevitable in mountainous regions, which obligates society to implement effective management to contribute to DRR (Kobiyama et al. 2018). The eastern part of the Brazilian territory is characterized by coastal and mountainous environments. Most of this territory contains the remaining areas of the Atlantic Forest which must be protected. This region possesses a lot of major Brazilian cities and tourism spots, too. Due to these human and social dynamics, there is a high demand for water, energy, and food security, requiring a Nexus approach whose concept can be seen in Leck et al. (2015) and Purwanto et al. (2021).

In the above-mentioned region, mass movement phenomena have historically been concentrated. Figure 11 shows municipalities with deaths records due to landslides during the period from 1988 to March 2011 and locations where major debris flow disasters occurred from 1900 to 2014.

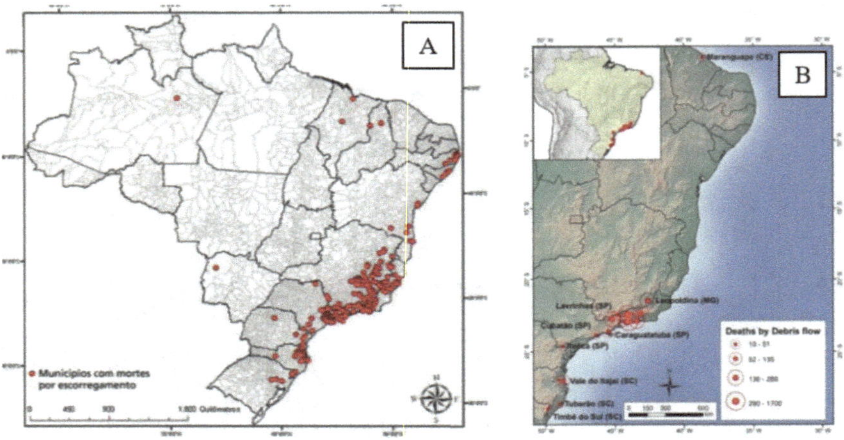

Fig. 11. Distribution of sediment-related disasters in Brazil: (a) Municipalities with deaths records due to landslides from 1988 to March 2011 2016 (Source: Carvalho and Galvão); (b) Locations of major debris flow disasters in Brazil during the period 1900–2014 (Source: Kobiyama et al. 2019).

Therefore, especially the eastern Brazilian region needs to simultaneously consider the issues of disasters, water resources, energy, food, ecosystem, in other words, a holistic Nexus approach: Water-Energy-Food-Disasters-Ecosystem (Kobiyama et al. 2022). Thus, instead of managing each item separately, an integrated management of disasters, water resources, food, energy, environment and catchments is needed, because a catchment is the ideal unit to carry out thisintegrated management.

3.4 Education for Disaster Risk Reduction Through the Use of School Catchment and Citizen Science

Practicing "Education for DRR based on citizen participation" becomes more important at the catchment level. Therefore, school catchments can serve as a foundation for this and should be built collaboratively by community groups, universities, public agencies, etc. According to Kobiyama et al. (2024), a school catchment is defined as a geographical region that has various measuring instruments and is useful for scientific research. Furthermore, it should be a location for developing didactic activities, serving for the learning of sciences, education, and any type of intellectual training for all citizens. The school catchment sparks community interest in hydrology and, consequently, expands knowledge in this study area, enhancing the population's participation in an integrated management.

To minimize the damage caused by natural disasters, Lamontagne (2002) highlighted the importance of popularizing science. As natural disasters in Brazil occur mainly due to water dynamics, hydrology certainly plays an important role in reducing them. In addition to studying and demonstrating the triggering mechanisms of these disasters, hydrology also brings insight into the hydrological phenomena experienced daily and highlights the importance of water and integrated coexistence with nature. In this context, the implementation of a school catchment network must be an urgent action in Brazil in order to reduce natural disasters, especially hydrological ones.

By its definition, a school catchment is important for hydrologists, and this value also extends to communities as well. This importance can be consolidated by greater community's awareness about hydrology, which further promotes citizen science. Involving the community in generating new knowledge about the environment can be the core of citizen science (Buytaert et al. 2014).

Since the school catchment is built for all citizens, its use is very appropriate to promote citizen science. Thus, local residents are instructed about hydrology, and their participation in data collection activities is encouraged. Consequently, greater community involvement increases the amount of information available about the catchment in which they are located, increasing citizen's knowledge and assisting the scientific community in their research. Starkey et al. (2017) mentioned that such increase contributes to the growing popularity and relevance of citizen science in the hydrological field. Besides enhancing citizens' perception, this approach can help them to recognize natural disasters early (risk awareness). According to Buytaert et al. (2014), although the concept and potential of citizen science are recent themes, they have been intrinsic to the process of building any scientific knowledge.

Therefore, a kind of "School Catchment Program" should be implemented in each region such as a municipality or state. Each city hall in a risk area must urgently build a school catchment in order to initiate DRR with citizen participation. The introduction of citizen sciences into the School Catchment Program will undoubtedly improve the performance of this participatory education. Then, the school catchments are a central point. This action will certainly intensify the practice of citizen science in local communities. Education is the basis of everything in any society. Education about water, which naturally serves DRR, does not require a high cost for its execution.

As mentioned in the item 2.3, both public institutions and local residents have little knowledge regarding natural hazards related to water and sediments. The use of school catchments in these regions will strengthen administrative and local communities, reducing vulnerabilities and, consequently, the risk of hydrological disasters.

In the execution of training courses focused on DRR, it is recommended to use the data obtained from the school catchments in the classroom. It is also strongly recommended to conduct field visiting within the school catchments to better understand the soil, pedogenesis, slopes, river network, fluvial regimes, among other aspects. Field excursion let participants directly observe and touch them with their own bodies.

4 Final Remarks

So far, Brazil, similar to Japan's previous system, has disaster prevention education such as evacuation training, or just waiting for guidance. In today's Japan, after experiencing mega-disasters (the Great Hanshin-Awaji Earthquake in 1995 and the Great East Japan Earthquake in 2011) that surpassed the limits of administrative response, society is expanding a perspective of self-support and mutual assistance based on disaster prevention education by protecting one's own life and education by helping others protect lives, respectively. Therefore, society is extensively working on comprehensive DRR learning to acquire more specific knowledge and skills, both in formal and informal education (Sato et al. 2021, Toyoda et al. 2021). With the experience of three devastating disasters reported in this study, the population and municipal, state and federal governments in Brazil should alter their actions, paying more attention to integrated DRR education.

Based on Sakaue and Murata (2019) and Kondo (2020), it can be said that the main objectives of DRR education should be: (i) to deepen the understanding of the current situation of risks and disasters, the causes and mitigation of natural disasters, etc., and to empower citizens to make decisions based on selecting appropriate actions and accurate thinking and assessment about disasters faced at the present and in the future; (ii) to understand and predict the hazards associated with natural phenomena, to be able to act to ensure their own safety, and to be able to prepare themselves on a daily basis; and (iii) to respect one's own life and that of others, to recognize the importance of creating a safe society, and to be willing to participate, cooperate and contribute to safety activities at school, at home and in local communities. In the case of Brazil, natural hazards are predominantly related to water dynamics. Therefore, the use of the school catchment with citizen science can be the starting point to begin strengthening DRR education.

Acknowledgments. This work was partially supported by the Brazilian agency CNPq (Chamada CNPq/MCTI Nº 10/2023 - Faixa B - Grupos Consolidados) (Processo: 403540/2023-2). The authors thank the CNPq and CAPES for research scholarships.

References

Aguir P (2022) Mulheres são maioria das vítimas dos temporais e soterramentos lideram causas de mortes; veja áreas onde ocorreram 130 óbitos. In: G1 Pernambuco. Recife, 10 jun. 2022. https://g1.globo.com/pe/pernambuco/noticia/2022/06/10/mulheres-sao-maioria-das-vitimas-dos-temporais-e-soterramentos-lideram-causas-de-mortes-veja-areas-onde-ocorreram-129-obitos.ghtml . Accessed 1 Aug 2022

Alves, P.: Mortes por chuvas no Grande Recife chegam a 109 e tragédia ultrapassa total de vítimas da cheia de 1975. 2022 (2022). https://g1.globo.com/pe/pernambuco/noticia/2022/06/01/mortes-por-chuva-no-grande-recife-total-de-vitimas-da-cheia-de-1975.ghtml. Accessed 17 Jun 2022

Amri, A., Lassa, J.A., Tebe, Y., Hanifa, N.R., Kumar, J., Sagala, S.: Pathways to Disaster Risk Reduction Education integration in schools: insights from SPAB evaluation in Indonesia. Int. J. Disaster Risk Reduction **73**, 102860 (2022). https://doi.org/10.1016/j.ijdrr.2022.102860

Below, R., Wirtz, A., Guha-Sapir, D.: Disaster Category - Classification and peril Terminology for Operational Purposes. Brussels: CRED/Munich: MunichRe Foundation (2009)

Brasil: Lei No. 12. 608, de 10 de abril de 2012. Institui a Política Nacional de Proteção e Defesa Civil - PNPDEC; dispõe sobre o Sistema Nacional de Proteção e Defesa Civil - SINPDEC e o Conselho Nacional de Proteção e Defesa Civil - CONPDEC; autoriza a criação de sistema de informações e monitoramento de desastres; altera as Leis nos 12.340, de 1o de dezembro de 2010, 10.257, de 10 de julho de 2001, 6.766, de 19 de dezembro de 1979, 8.239, de 4 de outubro de 1991, e 9.394, de 20 de dezembro de 1996; e dá outras providências. Brasília (2012). http://www.planalto.gov.br/ccivil_03/_Ato20112014/2012/Lei/L12608.htm. Accessed 20 Apr 2019

Buytaert, W., et al.: Citizen science in hydrology and water resources: opportunities for knowledge generation, ecosystem service management, and sustainable development. Front. Earth Sci. **2**, 1–21 (2014). https://doi.org/10.3389/feart.2014.00026

Candal, L., Tadeu, V.: Ocupação desordenada contribuiu com tragédia de Petrópolis, diz meteorologista. In: CNN Brasil. São Paulo, 16 Feb. 2022 (2022). https://www.cnnbrasil.com.br/nacional/ocupacao-desordenada-contribuiu-com-tragedia-de-petropolis-diz-meteorologista/. Accessed 10 Jul 2022

Carvalho, C.S.: Galvão T (2016) Prevenção de riscos de deslizamentos em encostas em áreas urbanas. In: Morais, M.P., Krause, C., Lima Neto, V.C. (eds.) Caracterização e tipologia de assentamentos precários: estudos de caso brasileiros, pp. 169–185. IPEA, Brasília (2016)

Casemiro, P., Matos, T.: Governo de SP e Prefeitura de São Sebastião foram avisados de risco de desastre 2 dias antes, diz diretor de órgão nacional de monitoramento. In: G1 São Paulo. São Paulo, 22 Feb. 2023 (2023). https://g1.globo.com/sp/sao-paulo/noticia/2023/02/22/governo-de-sp-e-prefeitura-de-sao-sebastiao-foram-avisados-de-risco-de-desastre-2-dias-antes-diz-diretor-de-orgao-nacional-de-monitoramento.ghtml. Accessed 1 Apr 2023

Chaudhary, M.T., Piracha, A.: Natural disasters – origins, impacts, management. Encyclopedia **1**, 1101–1131 (2021). https://doi.org/10.3390/encyclopedia1040084

Espinosa, L.A., Portela, M.M., Matos, J.P., Ghasbia, S.: Climate change trends in a european coastal metropolitan area: rainfall, temperature, and extreme events (1864–2021). Atmosphere **13**, 1995 (2022). https://doi.org/10.3390/atmos13121995

Frank, B., Sevegnani, L. (orgs.): Desastre de 2008 no Vale do Itajaí: Água, gente e política. Blumenau: Agência de Água do Vale do Itajaí, 192p. (2009)

Garnier, E., Lahournat, F.: Japanese stone monuments and disaster memory – perspectives for DRR. Disaster Prev. Manag. **31**(6), 1–12 (2022). https://doi.org/10.1108/DPM-03-2021-0089

Groisman, P., Knight, P., Easterling, D., Karl, T., Hegerl, G., Razuvaek, V.: Trends in intense precipitation in the climate record. J. Clim. **18**, 1326–1350 (2005). https://doi.org/10.1175/JCLI3339.1

Guimaraens, R.: A Enchente de 41. Porto Alegre: Libretos (2009)

IPCC - Intergovernmental Panel on Climate Change Climate Change (2007) Impacts, Adaptation and Vulnerability. Contribution of Working Group II to the Fourth Assessment Report of the Intergovernmental Panel on Climate Change. Cambridge: Cambridge University Press

Kobiyama, M., Chaffe, P.L.B., Goerl, R.F., Giglio, J.N., Reginatto, G.M.P.: Hydrological disasters reduction: lessons from hydrology. In: Sens ML, Mondardo RI (org.) Science and Technology for Environmental Studies: Experiences from Brazil, Portugal and Germany. Florianópolis: UFSC, pp. 49–72 (2010)

Kobiyama, M., Goerl, R.F., Fan, F.M., Corseuil, C.W., Michel, G.P., Dulac, V.F.: Abordagem integrada para gerenciamento de desastres em região montanhosa com ênfase no fluxo de detritos. Revista Gestão & Sustentabilidade Ambiental **7**(esp), 31–65 (2018). https://doi.org/10.19177/rgsa.v7e0201831-65

Kobiyama, M., Michel, G.P., Goerl, R.F.: Proposal of debris flow disasters management in Brazil based on historical and legal aspects. Int. J. Erosion Control Eng. **11**(3), 85–93 (2019). https://doi.org/10.13101/ijece.11.85

Kobiyama, M., et al.: Uso da bacia-escola na redução do risco de desastres: uma abordagem socio-hidrológica. In: Magnoni Junior L, Freitas CM, Lopes ESS, Castro GRB, Barbosa HA, Londe LR, Magnoni MGM, Silva RS, Teixeira T, Figueiredo WS (orgs.) Redução do risco de desastres e a resiliência no meio rural e urbano. 2. ed. São Paulo: Centro Paula Souza, pp. 560–583 (2020)

Kobiyama, M., Campagnolo, K., Fagundes, M.R.: Ruralization for water resources management in urban area revisited. Revista Geográfica Acadêmica **15**(2), 68–88 (2021)

Kobiyama, M., Fagundes, M.R., Brighenti, T.M., Stone, T.F., Corseuil, C.W.: Integrative approach for risk and disaster reduction: The Water-Energy-Food-Disaster-Ecosystem Nexus. In: Magnoni Junior, L., (orgs.) Ensino de geografia e a redução do risco de desastres em espaços urbanos e rurais. 1. ed. São Paulo: Centro Paula Souza, pp. 434–464 (2022)

Kobiyama, M., Franck, A.F., González-Ávila, I., Maciel, J.G.F.: Socio-Hydrogeomorphology and School Catchment as Useful Science and Tool for Disaster Risk Reduction. In: Hatano, H., Kovacs, P. (eds.) Proceedings of the 5th Global Summit of the Global Alliance of Disaster Research Institutes (GADRI). Springer, Berlin, Germany (2024). (in press)

Kondo, S.: Theoretical consideration on disaster risk reduction education from the viewpoint of the contingency. Res. J. Disaster Educ. **1**(1), 31–41 (2020). https://doi.org/10.51004/rjde.1.1_31

Lamontagne, M.: An overview of some significant eastern Canadian earthquakes and their impacts on the geological environment, buildings and the public. Nat. Hazards **26**, 55–67 (2002). https://doi.org/10.1023/A:1015268710302

Leão, E.B.S., Andrade, J.C.S., Nascimento, L.F.: Recife: a climate action profile. Cities **116**, 103270 (2021). https://doi.org/10.1016/j.cities.2021.103270

Leck, H., Conway, D., Bradshaw, M., Rees, J.: Tracing the water-energy-food nexus: description, theory and practice. Geogr. Compass **9**, 445–460 (2015). https://doi.org/10.1111/gec3.12222

Lenharo, M.: WHO declares end to COVID-19's emergency phase. Nature, News, May 5, 2023 (2023). https://doi.org/10.1038/d41586-023-01559-z. Accessed 20 Aug 2023

Lins, A., Aguiar, P., Novelino, R.: Garoto morreu em comunidade que pediu socorro 14 vezes desde 2015; 'tem que ter vontade para estabelecer política habitacional', diz especialista. In: G1 Pernambuco. Recife, 7 Jun. 2022 (2022). https://g1.globo.com/pe/pernambuco/noticia/2022/06/07/garoto-morreu-em-comunidade-que-pediu-socorro-14-vezes-desde-2015-tem-que-ter-vontade-para-estabelecer-politica-habitacional-diz-especialista.ghtml. Accessed 1 Aug 2022

Machado, C.C.: Tubarão 1974: Fatos e Relatos da Grande Enchente. Tubarão: Ed. Unisul (2005)

Madeiro, C.: Recife só acionou plano contra tragédia 2 dias após alerta de alto risco. 2022 (2022). https://noticias.uol.com.br/colunas/carlos-madeiro/2022/05/30/recife-so-acionou-plano-contra-tragedia-2-dias-apos-alerta-de-alto-risco.htm Accessed 17 Jun 2023

Marengo, J.A., Jones, R., Alvesa, L.M., Valverde, M.C.: Future change of temperature and precipitation extremes in South America as derived from the PRECIS regional climate modeling system. Int. J. Climatol. **29**, 2241–2255 (2009). https://doi.org/10.1002/joc.1863

Marques, C., Baesso, D.C.: Desastres e vulnerabilidade na Região Serrana do Rio de Janeiro (RSRJ). Idéias **12**, e021019 (2021). https://doi.org/10.20396/ideias.v12i00.8665127

Matsuo, P.M., Silva, R.L.F.: Desastres no Brasil? Práticas e abordagens em educação em redução de riscos e desastres. Educar em Revista **37**, e78161 (2021). https://doi.org/10.1590/0104-4060.78161

METSUL: Petrópolis enfrenta seu maior desastre desde a chuva de 1988 (2022). https://metsul.com/petropolis-enfrenta-seu-maior-desastre-desde-a-chuva-de-1988/ Accessed 17 Jun 2023

Michel, G.P., et al.: Relatório técnico dos desastres de dezembro de 2020 nos municípios de Presidente Getúlio, Ibirama e Rio do Sul – SC. Volume I. Porto Alegre: GPDEN/IPH/UFRGS (2021)

Min KS, Tsuboki K, Yoshioka MK, Moroda Y, Kanada S. Formation Mechanism of a Stationary Line-Shaped Precipitation System in the Kinki District, Japan —Case Study on 1: Event. J. Meteorol. Soc. Jpn **99**, 357–377 (September 2015). https://doi.org/10.2151/jmsj.2021-017

Modesto, F.B.F., Kuhlmann, L.G., Jacques, P.D., Ribeiro, R.S., Santos, T.D. (orgs.) Avaliação técnica pós-desastre, Petrópolis, RJ. Rio de Janeiro: CPRM (2022). https://rigeo.cprm.gov.br/handle/doc/22668 Accessed 10 Jul 2022

Nakano, G., Yamori, K.: Disaster risk reduction education that enhances the proactive attitudes of learners: a bridge between knowledge and behavior. Int. J. Disaster Risk Reduction **66**, 102620 (2021). https://doi.org/10.1016/j.ijdrr.2021.102620

Petal, M.: Disaster Risk Reduction Education. In: Shaw, R., Krishnamurty, R. (eds.) Disaster Management: Global Challenges and Local Solutions, p. 2010. Universities Press, Hyderabad (2010)

Petroliagkis, T.I., Alessandrini, A.: Screening and Selecting Climate Change Impact Parameters as Potential Drivers of Migration. Luxembourg: European Union (JRC123303) (2021). https://doi.org/10.2760/455010

Pinto, O.S., Iory, N.: Petrópolis cortou gasto com Defesa Civil em 2021. In: Poder360. Brasília, 16 Feb. 2022 (2022). https://www.poder360.com.br/brasil/petropolis-cortou-gasto-com-defesa-civil-em-2021/ Accessed 10 Jul 2022

Purwanto, A., Susnik, J., Suryadi, F.X., Fraiture, C.: Water-energy-food nexus: critical review, practical applications, and prospects for future research. Sustainability **13**, 1919 (2021). https://doi.org/10.3390/su13041919

Rädler, A.T.: Invited perspectives: how does climate change affect the risk of natural hazards? challenges and step changes from the reinsurance perspective. Nat. Hazard. **22**, 659–664 (2022). https://doi.org/10.5194/nhess-22-659-2022

Sakaue, H., Murata, S.: Development of disaster education in Japanese school education and its characteristics: From the viewpoint of the aftermath of the Great Hanshin and the Great East Japan Earthquake. Hyogo University Teacher Educ. J. **55**, 141–151 (2019)

Salata, A.R., Ribeiro, M.G.: Boletim Desigualdade nas Metrópoles. n. 09. Edição especial: dados anuais (2012–2021). Porto Alegre: PUCRS (2022). https://www.observatoriodasmetropoles.net.br/. Accessed 1 Aug 2022

Saldanha, C.B., Collischonn, W., Marques, M., Steinmetz, S., Almeida, I.R., Reisser Júnior, C.: O Evento de Chuva Intensa de Janeiro de 2009 Sobre a Região de Pelotas-RS. Revista Brasileira de Recursos Hídricos **17**, 255–265 (2012)

Sato, T., Sakurai, A., Oda, T., Hayashida, Y., Murayama, Y., Yamori, K.: Propulsive model of school safety for disaster in community school - Case report of Kita-Tsunashima Elementary School in Yokohama city. J. Japan Soc. Natural Disaster Sci. **40**, 175–190. https://doi.org/10.24762/jndsj.40.2_175

Schmidt, T.: Em 10 anos, São Sebastião recebeu ao menos 4 alertas sobre risco de tragédia. In: Brasil de Fato. São Paulo, 26 Feb. 2023 (2023). https://www.brasildefato.com.br/2023/02/26/em-10-anos-sao-sebastiao-recebeu-ao-menos-4-alertas-sobre-risco-de-tragedia Accessed 1 Apr 2023

Schumacher, R.S., Johnson, R.H.: Mesoscale processes contributing to extreme rainfall in a mid-latitude warm-season flash flood. Mon. Weather Rev. **136**, 3964–3986 (2008). https://doi.org/10.1175/2008MWR2471.1

Selby, D., Kagawa, F.: Disaster Risk Reduction in School Curricula: Case Studies from Thirty Countries. Paris: UNESCO, Geneva: UNICEF (2012). Disponível em <https://unesdoc.unesco.org/ark:/48223/pf0000217036>. Acesso em 30 jul. 2023

Silva, A.R.C., Kobiyama, M., Panceri, R.: "Programa defesa civil na escola" nos municípios de Imbituba e Lauro Müller/SC: avaliação com alunos e professores. Revista de Geografia (Recife) **38**(3), 154–179 (2021). https://doi.org/10.51359/2238-6211.2021.248819

Silva, A.R.C., Kobiyama, M., Vanelli, F.M.: Interfaces entre a política nacional de proteção e defesa civil e a política nacional de educação ambiental. Ciência e Natura **43**, e60 (2021). https://doi.org/10.5902/2179460X43612

Silveira, W.N., Kobiyama, M., Goerl, R.F., Brandenburg, B.: História das inundações em Joinville, 1851–2008. Curitiba: Organic Trading (2009)

Souza, S.O., Oliveira, R.C., Lupinacci, C.M.: Mapeamento da declividade enquanto instrumento de planejamento do litoral norte paulista. In: XII Simpósio Nacional de Geomorfologia, 2018, Crato. Proceedings (2018). https://www.sinageo.org.br/2018/trabalhos/9/9-5-546.html Accessed 1 Mar 2023

Starkey, E., Parkin, G., Birkinshaw, S., Large, A., Quinn, P., Gibson, C.: Demonstrating the value of community-based ('citizen science') observations for catchment modelling and characterization. J. Hydrol. **548**, 801–817 (2017). https://doi.org/10.1016/j.jhydrol.2017.03.019

Thomas, V., López, R.: Global Increase in Climate-Related Disasters. Manila: Asian Development Bank. (ADB Economics Working Paper Series No. 466) (2015)

Toyoda, Y., Muranaka, A., Kim, D., Kanegae, H.: Framework for utilizing disaster learning tools classified by real and virtual aspects of community space and social networks: Application to community-based disaster risk reduction and school disaster education on earthquakes in Japan for during- and post-COVID-19 periods. Progress Disaster Sci. **12**, 100210 (2021). https://doi.org/10.1016/j.pdisas.2021.100210

Unuma, T., Takemi, T.: Characteristics and environmental conditions of quasi-stationary convective clusters during the warm season in Japan. Q. J. R. Meteorol. Soc. **142**, 1232–1249 (2016). https://doi.org/10.1002/qj.2726

UN – United Nations: Transforming our World: The 2030 Agenda for Sustainable Development. New York: UN. (A/RES/70/1) (2015)

UNESCO; UNICEF Regional Office for East Asia and the Pacific,: Disaster Risk Reduction (DRR) in education: an imperative for education policymakers. UNESCO-UNICEF Office, Bangkok (2011)

UNISDR – United Nations International Strategy for Disaster Reduction: 2009 UNISDR Terminology on Disaster Risk Reduction, p. 2009. UNISDR, Geneva (2009)

UNISDR – United Nations Office for Disaster Risk Reduction: Sendai Framework for Disaster Risk Reduction 2015–2030. UNISDR, Geneva (2015)

UOL (2023) Governo de SP e Prefeitura de São Sebastião foram avisados de risco de desastre 2 dias antes, diz diretor de órgão nacional de monitoramento. In: UOL. São Paulo, 24 Feb. 2023. https://noticias.uol.com.br/cotidiano/ultimas-noticias/2023/02/24/o-que-poderia-ter-sido-feito-evitado-tragedia-litoral-sp.htm Accessed 1 Apr 2023

Van Aalst, M.K.: The impacts of climate change on the risk of natural disasters. Disasters **30**(1), 5–18 (2006). https://doi.org/10.1111/j.1467-9523.2006.00303.x

Veloso, J.A.V.: Tremeu a Europa e o Brasil também. Chiado Editora, Rio de Janeiro (2015)

WEF - World Economic Forum (2023) The Global Risks Report 2023. 18 ed. Insight Report. Geneva: WEF

Wynes, S., et al.: Climate action failure highlighted as leading global risk by both scientists and business leaders. Earth's Future **10**, e2022EF002857. https://doi.org/10.1029/2022EF002857

Xavier, P.: Diversos fatores contribuíram para tragédia de Petrópolis, apontam especialistas. In: Band-Uol. Rio de Janeiro, 14 Mar. 2022 (2022). https://www.band.uol.com.br/bandnews-fm/rio-de-janeiro/noticias/diversos-fatores-contribuiram-para-tragedia-de-petropolis-apontam-especialistas-16500015 Accessed 10 Jul 2022

Xavier, C.: Governos de SP e São Sebastião ignoram alertas de desastre. In: Vermelho. Brasília, 22 Feb. 2023 (2023). https://vermelho.org.br/2023/02/22/governos-de-sp-e-sao-sebastiao-ignoram-alertas-de-desastre-2-dias-antes/ Accessed 1 Apr 2023
Zanandrea, F., Kobiyama, M., Michel, G.P., Fleischmann, A.S., Collischonn, W. (orgs.): Desastres e água: eventos históricos no Brasil. Porto Alegre: ABRHidro (2023)

Cognizance of Climate Change Within the United Kingdom

Mark Ashley Parry(✉) and Andrew E. Collins

Northumbria University, Newcastle Upon Tyne, UK
ashley2.parry@northumbria.ac.uk

Abstract. This paper discusses the multifaceted perceptions of climate change within the United Kingdom. Whilst the understanding of anthropogenic climate change has gained prominence in more recent decades, especially in relation to greenhouse gas processes, its conceptual roots can be traced back to ancient Greek observations. Today, climate change is recognised as a significant environmental, societal, and economic challenge. This paper draws from both primary and secondary data to highlight how the perception of climate change varies widely amongst the British public. It is influenced by different factors such as political identity, gender, age, and other socio-demographic influences. Data obtained indicates that politically left-leaning voters and females are generally more likely to acknowledge the severity of climate change compared to their right-leaning and male counterparts. A critical focus is placed on the youth, as their perceptions and engagement are pivotal for long-term climate action. Despite this, there remains a gap in research concerning youth involvement with climate change. Therefore, it can be underscored that enhanced climate education should empower the younger next generation and promote active engagement in climate change action. Meanwhile, the terminology which is used in climate change discussions can significantly alter youth and wider public engagement, with terms such as "global warming" and "climate emergency" evoking varying levels of concern. Overall, these findings emphasise the importance of integrating scientific knowledge into public perception of climate change to develop both effective and sustainable climate change policies that can be acted upon more widely.

1 Introduction and Background

Whilst understanding of anthropogenic climate change is often regarded as a relevantly new concept for most people and has only been known about for a few decades, it is worth noting that the theory of humans altering the climate has its origins in ancient Greek times. Theophrastus observed that the draining of marshes in specific locations made them more susceptible to freezing and believed that land became warmer when cleared of forests (Neumann, 1985).

Climate change is viewed as one of the great environmental, societal, and economic issues challenging global security and posing societal risks (Stern, 2006; Brown, Hammill and McLeman, 2007; Butler, 2018; UKHSA, 2023; WHO, 2023a). The hard evidence is that global surface temperatures have risen by between 1.1 °C and 1.2 °C

H. Tatano et al. (Eds.): GSRIDRR 2023, DRRGADRIBS, pp. 292–326, 2025.
https://doi.org/10.1007/978-981-96-6609-6_22

above pre-industrial times (IPCC, 2023; WHO, 2023b). This rate of warming has been increasing since 1990 (Samset et al., 2023). In 2015, world leaders agreed on the Paris Agreement, which aimed to limit global temperature increases to 1.5 °C, and no more than 2 °C (UNFCCC, 2015), to avoid the worst of the impacts of climate change. It should be noted that the world has observed days in 2023 and 2024 that were above 2 °C (Le Page, 2023), and the planet has experienced 12 months where the average was above 1.5 °C (Poynting, 2024). The main reason for this is due to a strong El Niño that took place between 2023 and 2024 in the South Pacific (WMO, 2024).

In the United Kingdom, the decade from 2014 to 2023 was 1.25 °C warmer compared to the 1961–1990 average, with 2023 being 1.66 °C warmer compared to the 1961–1990 average (Kendon et al., 2024). Additionally, the warmest days during this decade were annually 2.8 °C warmer (Kendon et al., 2023). There have been numerous extreme weather events in the last decade, such as flooding (Quinn, 2015; Gillespie, 2019; Tapper and Walker, 2019; Gregory and Wright, 2024; Khalil, 2024), heatwaves (Met Office, 2018; Gregory, 2023), storms (Sale, 2021; Mathers, 2023; Robinson, 2023; Kirka, 2024), and wildfires (Walker and Findlay, 2019; McGonagle, 2020). These events can be seen as a part of the natural cycle, and similar events have previously occurred in the United Kingdom, such as the North Sea Flood of 1953 (Hickey, 2001; Baxter, 2005), Big Freeze of 1963 (Royal Meteorological Society, 2013), Great Glasgow Storm of 1968 (Tranter and Galvin, 2018), heatwave of 1976 (Baker, Shaffrey and Hawkins, 2021), 1990 (Burt, 1992), and 2003 (Met Office, 2012), October Storm of 1987 (Burt and Mansfield, 1988; Met Office, No Date), Burn's Day Storm of 1990 (McCallum, 1990), and 1992 New Year's Day Storm (Fogg et al., 1992; Johnson, 1992). However, there is a growing understanding that these are being made more extreme due to climate change (Zachariah *et al.,* 2023; Kew *et al.,* 2024).

In terms of its causation, there is a consensus within the scientific community that the current climate change is caused by anthropogenic activity. While the exact percentage is still up for debate, with figures of between 90 and 100% (Oreskes, 2004; Schulte, 2008; Doran and Zimmerman, 2009, Anderegg et al., 2010, Rosenberg et al., 2010; Powell, 2015; 2019; Lynas, Houlton and Perry, 2021; Myers et al., 2021), the commonly quoted figures reviewed by Cook et al. (2013; 2016) indicate that 97% of published scientific literature supports the theory of anthropogenic climate change. Despite studies demonstrating that most of the scientific community believe that current climate change is caused by anthropogenic activity, there are some that criticise the accuracy or relevance of the high values. For example, Legates et al. (2015) suggest that Cook et al.'s (2013) statement of 97% of academic literature supporting anthropogenic climate change is misleading, as 66.39% of the surveyed academic papers did not state a firm position. A recent study in late 2019 by James Powell highlights a 100% scientific consensus, as no research in the first half of 2019 was published that rejects anthropogenic climate change (Powell, 2019). Whilst plenty of research indicates that the scientific consensus is in the mid to late 90%, it is questionable whether there can ever truly be a 100% consensus on scientifically based analysis applied to climate change whilst various sectors of business, politics and society consider it 'controversial'.

Smaller percentages of the population and scientific community either do not believe in the notion of anthropogenic climate change or are sceptical about its causation or impacts on society; these are known as climate deniers and climate sceptics, respectively. Examples of these have included British politicians such as Andrew Bridgen (Bridgen, 2024); Neil Hamilton (Hayward, 2019), Andrea Jenkyns (Jenkyns, 2023), Nigel Lawson (Lawson, 2008; Carrington, 2017), Rupert Lowe (Lowe, 2020), Christopher Monckton (Monbiot, 2010; Monckton, 2015), Owen Paterson (Watt, 2012), John Redwood (Redwood, 2008), Matt Ridley (Ridley, 2023), Robert Rowland (Rowland, 2020), Richard Tice (Tice, 2023), Sammy Wilson (Bradfield, 2021), and Ann Widdecombe (Widdecombe, 2014). Additionally, there are researchers who are sceptical about either climate change or its causation, such as, John Abbot (Abbot and Marohasy, 2017), Robert Balling (Michaels et al., 2000), Craig Idso (Idso et al., 2009), Jennifer Marohasy (Abbot and Marohasy, 2017; Mann, 2021), Ian Plimer (Plimer, 2009), Frederick Seitz (Oreskes and Conway, 2010; Mann, 2021), Fred Singer (Singer and Avery, 2007; Idso et al., 2009; Oreskes and Conway, 2010), Roy Spencer (Spencer, 2007, 2008, 2010) and Valentina Zharkova (Nuccitelli, 2018).

1.1 Climate Change Perception

Perception studies are important because climate change perception will influence whether an individual will engage with the issue. Amongst the public, studies about perception of climate change have been undertaken for roughly thirty years (Brechin, 2010) and more regularly throughout the 2000s and 2010s, both by research groups and academia within the United Kingdom (Steentjes et al., 2017; Poortinga et al., 2019) and internationally (Steentjes et al., 2017; Funtatsu *et al.,* 2019; Lorencová, Loučková and Vačkářů, 2019; Poortinga et al., 2019; Leiserowitz et al., 2022; 2023).

Perception is defined by Mayhew (2009) as a way in which an individual interprets the world around them through senses, which is shaped by their individual upbringing. Also, as climate change has emerged as an interdisciplinary subject, its perception and interpretation can be defined in different ways. A way of defining perception with an environmental lens is "the way in which an individual perceives the environment; the process of evaluating and storing information received about the environment" (Mayhew, 2009, online). Pidgeon *et al.,* (1992, p.89) expands this definition by combining psychological aspects of perception with risks, including climate; by stating that "risk perception involves people's beliefs, attitudes, judgements and feelings, as well as the wider social or cultural values and dispositions that people adopt towards hazards and their benefits". As highlighted by Dessai et al. (2004, p.14) public perception of climate change is "based on psychological, social, moral, institutional and cultural processes".

It should be noted that the United States has been the most researched country in terms of climate change perception; however, there have also been studies undertaken within the United Kingdom. This research has been undertaken from within the academic community, other research groups and government organisations orientated to the British's public perception about subjects related to climate change, including, but not limited to, energy issues, climate terminology and climate imagery. In 2023, it was found that a majority of the British public tend to believe that anthropogenic activity is playing some role in some form in changing the world's climate (79%) (Leiserowitz et al., 2023). However, within the same study, it is demonstrated that the British public is behind in its understanding of the issue compared to some countries in Europe, such as

Hungary (90%), Greece (86%) and Denmark (85%); but is ahead of others, such as Germany (76%), Austria (74%), and Albania (67%) (Leiserowitz et al., 2023). Whilst, this demonstrates a level of belief that anthropogenic activity plays a role in climate change, it represents a year-on-year decline within the United Kingdom of eight% (Leiserowitz et al., 2022); and this 2022 figure is in line with other previous studies indicating 91% in 2005 (Poortinga, Pidgeon and Lorenzoni, 2006), and 86% in 2017 (Steentjes et al., 2017).

Research has also shown that the level and type of perception of climate change fluctuates over time (Nisbet and Myers, 2007; Ballew et al., 2019). This fluctuation could be due to media coverage (Mazur and Lee, 1993; Happer and Philo, 2016) or to the frequency and/or magnitude of extreme meteorological events in recent years (Brooks et al., 2014; Shao et al., 2014). Continuous studies of the public would be required to monitor how the public responds to individual events. Perception of climate change is not constant amongst the population, with variance amongst people with different worldviews, such as for example the contrasts between those who have a left-wing political ideology compared to those with a right-wing one (McCright, Dunlap and Marquart-Pyatt, 2016). However, a common gap is the limited attention to youth (Wray-Lake et al., 2010; MacDonald et al., 2013; Walsh and Cordero, 2019; Parry, 2020) in comparison to gender and political identity.

In general, an increased perception that climate change is occurring and is a threat to society results in increased levels of engagement including behaviour to try to reduce emissions and promote environment (Reser, Bradley and Ellul, 2014). However, in some extreme cases, researchers have observed eco-anxiety (Gifford and Gifford, 2016), which can cause disengagement due to a feeling of hopelessness about the future.

1.2 Factors Influencing Climate Change Perception

Existing research demonstrated that public perception of climate change is influenced by factors including ideologies and socio-demographics (Weber, 2010; 2016; Cipriani et al., 2024), as well as lived experience (Li, Johnson and Zaval, 2011). The following sub-sections examine this further.

1.2.1 Political Identity

Left-wing voters are more likely to believe anthropogenic climate change compared to right-wing voters in western countries (Leiserowitz, 2006; McCright and Dunlap, 2011; Whitmarsh, 2011; Unsworth and Fielding, 2014; Milfont et al., 2015; Hornsey et al., 2016; Poortinga et al., 2019; Ballew et al., 2020). Ballew et al. (2020) highlights that political differences between left and right voters in belief of climate change within the United States increases with higher levels of education and/or income. However, it has been found that the United Kingdom and Europe is much less divided compared to the United States (Poortinga et al., 2019). Also, within western countries, as individuals become older, the more conservative they become, in both nature and politically (Cornelis et al., 2009) and this produces a combined political and age-related influence on the distribution of beliefs about climate change. Traditionally, within the United Kingdom, the Labour Party is a left or centre-left wing party; whereas, the Conservative party are centre-right to right wing. However, in the 2019 General Election traditionally Labour voting areas changed to Conservative, which was in part due to lack of implementation of Brexit (Sabbagh, 2019). It is not known how this shift impacted on acceptance or rejection of actions that would address anthropogenic climate change.

1.2.2 Gender

During the 1960s and 1970s, there was no clear evidence that there was a relationship between gender and environmental concerns, including climate change (Van Liere and Dunlap, 1980; McCright, 2010). However, since the 1980s there has been a growing body of literature, which is increasingly suggesting that women are more concerned about the environment and therefore exercise a greater level of pro-environmental values and attitudes than their male counterparts (Greenbaum, 1995; Bord and O'Connor, 1997; Zelezny, Chua and Aldrich, 2000; Dietz, Kalof and Stern, 2002; McCright, 2010; Xiao and McCright, 2012; McCright and Xiao, 2014; McCright and Xiao, 2015). However, some research shows that in rare cases, men have been found to be more environmentally concerned than their female counterparts, such as in China (Xiao and Hong, 2010). Further, in a study by Chan, Pong and Tam (2019) it was found that environmental concerns vary depending on gender inequality.

One of the main theories that have been brought forward for this difference in beliefs is 'gender socialisation theory'. This theory argues that boys and girls are taught from an early age the differences in values and expectations between the genders in their societal culture (Wharton, 2009; Xiao and McCright, 2012). For example, in the western world, boys are taught about behaviour in a masculine way, which means that men are expected to show leadership and be independent, with traits of courage, assertiveness, physical strength and being unemotional (Franklin, 1984). In contrast, girls are taught about behaviour in a feminine way, which means that females are expected to show sensitivity (Windsor, 2015), warmth (Burke and Strets, 2009), empathy (Vetterling-Braggin, 1982), and be more emotional and understanding (Kite, 2001).

However, the expectation of masculinity and femininity are evolving with changing attitudes amongst civil society in the United Kingdom and other western nations (Anderson, 2009; Anderson and McCormack, 2015). For example, this change can be demonstrated with the increase in metrosexual males and more recently the spornosexual males within both Generation Y and Generation Z (Wolfman, 2017). Both types of men demonstrate a more liberal attribute, which is outgoing emotionally, inclusive and non-judging (Trubo, 2003; Adams, 2011).

In addition, it is likely that gender socialisation theory is having an impact on the level of knowledge that females have on climate change. Female people in Western countries are more likely to have misconceptions about climate change compared to their male counterparts (Ballew et al., 2018). They are also less likely to study sciences and mathematics where this is a choice for higher level study, including at college and university (Delaney and Devereux, 2019); this is despite females scoring similar levels to male counterparts in school (O'Dea et al., 2018). There are several potential explanations for this. Firstly, gender socialisation theory highlights that traditionally, hard science is perceived by school students to be masculine (Kelly, 1985). Another possibly is the lack of role models in science, technology, engineering and mathematics (STEM) subjects for girls to be inspired (Cheryan et al., 2011). A final potential reason is that female people tend to underestimate their ability in STEM subjects and their likelihood to succeed within these types of subjects (Meece et al., 1982; Sax, 1994; Correll, 2001; Ehrlinger and Dunning, 2003; Moakler and Kim, 2014).

Overall, attitudes attributed to masculinity and femininely are changing amongst today's youth. How gender intersectionality on the one hand and stereotyping on the other mixes up trends in climate related perception and engagement is unknown.

1.2.3 Youth

It is important to note that the terms "youth", "adolescent" and "young person" are often interchanged. This leads to the debate whether the term 'youth' or 'adolescent' is more appropriate in climate related awareness. Ostensibly, the two terms are identical, with the definition of 'youth', within a western context, being the stage of human life between childhood and adulthood (Kehily, 2007). The definition of 'adolescence' is the transitional period of a human's life between late childhood and the beginning of adulthood (Choudhury, Blakemore and Charman, 2006; Andersen and Vandehey, 2011). However, psychology researchers go further and state that the term also incorporates the social, mental and physical changes that virtually all humans undergo (Roche et al., 2004; Eaton et al., 2006; Ernst, Pine and Hardin, 2006). Consequently, adolescence deals with the biological and psychological aspects of a young person of a certain age, whilst youth is a socially constructed term. When exploring young people's relationship with climate change, these two terms should not be mutually excluded from each other, since during the adolescent period, young people's cognitive development is an ongoing process (Paus, 2005).

There are two types of brain development during adolescence. The first is in the regions of the brain that are important for the regulation of emotion, behaviour and the perception, whilst the second is in those linked to the evaluation of both risks and rewards (Steinberg, 2005). For example, it is estimated that within the United States, seventy-one per cent of all adolescence deaths each year are due to car accidents, other unintentional injuries, homicide, and suicide (Eaton et al., 2006). This is relevant to understanding climate change perception, since young people are increasingly likely to have this invincibility feeling, that the worst of the effects "won't-happen-to-me". This could foster behaviours which may not be prudent with environmental conscious decisions, as they do not see the increased environmental risks of the future. The other type of youth brain development flagged here is the still developing area of analytical skills and the relationship of this to world views (Vollerberg et al., 2001). Consequently, young people start to ask questions about the world around them, rather than accepting the beliefs of 'adults' (Lerner et al., 2001), and they start developing their own unique beliefs through their interactions with different social and cultural environments (Lerner et al., 2001). As such, it is important to provide this age group with new ideas, opinions and ways of understanding to allow engagement with societal issues and ultimately to help in addressing them.

1.3 Methodology

A predominantly pragmatism-based approach orientated the research that led to the findings of this paper, predominantly to draw from varied sources and perspectives to understand multi-causal and interrelated climate issues. This research paradigm was also useful in allowing mixed methods, which "focuses instead on 'what works' as the truth regarding the research questions under investigation" (Tashakkori and Teddlie 2003, p. 713). In addition, an objective approach balanced the often critical and emotive climate change perspectives. The research technique aimed at being value-free, described by McLean (2011, p. 316) as "research that has been produced by a completely impartial and dispassionate researcher".

1.3.1 Questionnaire Derived Data

Questionnaire responses from between March and September 2017 provide the contextual overview of people's perception of climate change. Unlike much previous research on climate change perception the use of open questions, as well as closed, allowed the respondent to respond to the question more freely, uninfluenced by the researcher (Foody, 1993), providing explanation to responses to closed questions and to better understand trends observed. The data used within this chapter comes from the open questions. A mix of both online and face-to-face approaches were used, noting that electronic questionnaires are increasingly recognised Van Selm and Jankowski, 2006; Dolnicar et al., 2009), and are cost effective (Mavis and Brocato, 1998; Kaplowitz, Hadlock and Levine, 2004), albeit potentially raising issues of low response rates(Murphy et al. 2020) in part through lack of trust in sending information over the internet Scott et al. (2011). According to ONS (2015) only 14% of households within the United Kingdom do not have access to the internet. This is a decline from 2006, when the figure was 43%. With more people using the internet for their everyday activities, it was observed that for households that only have one or more adults over the age of 65, this figure increases to 51%. This demonstrates that more people are using the internet, but that a majority of people over the age of 65 still do have access to the internet. The study attributes this to cost and lack of skills.

The questionnaire was posted to local community groups via the social media platform Facebook, such as for example the "Isle of Man, advice, help & local events". Facebook was used since the large proportion of the United Kingdom population use it regularly, Data gathered from the ONS (2020) suggests that in 2020, roughly 70% of the United Kingdom population used one of these sites within the last three months of being surveyed. Nonetheless, this approach raised the potential for self-selection bias. To reduce the potential impact of self-selection bias, the social demographic data was regularly monitored. In the case that the researcher discerned that there was an underrepresentation of a certain social demographic group, the researcher would attempt to seek out people from these underrepresented groups via the use of targeted emails to public body organisations. Overall, 1,134 questionnaires were completed by members of the public, and a full breakdown of the socio demographics of the respondents compiled.

The research supporting this chapter also used questionnaire data collected via the Department of Energy and Climate Change [DECC] until 2016 and then by the Department of Business, Energy & Industrial Strategy [BEIS]. These questionnaires about the

British public energy and climate change perceptions are collected roughly every three months.

2 Perception of Climate Change in the United Kingdom

A consideration in this research was that the majority of British people believe that the climate is currently changing variously through anthropogenic [burning of fossil fuels] and natural [solar activity, volcanic activity, etc.] processes. Where the main cause of current climate change is considered natural respondents would be less likely to believe that individual or societal actions are going to make much difference. As such, some educational institutions may have only been teaching about the human processes, to avoid confusing their students (Seow and Ho, 2016), resulting in young people not fully understanding or appreciating the complexity of the climate system.

Figure 1 presents the BEIS survey results for the question "thinking about the causes of climate change, which, if any, of the following best describe your opinion?", asked each year in March between 2013 and 2020. Overall, it demonstrates that there has been a general decline in those indicating that the climate changed due to natural processes or that there is no such thing as climate change between 2013 and 2020. Therefore, it would be reasonable to postulate that increasingly more believe that climate change is a process in which there is collective responsibility, and that society can limit it.

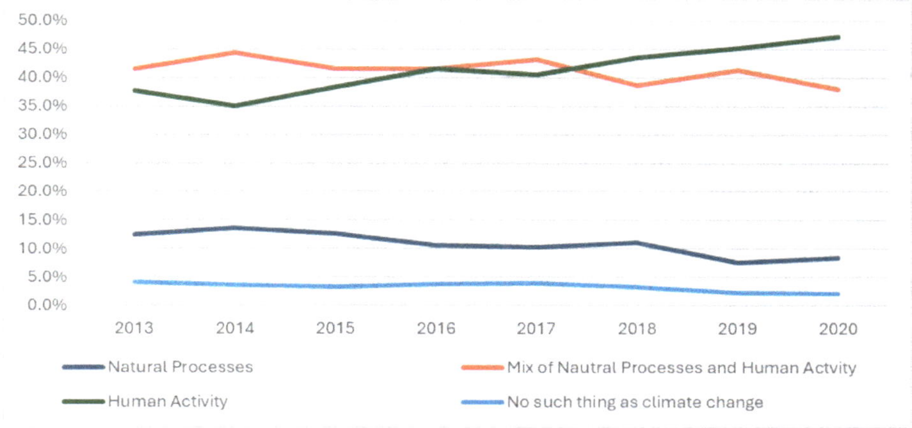

Fig. 1. Perception of Cause of Current Climate Change between March 2013 and March 2020 (Data Adapted from BEIS, 2017; 2018; 2019; 2020; DECC, 2013; 2014; 2015; 2016).

Persistent scepticism regards the non-natural process is however prevalent. Respondent 'OLA181' from our own survey highlighted; "climate change is a natural process and all this hype is for other companies to get money out of us including the government" [Female, 45–54, Northern Ireland]. This was backed up by respondent 'OLC173' who indicated that; "'climate change' is a natural phenomenon which is hyped up by scientists, government and the media to justify excessive additional taxation." [Male, 65+,

South East England]. These two quotes exemplify how some believe climate change is no more than fiscal policy to generate additional taxation for the treasury. This interpretation has previously been identified as a concern by the Grantham Research Institute on Climate Change and the Environment, who highlight that there is distrust, such that climate change is a clandestine way to increase taxation, rather than incentive to reduce emissions (Carattini et al., 2017). The United Kingdom has employed taxation methods linked to climate change that have proven to encourage people and companies to reduce their carbon emissions. One of the most known is the Hydrocarbon oil duty, more commonly known as fuel duty, which is a fuel tax which is levied on fuel which has a carbon origin (Hydrocarbon Oil Duties Act 1979, c.5). Whilst Fuel duty was first introduced in the United Kingdom in 1909 with the passing of the 'Financial Act 1908' (Mackett et al., 2013), the way in which the base value of the tax is calculated was altered in 1993, with the introduction of the 'road fuel escalator', which is also known as the 'fuel price escalator' (Seely, 2011). The initial rate of the escalator was set at three percent plus inflation and has been adjusted at various times since. There were varied reasons why the fuel price escalator was introduced and increased, but one was as it is a technique for the United Kingdom to meet its carbon dioxide targets set at the 1992 United Nations Conference on Environment and Development (AA, 2008). It would force manufacturers to make more fuel-efficient vehicles and discourage the public in making unnecessary trips (Gray et al., 2001). The OECD (2001) estimated that the escalator had helped reduce between two and five million tonnes of carbon dioxide. However, the significant increase in fuel duty in a short period of time resulted in growing discontent amongst the public, which lead to a mass fuel protest in September 2000. Consequently, the then Labour Government decided to freeze fuel duty until 2002, and since it has only been raised twice, only in line with inflation, and decreased in 2022.

The second disbelief in climate change according to comments received related to conspiracy and claims, specifically misleading and controversial claims. For example, respondent OLB010 stated that a 'credulous public believes big-business conspiracy that climate change is a hoax' [Male, 45–54, North East England]. Misinformation campaigns have been waged against climate change concerns by climate-denier groups and individuals. Examples of these groups are 'Global Warming Policy Foundation', 'Exxon', and 'Total' (Monbiot, 2006; Supran and Oreskes, 2017; Bonneuil, Choquet, and Franta, 2021). It is not possible to list all manner of changing climate denial beliefs here. However, in recent years, due to the escalation of social media, a new style denialism relating to both climate change and science in general is very detectable. Examples are famously provided by Donald Trump who tweeted that "the concept of global warming was created by and for the Chinese in order to make U.S. manufacturing non-competitive" (Trump, 2012, Online). This is an example of a conspiracy claim about climate change, which will have been seen by millions of Twitter users and has been reported by numerous new reports in the intervening years, giving this conspiracy claim large exposure.

These types of conspiracies are not mentioned in Mann's (2012) six stage of denial. However, Hoofnagle (2009) had a different take on science denial indicating the following five types for which some examples are included:

1. The use of forceful statements to subdue the truth.

2. Being selective of the data. An example of this is when climate sceptics (Eastbrook, Oliver and Carter, 2013) state that global temperatures for most of the last 10,000 years have been warmer than they are presently based on work by Alley (2000). However, the last recorded temperature was from 1855, and when the temperatures since then are included, results show that current temperatures are above much of the past 10,000 years.
3. The use of fake experts. An example of this occurred when a letter by self-proclaimed 'prominent scientists' sent a letter to António Guterres, the UN Secretary-General stating that there is no climate emergency. When observing the list, most had a history of climate denialism and little to no expertise in climate change science (Walker and Leviston, 2019).
4. Either moving the goalposts on expectations or having implausible expectations. This is when denialists refuse to concede when their challenges have been addressed by climate scientists and start making new claims.
5. Logical paradox. This is along the lines of Betz (2020) who called out a number of royal family members and celebrities who have used private jets as 'hypocrites'.

In addition to these arguments, there are some whose scepticism about climate change is due to their historical engagement with the issue being earlier related to global cooling. This is demonstrated with respondent 'OLC173' who recalls that they are "old enough to remember when we were being told that we would all have to move to the Sahara because of the new 'ice age'" [Male, 65+, South East England]. This perception about climate change has its origins in the 1970s, when there were stories within the media, especially within the United States, leading the public to the belief that the climate was cooling. Some examples of these are shown in Table 1. This occurred despite growing scientific research at the time, that greenhouse gas emissions were causing the temperature to rise.

Table 1. Some global cooling news from the 1970s

Headline	Publication	Reference
"Colder Winters Herald Dawn of New Ice Age: Scientists See Ice Age In the Future"	Washington Post 11th January 1970	Boldt (1970, p. A1)
"Science: Another Ice Age?"	Time Magazine 13th November 1972	Time Magazine (1972, Online)
"Scientists Ask Why World Climate Is Changing ... A Major Cooling Widely Considered to Be Inevitable"	New York Times 21st May 1975	Sullivan (1975a, p. 92)
"WARMING TREND SEEN IN CLIMATE: Two Articles Counter View That Cold Period is Due"	New York Times 14th August 1975	Sullivan (1975b, p. 24)

Source: Author

The former climate cooling alert leads to some older respondents' scepticism about current concerns as they are more likely to believe that climate change is natural due to having seen the media discussing global cooling and then global warming. This aside, and despite the overwhelming consensus on climate change as currently known about more comprehensively, there are still stories in the media that try to promote the natural phenomena sometimes including global cooling. Examples of these for the past fifteen years are demonstrated in Table 2.

3 The Problem of Climate Change Impact as Considered by People in the United Kingdom

Extensive changes in the United Kingdom's climate are anticipated in the coming years (IPCC, 2023; EA, 2018; Lowe *et al.*, 2018; Brown, 2020). However, public awareness of these changes and their impacts l remain questionable with the data from BEIS conducted in March 2019 and March 2020, indicating only a slight increase in the belief that climate change is already affecting the UK (from 65.4 to 65.9%). The perception of impact is being maintained by recent extreme weather events, such as heatwaves and severe winter flooding where personal experiences, media exposure, and scientific education have an influence.

In terms of current impacts of climate change within the United Kingdom, the survey data reveals that the British public perceives climate change impacts more significantly in other countries compared to the United Kingdom. This perception increased by 4.2% between March 2019 and March 2020, which is likely influenced by media coverage of severe weather events, such as the Australian wildfires between September 2019 and February 2020. Within the United Kingdom, 25.3% of respondents in 2020 believe climate change impacts were a great deal, with a further 53.1% acknowledging there were impacts to some extent.

When asked about the specific impacts within the United Kingdom, flooding and rising sea levels were the most noted changes between March 2019 and March 2020, with an 18.9% increase, likely due to recent widespread flooding during the winter of 2019–2020. Fewer respondents in 2020 identified rising temperatures as a significant issue, likely due to a warm summer in 2018 and the late winter heatwave of 2019. Younger respondents (16–24) are less likely to acknowledge current impacts, potentially due to their shorter lifespan observation period.

In terms of future impacts of climate change within the United Kingdom, it is highlighted from the BEIS data that rising sea levels and increased flooding are of greatest concern, with 60.6% of respondents in 2020 identifying these as likely future issues, marking a significant increase from 2019. However, the overall concern for other impacts declined, possibly influenced by the COVID-19 pandemic[1] diverting public attention and worry. The 'finite pool of worry' theory suggests that the immediate crisis of the pandemic and its impact could overshadow future climate concerns. The data also indicates that the younger individuals are less likely to anticipate severe future impacts, possibly due to their limited exposure to past climate/weather variability.

Analysis across age groups showed that there were significant differences, with older individuals being more likely to anticipate different future impacts of climate

[1] The UK experienced a marked number of cases of COVID-19 in March 2020, which resulted in the first national lockdown on the 23rd March.

Table 2. Global cooling news articles since 2010 in the United Kingdom

Headline	Publication	References
"The mini ice age starts here"	Daily Mail 9th January 2010	Rose (2010, Online)
"Dawn of a new ice age"	Daily Express 21st December 2010	Warren (2010, Online)
"So much for global warming as Planet Earth gets colder"	Daily Express 5th July 2011	Ingham (2011, Online)
"Britain faces a mini 'ice age'	Daily Express 10th October 2011	Caroe (2011, Online)
"Forget global warming – it's Cycle 25 we need to worry about (and if NASA scientists are right the Thames will be freezing over again)"	Daily Mail 28th January 2012	Rose (2012, Online)
"And now it's global COOLING! Return of Arctic ice cap as it grows by 29% in a year"	Daily Mail 8th September 2013	Rose (2013, Online)
"Scientists warn the sun will 'go to sleep' in 2030 and could cause temperatures to plummet"	Daily Mail 10th July 2015	Prigg (2015, Online)
"Earth heading for 'mini ice age' within 15 years"	Daily Mail 11th July 2015	Hyde (2015, Online)
"World is on brink of new ice age and Britain could be hit the hardest"	Daily Mirror 16th October 2015	Carr (2015, Online)
"GLOBAL COOLING: Decade long ice age predicted as sun 'hibernates'"	Daily Express 1st December 2015	Austin (2015, Online)
"The silent sun: Eerie image revealed as solar activity remains the quietest it has been in more than a century – and some clims it could even trigger a mini ice age"	Daily Mail 12th February 2016	Zolfagharifard (2016, Online)
"Freak climate changes could spark 'mini ice-age' in Britain from 2017"	Daily Express 13th August 2016	O'Brien (2016, Online)
"Three decade ICE AGE to freeze Earth from 2030 – scientists make '97% accurate' prediction"	Daily Star 27th October 2016	Nevett, (2016, Online)
"Plummeting temperatures could send the world into a 'mini ice age' in 2030 and could OVERRIDE global warming, claim mathematicians"	Daily Mail 27th December 2017	Weston (2017, Online)

(continued)

Table 2. *(continued)*

Headline	Publication	References
"Could we face a mini ice age in the next 30 years? Scientists make extraordinary prediction based on the sun's natural cycles…and it would even reverse global warming!"	Daily Mail 1st January 2018	Naish (2018, Online)
"Extreme cold winter weather conditions such as the 'Beast from the East' are linked to a weak solar cycle that could trigger a MINI ICE AGE by 2050, experts warn"	Daily Mail 20th March 2018	Pettit (2018, Online)
"'Mini ice age' on Earth could cause -50C temperatures and food shortage, experts warn"	Daily Star 2nd February 2020	Taylor (2020, Online)

Source: Author

change. Social grade and education levels have been demonstrated to also play a role, as higher social grades and education levels correlate with greater concern for future climate impacts. In addition, females are generally more likely to believe different future impacts of climate change will happen, though the differences are not always statistically significant. Urban residents and ethnic minorities also show varying levels of concern, particularly regarding water availability and pollution. An implication here is that demographic differences require tailoring communication of climate change messages effectively across the population.

The results show that the British public generally think that the United Kingdom is already being impacted by climate change, albeit with varying views on how that has been occurring and how it will occur in the future. This raises the further question as to whether, although they think that it is already happening, they think that climate change is a serious problem.

The Eurobarometer has been recording how serious a problem civil society across the European Union has viewed climate change between 2008 and 2019. The respondents were asked "how serious a problem do you think climate change is at this moment?". These results were collected using a Likert Scale between one and 10 where 1 indicated not at all a serious problem, and 10 an extremely serious problem. Table 3 indicates the top five countries overall within the European Union by levels of how serious civil society in each view climate change for the previous two Eurobarometer surveys of 2017 and 2019. It shows that that for both dates, the United Kingdom scored lower concern scores compared to several other European Union countries. The April 2019 survey shows that the United Kingdom was seventeenth out of twenty-eight countries, and averaging a mean score of 1.07 behind Spain, the highest scoring country. This gap is quite constant, as the gap in 2017 was 1.02; however, the United Kingdom gained seven positions in the intervening period with an expanding heterogeneity in how serious a problem climate change is viewed across the European Union.

The United Kingdom based perception of how serious a problem climate change is currently has not been constant. This is shown in both Table 3 and then in Fig. 2, the latter of which demonstrates that between 2008 and 2019 there are two main stages to the levels of seriousness with which the British respondents considered climate change. The first stage is the decline from the start of the survey in April 2008 to December 2013, where the mean score dropped from 7.01 to 6.09. The next stage is the recovery of the perception score from December 2013, with the score rising from 6.09 to 7.83 in April 2019. This data is consistent with the likelihood of at least two 'social tipping points' in the last decade and half.

Table 3. Top five EU countries and the United Kingdom ranking for 2017 and 2019 of climate change perception as a serious problem

2017		
Rank	**Country**	**Mean**
1	Spain	8.21
2	Greece	8.19
3	Italy	8.12
4	France	8.08
	Hungary	
24	United Kingdom	7.19
-	European Union	7.59

2019		
Rank	**Country**	**Mean**
1	Malta	8.90
2	Spain	8.55
3	Greece	8.48
4	Cyprus	8.33
5	Hungary	8.31
17	United Kingdom	7.83
-	European Union	7.88

Source: Data based on Eurobarometer (2017 and 2019)

Within climate change studies, tipping points are regularly and increasingly talked about in terms of an environmental issue. Milkoreit et al. (2018, p. 10) defines social tipping points as "self-reinforcing positive feedback mechanisms, that inevitably and often irreversibly lead to a qualitatively different state of the social system". The UK study reported in this chapter suggests this in that the data is consistent with a self-reinforcing feedback mechanism that can be controlled by numerous external focuses, including extreme meteorological events, controversy, activism or media coverage. It demonstrates that changes in perception are inevitable.

In relation to the trend line indicated in Fig. 2, the first tipping point occurred at some point before 2009. Though this is before the available data, this social tipping point in climate change perspectives has been observed in numerous previous studies from roughly this time period, when there was an increase in climate scepticism, especially within Australia (Connor and Higginbotham, 2013), Germany (Ratter, Philipp and von Storch, 2012), the United Kingdom (Spence et al., 2010; Corner et al., 2011), and the United States (Shwartz, 2010; Leiserowitz et al., 2014). Other studies speculate this occurred due to extremely cold and snowy weather within the United States and Europe (Perkins, 2010; Moser and Dilling, 2011; Capstick and Pidgeon, 2014). A further potential cause was the state of the economy; past research reported by Kahn and Kotchen (2010) demonstrated that higher levels of unemployment have a positive association with lower levels of environmental concerns. This is also linked to the 'finite pool of worry' theory as backed up by Weber (2010). However, as highlighted earlier in this chapter, there are some questions as to whether the 'finite pool of worry' effect is strong.

Despite the aftermath of Brexit and accompanying economic uncertainty, and in addition to the increased number of terrorist attacks within the United Kingdom and Europe in the mid-2010s, the level of concern with climate, as observed within Fig. 2, continued to rise following its earlier decline. A further theory is the controversy that surrounded 'climategate', but researchers believed that this impact was only short-lived (Anderegg and Goldsmith, 2014). The final theory is the lower levels of media coverage during that period (Leiserowitz et al., 2013). Boykoff (2011) demonstrated that apart from one month in 2009, the time of the Copenhagen COP conference, there was a decline in the amount of media coverage that climate change received in both the United Kingdom and the United States from 2007. This would have an impact since the institutional media has previously been the most important way to disseminate scientific information about climate change. Whilst it is becoming less important with the advent of social media, it remains an important tool in communicating climate science with the public. The current trajectory is suggesting that climate change scepticism is still declining, but it is an uncertain trajectory since events can play an influential role in forming the national psyche towards climate change perception and engagement.

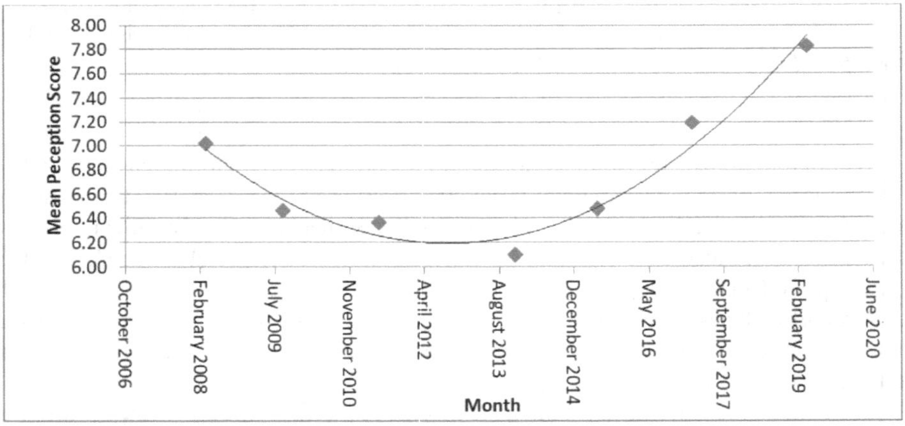

Fig. 2. Mean Perception Score for the British Respondents to the question "How Serious a Problem is Climate Change?" Diamonds = Observation Data. Source: Data Adapted from Eurobarometer (2008, 2009, 2011, 2013, 2015, 2017, 2019)

Whilst there are differences in the perception of how serious the public of each nation believe climate change is currently, this difference can be more localised within the borders of a country, as demonstrated clearly in the United States. Figure 3 demonstrates the perception score on the seriousness of climate change in the continent of Europe using the NUTS1 (*Nomenclature d'Unités Territoriales Statistiques*) break down of regions. Within the United Kingdom, there are regions, such as Cornwall and Northern Ireland, which are ranked as the top category on believing there is a problem. In contrast, North East of England and 'Yorkshire and the Humber' are ranked much lower having similar scores to those recorded in Eastern Europe and the most northern extremes of Europe. The map indicates contrasting differences are often for regions next to each other. This is consistent with the finding that each area has a unique balance of factors that influences perception in that area, suggesting that one mode of climate change communication for all is likely missing specificity of perception influencers within regions.

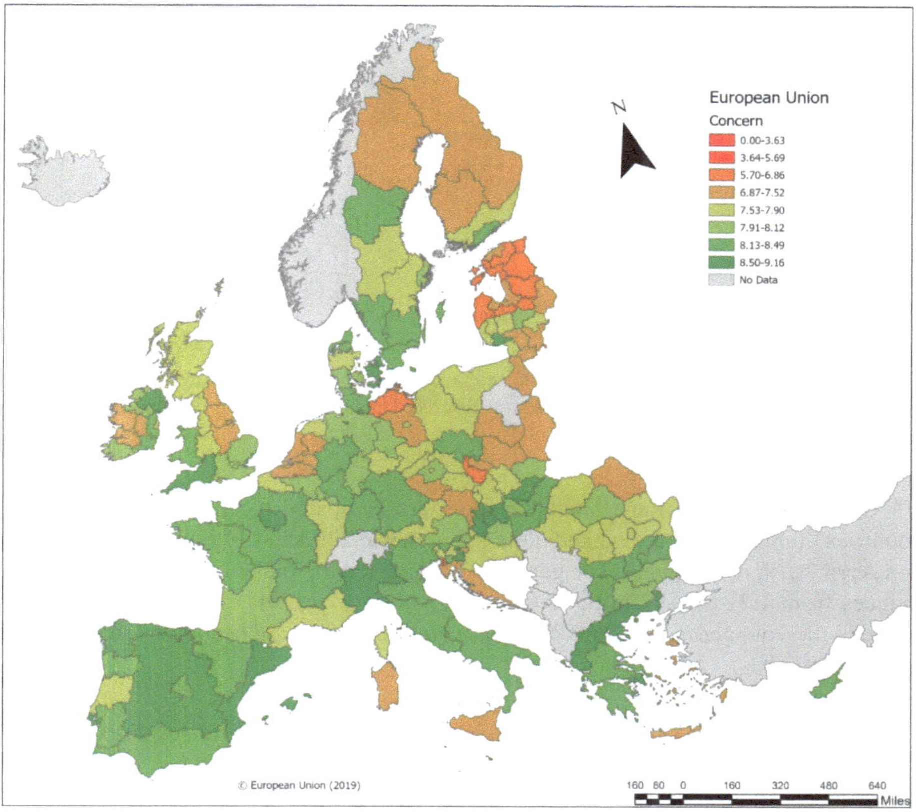

Fig. 3. Seriousness of the Climate Change problem as viewed Across the European Union in April 2019 at NUTS1 (Nomenclature d'Unités Territoriales Statistiques) Level. Source: Data Adapted from Eurobarometer (2019)

Figure 4 demonstrates the difference in perception by age groups between April 2008 and April 2019. The data shows that young people [15–24] have throughout the eleven years period been the third out of the four age groups to be more concerned about climate change within the United Kingdom. This demonstrates whilst young people are concerned about climate change, compared to the two age groups above them they have not constantly viewed it as a serious problem. Constant with the results from the earlier findings referred to in this chapter the most elderly respondents within the survey are typically least likely to think that climate change is a serious problem compared to other age groups. However, the gap observed pre-2013 is not as large by 2018. The overall difference between age group perception of a problem has reduced in recent years, now standing being just 0.19 compared to 1.32 in April 2008.

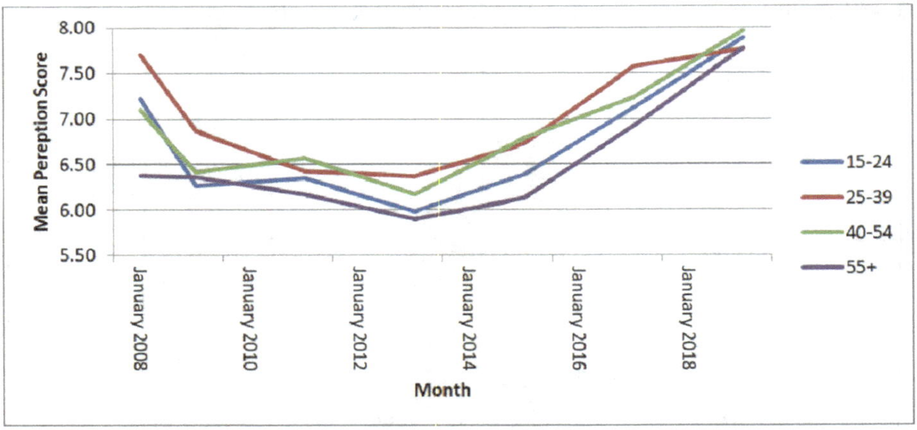

Fig. 4. Mean Perception Score for British Respondents to the question "How Serious a Problem is Climate Change?" based upon Age Groups. Source: Data Adapted from Eurobarometer (2008, 2009, 2011, 2013, 2015, 2017, 2019)

When the youth of the United Kingdom are compared to other European Union countries (Table 4) they are ranked quite low down at 22nd and 17th place in June 2017 and April 2019, respectively. The gap below Malta, the top ranked country at both dates reduces from 1.21 to 0.94 in the intervening two-year period. As such the difference between the younger respondents and the overall responses are becoming quite like the other countries.

Table 4. Top five EU countries and the United Kingdom ranking for 2017 and 2019 of climate change as a serious problem for respondents aged 15 to 24 years old

2017		
Rank	Country	Mean
1	Malta	8.37
2	Greece	8.29
3	Italy	8.26
4	Spain	8.23
5	Sweden	8.16
22	United Kingdom	7.16
-	European Union	7.51

2019		
Rank	Country	Mean
1	Malta	8.82
2	Italy	8.71
3	Spain	8.62
4	Hungary	8.52
5	Denmark	8.49
17	United Kingdom	7.88
-	European Union	7.90

Source: Data Adapted from Eurobarometer (2017, 2019)

The Eurobarometer data also indicates that females within the United Kingdom were more likely to think that climate change was a serious problem compared to their male counterparts (7.95 versus 7.73). This is consistent with earlier indications that female respondents are showing to be generally more concerned with environmental issues, such as climate change, albeit some of the data in the research showed this to be a barely significant difference. In terms of political identification, it was found that the further left leaning the voter, the higher they would score in terms of acknowledging the seriousness of climate change, with the average for a left-wing voter scoring 8.21 compared to 7.09 for a right-wing voter. These results are consistent with past research (Whitmarsh, 2011; Unsworth and Fielding, 2014; Milfont et al., 2015; Poortinga et al., 2019; Ballew et al., 2020), which suggested that there is some kind of political polarisation in climate change perception within the United Kingdom.

4 The Role of Climate Change Description in How It is Perceived

The ambiguities in the above research findings in terms of variation between subgroups, also raised questions about the term "climate change", which has its origins dating back to at least the 1920s (Willis, 1925). Since the 1980s, the term 'global warming' has predominantly been used to describe the impacts of the increase greenhouse gas emissions, due to anthropogenic activity, and its effects on current and future climates (Whitmarsh, 2009). It has been highlighted by Houghton (2004) that the metaphor of "warming" has been effective in capturing the public attention to the issue. However, it does not demonstrate the complexities of the climate system and potential global risks that climate change is expected to trigger.

The two terms are part of a nomenclature, or terminology, with varying technical definitions. The term 'global warming' only refers to the increase in global surface temperatures (Jacobs et al., 2016). The term 'climate change' does not just refer to the increase of temperatures both at a local and global scale, but also describes the changes in climatological features and/or processes due to these temperature changes, including the prevalence of extreme climatic conditions (Meehl et al., 2000). For example, the alteration of spatial distribution of precipitation globally (Trenberth, 1998; Sorribas et al., 2016); the distribution and frequency of droughts (Dai, 2013); and the location, frequency and intensity of tropical storms (Knutson et al., 2010).

Climate researchers highlight that civil society views these two terms in completely different ways. Whitmarsh (2009) found that 79.8% of the British civil society thought that global warming was an important issue compared to 66.6% for the case of climate change. This seems to suggest that there is a difference in which people in Britain cognitively frame each issue. This study is now 11 years old and with increased usage of climate change, it is possible that these differences might not be as large or remain statistically significant now. Also, these are not the only two terms that have been used regularly to describe climate change. The term 'climate emergency' has increased in frequency within the last couple of years. This is demonstrated by the term being declared the word of the year in 2019 for Oxford Dictionaries (Zhou, 2019), as they said that the usage of the word increased by one hundred-fold compared to the previous twelve months.

Figure 5 demonstrates the popularity of 'climate change', 'global warming', 'global cooling', and 'climate emergency' terms between January 2004 and July 2020 within the United Kingdom according to Google Trends, which looks at the history of Google searches. The inclusion of 'climate emergency' is due to the increased discussion of this term in 2019 by prominent news agencies. The values in Fig. 5 demonstrate the relative amount of use of the terms in searches within the United Kingdom. In the early years the term 'global warming' was for most months the most searched term about the change in the climate. The term 'climate change' was searched more frequently than 'global warming' only five times in the first five years between 2004 and 2008, with all of these been observed in the month of August. The change of the popularly used terms starts to be observed in 2009 onwards, when the search terms were similarly searched until 2014. At this stage, it becomes clearer that the term 'climate change' was searched more regularly than 'global warming'.

Overall, the British public searches of these terms followed a general decline throughout the late 2000s and early to the mid-2010s for 'climate change' and 'global warming'. This coincides with what is considered the declining concern with climate change as perceived amongst people in the United Kingdom and a decline in media coverage (Boykoff, 2011). However, in recent years there has been an increase in concern with climate change amongst people in Britain reflected by the increased searches for the topic on Google. It is likely that whilst high levels of climate change can drive concerns about it, online searches reflects both changing concerns and changing perceptions about how climate change is interpreted.

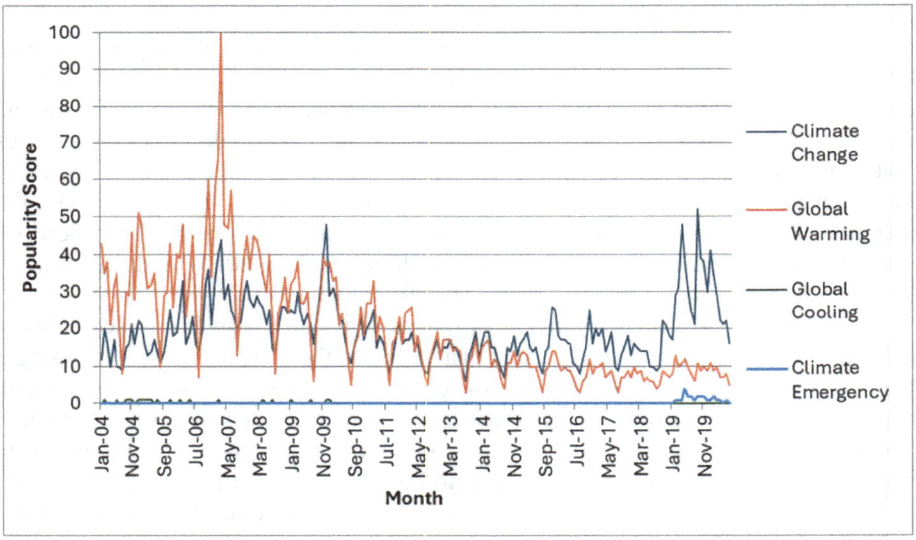

Fig. 5. Google Web Searches within the United Kingdom for Terms Climate Change, Global Warming, Global Cooling and Climate Emergency between January 2004 and July 2020. Source: Data Adapted from Google Trends (2020).

In addition to the usage and consequent understanding of climate change throughout this period, the research allowed analysis of when climate change was most talked about within the United Kingdom. This is demonstrated with the large spike for communications about 'global warming' in March 2007. The most likely explanation of this spike is the release of a polemical documentary film called "The Great Global Warming Swindle" by Channel 4. In addition, there was another spike in the previous month, due to the release of Al Gore's documentary film "An Inconvenient Truth" winning two awards at the Academy Awards.

To further demonstrate how events can result in engagement of the public to search for climate change related terms, there were two spikes for the term 'climate change' in November/December 2015 and November 2016. Both spikes are likely to be due to the Agreement and Effective of the Climate Change Paris Agreement, respectively. In addition, it can be observed that since late 2018, there has been a large increase in the amount of searching of the term 'climate change', and it is at its most sustained level since the late 2000s. This is likely to do with the increase of climate activism, notably amongst youth, that has occurred both within the United Kingdom and throughout the world, due to the rise of 'Extinction Rebellion', the 'School Strikes for Climate' and the activist Greta Thunberg.

The term 'global cooling' was rarely searched throughout the period, only a few times in sporadic intervals for which the term received a popularity score of 1. The most recent score of 1 was during the winter of 2009–10, which happened to coincide with the unusually snowy and cold winter that occurred across the whole of the United Kingdom. This likely demonstrates that extreme meteorological events, as well as activism and political events, can influence the number of times the public search for climate related terms.

Lastly, the term 'climate emergency', before 2019 was ranked with a popularity score of zero, indicating that virtually nobody was searching for this term on Google. However, there has since been an increase in its usage, but despite being named as the Oxford Dictionary Word, it only scored 4 at its peak in May 2019. This peak coincided with it's informal declaration by the UK Parliament. In addition, as with all the other terms, the search frequency of the term 'climate emergency' reduced back to a score of 1, returning nearly back to pre-2019 levels. Arguably, this demonstrates that despite the popular use of the term by activists and the media, the public are still using the terms that had earlier been set out via the science of climate change.

4.1 Conclusion

This chapter has explored some of the multifaceted aspects of climate change perception within the United Kingdom. The understanding of anthropogenic climate change, although a relatively new discourse until the 1980s, is today widely recognised as one of the most significant environmental, societal, and economic challenges. The perception of climate change among people in Britain varies widely, being influenced by political identity, gender, age and other socio-demographics. Left-leaning and female people are generally more likely to acknowledge the severity of climate change compared to their right-wing and male counterparts. However, perceptions of climate change are not static and can fluctuate over time due to current events, the level of media coverage and personal experiences of it.

Youth perception is particularly significant in the analysis since young people will face the long-term consequences of climate change for longer. Despite their increasing role in being aware and raising concerns, research focusing on youth perception and engagement remains limited. It will be increasingly imperative to enhance climate related education, to empower the next generation to participate actively in climate action and necessary changes to the way people live in the United Kingdom and elsewhere in the world. However, some data trends and our engagement in the education sector also shows that whilst a younger generation believe that climate change is more serious in comparison to the population as a whole, their perception of its impacts most often remain unclear.

Furthermore, the terminology used in discussing climate change significantly affects public perception and engagement with the issue. Terms such as "global warming" and "climate emergency" evoke different levels of concern and urgency among the people who hear them. In recent years, people represented in this research are more likely to use the term "climate change" as scientifically correct but this may not evoke such a strong reaction.

In summary, addressing climate change will require an inclusive approach that considers a diverse perception of what climate change is all about across different demographics throughout the United Kingdom and elsewhere. If follows that continuous and improved education, targeted communication and involvement of all societal groups, in particular youth, are essential steps for developing a climate sentient, resilient and informed society. The chapter therefore ends by emphasising the importance of integrating scientific knowledge with public perception, to encourage and enable governments and other stakeholders to develop effective and sustainable policies for tackling climate change.

Declaration. A significant portion (roughly 75%) of this book chapter's content is derived from the doctoral thesis of the lead author, Dr. Mark Ashley Parry, available here: https://core.ac.uk/download/467108101.pdf. This material has not been previously published and will not be republished elsewhere in the future.

References

AA (Automotive Association): Taxation and investment – 1993–2008. Automotive Association, London, United Kingdom (2008)

Abbot, J., Marohasy, J.: The application of machine learning for evaluating anthropogenic versus natural climate change. GeoResJ **14**, 36–46 (2017). https://doi.org/10.1016/j.grj.2017.08.001

Adams, A.: "Josh Wears Pink Cleats': inclusive masculinity on the soccer field. J. Homosex. **58**(5), 579–596 (2011). https://doi.org/10.1080/00918369.2011.563654

Alley, R.B.: The Younger Dryas cold interval as viewed from central Greenland. Quatern. Sci. Rev. **19**(1–5), 213–226 (2000). https://doi.org/10.1016/S0277-3791(99)00062-1

Anderegg, W.R.L., Goldsmith, G.R.: Public interest in climate change over the past decade and the effects of the 'climategate' media event. Environ. Res. Lett. **9**, 111004 (2014). https://doi.org/10.1088/1748-9326/9/5/054005

Anderegg, W.R.L., Prall, J.W., Harold, J., Schneider, S.H.: Expert credibility in climate change. Proc. Natl. Acad. Sci. **107**(27), 12107–12109 (2010). https://doi.org/10.1073/pnas.1003187107

Andersen, P., Vandehey, M.: Career Counseling and Development in a Global Economy. Cengage Learning, Boston, Massachusetts, United States (2011)

Anderson, E.: Inclusive Masculinity: The Changing Nature of Masculinities. Routledge, New York City, New York, United States (2009). https://doi.org/10.4324/9780203871485

Anderson, E., McCormack, M.: Cuddling and spooning: heteromasculinity and homosocial tactility among student-athletes. Men Masculinities **18**(2), 214–230 (2015). https://doi.org/10.1177/1097184X14523433

Austin, J.: GLOBAL COOLING: Decade long ice age predicted as sun "hibernates". Daily Express, 1st December (2015). https://www.express.co.uk/news/science/616937/GLOBAL-COOLING-Decade-long-ice-age-predicted-as-sun-hibernates. Accessed 5 May 2020

Baker, L., Shaffrey, L., Hawkins, E.: Has the risk of a 1976 north-west European summer drought and heatwave event increased since the 1970s because of climate change? Q. J. R. Meteorol. Soc. **147**(741), 4143–4162 (2021). https://doi.org/10.1002/qj.4172

Ballew, M.T., et al.: Climate change in the american mind: data, tools, and trends. Environ. Sci. Policy Sustain. Dev. **61**(3), 4–18 (2019). https://doi.org/10.1080/00139157.2019.1589300

Ballew, M.T., Marlon, H., Leiserowitz, A., Maibach, E.: Gender Differences in Public Understanding of Climate Change. Yale University and George Mason University. New Haven, Connecticut, United States: Yale Project on Climate Change Communication (2018). https://climatecommunication.yale.edu/publications/gender-differences-in-public-understanding-of-climate-change/. Accessed 5 Feb 2024

Ballew, M.T., Pearson, A.R., Goldberg, M.H., Rosenthal, S.A., Leiserowitz, A.: Does socioeconomic status moderate the political divide on climate change? The role of education, income, and individualism. Glob. Environ. Change **60**(102024), 1–12 (2020). https://doi.org/10.1016/j.gloenvcha.2019.102024

Baxter, P.J.: The east coast Big Flood, 31 January–1 February 1953: a summary of the human disaster. Phil. Trans. R. Soc. A **363**, 1293–1312 (2005). https://doi.org/10.1098/rsta.2005.1569

BEIS (Department for Business, Energy and Industrial Strategy): Energy and Climate Change Public Attitude Tracker: Wave 21 (2017). https://assets.publishing.service.gov.uk/government/uploads/system/uploads/attachment_data/file/611988/DISCLOSURE_CHECKED_PAT_-_Wave_21.xlsx. Accessed 24 July 2020

BEIS (Department for Business, Energy and Industrial Strategy): Energy and Climate Change Public Attitude Tracker: Wave 25 (2018). https://assets.publishing.service.gov.uk/government/uploads/system/uploads/attachment_data/file/702647/Wave_25_dataset.xlsx. Accessed 24 July 2020

BEIS (Department for Business, Energy and Industrial Strategy): BEIS Public Attitude Tracker: Wave 29 (2019). https://assets.publishing.service.gov.uk/government/uploads/system/uploads/attachment_data/file/800437/BEIS_Public_Attitudes_Tracker_-_Wave_29_-_excel_dataset.xlsx. Accessed 24 July 2020

BEIS (Department for Business, Energy and Industrial Strategy): BEIS Public Attitude Tracker: Wave 33 (2020). https://assets.publishing.service.gov.uk/government/uploads/system/uploads/attachment_data/file/883261/BEIS_PAT_W33_-_Excel_Dataset.xlsx. Accessed 24 July 2020

Betz, B.: Prince Harry, Leonard DiCaprio and other celebrities who are hypocritical on climate change. Fox News, 29th January (2020). https://www.foxnews.com/entertainment/celebrities-hypocritical-climate-change. Accessed 3rd May 2020

Boldt, D.R.: Colder Winters Held Dawn of New Ice Age: Scientists See Ice Age In the Future. Washington Post, 11th January, p. A1 (1970)

Bonneuil, C., Choquet, P.-L., Franta, B.: Early warnings and emerging accountability: total's responses to global warming, 1971–2021. Glob. Environ. Chang. **71**, 102386 (2021). https://doi.org/10.1016/j.gloenvcha.2021.102386

Bord, R., O'Connor, R.: The gender gap in environmental attitudes. Soc. Sci. Q. **789**(4), 830–840 (1997)

Boykoff, M.T.: Who Speaks for the Climate? Cambridge University Press, Cambridge, Cambridgeshire, United Kingdom (2011)

Bradfield, P.: COP26 Glasgow: DUP MP Sammy Wilson fires broadside at climate change science of global summit. News Letter, 18th November (2021). https://www.newsletter.co.uk/news/politics/sammy-wilson-fire-broadside-at-cop26-science-on-climate-change-3461789. Accessed 10 Mar 2024

Brechin, S.R.: Public opinion: a cross-national view. In: Lever-Tracy, C. (ed.) Routledge Handbook of Climate Change and Society, pp. 179–209. Routledge, Abingdon, Oxfordshire, United Kingdom (2010)

Bridgen, A. [@ABridgen]: Parliament is deceiving the public over the vaccine harms and excess deaths, the Net Zero #Climatescam, the plans of the Who, Digital ID, and CBDCs. [Twitter/X] 19th March (2024). https://twitter.com/i/web/status/1770010120333951225. Accessed 19th Mar 2024

Brooks, J., Oxley, D., Vedlitz, A., Zahran, S., Lindsey, C.: Abnormal daily temperature and concern about climate change across the United States. Rev. Policy Res. **31**(3), 199–217 (2014). https://doi.org/10.1111/ropr.12067

Brown, K.: How much more climate change is inevitable for the UK? (2020). https://www.theccc.org.uk/2020/04/21/how-much-more-climate-change-is-inevitable-for-the-uk/. Accessed 5 July 2020

Brown, O., Hammill, A., McLeman, R.: Climate change as the 'new' security threat: implications for Africa. Int. Aff. **83**(6), 1141–1151 (2007)

Burke, P.J., Strets, J.E.: Identity Theory. Oxford University Press, Oxford, Oxfordshire, United Kingdom (2009)

Burt, S.: The exceptional hot spell of early August 1990 in the United Kingdom. Int. J. Climatol. **12**(6), 547–567 (1992). https://doi.org/10.1002/joc.3370120603

Burt, S.D., Mansfield, D.A.: The great storm of 15–16 October 1987. Weather **43**(3), 90–100 (1988). https://doi.org/10.1002/j.1477-8696.1988.tb03885.x

Butler, C.D.: Climate change, health and existential risks to civilization: a comprehensive review (1989–2013). Int. J. Environ. Res. Publ. Health **15**(1), 2266 (2018). https://doi.org/10.3390/ijerph15102266

Capstick, S.B., Pidgeon, N.F.: Public perception of cold weather events as evidence for and against climate change. Clim. Change **122**, 695–708 (2014). https://doi.org/10.1007/s10584-013-1003-1

Carattini, S., Carvalho, M., Fankhauser, S.: How to make carbon taxes more acceptable. London, England, United Kingdom: Grantham Research Institute on Climate and the Environment and Centre for Climate Change Economics and Policy, London School of Economics and Political Science (2017)

Caroe, L.: Britain face a mini 'ice age'. Daily Express, 10th October (2011). https://www.express.co.uk/news/uk/276516/Britain-faces-a-mini-ice-age. Accessed 5 May 2020

Carr, S.: World is on brink of new ice age and Britain could be hit the hardest. Daily Mirror, 16th October (2015). https://www.mirror.co.uk/news/uk-news/world-brink-new-ice-age-6646041. Accessed 5 May 2020

Carrington, D.: BBC apologies over interview with climate denier Lord Lawson. The Guardian, 24th October (2017). https://www.theguardian.com/environment/2017/oct/24/bbc-apologises-over-interview-climate-sceptic-lord-nigel-lawson. Accessed 29 July 2018

Chan, H.-W., Pong, V., Tam, K.-P.: Cross-national variation of gender differences in environmental concern: testing the sociocultural hindrance hypothesis. Environ. Behav. **51**(1), 81–108 (2019). https://doi.org/10.1177/0013916517735149

Cheryan, S., Siy, J.O., Vichayapai, M., Drury, B.J., Kim, S.: Do female and male role modesl who embody STEM stereotypes hinder women's anticipated success in STEM? Personal. Sci. **2**(6), 656–664 (2011). https://doi.org/10.1177/1948550611405218

Choudhury, S., Blakemore, S.-J., Charman, T.: Social cognitive development during adolescence. Soc. Cogn. Affect. Neurosci. **1**(3), 165–174 (2006). https://doi.org/10.1093/scan/nsl024

Cipriani, E., Frumento, S., Grassini, S., Gemignani, A., Menicucci, D.: Do individual differences in perception affect awareness of climate change? Brain Sci. **14**(3), 266 (2024). https://doi.org/10.3390/brainsci14030266

Connor, L.H., Higginbotham, N.: "Natural cycles" in lay understandings of climate change. Glob. Environ. Chang. **23**(6), 1852–1861 (2013). https://doi.org/10.1016/j.gloenvcha.2013.07.002

Cook, J., et al.: Quantifying the consensus on anthropogenic global warming in the scientific literature. Environ. Res. Lett. **8**, 1–7 (2013). https://doi.org/10.1088/1748-9326/8/2/024024

Cook, J., et al.: Consensus on consensus: a synthesis of consensus estimates on human-caused global warming. Environ. Res. Lett. **11**, 048002 (2016)

Cornelis, I., van Hiel, A., Roets, A., Kossowska, M.: Age differences in conservatism: evidence on the mediating effects of personality and cognitive style. J. Pers. **77**(1), 51–88 (2009). https://doi.org/10.1111/j.1467-6494.2008.00538.x

Corner, A., Venables, D., Spence, A., Poortinga, W., Demski, C., Pidgeon, N.: Nuclear power, climate change and energy security: exploring British public attitudes. Energy Policy **39**(9), 4823–4833 (2011). https://doi.org/10.1016/j.enpol.2011.06.037

Correll, S.J.: Gender and the career choice process: the role of biased self-assessments. Am. J. Sociol. **106**, 1691–1730 (2001)

Dai, A.: Increasing drought under global warming in observations and models. Nat. Clim. Chang. **3**, 52–58 (2013). https://doi.org/10.1038/nclimate1633

DECC (Department of Energy and Climate Change): Public attitudes tracking survey: wave 5 (2013). https://assets.publishing.service.gov.uk/government/uploads/system/uploads/attachment_data/file/193082/Wave_5_Excel_dataset.xlsx. Accessed 24 July 2020

DECC (Department of Energy and Climate Change): Public attitudes tracking survey: wave 9 (2014). https://assets.publishing.service.gov.uk/government/uploads/system/uploads/attachment_data/file/306901/wave_9_excel_data.xlsx. Accessed 24 July 2020

DECC (Department of Energy and Climate Change): Public attitudes tracking survey: wave 13 (2015). https://assets.publishing.service.gov.uk/government/uploads/system/uploads/attachment_data/file/424516/DECC_Public_Attitudes_Tracker_wave_13_excel_data.XLSX. Accessed 24 July 2020

DECC (Department of Energy and Climate Change): Public Attitudes Tracking Survey: Wave 17 (2016). https://assets.publishing.service.gov.uk/government/uploads/system/uploads/attachment_data/file/519439/Dataset-_Wave_17.xlsx. Accessed 24 July 2020

Delaney, J.M., Devereux, P.J.: Understanding gender differences in STEM: evidence from college applications. Econ. Educ. Rev. **72**, 219–238 (2019). https://doi.org/10.1016/j.econedurev.2019.06.002

Dessai, S., Adger, W.N., Hulme, M., Turnpenny, J., Köhler, J., Warren, R.: Defining and experiencing dangerous climate change. Clim. Change **64**(1–2), 11–25 (2004). https://doi.org/10.1023/B:CLIM.0000024781.48904.45

Dietz, T., Kalof, L., Stern, P.C.: Gender, values, and environmentalism. Soc. Sci. Q. **83**, 353–364 (2002)

Dolnicar, S., Laesser, C., Matus, K.: Online versus paper: format effects in tourism surveys. J. Travel Res. **47**(3), 295–316 (2009). https://doi.org/10.1177/0047287508326506

Doran, P.T., Zimmerman, M.K.: Examining the scientific consensus on climate change. Eos **90**(3), 22–23 (2009). https://doi.org/10.1029/2009EO030002

Eastbrook, D.J., Oliver, C.D., Carter, R.M.: Chapter 5 – Observation: The Cryosphere. In: Best, D.C., Karnick, S.T. (eds.) Climate Change Reconsidered II: Physical Science, Arlington Heights, Illinois, United States. Nongovernmental International Panel on Climate Change, The Heartland Institute, pp. 629–712 (2013)

EA (Environment Agency): Climate change impacts and adaptation (2018). https://assets.publishing.service.gov.uk/government/uploads/system/uploads/attachment_data/file/758983/Climate_change_impacts_and_adaptation.pdf. Accessed 5 July 2020

Eaton, D.K., et al.: Youth risk behavior surveillance – United States, 2005. Morbidity Mortal. Week. Surveillan. Summary **55**(5), 1–108 (2006)

Ehrlinger, J., Dunning, D.: How chronic self-views influence (and potentially mislead) estimates of performance. J. Pers. Soc. Psychol. **84**(1), 5–17 (2003). https://doi.org/10.1037/0022-3514.84.1.5

Ernst, M., Pine, D.S., Hardin, M.: Triadic model of neurobiology of motivated behaviour in adolescence. Psychol. Med. **36**(3), 299–312 (2006). https://doi.org/10.1017/S0033291705005891

Eurobarometer: Eurobarometer 69.2 (Mar-May 2008). Brussels, Belgium: European Union (2008). https://dbk.gesis.org/dbksearch/sdesc2.asp?no=4744&db=e. Accessed 5 Jan 2020

Eurobarometer: Special Eurobarometer 322: Europeans' attitudes towards climate change. Brussels, Belgium: European Union (2009). https://data.europa.eu/euodp/en/data/dataset/S703_72_1_EBS322. Accessed 5 Jan 2020

Eurobarometer: Special Eurobarometer 372: Social Climate. Brussels, Belgium: European Union (2011). https://data.europa.eu/euodp/en/data/dataset/S1007_75_4_EBS372. Accessed 5 Jan 2020

Eurobarometer: Special Eurobarometer 409: Climate Change. Brussels, Belgium: European Union (2013). https://data.europa.eu/euodp/en/data/dataset/S1084_80_2_409. Accessed 5 Jan 2020

Eurobarometer: Special Eurobarometer 435: Climate Change. Brussels, Belgium: European Union (2015). https://data.europa.eu/euodp/en/data/dataset/S2060_83_4_435_ENG. Accessed 5 Jan 2020

Eurobarometer: Special Eurobarometer 459: Climate Change. Brussels, Belgium: European Union (2017). https://data.europa.eu/euodp/en/data/dataset/S2140_87_1_459_ENG. Accessed 5 Jan 2020

Eurobarometer: Special Eurobarometer 490: Climate Change. Brussels, Belgium: European Union (2019). https://data.europa.eu/euodp/en/data/dataset/S2212_91_3_490_ENG. Accessed 5 Jan 2020

Foody, W.: Constructing Questions for Interviews and Questionnaires: Theory and Practice in Social Research. Cambridge University Press, Cambridge, Cambridgeshire, United Kingdom (1993)

Fogg, R., Munro, A., Setrice, J.-B.: Unhappy New Year as homes are blown apart. The Shetland Times, 3rd January, p. 1 (1992)

Franklin, C.W., II.: The Changing Definition of Masculinity. Plenum Press, New York City, New York, United States (1984)

Funatsu, B.M., Dubreuil, V., Racapé, A., Debortoli, N.S., Nasuti, S., Le Tourneau, F.-M.: Perception of climate and climate change by Amazonian communities. Glob. Environ. Chang. **57**, 101923 (2019). https://doi.org/10.1016/j.gloenvcha.2019.05.007

Gifford, E., Gifford, R.: The largely unacknowledged impact of climate change on mental health. Bull. Atom. Sci. **72**(5), 292–297 (2016). https://doi.org/10.1080/00963402.2016.1216505

Gillespie, T.: UK weather: more rain falls overnight as floods force 1,200 homes to evacuate. Sky News, 15th November (2019). https://news.sky.com/story/uk-weather-1-200-homes-evacuated-due-to-flooding-with-more-rain-on-the-way-11861270. Accessed 20 Mar 2024

Gray, D., Farrington, J., Shaw, J., Martin, S., Roberts, D.: Car dependence in rural Scotland: transport policy, devolution and the impact of the fuel duty escalator. J. Rural. Stud. **17**(1), 113–125 (2001). https://doi.org/10.1016/S0743-0167(00)00035-8

Greenbaum, A.: Taking stock of two decades of research on the social bases of environmental concern. In: Mehta, M.D., Ouellet, E. (eds.) Environmental Sociology, pp. 125–152. Captus Press, North York, Ontario, Canada (1995)

Gregory, J.: September heatwave sets new record. BBC News, 7th September (2023). https://www.bbc.co.uk/news/uk-66743788. Accessed 20 Mar 2024

Gregory, J., Wright, G.: England flooding: hundreds of homes evacuated after heavy rain. BBC News, 5th January (2024). https://www.bbc.co.uk/news/uk-67888959. Accessed 20 Mar 2024

Google Trends: Compare (2020). https://trends.google.com/trends/explore?date=all&geo=GB&q=Climate%20Change,global%20warming,global%20cooling,climate%20emergency. Accessed 14 Aug 2020

Happer, C., Philo, G.: New approaches to understanding the role of the news media in the formation of public attitudes and behaviours on climate change. Eur. J. Commun. **31**(2), 136–151 (2016). https://doi.org/10.1177/0267323115612213

Hayward, W.: A politician posted a bizarre video denying climate change – and got a lot of things wrong. WalesOnline, 7th May (2019). https://www.walesonline.co.uk/news/wales-news/politician-posted-bizarre-video-denying-16237701. Accessed 5 Oct 2020

Hickey, K.R.: The storm of 31 January to 1 February 1953 and its impact on Scotland. Scott. Geogr. J. **117**(4), 283–295 (2001). https://doi.org/10.1080/00369220118737129

Hoofnagle, M.: Climate change deniers: failsafe tips on how to spot them. The Guardian, 11th March (2009). https://www.theguardian.com/environment/blog/2009/mar/10/climate-change-denier. Accessed 29 Mar 2020

Hornsey, M.J., Harris, E.A., Bain, P.G., Fielding, K.S.: Meta-analyses and outcomes of belief in climate change. Nat. Clim. Chang. **6**, 622–626 (2016). https://doi.org/10.1038/nclimate2943

Houghton, J.: Global Warming: The Complete Briefing, 3rd edn. Cambridge University Press, Cambridge, Cambridgeshire, United Kingdom (2004)

Hyde, D.: Earth heading for 'mini ice age' within 15 years. The Telegraph, 11th July (2015). https://www.telegraph.co.uk/news/science/11733369/Earth-heading-for-mini-ice-age-within-15-years.html. Accessed 5 May 2020

Hydrocarbon Oil Duties Act 1979, c. 5. http://www.legislation.gov.uk/ukpga/1979/5. Accessed 22 Nov 2017

Idso, C., et al.: Climate Change Reconsidered: 2009 Report of the Nongovernmental International Panel on Climate Change (NIPCC). Chicago, Illinois, United States: The Heartland Institute (2009)

Ingham, J.: So much for global warming as Planet Earth gets colder. Daily Express, 5th July (2011). https://www.express.co.uk/news/uk/256855/So-much-for-global-warming-as-Planet-Earth-gets-colder. Accessed 1 Oct 2020

IPCC [Intergovernmental Panel on Climate Change]: Summary for Policymakers. In: Core Writing Team, Lee, H., Romero, J. (eds.) Climate Change 2023: Synthesis Report. Contribution of Working Groups I, II and III to the Sixth Assessment Report of the Intergovernmental Panel on Climate Change, pp. 1–34. IPCC, Geneva, Switzerland (2023). https://doi.org/10.59327/IPCC/AR6-9789291691647.001

Jacobs, P.H., Jokimäki, A., Rice, K., Green, S.A., Winkler, B.: Polluted discourse: communication and myths in a climate of denial. In: Drake, J.L., Kontar, Y.Y., Eichelberger, J.C., Rupp, T.S., Taylor, K.M. (eds.) Communicating Climate-Change and Natural Hazard risk and Cultivating Resilience. Springer, Cham, Switzerland (2016)

Jenkyns, A.: Climate Targets are Bureaucratic Nonsense! Finally we're Starting to Listen to the British People, says Dame Andrea Jenkyns (2023). https://www.andreajenkyns.co.uk/climate-targets-are-bureaucratic-nonsense-finally-were-starting-to-listen-to-the-british-people-says-dame-andrea-jenkyns/. Accessed 11 Mar 2023

Johnson, P.: Aftermath: time to count. The Shetland Times, 10th January. p. 2 (1992)

Kahn, M.E., Kotchen, M.J.: Environmental Concern and the Business Cycle: The Chilling Effect of Recession. Cambridge, Massachusetts, United States: National Bureau of economic Research. Working Paper 16241 (2010). https://doi.org/10.3386/w16241, https://www.nber.org/papers/w16241. Accessed 4 Mar 2020

Kaplowitz, M.D., Hadlock, T.D., Levine, R.: A comparison of web and mail survey response rates. Public Opin. Q. **68**(1), 94–101 (2004). https://doi.org/10.1093/poq/nfh006

Kehily, M.J.: Understanding Youth: Perspectives, Identities & Practices. SAGE Publications Ltd., London, United Kingdom (2007)

Kelly, A.: The Construction of Masculine Science. Br. J. Sociol. Educ. **6**(2), 133–154 (1985)

Kendon, M., et al.: State of the UK Climate 2023. Int. J. Climatol. **44**(s1), 1–117 (2024). https://doi.org/10.1002/joc.8553

Kendon, M., et al.: State of the UK Climate 2022. Int. J. Climatol. **43**(s1), 1–82 (2023). https://doi.org/10.1002/joc.8167

Kew, S.F., et al.: Autumn and Winter storms over UK and Ireland are becoming wetter due to climate change. Grantham Institute for Climate Change London, United Kingdom (2024). https://doi.org/10.25561/111577

Khalil, H.: Cars and rail line submerged as flood risk persists. BBC News, 27th September (2024). https://www.bbc.co.uk/news/articles/cn8j6le2498o. Accessed 30 Sept 2024

Kirka, D.: Britain on alert after 2nd major storm of the season batters country, leaving at least 2 dead. AP News, 26th November (2024). https://apnews.com/article/flooding-britain-wales-bert-storm-wind-rivers-f955134de835d1ff591c6be29325e656. Accessed 2 Dec 2024

Kite, M.E.: Changing Time, Changing Gender Roles:@ Who De We Want Women and Men to Be? In: Unger, R. (ed.) The Handbook of the Psychology of Women and Gender, pp. 215–227. Wiley, New York City, New York, United States (2001)

Knutson, T.R., et al.: Tropical cyclones and climate change. Nat. Geosci. **3**, 157–163 (2010). https://doi.org/10.1038/ngeo779

Lawson, N.: An Appeal to Reason: A Cool Look at Global Warming. Duckworth Overlook, London, United Kingdom (2008)

Le Page, M.: Earth passes 2 °C of warming on hottest day ever recorded. New Scientist, 20th November (2023). https://www.newscientist.com/article/2403743-earth-passes-2c-of-warming-on-hottest-day-ever-recorded/. Accessed 10 Mar 2024

Legates, D.R., Soon, W., Briggs, W.M., Monkton, C.: Climate consensus and 'misinformation': a rejoinder to agnotology, scientific consensus, and the teaching and learning of climate change. Sci. Educ. **24**(3), 299–318 (2015). https://doi.org/10.1007/s11191-013-9647-9

Leiserowitz, A.: Climate change risk perception and policy preferences: the role of affect, imagery and values. Clim. Change **77**, 45–72 (2006). https://doi.org/10.1007/s10584-006-9059-9

Leiserowitz, A., et al.: International Public Opinion on Climate Change, 2022. New Haven, Connecticut, United States: Yale Program on Climate Change Communication and Data for Good at Meta (2022)

Leiserowitz, A., Maibach, E., Roser-Renouf, C., Feinberg, G., Howe, P.: Extreme Weather and Climate Changes in the American Mind: April 2013. Yale University and George Mason University. New Haven, Connecticut, United States: Yale Program on Climate Change Communication (2013)

Leiserowitz, A., Maibach, E., Roser-Renouf, C., Feinberg, G., Rosenthal, S., Marlon, J.: Climate Change in the American Mind: October 2014. Yale University and George Mason University.

New Haven, Connecticut, United States: Yale Program on Climate Change Communication (2014)
Leiserowitz, A., et al.: International Public Opinion on Climate Change, 2023. New Haven, Connecticut, United States: Yale Program on Climate Change Communication and Data for Good at Meta (2023)
Lerner, J.V., Lerner, R.M., Finkelstein, J.: Adolescence in America: An Encyclopaedia. Santa Barbara, California, United States: ABC-CLIO (2001)
Li, Y., Johnson, E.J., Zaval, L.: Local warming: daily temperature change influences belief in global warming. Psychol. Sci. **22**(4), 454–459 (2011). https://doi.org/10.1177/095679761140 0913
Lorencová, E.K., Loučková, B., Vačkářů, D.: Perception of climate change risk and adaptation in the Czech Republic. Climate **7**(5), 61 (2019). https://doi.org/10.3390/cli7050061
Lowe, J.A., et al.: UKCP18 Science Overview Report. Exeter, United Kingdom: Met Office Hadley Centre (2018)
Lowe, R.: Devastating Bushfires in Australia and other extreme weather events as a consequence of Climate Change (debate) (2020). https://www.europarl.europa.eu/doceo/document/CRE-9-2020-01-13-INT-1-062-0000_EN.html. Accessed 25 Feb 2024
Lynas, M., Houlton, B.Z., Perry, S.: Greater than 99% consensus on human caused climate change in the peer-reviewed scientific literature. Environ. Res. Lett. **16**, 114005 (2021). https://doi.org/10.1088/1748-9326/ac2966
MacDonald, J.P., Harper, S.L., Willox, A.C., Edge, V.L., Rigolet Inuit Community Government: A necessary voice: climate change and lived experiences of youth in Rigolet, Nunatsiavut, Canada. Glob. Environ. Change **23**(1), 360–371 (2013). https://doi.org/10.1016/j.gloenvcha.2012.07.010
Mackett, R.L., May, A.D., Kii, M., Pan, H.: Sustainable Transport for Chinese Cities. Emerald Group Publishing, Bingley, West Yorkshire, United Kingdom (2013)
Mann, M.E.: The Hockey Stick and the Climate Wars: Dispatches from the Front Lines. Columbia University Press, New York City, New York, United States (2012)
Mann, M.E.: The New Climate War: The Fight to Take Back Our Planet. Scribe Publications, London, United Kingdom (2021)
Mathers, M.: Storm Babet death toll rises to seven as women, 80, found dead at home. The Independent, 22nd October (2023). https://www.independent.co.uk/news/uk/home-news/storm-babet-death-toll-floods-b2433930.html. Accessed 20 Mar 2024
Mavis, B.E., Brocato, J.J.: Postal surveys versus electronic mail surveys: the tortoise and the hare revisited. Eval. Health Prof. **21**(3), 395–408 (1998). https://doi.org/10.1177/016327879802 100306
Mayhew, S.: A Dictionary of Human Geography, 4th edn. Oxford University Press, Oxford, Oxfordshire, United Kingdom (2009)
Mazur, A., Lee, J.: Sounding the global alarm: environmental issues in the US National News. Soc. Stud. Sci. **23**(4), 681–720 (1993). https://doi.org/10.1177/030631293023004003
McCallum, E.: The burns' day storm, 25 January 1990. Weather **45**(5), 166–173 (1990). https://doi.org/10.1002/j.1477-8696.1990.tb05607.x
McCright, A.M.: The effects of gender on climate change knowledge and concern in the American public. Popul. Environ. **32**(1), 66–87 (2010). https://doi.org/10.1007/s11111-010-0113-1
McCright, A.M., Dunlap, R.E.: Cool dudes: the denial of climate change among conservative white males in the United States. Glob. Environ. Chang. **21**(4), 1163–1172 (2011). https://doi.org/10.1016/j.gloenvcha.2011.06.003
McCright, A.M., Dunlap, R.E., Marquart-Pyatt, S.T.: Political ideology and views about climate change in the European Union. Environ. Polit. **25**, 338–358 (2016). https://doi.org/10.1080/09644016.2015.1090371

McCright, A.M., Xiao, C.: Gender and environmental concern: insights from recent work and for future research. Soc. Nat. Resour. **10**, 1109–1113 (2014). https://doi.org/10.1080/08941920.2014.918235

McCright, A.M., Xiao, C.: Gender differences in environmental concern: revisiting the institutional trust hypothesis in the USA. Environ. Behav. **47**(1), 17–37 (2015). https://doi.org/10.1177/0013916513491571

McGonagle, S.: Fire service fights 114 gorse and wildfires in 48 hours. The Irish News, 3rd June (2020). https://www.irishnews.com/news/northernirelandnews/2020/06/04/news/fire-service-called-to-more-than-100-gorse-and-wildfires-across-northern-ireland-this-week-alone-1962541/. Accessed 20 Mar 2024

McLean, C.: Value-free research. In: Jupp, V. (ed.) The SAGE Dictionary of Social Research Methods, p. 316. SAGE Publications, Ltd., London, United Kingdom (2011)

Meece, J.L., Parson, J.E., Kaczala, C.M., Goff, S.B.: Sex differences in math achievement: toward a model of academic choice. Psychol. Bull. **91**(2), 324–348 (1982). https://doi.org/10.1037/0033-2909.91.2.324

Meehl, G.A., Zwiers, F., Evans, J., Knutson, T., Mearns, L., Whetton, P.: Trends in extreme weather and climate events: issues related to modelling extremes in projections of future climate change. Bull. Am. Meteor. Soc. **81**(3), 427–436 (2000). https://doi.org/10.1175/1520-0477(2000)081%3c0427:TIEWAC%3e2.3.CO;2

Met Office: Hot spell – August 2003 (2012). https://www.metoffice.gov.uk/binaries/content/assets/metofficegovuk/pdf/weather/learn-about/uk-past-events/interesting/2003/hot-spell---august-2003---met-office.pdf. Accessed 20 Mar 2024

Met Office: Was summer 2018 the hottest on record? (2018). https://www.metoffice.gov.uk/about-us/news-and-media/media-centre/weather-and-climate-news/2018/end-of-summer-stats. Accessed 20 Mar 2024

Met Office (No Date) The Great Storm of 1987. https://www.metoffice.gov.uk/weather/learn-about/weather/case-studies/great-storm. Accessed 20 Mar 2024

Michaels, P.J., Balling, R.C., Balling Jr., R.C.: The Satanic Gases: Clearing the Air about Global Warming. Cato Institute, Washington D.C., United States (2000)

Milkoriet, M., et al.: Defining tipping points for ecological-ecological systems scholarship – an interdisciplinary literature review. Environ. Res. Lett. **13**, 033005 (2018). https://doi.org/10.1088/1748-9326/aaaa75

Milfont, T.L., Milojev, P., Greaves, L.M., Sibley, C.G.: Socio-structural and psychological foundations of climate change beliefs. N. Z. J. Psychol. **41**(1), 17–30 (2015)

Moakler, M.W., Jr., Kim, M.M.: College major choice in STEM: revisiting confidence and demographic factors. Career Dev. Q. **62**(2), 128–142 (2014). https://doi.org/10.1002/j.2161-0045.2014.00075.x

Monbiot, G.: The denial industry. The Guardian, 19th September (2006). https://www.theguardian.com/environment/2006/sep/19/ethicalliving.g2. Accessed 6 July 2020

Monbiot, G.: Monckton climate denial is a gift to those who take the science seriously. The Guardian, 8th January (2010). https://www.theguardian.com/environment/georgemonbiot/2010/jun/08/monckton-gift-climate-denial. Accessed 28 June 2018

Monckton, C.: Was 2014 the warmest Year on Record? No, It Wasn't…. Haymarket, Virginia, United States: Science & Public Policy Institute (2015). http://www.scienceandpublicpolicy.org/images/stories/papers/reprint/monckton_2014_not_hot.pdf. Accessed 29 Jan 2018

Moser, S.C., Dilling, L.: Communicating climate change: Closing the science-action gap. In: Dryzek, J., Norgaard, R.B., Schlosberg, D. (eds.) The Oxford Handbook of Climate Change and Society, pp. 161–176. Oxford University Press, Oxford, Oxfordshire, United Kingdom (2011)

Murphy, C.C., et al.: A randomized trial of mail and email recruitment strategies for a physician survey on clinical trial accural. BMC Med. Res. Methodol. **20**, 123 (2020). https://doi.org/10.1186/s12874-020-01014

Myers, K.F., Doran, P.T., Cook, J., Kotcher, J.E., Myers, T.A.: Consensus revisited: quantifying scientific agreement on climate change and climate expertise among Earth scientists 10 years later. Environ. Res. Lett. **16**(10), 104030 (2021). https://doi.org/10.1088/1748-9326/ac2774

Naish, J.: Could we face a mini ice age in the next 30th years? Scientists make extraordinary prediction based on the sun's natural cycles…and it would even reverse global warming!. Daily Mail, 1st January (2018). https://www.dailymail.co.uk/news/article-5227073/Could-face-mini-ice-age-30-years.html. Accessed 5 May 2020

Neumann, J.: Climatic change as a topic in the classical Greek and Roman literature. Clim. Change **7**(4), 441–454 (1985). https://doi.org/10.1007/BF00139058

Nevett, J.: Three decade ICE AGE to freeze Earth from 2030 – scientists make '97% accurate' prediction. Daily Star, 27th October (2016). https://www.dailystar.co.uk/news/latest-news/three-decade-ice-age-2030-17255953. Accessed 5 May 2020

Nisbet, M.C., Myers, T.: Trends: twenty years of public opinion about global warming. Publ. Opin. Q. **71**(3), 444–470 (2007)

Nuccitelli, D.: The 'imminent mini ice age' myth is back, and it's still wrong. The Guardian, 9th January (2018). https://www.theguardian.com/environment/climate-consensus-97-per-cent/2018/jan/09/the-imminent-mini-ice-age-myth-is-back-and-its-still-wrong. Accessed 6 Apr 2020

O'Brien, Z.: Freak climate changes could spark 'mini ice-age' in Britain from 2017. Daily Express, 13th August (2016). https://www.express.co.uk/news/uk/699681/FREAK-climate-changes-spark-min-ice-age-Britain-2017. Accessed 5 May 2020

O'Dea, R.E., Lagisz, M., Jennions, M.D., Nakagawa, S.: Gender differences in individual variations in academic grades fail to fit expected patterns for STEM. Nat. Commun. **9**(3777), 1–8 (2018). https://doi.org/10.1038/s41467-018-06292-0

OECD (Organisation for Economic Co-Operation and Development): Environmental Related Taxes on OECD Countries: Issues and Strategies. OECD, Paris, France (2001)

ONS (Office for National Statistics): Internet Access – Households and Individuals: 2015 (2015). http://www.ons.gov.uk/peoplepopulationandcommunity/householdcharacteristics/homeinternetandsocialmediausage/bulletins/internetaccesshouseholdsandindividuals/2015-08-06. Accessed 25 May 2016

ONS (Office of National Statistics): Internet access – households and individuals: 2020 (2020). https://www.ons.gov.uk/file?uri=/peoplepopulationandcommunity/householdcharacteristics/homeinternetandsocialmediausage/datasets/internetaccesshouseholdsandindividualsreferencetables/2020/internetaccesspublicationtables2020final1.xlsx. Accessed 25 Feb 2021

Oreskes, N.: Beyond the ivory tower: the scientific consensus on climatic change. Science **306**(5702), 1686 (2004). https://doi.org/10.1126/science.1103618

Oreskes, N., Conway, E.M.: Merchants of Doubt: How a Handful of Scientists Obscured the Truth on Issues from Tobacco Smoke to Global Warming'. Bloomsbury Publishing, London, United Kingdom (2010)

Parry, M.A.: Climate change perception, engagement and response in the United Kingdom and the Crown Dependencies: the case of youth participation. Doctoral thesis, Northumbria University (2020). https://nrl.northumbria.ac.uk/id/eprint/46769/. Accessed 8 Aug 2023

Paus, T.: Mapping brain maturation and cognitive development during adolescence. Trends Cogn. Sci. **9**(2005), 60–68 (2005). https://doi.org/10.1016/j.tics.2004.12.008

Pettit, H.: Extreme cold winter weather conditions such as the 'Beast from the East' are linked to a weak solar cycle that could trigger a MINI ICE AGE by 2050, experts warn. Daily Mail, 20th March (2018). https://www.dailymail.co.uk/sciencetech/article-5522707/Extreme-weather-Beast-East-linked-solar-cycle.html. Accessed 27 Mar 2024

Perkins, S.: Atmospheric science: the cold facts. Nat. Clim. Chang. (2010). https://doi.org/10.1038/nclimate1008

Pidgeon, N.F., Hood, C., Jones, D., Tucker, B., Gibson, R.: Risk perception. In: Royal Society Study Group (ed.) Risk Analysis, Perception and Management: Report of a Royal Society Group, pp. 89–134. The Royal Society, London, United Kingdom (1992)

Plimer, I.: Heaven and Earth: Global Warming – The Missing Science. Connor Court Publishing, Brisbane, Queensland, Australia (2009)

Poortinga, W., Pidgeon, N.F., Lorenzoni, I.: Public perceptions of nuclear power, climate change and energy options in Britain: summary findings of a survey conducted during October and November 2005. In: Understanding Risk Working Paper 06-02. Norwich, Norfolk, United Kingdom: Centre for Environmental Risk (2006)

Poortinga, W., Whitmarsh, L., Steg, L., Böhm, G., Fisher, S.: Climate change perceptions and their individual-level determinants: a cross-European analysis. Glob. Environ. Chang. **55**, 25–35 (2019). https://doi.org/10.1016/j.gloenvcha.2019.01.007

Powell, J.L.: Climate scientists virtually unanimous: anthropogenic global warming is true. Bull. Sci. Technol. Soc. **35**(5–6), 121–124 (2015)

Powell, J.L.: Scientists reach 100% consensus on anthropogenic global warming. Bull. Sci. Technol. Soc. **37**(4), 183–184 (2019). https://doi.org/10.1177/0270467619886266

Poynting, M.: World's first year-long breach of key 1.5C warming limit. BBC News, 8th February (2024). https://www.bbc.co.uk/news/science-environment-68110310. Accessed 10 Mar 2024

Prigg, M.: Scientists warn the sun will 'go to sleep' in 2030 and could cause temperatures to plummet. Daily Mail, 10th July (2015). https://www.dailymail.co.uk/sciencetech/article-3156594/Is-mini-ICE-AGE-way-Scientists-warn-sun-sleep-2020-cause-temperatures-plummet.html. Accessed 5 May 2020

Quinn, B.: Cumbrian village deals with more devastation as river bursts banks again. The Guardian, 10th December (2015). https://www.theguardian.com/uk-news/2015/dec/09/glenridding-defences-fail-as-flood-waters-rise-in-isolated-cumbria-town. Accessed 27 Mar 2024

Ratter, B.M.W., Philipp, K.H.I., von Storch, H.: Between hype and decline: recent trends in public perception of climate change. Environ. Sci. Policy **18**(1), 3–8 (2012). https://doi.org/10.1016/j.envsci.2011.12.007

Redwood, J.: UK human CO2 output represents 0.0025% of greenhouse gases. http://johnredwoodsdiary.com/2008/04/10/uk-human-co2-output-represents-0-0025-of-greenhouse-gases/. Accessed 10 Mar 2024

Reser, J.P., Bradley, G.L., Ellul, M.C.: Encountering climate change: 'seeing' is more than 'believing.' WIREs Clim. Change **5**(4), 521–537 (2014). https://doi.org/10.1002/wcc.286

Ridley, M.: The Public Isn't Being Told the Full Truth about the Climate Threat – Matt Ridley (2023). https://www.mattridley.co.uk/blog/the-public-isnt-being-told-the-full-truth-about-the-climate-threat/. Accessed 21 Mar 2024

Robinson, J.: Storm Ciaran: Record-breaking storm batters Channel Islands with 104mph gusts – as southern England hit by flooding', Sky News, 2nd November (2023). https://news.sky.com/story/storm-ciaran-red-warning-in-channel-islands-85mph-gusts-and-ferries-cancelled-as-wet-weather-sweeps-towards-uk-12997830. Accessed 20 Mar 2024

Roche, J., Tucker, S., Flynn, R., Thomson, R.: Youth in Society: Contemporary Theory, Policy and Practice. SAGE Publications Ltd., London, United Kingdom (2004)

Rose, D.: The mini ice age starts here', Daily Mail, 9th January (2010). https://www.dailymail.co.uk/sciencetech/article-1242011/DAVID-ROSE-The-mini-ice-age-starts-here.html. Accessed 5 May 2020

Rose, D.: Forget global warming – it's Cycle 25 we need to worry about (and if NASA scientists are right the Thames will be freezing over again). Daily Mail, 29th January (2012). https://www.dailymail.co.uk/sciencetech/article-2093264/Forget-global-warming--Cycle-25-need-worry-NASA-scientists-right-Thames-freezing-again.html. Accessed 27 Mar 2024

Rose, D.: And now it's global COOLING! Return of Arctic ice cap as it grows by 29% in a year. Daily Mail, 8th September (2013). https://www.dailymail.co.uk/news/article-2415191/And-global-COOLING-Return-Arctic-ice-cap-grows-29-year.html. Accessed 5 May 2020
Rosenberg, S., Vedlitz, A., Cowman, D.F., Zahran, S.: Climate change: a profile of US climate scientists' perspectives. Clim. Change **101**(3), 311–329 (2010). https://doi.org/10.1007/s10584-009-9709-9
Rowland, R.: Devastating Bushfires in Australia and other extreme weather events as a consequence of Climate Change (debate) (2020). https://www.europarl.europa.eu/doceo/document/CRE-9-2020-01-13-INT-1-071-0000_EN.html. Accessed 25 Feb 2024
Royal Meteorological Society: The Big Freeze of 1963: One of the coldest winters on record (2013). https://www.rmets.org/metmatters/big-freeze-1963-one-coldest-winters-record. Accessed 20 Mar 2024
Sabbagh, D.: Election results signifies realignment of UK politics. The Guardian, 13th December (2019). Local Warming: Daily Temperature Change Influences Belief in Global Warming. Accessed 28 Dec 2019
Sax, L.J.: Predicting gender and major-field differences in mathematical self-concept during college. J. Women Minor. Sci. Eng. **1**, 291–307 (1994). https://doi.org/10.1615/jwomenminorscieneg.v1.i4.30
Sale, I.: Storm arwen: thousands of homes still without power after some of the 'worst' disruption in 20 years. Sky News, 29th November (2021). https://news.sky.com/story/storm-arwen-thousands-of-homes-still-without-power-after-some-of-the-worst-disruption-in-20-years-12482272. Accessed 20 Mar 2024
Samset, B.H., Zhou, C., Fuglestvedt, J.S., Lund, M.T., Marotzke, J., Zelinka, M.D.: Steady global surface warming from 1973 to 2022 but increased warming rate after 1990. Commun. Earth Environ. **4**, 400 (2023). https://doi.org/10.1038/s43247-023-01061-4
Schulte, K.-M.: Scientific consensus on climate change? Energy Environ. **19**(2), 281–286 (2008). https://doi.org/10.1260/095830508783900744
Scott, A., et al.: A randomised trial and economic evaluation of the effect of response mode on response rate, response bias, and item non-response in a survey of doctors. BMC Med. Res. Methodol. **11**(126), 1–12 (2011). https://doi.org/10.1186/1471-2288-11-126
Seely, A.: Taxation of Road Fuels: The Road Fuel Escalator (1993–2000). House of Commons Library, London, United Kingdom (2011)
Seow, T., Ho, L.-C.: Singapore teachers' beliefs about the purpose of climate change education and student readiness to handle controversy. Int. Res. Geograph. Environ. Educ. **25**(4), 358–371 (2016). https://doi.org/10.1080/10382046.2016.1207993
Shao, W., Keim, B.D., Garand, J.C., Hamilton, L.C.: Weather, climate, and the economy: explaining risk perceptions of global warming, 2001–10. Am. Meteorol. Soc. **6**(1), 119–134 (2014). https://doi.org/10.1175/WCAS-D-13-00029.1
Shwartz, M.: Majority of Americans Continue to Believe that Global Warming is Real. Stanford University, Stanford, California, United States (2010)
Singer, S.F., Avery, D.T.: Unstoppable Global Warming: Every 1,500 Years. Rowman & Littlefield Publishers Inc., Lanham, Maryland, United States (2007)
Spence, A., Venables, D., Pidgeon, N., Poortinga, W., Demski, C.: Public Perceptions of Climate Change and Energy Futures in Britain: Summary Findings of a Survey Conducted in January-March 2010. Understanding Risk Working Paper 10-01. Cardiff, Wales, United Kingdom: School of Psychology, Cardiff University (2010)
Spencer, R.W.: Not that simple. New York Post, 26th February (2007). https://nypost.com/2007/02/26/not-that-simple/. Accessed 30 Sept 2020
Spencer, R.W.: Climate Confusion: How Global Warming Hysteria Leads to Bad Science, Pandering Politicians and Misguided Policies That Hurt the Poor. Encounter Books, New York City, New York, United States (2008)

Spencer, R.W.: The Great Global Warming Blunder: How Mother Nature Fooled the World's Top Climate Scientists. Encounter Books, New York City, New York, United States (2010)

Sorribas, M.V., et al.: Projections of climate change effects on discharge and inundation in the Amazon basin. Clim. Change **136**, 555–570 (2016). https://doi.org/10.1007/s10584-016-1640-2

Steentjes, K., et al.: European Perception of Climate Change: Topline Findings of a Survey Conducted in Four European Countries in 2016. Cardiff University, Cardiff, Wales, United Kingdom (2017)

Steinberg, L.: Cognitive and affective development in adolescence. Trends Cogn. Sci. **9**(2), 69–74 (2005). https://doi.org/10.1016/j.tics.2004.12.005

Stern N.: Stern Review: The Economics of Climate Change. HM Treasury, London, United Kingdom (2006)

Sullivan, W.: Scientists ask why world climate is changing; major cooling may be ahead: scientists ponder why world's climate is changing; a major cooling widely considered to be inevitable. The New York Times, 21st May, p. 92 (1975a)

Sullivan, W.: Warming trend seen in climate: two articles counter view that cold period is due. The New York Times, 14th August, p. 24 (1975b)

Supran, G., Oreskes, N.: Assessing ExxonMobil's climate change communications (1977–2014). Environ. Res. Lett. **12**(8), 084019 (2017). https://doi.org/10.1088/1748-9326/aa815f

Tapper, J., Walker, A.: Lincolnshire flooding: hundreds of homes evacuated. The Guardian, 15th June (2019). https://www.theguardian.com/uk-news/2019/jun/15/lincolnshire-flooding-homes-evacuated-river-steeping-wainfleet. Accessed 20 Mar 2024

Tashakkori, A., Teddlie, C.: Handbook of Mixed Methods in Social & Behavioral Research. SAGE Publications Inc., Thousand Oaks, California, United States (2003)

Taylor, J.: 'Mini ice age' on Earth could cause −5°C temperatures and food shortage, experts warn. Daily Star, 2nd February (2020). https://www.dailystar.co.uk/news/mini-ice-age-earth-could-21414608. Accessed 5 May 2020

Tice, R.: CO_2 isn't a poison. It's plant food! We need to challenge the climate change nonsense! #ticetake #climatechange #co2 #plantfood' [X] 25th November (2023). https://twitter.com/TiceRichard/status/1728511408784953512. Accessed 10 Mar 2024

Time Magazine: Science: another Ice Age?. Time Magazine, 13th November (1972). http://content.time.com/time/magazine/article/0,9171,910467,00.html. Accessed 3 May 2020

Tranter, P., Galvin, J.: Fifty years on from the Great Glasgow Storm… What have we learned from powerful storms since then? Weather **73**(2), 39–45 (2018). https://doi.org/10.1002/wea.3218

Trenberth, K.E.: Atmospheric moisture residence times and cycling: implications for rainfall rates and climate change. Clim. Change **39**, 667–694 (1998). https://doi.org/10.1023/A:1005319109110

Trubo, R.: Metrosexuals: It's a Guy Thing!. WebMD, 28th July (2003). https://www.webmd.com/sex-relationships/features/metrosexuals-its-guy-thing#1. Accessed 1st Dec 2019

Trump, D.J. [realDonaldTrump]: The concept of global warming was created by and for the Chinese in order to make U.S. manufacturing non-competitive [Twitter/X] (2012). https://x.com/realDonaldTrump/status/265895292191248385. Accessed 3 Feb 2024

UKHSA [UK Health Security Agency]: Health Effects of Climate Change (HECC) in the UK. UK Health Security Agency, London, United Kingdom (2023)

UNFCCC [United Nations Framework Convention on Climate Change]: Adoption of the Paris agreement. United Nations Framework Convention on Climate Change, Born, North Rhine-Westphalia, Germany (2015)

Unsworth, K.L., Fielding, K.S.: It's political: how the salience of one's political identity change climate beliefs and policy support. Glob. Environ. Chang. **27**, 131–137 (2014). https://doi.org/10.1016/j.gloenvcha.2014.05.002

Van Liere, K.D., Dunlap, R.E.: The social bases of environmental concern: a review of hypotheses, explanations and empirical evidence. Publ. Opinion Q. **44**(2), 181–197 (1980)

Van Selm, M., Jankowski, N.W.: Conducting online surveys. Qual. Quant. **40**(3), 435–456 (2006). https://doi.org/10.1007/s11135-005-8081-8

Vetterling-Braggin, M.: Femininity, Masculinity and Androgyny: A Modern Philosophical Discussion. Rowman & Littlefield, Lanham, Maryland, United States (1982)

Vollerberg, W.A.M., Iedema, J., Raaijmakers, Q.A.W.: Intergenerational transmission and the formation of cultural orientations in adolescence and young childhood. J. Marriage Family **63**(4), 1185–1198 (2001). https://doi.org/10.1111/j.1741-3737.2001.01185.x

Walker, D., Findlay, S.: Fire crews battle through the night as major wildfire threatens Moray wind farm. Press J. 23rd April 2019 (2019). https://www.pressandjournal.co.uk/fp/news/moray/1731320/major-wildfire-threatens-wind-farm-in-moray/. Accessed 20 Mar 2024

Walsh, E.M., Cordero, E.: Youth science expertise, environmental identity, and agency in climate action filmmaking. Environ. Educ. Res. **25**(5), 656–677 (2019). https://doi.org/10.1090/12504622/2019/1569206

Warren, J.: Dawn of a new ice age. Daily Express, 21st December (2010). https://www.express.co.uk/expressyourself/218537/Dawn-of-a-new-ice-age. Accessed 5 May 2020

Watt, H.: Owen Paterson becomes new environment secretary. The Telegraph, 4th September (2012). https://www.telegraph.co.uk/news/politics/9520174/Owen-Paterson-becomes-new-Environment-Secretary.html. Accessed 30 Sept. 2020

Weber, E.U.: What shapes perceptions of climate change? WIREs Clim. Change **1**(3), 332–342 (2010). https://doi.org/10.1002/wcc.41

Weber, E.U.: What shapes perceptions of climate change? New research since 2010. WIREs Clim. Change **7**(1), 125–134 (2016). https://doi.org/10.1002/wcc.377

Weston, P.: Plummeting temperatures could send the world into a 'mini ice age' in 2030 and could OVERRIDE global warming, claim mathematicians. Daily Mail, 27th December (2017). https://www.dailymail.co.uk/sciencetech/article-5215575/Plummeting-temperatures-cause-mini-ice-age-2030.html. Accessed 5 May 2020

Wharton, A.S.: The Sociology of Gender: An Introduction to Theory and Research. Blackwell Publishing, Oxford, Oxfordshire, United Kingdom (2009)

Whitmarsh, L.: What's in a name? Commonalities and differences in public understanding of "climate change" and "global warming." Publ. Underst. Sci. **18**, 401–420 (2009). https://doi.org/10.1177/0963662506073088

Whitmarsh, L.: Scepticism and uncertainty about climate change, dimensions, determinants and change over time. Glob. Environ. Chang. **21**(2), 690–700 (2011). https://doi.org/10.1016/j.gloenvcha.2011.01.016

WHO [World Health Organization]: Climate change (2023a). https://www.who.int/news-room/fact-sheets/detail/climate-change-and-health#:~:text=Climate%20change%20affects%20food%20availability,700%20000%20annually%2C%20may%20rise. Accessed 25 Mar 2024

WHO [World Health Organization]: State of the Global Climate 2022. World Health Organization, Geneva, Switzerland (2023)

Widdecombe, A.: 'Fighting the nonsense' The climate change rebels were spot on, says ANN WIDDECOMBE. Daily Express, 29th October (2014). https://www.express.co.uk/comment/columnists/ann-widdecombe/528569/Ann-Widdecombe-on-Climate-change-Ched-Evans-and-Fiona-Woolf. Accessed 19 Feb 2020

Willis, R.: Physiography of the California coast ranges. Geol. Soc. Am. Bull. **35**(4), 641 (1925). https://doi.org/10.1130/GSAB-36-641

Windsor, E.J.: Femininities. In: Wright, J.D. (ed.) International Encyclopaedia of the Social & Behavioral Sciences, pp. 893–897. Elsevier, Amsterdam, North Holland, Netherlands (2015)

WMO [World Meteorological Organization]: El Niño weakens but impacts continue (2024). https://wmo.int/news/media-centre/el-nino-weakens-impacts-continue. Accessed 10 Mar 2024

Wolfman, G.: Metrosexual, hipster, spornosexual: why do we keep redefining men?. The Conversation, 5th May (2017). https://theconversation.com/metrosexual-hipster-spornosexual-why-do-we-keep-redefining-men-72444. Accessed 25 Nov 2019

Wray-Lake, L., Flanagan, C.A., Osgood, D.W.: Examining trends in adolescent environmental attitudes, beliefs, and behaviors across three decades. Environ. Behav. **42**(1), 61–85 (2010). https://doi.org/10.1177/0013916509335163

Xiao, C., Hong, D.: Gender differences in environmental behaviors in China. Popul. Environ. **32**, 88–104 (2010). https://doi.org/10.1007/s11111-010-0115-z

Xiao, C., McCright, A.M.: Explaining gender difference in concern about environmental problems in the United States. Soc. Nat. Resour. **25**, 1067–1084 (2012). https://doi.org/10.1080/08941920.2011.651191

Zachariah, M., et al.: Extreme heat in North America, Europe and China in July 2023 made much more likely by climate change. Grantham Institute for Climate Change London, United Kingdom (2023). https://doi.org/10.25561/105549

Zelezny, L.C., Chua, P.-P., Aldrich, C.: Elaborating on gender differences in environmentalism. J. Soc. Issues **56**, 443–457 (2000). https://doi.org/10.1111/0022-4537.00177

Zolfagharifard, E.: The silent sun: Eerie image revealed as solar activity remains the quietest it has been in more than a century - and some claims it could even trigger a mini ice age. Daily Mail, 12th February (2016). https://www.dailymail.co.uk/sciencetech/article-3444633/What-happened-sun-Solar-activity-remains-quietest-century-trigger-mini-ice-age.html. Accessed 15 Feb 2024

Zhou, N.: Oxford Dictionaries declares 'climate emergency' the word of 2019. The Guardian, 21st November (2019). https://www.theguardian.com/environment/2019/nov/21/oxford-dictionaries-declares-climate-emergency-the-word-of-2019. Accessed 4 Jan 2020

Application of GIS and Remote Sensing-Based GeoSM-NatE Tool and PAR Model to Delineate Urban Flood Hotspot Using: A Case of Chaktai Khal, Chattogram, Bangladesh

Archita Saha(✉), Mahua Mukherjee, Atul Kumar(✉), and Kumar Abhinay

Department of Architecture and Planning, Indian Institute of Technology Roorkee, Haridwar 247667, India
{archita_s,akumar5,kumar_a}@ar.iitr.ac.in, mahuafap@iitr.ac.in

Abstract. The study selected the historic commercial waterway Chaktai khal that exits the city into the Karnaphuli River as a study area. Monsoons flood this watershed annually, affecting livability and the economy. Identifying flood risk hotspot areas in response to these patterns is a crucial first step in developing an integrated system for managing and mitigating urban flood risk. The study adopted the approach of PAR model (Pressure and Release Model), under this model the risk pressure has been quantified using Geo-Spatial Mapping of Natural Ecosystem (GeoSM-NatE tool) and finally under the release model a strategic approach of Blue-Green Infrastructure (BGI) has been discussed. Multi-temporal remote sensing analysis using GeoSM-NatE tool can detect hotspot regions for urban floods, UHI, and pollution utilizing secondary data, RS and GIS analysis. The first part of GeoSM-NatE tool describes the pressure part of PAR model whereas its second part discusses the release part. The research considered 1 km by 1 km grids delineation in Chattogram city corporation (CCC) boundary then using GeoSM-NatE tool from it extracted the grids that are urban flood risk hotspots in Chaktai Khal catchment area. Hotspots have been delineated using five input layers—NDBI, NDVI, NDWI, stream density, and slope delineated on Arc-GIS platform. Sentinel-02 with a resolution of 30 m was used to map NDBI, NDVI, and NDWI, while Cartosat DEMs provided from NASA's Shuttle Radar Topography Mission (SRTM) was utilized for generating hydrological maps such as elevation and stream density. Finally, the study identified major urban flood risk hotspots in Chattogram city corporation and four under the Chaktai Khal watershed. Two were in the city's densest wards—4, 6, 7, 8, 16, 17, and 19—at the highest risk. The result has been validated with the flood location reported in multiple sources of government and non-government report such as Chaktai Khal catchment area covers the 10 most flood-prone city wards: 04, 06, 07, 08, 16, 17, 19, 20, 34 and 35. After considering all risk factors, the paper recommends NbS implementation for Chattogram's urban flood problem in Bangladesh's Chaktai Khal watershed.

Keywords: Urban flood Risk · GeoSM-NatE · Pressure and Release Model (PAR): Hotspots · GIS

H. Tatano et al. (Eds.): GSRIDRR 2023, DRRGADRIBS, pp. 327–347, 2025.
https://doi.org/10.1007/978-981-96-6609-6_23

1 Introduction

Unplanned growth is synonymous with urbanization (Reddy, 2019). Urbanization refers to the phenomenon of a growing number of individuals relocating from rural to urban regions (Rashid et al., 2007). According to the World Bank, the urban population will exceed 5 billion by 2030, accounting for about two-thirds of the global population (The World Bank, 2009). This increase is 90% in developing countries (The World Bank, 2009). In South Asia Bangladesh's urban population has higher growth rate of 3.7% each year than any other country except Nepal (H. Z. Rahman, 2014). As per the data provided by the United Nations, the urban population of Bangladesh has reached 35 million, which accounts for 25% of the total population. This figure has seen a significant increase from the 8% recorded at the time of the country's independence (UN-Habitat, 1996; M. S. Rahman et al., 2012). It will reach 80 million by 2030 (M. S. Rahman et al., 2012). Dhaka, Chattogram, Khulna, and Rajshahi are Bangladesh's four largest cities, and they are home to nearly 60% of the country's total urban population (Rana, 2011). Dhaka, a rapidly expanding megacity, accounts for 35% of the total population in its region (Rana, 2011) while Chattogram is home to nearly 30% (Samad, 2015). However, the growing trend of urbanization around the world is causing the problem of climate change (Aalst, 2006; Chan et al., 2022; McCarthy et al., 2010; D. Mukherjee, 2016; Sue Grimmond, 2019). Climate change is the most common sign of environmental damage, which can be estimated by carbon dioxide (CO2) pollution levels (Aalst, 2006; Chan et al., 2022; McCarthy et al., 2010; D. Mukherjee, 2016; Sue Grimmond, 2019). Cities produce 70% of global greenhouse gas emissions while covering just 2% of the planet, according to UN Habitat (Aalst, 2006; Chan et al., 2022; McCarthy et al., 2010 D. Mukherjee, 2016; Sue Grimmond, 2019). Climate change caused by urbanization is making people more vulnerable to extreme weather events in cities in recent years for example flooding, heat waves, UHI (urban heat island effect), tropical cyclone etc. (Aalst, 2006; M. S. Rahman et al., 2012). The expansion of impervious surfaces due to urbanization leads to a rise in regional flood vulnerability as it decelerates the hydrological response time (Feng et al., 2021). The number of persons impacted by flooding, as well as financial, economic, and insurance losses, has grown (S. Yu. Schreider, 2012; Stevens, 2012). From 1995–2015, 47% of all natural catastrophes happened as a result of flooding, impacting 2.3 billion people and costing an estimated $662 billion in economic losses (Matheswaran et al., 2019). An estimated 70 million individuals are at risk of flooding every year, out of a total global population of 800 million (Matheswaran et al., 2019). Floods harmed 178 million people worldwide in 2010 alone (S. Yu. Schreider, 2012; Stevens, 2012). In extreme years such as 1998 and 2010, total losses topped $40 billion (S. Yu. Schreider, 2012; Stevens, 2012). Since the 1960s, flood disaster events have increased, and since the early 1990s, they have increased even more (Stevens, 2012; Zevenbergen et al., 2014). In developing nations, such events have increased economic losses (Stevens, 2012; Zevenbergen et al., 2014). Population expansion has increased the number of people living in flood-prone regions, reducing their ability to cope (Stevens, 2012; Zevenbergen et al., 2014). For these reasons in Bangladesh also flooding is by far the most common and damaging kind of natural disaster (Rashid et al., 2007). The commercial capital of Bangladesh, Chattogram, also has to deal with flash floods during every rainy season (Hasan Tarek et al., 2017). While the city's 60-year rainfall history and total amount have remained unchanged, the city's physical environment has undergone significant changes (Hasan Tarek et al., 2017). City areas like Chawkbazar, Chaktai, Sholoshahar, and Bahaddarhat frequently flood during heavy rainfall (Hasan Tarek et al., 2017). The economic and social consequences of this are devastating (Hasan Tarek et al., 2017). In 2017, three separate waterlogging events in Chattogram, the most prosperous economic area in Southern Bangladesh, resulted in the region's largest recorded economic loss of 6,000 million BDT (Islam and Raja, 2021). The lack of planned and inadequate maintenance of drainage systems results in significant flooding in regions with low elevation within the urban area, leading to considerable adversity for a substantial population (Akter et al., 2017; Tanim and Akter, 2016). With more water to drain, the city's drainage system becomes clogged, especially in low-lying areas (Hasan Tarek et al., 2017). Paved surfaces and water blockage have increased runoff due to rapid development in Chattogram (Tuhin et al., 2021). The urban area is endowed with multiple natural khals (canals), which serve a crucial function in facilitating the discharge of surface water into the Karnaphuli river (Tuhin et al., 2021). Due to unplanned urban growth, encroachment and congestion of channels, and urban development of local water retention bodies, these khals are unable to discharge excess surface runoff water (Chowdhury and Akter, 2021; Mahmood and Matin, 2018). Further, Chattogram has not yet begun implementing its drainage masterplan (Tuhin et al., 2021). The city's drainage infrastructure haven't been able to handle the surge, as seen by the regular drainage overflow and urban flooding (Tuhin et al., 2021). Proper assessment of flood prone areas is required to solve this issue (Matheswaran et al., 2019) though due to a lack of flood hazard maps, there is an information gap regarding 'flood risk hotspots' within the city (Chowdhury and Akter, 2021; Mahmood and Matin, 2018). In the realm of disaster management, the term 'risk' has conventionally been construed as the probability of prospective detriment stemming from three key determinants: hazard (the probability of an impending detrimental physical occurrence, such as a cyclone, earthquake, or drought); exposure (the populace or resources, encompassing human habitation, infrastructure, crops, and livestock, that are susceptible to the hazard); and vulnerability (the proneness to loss, which is associated with a gamut of physical, societal, economic, and other characteristics of the exposed entities) (Bijan Khazai et al., 2013; Roxana L. Ciurean, 2016; Serje, 2010). The term "hotspot" is used in a number of fields (including urban planning, economics, and criminology) to describe areas where a particular phenomenon is particularly concentrated or intense areas (De Risi et al., 2017; Jalayer et al., 2014). According to scientists, hotspots are places where natural disasters are more likely to occur (De Risi et al., 2017; Dilley et al., 2005; Jalayer et al., 2014). In a similar vein, "flood risk hotspots" are areas where flooding is especially likely to occur (De Risi et al., 2017; Jalayer et al., 2014). The identification of high-risk areas is considered to be one of the primary steps in an integrated methodology for managing and mitigating flood risks in urban areas (De Risi et al., 2017; Jalayer et al., 2014). Urban planning and risk management begin with identifying urban flooding areas (Jalayer et al., 2014). Mapping urban flooding risk hotspots helps urban planners identify areas that need urgent or long-term attention (Jalayer et al., 2014). These efforts may include more accurate small-scale risk assessment and preventive initiatives (Jalayer et al., 2014). Preventive methods include design for flood-resistant buildings, relocation (if necessary), territorial limitation, and public awareness campaigns (Jalayer et al., 2014). Researchers used several numerical models, hydrological and hydraulic simulations for urban flood risk hotspot delineation (Chowdhury and Akter, 2021). The integration of GIS and Remote Sensing (RS) data has been a popular method for tackling this problem in recent years (De Risi et al., 2017; Jalayer et al., 2014; Samela et al., 2016). This study used the framework of GeoSM-NatE tool for identifying urban flood risk hotspot areas in a selected study area. The GeoSM-NatE tool is a technology-based spatial decision-support system that intends to identify urban blue and green zones ideal for decreasing catastrophe risks using nature-based solutions through GIS analysis and Remote Sensing methods (P. M. Mukherjee et al., n.d.). It's an urban risk resilience toolbox for strategic integration of BGI (Blue-Green Infrastructure) for Urban Risk Reduction, developed in the 'Climatology Lab,' IIT Roorkee, India. The GeoSM-NatE tool can detect long-term changes in ecosystem coverage and ecological degradation in urban settings through multi-temporal analysis with remote sensing techniques (P. M. Mukherjee et al., n.d.). It also identifies high-risk hot spots, for instance, UHI, urban flooding, or pollution using secondary data and GIS analysis (P. M. Mukherjee et al., n.d.). Finally, it selects potential areas for nature-based intervention to mitigate risk. The toolbox primarily deploys GIS analysis by generating geospatial data layers from RS and overlaying operations. Linkages between ecosystem loss in urban areas and risk exposure can be analyzed through temporal earth observation data and comparing trends in a GIS environment (P. M. Mukherjee et al., n.d.). The study adopted the approach of PAR model (Pressure and Release Model) under Geo-Spatial Mapping of Natural Ecosystem or GeoSM-NatE tool. The PAR approach placed an emphasis on the underlying factors, changing pressures, and harmful environments that put individuals at risk (Wisner et al., 2004). The distribution of power is reflected in demographic, economic, and political dynamics, which are the root causes (Wisner et al., 2004). Growing populations, more urbanization, tighter budgets, deforestation, and climate change all provide new challenges (Wisner et al., 2004). When people don't have the means, fortitude, and tenacity to deal with their exposure, unsafe situations develop (Wisner et al., 2004). Dynamic pressures direct primary causes into precarious situations, which then interact with dangers to produce an undesirable outcome (Wisner et al., 2004). The release part of this model is about safety progression to ensure risk reduction and reduced hazard impacts. This part involves addressing root causes, reducing pressures, and improving conditions. For example, hazards like urban flooding can be mitigated by implementation of nature-based solution (Ferreira et al., 2021). The first part of GeoSM-NatE tool describes the pressure part of PAR model whereas its second part discusses the release part. This study used GeoSM-NatE tool and focused only on the pressure part of PAR model to detect long-term changes in urban ecosystem coverage and ecological degradation using multi-temporal analysis and remote sensing and thus identified urban flood risk hotspots in Chaktai Khal neighborhood. By superimposing regions of high flooding susceptibility (as a proxy for regions with high flooding hazard) with regions of high exposure to flooding risk, the method detailed in this paper identifies flood risk hotspot areas (Jalayer et al., 2014). Further risk indicators investigation and flood risk assessment may be conducted for the study's highlighted areas. The elimination of flora and the enlargement of impermeable surfaces in metropolitan regions also disturbs the hydrologic cycle, leading to a decrease in precipitation interception, evapotranspiration, and infiltration, and consequently, an increase in runoff (Ferreira et al., 2021; Huang et al., 2020; Majidi et al., 2019; Oral et al., 2020; Vojinovic et al., 2021). Nature-based solutions (NBS) offer a promising strategy for managing urban water resources, as they strive to maintain the urban hydrological cycle in a way that closely emulates natural circumstances (Ferreira et al., 2021). Natural-based solutions (NBS) have the potential to enhance both mitigation and adaptation to global changes, encompassing land-use and climate change, while simultaneously mitigating floods (Ferreira et al., 2021). Additionally, NBS offer valuable services for the preservation and rehabilitation of ecological functions (Ferreira et al., 2021). Thus, this study concludes by examining the potential of nature-based solutions (NBS) within the context of the release model of the Pressure and Release (PAR) model modified from Wisner et al., 2004 for the purpose of mitigating urban flood events in the study area.

2 Data Preparation

2.1 Study Area

The research was carried out within the jurisdiction of Chittagong City Corporation, as illustrated in Fig. 1. The area is situated between 22°14 to 22°22 N latitude and 91°46 to 91°51 E longitude and encompasses a total of 41 urban wards (Mia et al., 2015). Bangladesh's commercial center and largest seaport city, Chattogram (formerly Chittagong), adjacent to the Karnaphuli River borders the Bay of Bengal in the southeast (Ashraf and Chowdhury, 2009; Mia et al., 2015). With an area of 157 km^2, this city ranks as the second-largest in the country (Ashraf and Chowdhury, 2009; Mia et al., 2015). Chattogram city flanked by the Karnaphuli river and the Bay of Bengal, has a coastline hilly landscape due to water resources including streams, canals, ponds, and lakes (Ashraf and Chowdhury, 2009; Mia et al., 2015). The northern hills vary in height from 30 to 140 m above sea level, whereas the lowlands average just around 7 m above sea level (Ashraf and Chowdhury, 2009; Mia et al., 2015). The area has a tropical monsoon climate characterized by hot, humid summers and mild, dry winters. Summertime highs may exceed 32 °C, while wintertime lows seldom fall below 13° (Ashraf and Chowdhury, 2009; Mia et al., 2015). Since this place is in the tropical zone, the summer has high temperatures, heavy rain, and high humidity. The city gets between 2100 mm and 3800 mm of rain a year, with 2400 mm falling only during the monsoon season (Ashraf and Chowdhury, 2009; Mia et al., 2015). The rainy season—June to September—accounts for 60% of this rainfall (Ashraf and Chowdhury, 2009; Mia et al., 2015). Only 60% of the drains that had been realized from the 1995 master plan remained in the 2006 drainage network model (Chowdhury and Akter, 2021). With fewer places to live and more frequent flooding, urban areas are struggling to accommodate a growing population (Chowdhury and Akter, 2021). Many areas of the city have become more prone to flooding as the urbanization rate has risen by nearly 3 percentage points and land use has changed accordingly (Chowdhury and Akter, 2021). Chaktai Khal, Rajakhali Khal, Moheshkhal, Jamalkhan khal, and Nasir khal are some of the most important khals that run through the Chattogram metropolitan area (Mahmood and Matin, 2018). Through a network of tributaries, all these khals eventually empty into the southernmost Karnaphuli river area (Mahmood and Matin, 2018). Due to rapid population growth, there isn't much land left. Because of this, many canals are illegally occupied and used to build physical infrastructure (like houses and shops) (Mahmood and Matin, 2018). Also, poor land use has caused a lot of silting, which makes it hard for water to flow from the city to the Karnaphuli River and the Bay of Bengal (Nishat et al., 2022). Because of this, flooding has become a very big problem (Nishat et al., 2022). Also, the lack of a well-connected underground drainage system in the city has made flooding worse (Ashraf and Chowdhury, 2009; Mia et al., 2015). Population growth, economic concentration, and illegal, unchecked growth have all contributed to the worsening of this problem over the past decade or so (Ashraf and Chowdhury, 2009; Mia et al., 2015). Here is where Chattogram city corporation (CCC) has the hardest time managing unauthorized channel occupancy and maintaining the drainage system (Ashraf and Chowdhury, 2009; Mia et al., 2015). The city's drainage system relies majorly on the Chaktai Khal (Chowdhury and Akter, 2021; Hasan Tarek et al., 2017; Mahmood and Matin, 2018). From the Mughal dynasty, this river channel was used for commercial transactions and inner-land travel (Ashraf and Chowdhury, 2009; Mia et al., 2015). From the 19th century, the area surrounding the Chaktai Khal saw population growth and the development of 'gudams' and 'golas', or 'warehouses' (Ashraf and Chowdhury, 2009; Mia et al., 2015). The wholesale market next to the khal sells a variety of foods and building materials, such as C. I. Sheet and paint in addition to rice, lentils, onions, ginger, powdered milk, sugar, edible oil and dried fish (Ashraf and Chowdhury, 2009; Mia et al., 2015). Trucks loading and unloading goods at wayside warehouses and yards congest Chaktai's highways (Ashraf and Chowdhury, 2009; Mia et al., 2015). Chaktai is a popular pick-up spot for Chattogram retailers (Ashraf and Chowdhury, 2009; Mia et al., 2015). Though Chaktai no longer controls the nation's food industry, it remains a regional commercial powerhouse (Ashraf and Chowdhury, 2009; Mia et al., 2015). It boosts our national economy. Waste disposal, plastic dumping, untreated sewage, illegal building across the channel, and other issues prevent the khal from handling increased surface run-off (Ashraf and Chowdhury, 2009; Mia et al., 2015). As a result, during heavy rains, some major areas in this catchment, Chawkbazar, Chaktai, Sholoshahar, and Bahaddarhat overflow (Ashraf and Chowdhury, 2009; Mia et al., 2015). Drainage and waterlogging hinder Chattogram's economic growth. The identification of urban flood risk hotspots in the Chaktai Khal catchment area will aid in identifying potential areas for the implementation of urban flood mitigation measures, which is currently a critical need (Chowdhury and Akter, 2021). For this research, the port city of Bangladesh Chattogram has been taken to explore the scope of a historically significant yet decaying watershed Chaktai Khal for identifying the major flood risk hotspot areas. For this study, the length of the Chaktai Khal was measured from Bahaddarhat Circle to the Chaktai Khal's mouth at Chaktai commercial area, approximately 6 km. The entire length of this khal passes through or right adjacent to ten city wards; ward number: 4-Chandgaon, 6-East Sholoshahar, 7-West Sholoshahar, 8-Solakbahar, 16-Chawkbazar, 17-West Bakalia, 19-South Bakalia, 20- Dewan Bazar, 34- Patharghata, and 35-Boxirhat.

2.2 Materials and Methodology

The study focused on the 'Chaktai Khal' catchment area and consisted primarily of two phases of data acquisition, i. Field survey and ii. GeoSM-NatE. The research makes use

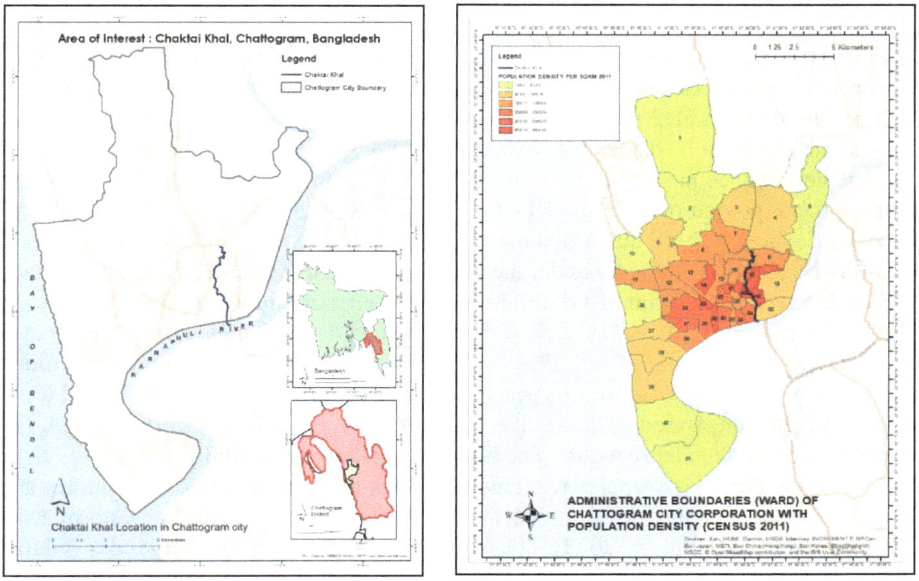

Fig. 1. Study Area (left) and Ward Boundaries of Chattogram with population density data (according to Bangladesh Population and Housing Census 2011, Administrative Report) (right), Source: Self

of both primary and secondary sources of information. A field study was carried out to ascertain the current challenges and threats of the Chaktai Khal catchment region. Photographs were used for the visual survey. An unstructured questionnaire survey was also conducted. The findings of this physical survey validated the results of the secondary data analysis. For secondary data analysis the Chattogram City Corporation (CCC) boundary area was taken as a study area. The identification of urban flood risk zones was carried out through the implementation of a conditional ranking system. This system involved the assessment of urban grids measuring 1 km^2, wherein hazard exposure was evaluated using metrics derived from the mapping process (Manfreda et al., 2011). Digital elevation models (DEMs) of terrain have extensive use in hydrological mapping (Manfreda et al., 2011). The study used Digital Elevation Model of 30 m (1 arc second) resolution from SRTM (shuttle radar topography mission) as an input data for hydrological parameters mapping (Samela et al., 2016). Topographic data covering over 80% of land areas on Earth was acquired by the SRTM, making it the first practical worldwide record of land elevations (Shuttle Radar Topography Mission (SRTM), Earth data). Sentinel-2 was also acquired for mapping indices such as NDBI, NDVI and NDWI (USGS, Sentinel-02, Kuc and Chormański, 2019). This study used 10 m and 20 m resolution cloud free sentinel 02 data of December 2021 (USGS, Sentinel-02, Kuc and Chormański, 2019). Parameters for urban flooding vulnerability in the Chaktai Khal catchment area of Chattogram have been investigated using a conceptual framework named the Pressure and Release (PAR) model. Disaster managers would benefit from having a conceptual model like the Pressure and Release (PAR) model (Wisner et al., 2004) as a basis for assessing vulnerability and taking steps to reduce it (Awal, 2015). It is argued in the PAR model that the intersection of natural hazards and vulnerability-generating processes is where disasters tend to occur (Awal, 2015). Disaster strikes when these two factors interact with one another (Awal, 2015). Pressure may come from any side, but in order to relieve it, vulnerability must be decreased (Awal, 2015). Figure 8 displays how the PAR model was adapted to account for urban flooding in Chattogram city, Bangladesh. The study adopted the approach of PAR model (Pressure and Release Model), under this model the risk pressure has been quantified using Geo-Spatial Mapping of Natural Ecosystem (GeoSM-NatE tool) and finally under the release section of the model strategic approach of Blue-Green Infrastructure (BGI) has been discussed. For ranking the delineated urban grids of 1 km in city corporation periphery, five urban flood risk indicators were selected based on the understanding from 'Pressure' part of PAR model. They are dense buildup areas, least vegetation, least surface waterbody, highest stream density, and lowest topographical elevation. Based on these five indicators five input layers were selected. For identifying the highly dense areas, 'The Normalized Difference Built-up Index (NDBI)' was created to determine what percentage of a given area is comprised of buildings and other man-made structures (Andriani et al., 2018; Bahadur, 2018; Kuc and Chormański, 2019; Sunarta and Saifulloh, 2022). The Normalized Difference Build-up Index (NDBI) identifies metropolitan areas with increased shortwave-infrared reflectance (SWIR) (Kuc and Chormański, 2019) (Figs. 2, 3, 4, 5, 6 and 7).

Fig. 2. Mouth of Chaktai Khal, Chaktai Commercial Area, source: self.

Fig. 3. Encroachment on Chaktai Khal Existing Situation, source: self.

Fig. 4. Solid Wastage Garbage Dumps in the khal, source: self.

Fig. 5. Adjacent roads of the Khal gets inundated during heavy rainfall, source: self.

Fig. 6. Concrete embankment, wastage dumping, illegal encroachment of Chaktai Khal, source: self.

Fig. 7. Untreated surface runoff water runs into the Chaktai Khal, source: self.

As a result, the NDBI was calculated as follows:

$$NDBI = \frac{SWIR - NIR}{SWIR + NIR} \tag{1}$$

Build-up regions are represented by NDBI indices that are closer to 1 than -1 (Dammalage and Jayasinghe, 2019). More positive values indicate more impervious

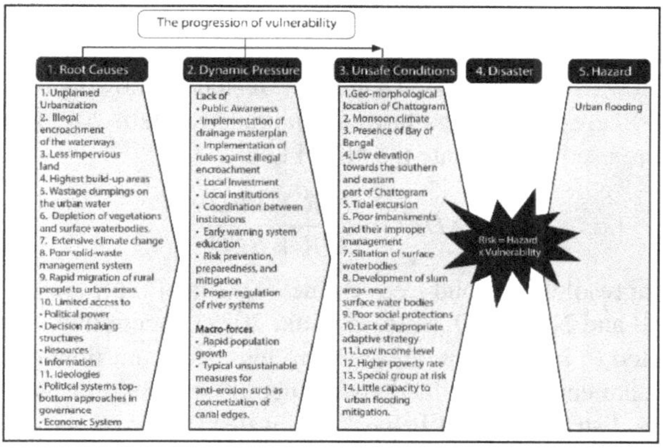

Fig. 8. Pressure and Release (PAR) model of Urban Flooding Disaster in Chattogram city, Bangladesh. (Modified from Wisner et al., 2004), Source: Self

land and less infiltration rate and thus more surplus surface runoff water (Kuc and Chormański, 2019; Mrekva et al., 2012). Secondly, 'The Normalized Difference Water Index (NDWI)' was delineated to identify the surface waterbodies in the study area (Bahadur, 2018; Hussain et al., 2018; Xu, 2006). The NDWI pictures were created by (2) (Dammalage and Jayasinghe, 2019; Yang et al., 2017). Water characteristics in the categorized image were detected using the NDWI index (Dammalage and Jayasinghe, 2019; Yang et al., 2017). In general, open water is represented by positive values while all other features by negative ones in data sets (Dammalage and Jayasinghe, 2019; Yang et al., 2017). Hence, NDWI can demonstrate if there is a scarcity of surface waterbodies in the research region, implying a lower capability to serve surplus surface water runoff and a higher likelihood of urban flooding (Yang et al., 2017). The NDWI was calculated as follows (Table 1):

$$NDWI = \frac{NIR - GREEN}{NIR + GREEN} \tag{2}$$

Table 1. The Data Used for Urban Flood Risk Mapping using GeoSM-NatE tool.

Input Data	Purpose
1. Administrative boundary and legal limits of the urban authority	Area of Interest (AOI)
2. Geospatial data	Indices of Urban features development and Urban risk mapping (NDBI, NDVI, NDWI)
3. Secondary data like DEM SRTM	Hydrological Mapping on GIS platform

Thirdly, 'The Normalized Difference Vegetation Index (NDVI)' was delineated using sentinel-2 data in same manner. A number of studies have looked into how parks and other green areas affect surface runoff (Kim et al., 2017). Less green areas in a study area indicate more impervious lands resulting in more vulnerability towards urban flooding (Barroca et al., 2006). There have been a number of studies showing that NDVI is useful for mapping out where floodwaters have accumulated (Atefi, 2022). Following is the standard procedure used to calculate the NDVI (3):

$$NDVI = \frac{NIR - RED}{NIR + RED} \tag{3}$$

10 m and 20 m resolution cloud free sentinel-02 data of 2021 were used for measuring NDBI, NDVI, and NDWI (Onačillová et al., 2022). Accordingly, a stream density map was delineated to identify the areas where higher stream flow or higher runoff is generating (Abdelkareem, 2017). The abundance of streams in a watershed is a representation of the distribution and behavior of water (EnviroAtlas, 2015). Density of streams, or drainage, varies from watershed to watershed due to differences in infiltration capacity (EnviroAtlas, 2015). Increased runoff and erodible land are characteristics of watersheds with a high stream density (EnviroAtlas, 2015). Higher stream densities occur on steeper slopes, which are influenced by topography and elevation (EnviroAtlas, 2015). It is defined as the total length of streams by drainage area of all orders (Hema et al., 2021). Concentrating flood flow in the basin is mostly determined by drainage density; hence, a higher drainage density indicates a larger flood danger (Dang and Nguyen, 2018). SRTM based DEM data of 30 m spatial resolution was used with ArcGIS in this regard (EnviroAtlas, 2015; Hema et al., 2021). The highest values of stream density indicate the highest accumulation of surface runoff. The stream density map was generated using publicly accessible DEM—Shuttle Radar Topography Mission (SRTM 30 m) data (EnviroAtlas, 2015; Hema et al., 2021). Via the website (https://earthexplorer.usgs.gov/), it was downloaded (EnviroAtlas, 2015; Hema et al., 2021). Finally, an elevation map of the study area was generated, taking into account the fifth parameter for identifying urban flood risk hotspots: the minimum elevation of a given topographical region (Dunn and Hickey, 1998). An important factor in establishing the infiltration/runoff relationship is the elevation (Abdelkareem, 2017; Manfreda et al., 2011; Reedevi et al., 2009). More infiltration and less runoff occur when the elevation is less steep, while the converse is true when the elevation is steeper (Abdelkareem, 2017; Manfreda et al., 2011; Reedevi et al., 2009). Many researchers suggested obtaining DEM using SRTM data and GIS based approach for delineating elevation map (Abdelkareem, 2017; Manfreda et al., 2011; Reedevi et al., 2009; Sreedevi et al., 2013). This study used SRTM DEM of 30m resolution from https://earthexplorer.usgs.gov/ and then imported it to the ArcGIS. After processing all the five maps, using the GeoSM-NatE tool, the severity of each parameter's risk factors was calculated by assigning a weight to the 1 square kilometer urban grids. In ArcGIS each grid had an individual object id. Thus approx. 200 grids were generated within Chattogram City Corporation boundary. For each map feature or index, the average value and its corresponding item ID were recorded. Mean values for each 1 km-by-1 km grid were extracted from the attribute tables using ArcGIS for each map. For elevation, NDVI, NDWI maps the extracts mean values with their object id were sorted in a descending order and the lowest value of them indicated lowest elevation area, least vegetation and least surface water respectively. On the other hand, the mean values extracted from NDBI and Stream density maps were sorted in an ascending

order and there the highest values of those indicated highest buildup areas and highest accumulation of surface water runoff respectively. Afterwards the sorted mean values were distributed into 10 classes. Each class consists of 20 mean values. The first 20 mean values from each five sorting were marked as 'risk rank 10' and then accordingly next 20 mean values as 'risk rank 9' in the same manner and thus risk ranks from 10 to 1 were sorted where each rank consists of 20 mean values orderly arranged. Then from each Table 10 to 5 risk ranks were assessed in a manner so that the grids which were commonly found for each parameter can be differentiated. The 1 km-by-1 km grids which were found 5 times from five parameter's risk assessment were marked as '5th order risk zones', which were found 4 times in a same manner were marked as '4th order risk zones' and then 3rd, 2nd risk zones were identified similarly. As the 5th and 4th order risk zones possess the highest potential among all, only those urban flood risk grids were extracted considering the metropolitan city's perimeter. Again then, only the urban grids encompassing the Chaktai Khal were extracted from them and thus four 1 km by 1 km urban flood hotspot grids were found only within which Chaktai Khal runs. Among these four grids two are from '5th order risk zones': the highest risk order zones and other two are from the '4th order risk zones': the second highest risk order zones (Fig. 9).

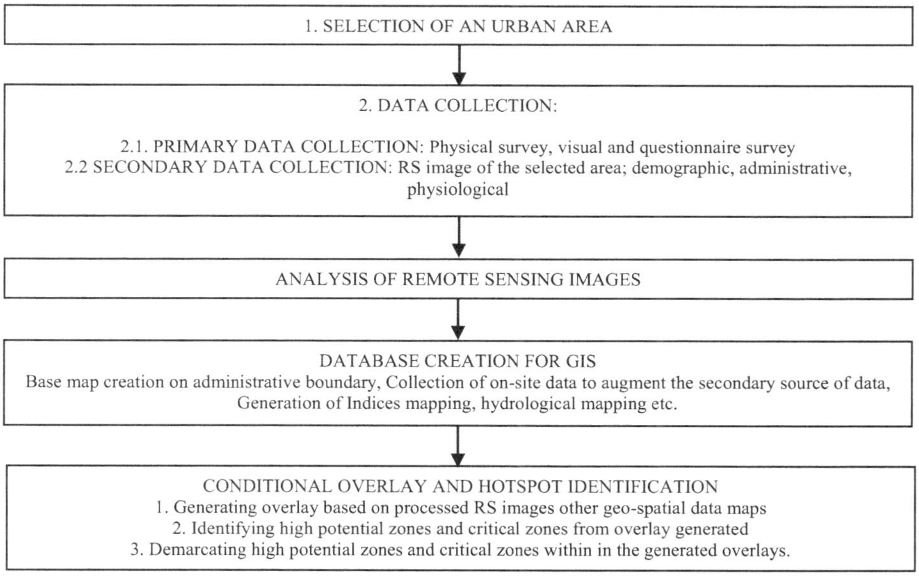

Fig. 9. Methodology Chart

3 Data Analysis

<u>a. Vegetation Mapping:</u> The present study employed NDVI (Normalized Difference Vegetation Index) mapping technique to map the vegetation in the Chattogram city corporation area. Figure 10 depicts the presence of vegetation within the city corporation region during the year 2021, as indicated by the NDVI map. From the visual interpretation of the NDVI map, the northern region of the city is characterized by the presence of densely vegetated patches. Again, upon comparison with the elevation map, it is evident that the upper stream regions of the city exhibit a high density of vegetation. Conversely, the downstream regions exhibit higher levels of development and a lower prevalence of vegetation patches. The present study centers on the catchment area of 'Chaktai Khal', which exhibits the lowest vegetation coverage in comparison to the overall vegetation coverage of the city. This suggests that the Khal area, the downstream regions of the city exhibit the lowest levels of permeability. Upon comparison with the basin map, it is evident that the area with the largest basin in the city exhibits the lowest quantity of vegetation. The basin area with the largest surface runoff in the city exhibits limited potential for infiltration due to the minimal presence of vegetation. Vegetation predominates on the outskirts of the city, to the north, south, east, and west. According to the pixel count of the NDVI map approx. 8155 ha of vegetation area was present in 2021 in Chattogram city.

b. Build-up Area Mapping: The research employed NDBI (Normalized Difference Build-up Index) mapping as a means of delineating the build-up areas within Chattogram city. The NDBI map's spatial distribution visually indicates that the built-up area, predominantly situated in the central region of the city, constituted the most significant portion of land use. Figure 11 illustrates that the 'Chaktai Khal' is situated in highly populated regions, which suggests that there has been unauthorized occupation of the khal region. The northern region of the city exhibits highly concentrated urban development as observed from the riverbank areas of Karnaphuli. As one moves towards the northern regions of the city, specifically the upper stream areas (as depicted in Fig. 13), the prevalence of built-up areas gradually decreases. According to the NDBI map 2021 for Chattogram city approx. 8815.95 ha of built-up area was present.

c. Waterbody Mapping: The NDWI (Normalized Difference Water Index) was employed in this study to cartographically represent the surface waterbodies that exist within the Chittagong metropolitan area. The analysis of Fig. 12 reveals a notable insufficiency of waterbodies in the central region of the city. A higher concentration of water bodies is observable in the Western region of the city, or in proximity to coastal regions. The catchment area of Chaktai Khal exhibits a limited presence of waterbodies, indicating a reduced capacity for runoff detention in the surrounding region.

d. Urban Hydrology Mapping: The present study employs the hydrology analysis module within the spatial analyst toolbox of ArcGIS software to investigate the inherent surface water flow dynamics of Chattogram city. According to the elevation map depicted in Fig. 13, the city exhibits a predominantly high elevation in its North-Western region. Conversely, a significant proportion of the urban landscape is situated in regions of reduced altitude. The regions along the banks of the Karnaphuli river and the coastal areas of the city are predominantly situated within lower elevation zones. Chaktai Khal is situated in low-lying regions, which presents opportunities for managing surface water runoff originating from upstream areas. The illustration depicted in Fig. 14 portrays the inherent mechanism of surface water runoff accumulation. The central region, eastern riverbank, and western coastal areas of the city exhibit higher natural stream densities.

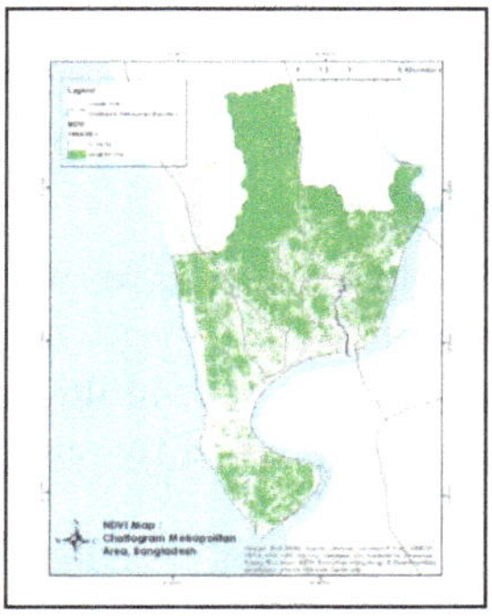

Fig. 10. NDVI Map of Chattogram City, 2021

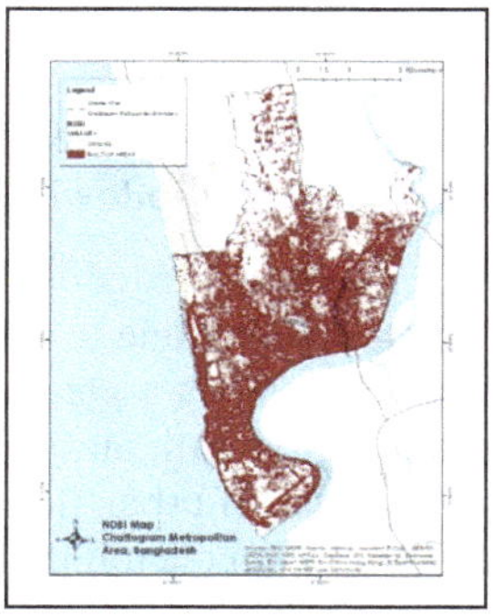

Fig. 11. NDBI Map of Chattogram City, 2021

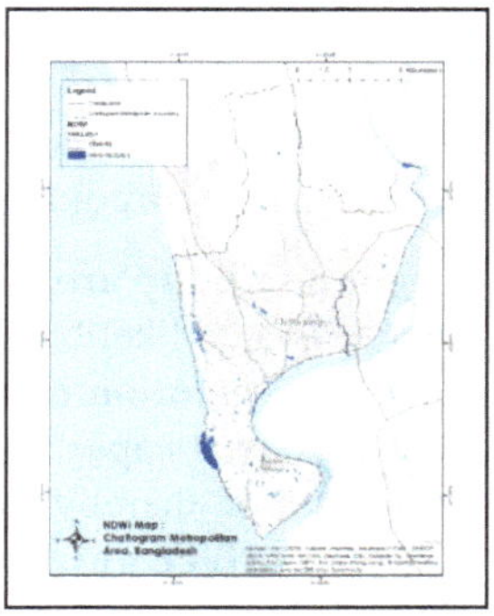

Fig. 12. NDWI Map of Chattogram City, 2021

The Chaktai khal is situated in an area where the concentration of streams is elevated, and it is oriented orthogonally to the direction of surface water flow. Thus, the construction of this man-made canal has the potential to effectively manage surface water flow from upstream areas to downstream areas. Furthermore, the existence of the khal in the two most expansive basins within the urban area suggests that a significant amount of surface water runoff can be directed towards this watercourse.

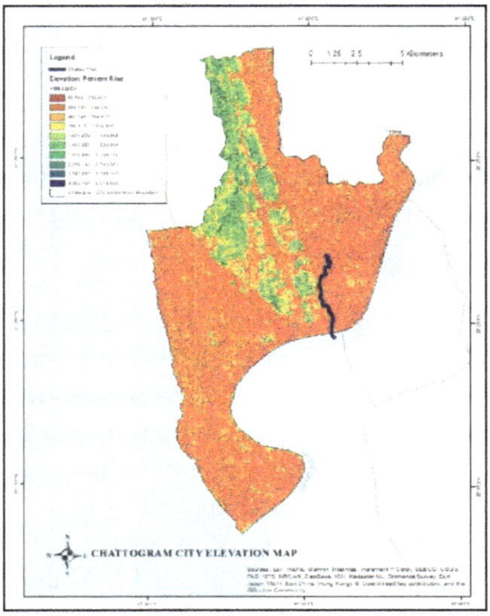

Fig. 13. Elevation Map Chattogram City

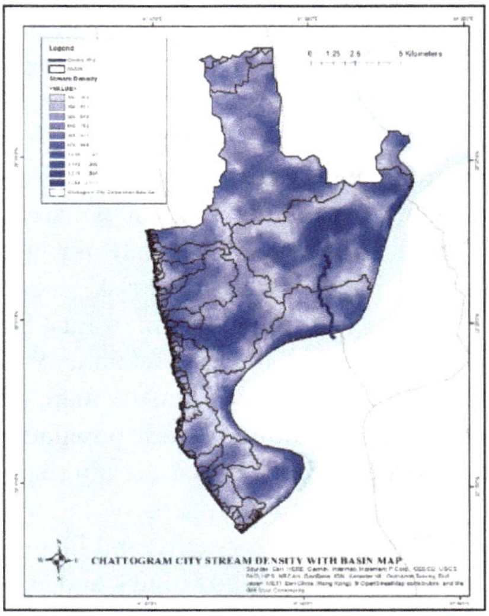

Fig. 14. Chattogram City Basin with Stream Density Map

4 Result and Discussion

This study adopted conditional approach for delineating urban flood risk hotspot in Chattogram city corporation area considering following 5 parameters –

- **Highest buildup areas** – which denote a greater degree of imperviousness, resulting in an elevated likelihood of urban flooding due to surface water runoff.
- **Lowest vegetation count** – which denotes a greater likelihood of poor infiltration rates, reduced capacity for retaining rainfall, and an increased potential for surface water runoff, thereby elevating the risk of urban flooding.
- **Lowest surface waterbodies** – which denote minimal opportunities for impeding the velocity of surface water runoff and minimal capacity for retention, thereby increasing the likelihood of generating the maximum amount of surface water runoff.
- **Lowest elevation area** – which denotes maximum chances of getting the surface water runoff flow than the higher elevation areas thus high risk for urban flooding.
- **Highest stream density** – which denotes a greater likelihood of inundation due to the topographic accumulation of surface water runoff.

The study considered these five parameters to be equally significant in delineating regions with increased urban flood risk. By superimposing five distinct maps for these five parameters, namely NDBI, NDVI, NDWI, elevation, and stream density, and applying the GeoSM-NatE tool's methodology, two categories of urban flood risk hotspots were identified. Type One refers to the fifth-order urban flood risk hotspots, which are defined as areas of one square kilometer where the five parameters coexist, resulting in the highest urban flood risk areas. The Type Two category pertains to urban flood risk hotspot areas of 1 square kilometer with the presence of any four parameters, thereby designating them as the second highest urban flood risk areas at the fourth order. The Fig. 16 displays two distinct categories of flood risk hotspot regions within the jurisdiction of Chattogram City Corporation. Figure 17 illustrates that the Chaktai Khal traverses four urban areas that are susceptible to flooding and among them two areas belong to the highest urban flood risk hotspot areas. The urban flood risk hotspot, including Chaktai Khal, is identified in Fig. 18 through the demarcation of city wards. The research indicates that the urban flood risk, in conjunction with the khal, pertains to specific city wards, namely 4- Chandgaon, 6- East Sholoshahar, 7- West Sholoshahar, 8-Solakbahar, 16-Chawkbazar, 17- West Bakalia, 19-South Bakalia, and 20-Dewan Bazar. Figure 01 illustrates the population density map, which reveals that the wards mentioned above are the localities with the highest population density within the city. Numerous academic studies have also identified certain regions as highly susceptible to urban flooding due to their high population density and unfavorable conditions such as low-lying terrain, obstructed drainage systems, and inadequate urban development, all of which can contribute to flooding (Chowdhury and Akter, 2021; Hasan Tarek et al., 2017; Mahmood and Matin, 2018; Nishat et al., 2022). The influence of topography on the incidence of flooding is significant. In instances of intense precipitation, a substantial volume of surplus water is transported downstream from elevated terrain to regions of lower elevation. The inadequacy of water carriers, namely canals, sewers, and drains, to effectively discharge a significant volume of water results in the occurrence of flooding in adjacent streets and residential properties (Chowdhury and Akter, 2021; Hasan Tarek et al., 2017; Mahmood and Matin, 2018; Nishat et al., 2022). Also, areas in proximity to Chaktai Khal are susceptible to inundation due to their low elevation. The flood-prone regions depicted in the flood maps suggest a correlation with the phenomenon of urbanization, thereby implicating it as a contributing factor to these incidents. The failure to implement appropriate urbanization measures is exacerbating the already rapid pace of urbanization. Moreover, most of the thoroughfares, edifices, and vicinities in proximity to these sites are constructed from concrete. The gradual reduction of the city's natural infiltration zone is evident. For the last few decades, the central region of the city has experienced a significant reduction in surface waterbodies and vegetation cover due to the rapid process of urbanization. Consequently, the infiltration rate in this region has decreased, resulting in the identification of a significant concentration of urban flood risk hotspots in this section of the city (refer to Fig. 16). Furthermore, based on the findings of the visual survey conducted in this study, it was ascertained that the Chaktai khal locality is experiencing unlawful encroachment because of swift urbanization. Most of the residents in this locality are situated within impervious concrete cover, including roads, streets, and densely populated commercial areas. The inhabitants of this region encounter swift surface water drainage during intense monsoon precipitation on an annual basis (Fig. 15).

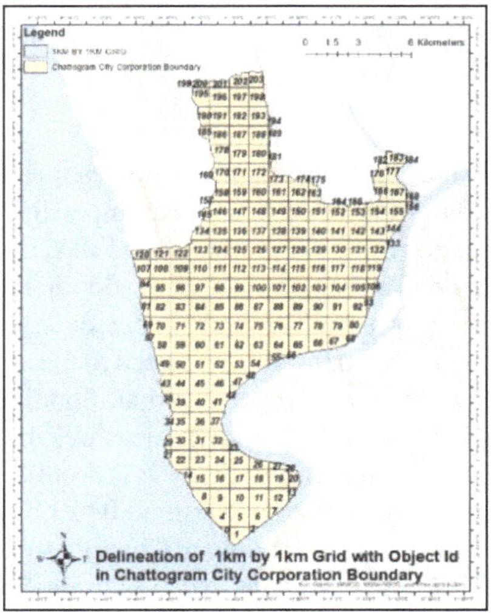

Fig. 15. 1 km by 1 km Grid Delineation in Chattogram City Corporation Boundary

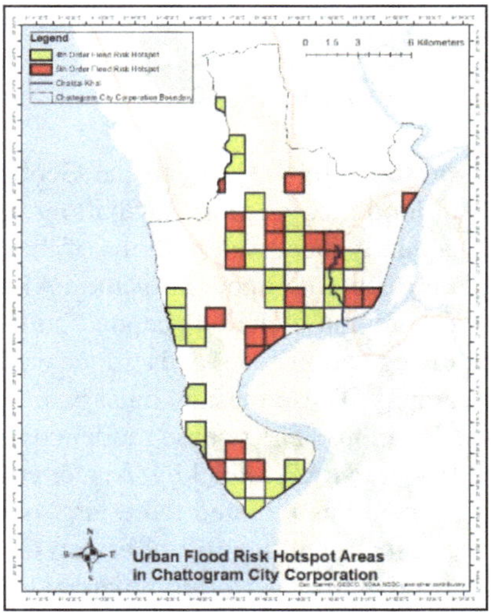

Fig. 16. Urban Flood Risk Hotspot Delineation in Chattogram City Corporation Boundary

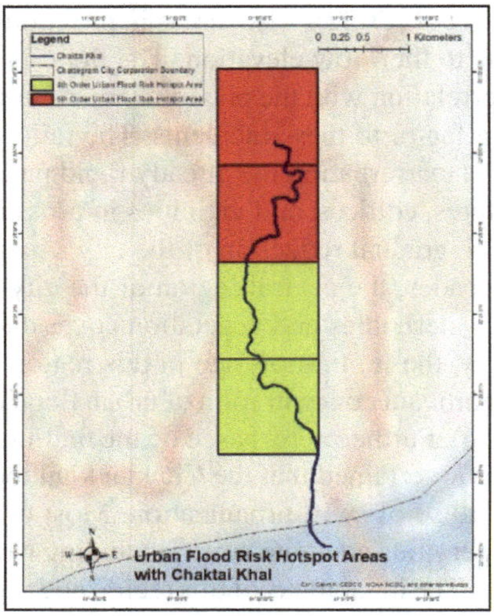

Fig. 17. Urban Flood Risk Hotspot Areas with Chaktai Khal

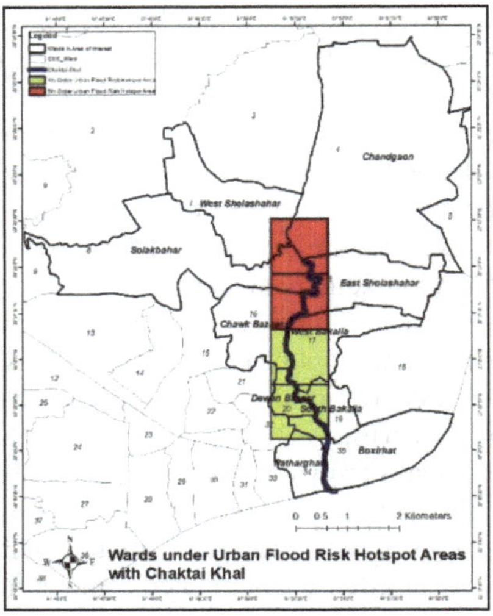

Fig. 18. Wards under Urban Flood Risk Hotspot Areas with Chaktai Khal

5 Conclusion

The present study employed the GeoSM-NatE tool and PAR model to investigate the urban flood risk hotspots in Chattogram city. The findings of this research paper were subsequently discussed in terms of the measures for mitigating urban flood risk, utilizing the release component of the PAR model. The key causes for urban floods in the research region include expanded built-up areas, limited infiltration scopes owing to a lack of vegetation cover and surface waterbodies, and vulnerabilities related to the area's topography. These reasons must be changed to reduce the danger of urban flooding in these locations. The release model considering urban flood mitigation measures, modified from the PAR model by Wisner et al., 2004, is shown in the Fig. 19. Presently, the government has initiated three large-scale projects valued at approximately 10 billion rupees aimed at mitigating the effects of rain, tidal water, and flooding in Chattogram city, widely recognized as the commercial hub (Arafat, 2015; *DailySun*, 2017; Mamun, 2018; Nishat et al., 2022). Following the expiration of the designated timeframe for the two projects, two additional one-year extensions have been granted (Arafat, 2015; *DailySun*, 2017; Mamun, 2018; Nishat et al., 2022). The construction project involving the development of a road and 12 regulators along the banks of Karnaphuli river to Chaktai canal via Kalurghat bridge has been granted a two-year extension, spanning from July to June 2024 (Arafat, 2015; *DailySun*, 2017; Mamun, 2018; Nishat et al., 2022). As of present, the construction of 75% of the sluice gates, specifically 9 out of 12, within the scope of our project has been finalized. As of now, 80% of the task has been accomplished (Arafat, 2015; *DailySun*, 2017; Mamun, 2018; Nishat et al., 2022). However, this study proposes the adoption of Nature-based solutions for mitigating urban floods, as its results indicate that the degradation of natural ecosystems is the primary cause of the urban flooding problem in the city and Chaktai khal area. NBS are novel technical, economic, and environmental planning principles that sustain and restore hydrological functioning in human and natural systems (Huang et al., 2020). NBS uses green infrastructures to solve issues like urban floods while preserving a city's hydrological systems (Huang et al., 2020). Low Impact Development (LID), Best Management Practices (BMPs), Water Sensitive Urban Design (WSUD), the Sustainable Urban Drainage System (SUDS), Green Stormwater Infrastructure/Green Infrastructure, and Sponge City are all regional terms (Huang et al., 2020). Enhancement of natural ecosystems can be possible in the Chattogram city and Chaktai Khal locality by implementing any of these nature-based strategies for urban flood mitigation. Due to its cost effectiveness, low maintenance and immense beneficial impacts NbS are suggested as quite best approaches by many researchers for urban flood mitigation (Krebs et al., 2013). The potential expansion of this research could encompass a quantitative examination of nature-based approaches to address urban flooding in the designated study location.

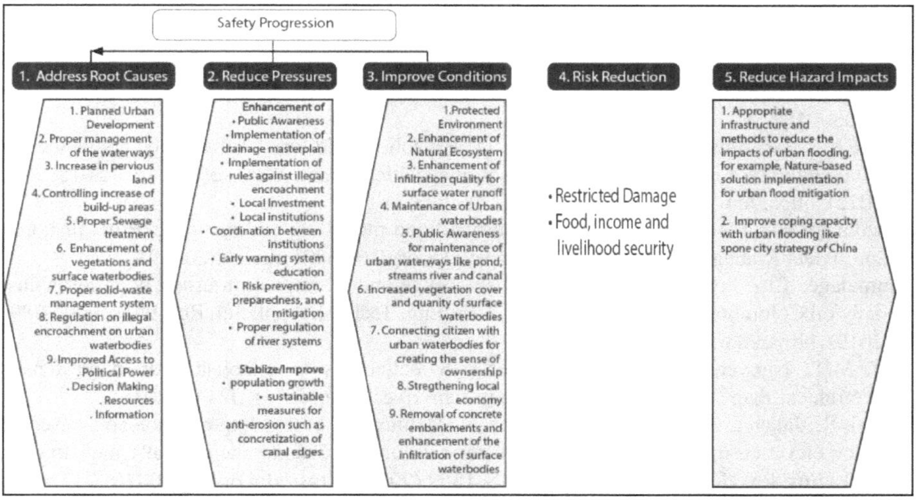

Fig. 19. The release part of PAR model (modified from Wisner et al., 2004) for Urban Flood Mitigation

References

van Aalst, M.K.: The impacts of climate change on the risk of natural disasters. Disasters **30**(1), 5–18 (2006). https://doi.org/10.1111/j.1467-9523.2006.00303.x

Abdelkareem, M.: Targeting flash flood potential areas using remotely sensed data and GIS techniques. Nat. Hazards **85**(1), 19–37 (2017). https://doi.org/10.1007/s11069-016-2556-x

Akter, A., Mohit, S.A., Chowdhury, M.A.H.: Predicting urban storm water-logging for Chittagong city in Bangladesh. Int. J. Sustain. Built Environ. **6**(1), 238–249 (2017). https://doi.org/10.1016/j.ijsbe.2017.01.005

Andriani, Putranto, D.D.A., Affandi, A.K., Ibrahim, E.: Interpretation of land use and land cover at lowland area using by NDVI and NDBI. Ecol. Environ. Conserv. **24**(2), 651–657 (2018)
Arafat, S.M.G.B.: The "sorrow of Chittagong City. The Daily Star 2021 (2015). http://www.thedailystar.net/op-ed/politics/the-"sorrow-chittagong-city"-105313
Ashraf, M.A., Chowdhury, M.S.A.: Drainage planning in the cities of Bangladesh: case study of drainage and water logging in Chaktai commercial area, Chittagong. J. Bangladesh Inst. Plan. **2**(December), 49–60 (2009). https://doi.org/10.3329/jbip.v2i0.9556
Atefi, M.R.: Detection of Flash Flood Inundated Areas Using Relative Difference in NDVI from Sentinel-2 Images: A Case Study of August 2020 (2022)
Awal, M.A.: Vulnerability to disaster: pressure and release model for climate change hazards in Bangladesh vulnerability to disaster: pressure and release model for climate change hazards in Bangladesh. Int. J. Environ. Monit. Protect. **2**(2), 15–21 (2015). http://www.openscienceonline.com/journal/ijemp
Bahadur: NDVI, NDBI & NDWI Calculation Using Landsat 7,8 (2018). https://www.linkedin.com/. Accessed 8 Sept 2018. https://www.linkedin.com/pulse/ndvi-ndbi-ndwi-calculation-using-landsat-7-8-tek-bahadur-kshetri
Barroca, B., Bernardara, P., Mouchel, J.M., Hubert, G.: Indicators for identification of urban flooding vulnerability. Nat. Hazard. **6**(4), 553–561 (2006). https://doi.org/10.5194/nhess-6-553-2006
Khazai, B., Bendimerad, F., Cardona, O.D., Carreño, M.-L., Barbat, A.H., Burton, C.G.: A Guide to Measuring Urban Risk Resilience Principles, Tools and Practice of Urban Indicators. In The Jerusalem Talmud, First order: Zeraim, Tractate Berakhot (2013). https://doi.org/10.1515/9783110800487.39
Chan, F.K.S., Griffiths, J., Sang, Y., Komori, D., Pezzoli, A.: Editorial: urban flood resilience and sustainable flood management strategies in megacities. Front. Water **3**(January), 2018–2022 (2022). https://doi.org/10.3389/frwa.2021.738493
Chowdhury, M.A.H., Akter, A.: City scale urban flood modelling and mapping. Proc. Inst. Civ. Eng. Water Manag. **176**, 1–18 (2021). https://doi.org/10.1680/jwama.20.00094
Dammalage, T.L., Jayasinghe, N.T.: Land-use change and its impact on urban flooding: a case study on Colombo district flood on May 2016. Eng. Technol. Appl. Sci. Res. **9**(2), 3887–3891 (2019). https://doi.org/10.48084/etasr.2578
Dang, M.T., Nguyen, D.B.: Application of GIS technology to establish a drainage density hierarchical map for flood hazard zoning in Lam river basin, **59**(6), 32–42 (2018)
De Risi, R., Jalayer, F., De Paola, F., Lindley, S.: Delineation of flooding risk hotspots based on digital elevation model, calculated and historical flooding extents: the case of Ouagadougou. Stoch. Env. Res. Risk Assess. **32**(6), 1545–1559 (2017). https://doi.org/10.1007/s00477-017-1450-8
Dilley, M., Chen, R.S., Deichmann, U., Lerner-Lam, L., Arnold, M.A.: Natural disaster hotspots a global risk analysis. In: The World Bank Hazard Management Unit 2005 Washington, D.C, vol. 19, no. 1 (2005). https://doi.org/10.1002/j.1834-4461.1948.tb00495.x
Dunn, M., Hickey, R.: The effect of slope algorithms on slope estimates within a GIS. Cartography **27**(1), 9–15 (1998). https://doi.org/10.1080/00690805.1998.9714086
EnviroAtlas: Stream Density, 1–2 (2015)
Feng, B., Zhang, Y., Bourke, R.: Urbanization impacts on flood risks based on urban growth data and coupled flood models. Nat. Hazards **106**(1), 613–627 (2021). https://doi.org/10.1007/s11069-020-04480-0
Ferreira, C.S.S., Potočki, KristinaMarijana Kapović-Solomun, A., Kalantari., Z.: Nature-Based Solutions for Flood Mitigation and Resilience in Urban Areas. May, pp. 59–78 (2021). https://doi.org/10.1007/698_2021_758
Hasan Tarek, M.Md.S., Kabir, I., Hassan, A.: Application of Swmm for Analysis of Flash Floods in Urban Areas: a Case Study for Chaktai Khal Watershed in Chittagong. March 2017 (2017)

Hema, H.C., Vittala, S.S., Govindaiah, S.: Quantitative morphometric infer in the hard rock Terrain based on SRTM - DEM and GIS - Chintamani Watershed, Chikkaballapur District, Karnataka, India. Sustain. Water Resourc. Manag. **7**(4), 1–12 (2021). https://doi.org/10.1007/s40899-021-00534-8

https://www.daily-sun.com/post/230621/Waterlogging-hits-life-in-Chittagong1/1.(2017).230621

Huang, Y., et al.: Nature-based solutions for urban pluvial flood risk management. Wiley Interdiscip. Rev. Water **7**(3), 1–17 (2020). https://doi.org/10.1002/WAT2.1421

Hussain, N., Islam, M.H., Firdaus, F.: Impact of Tidal River Management (TRM) for water logging: a geospatial case study on coastal zone of Bangladesh. J. Geosci. Environ. Protect. **06**(12), 122–132 (2018). https://doi.org/10.4236/gep.2018.612009

Islam, M.R., Raja, D.R.: Waterlogging risk assessment: an undervalued disaster risk in coastal urban community of Chattogram, Bangladesh. Earth **2**(1), 151–173 (2021). https://doi.org/10.3390/earth2010010

Jalayer, F., et al.: Probabilistic GIS-based method for delineation of urban flooding risk hotspots. Nat. Hazards **73**(2), 975–1001 (2014). https://doi.org/10.1007/s11069-014-1119-2

Kim, H.W., Kim, J.H., Li, W., Yang, P., Cao, Y.: Exploring the impact of green space health on runoff reduction using NDVI. Urban Forest. Urban Green. **28**(July), 81–87 (2017). https://doi.org/10.1016/j.ufug.2017.10.010

Krebs, G., Kokkonen, T., Valtanen, M., Koivusalo, H., Setälä, H.: A high resolution application of a stormwater management model (SWMM) using genetic parameter optimization. Urban Water J. **10**(6), 394–410 (2013). https://doi.org/10.1080/1573062X.2012.739631

Kuc, G., Chormański, J.: Sentinel-2 imagery for mapping and monitoring imperviousness in urban areas. Int. Arch. Photogram. Remote Sens. Spat. Inf. Sci. - ISPRS Arch. **42**(1/W2), 43–47 (2019). https://doi.org/10.5194/isprs-archives-XLII-1-W2-43-2019

Mahmood, F., Matin, M.A.: Assessment of Drainage Capacity of Chaktai and Rajakhali Khal in Chittagong City and Inundation Adjacent of Urban Areas. February, 1–13 (2018)

Majidi, A.N., et al.: Planning nature-based solutions for urban flood reduction and thermal comfort enhancement. Sustainability (Switzerland) **11**(22), 1–27 (2019). https://doi.org/10.3390/su11226361

Mamun, M.M.A.L.: Improve urban waterways management in Chattogram city - Op-Ed - observerbd.com, 1–6 (2018). https://www.observerbd.com/details.php?id=149108

Manfreda, S., Di Leo, M., Sole, A.: Detection of flood-prone areas using digital elevation models. J. Hydrol. Eng. **16**(10), 781–790 (2011). https://doi.org/10.1061/(asce)he.1943-5584.0000367

Matheswaran, K., Alahacoon, N., Pandey, R., Amarnath, G.: Flood risk assessment in South Asia to prioritize flood index insurance applications in Bihar, India. Geomat. Nat. Haz. Risk **10**(1), 26–48 (2019). https://doi.org/10.1080/19475705.2018.1500495

McCarthy, M.P., Best, M.J., Betts, R.A.: Climate change in cities due to global warming and urban effects. Geophys. Res. Lett. **37**(9), 1–5 (2010). https://doi.org/10.1029/2010GL042845

Mia, M.A., Nasrin, S., Zhang, M., Rasiah, R.: Chittagong, Bangladesh. Cities **48**, 31–41 (2015). https://doi.org/10.1016/j.cities.2015.05.011

Mrekva, L., Engi, Z., Tóth, G.: Urban flood risk and hydrology. Zeitschrift Fur Geomorphologie **56**(September), 021–035 (2012). https://doi.org/10.1127/0372-8854/2012/S-00083

Mukherjee, D.: Ijesrt International Journal of Engineering Sciences & Research Technology Effect of Urbanization on Flood - a review with recent flood in Chennai (India). Int. J. Eng. Sci. Res. Technol. **5**(7), 451–455 (2016). http://www.ijesrt.com

Mukherjee, P.M., Ravan, D.S., Satheesh, A., Kumar, A., Naorem, V.: GeoSM-NatE: urban risk resilience tool, using remote sensing and GIS for Nature-based risk mitigation (n.d.)

Nishat, T.A., Sunny, D.S., Khan, R.T., Mullick, M.R.A., Datta, P.: Urban waterlogging risk profiling: the case of Khatunganj wholesale commodity market, Chattogram. In: Water Management: A View from Multidisciplinary Perspectives, pp. 93–119 (2022). https://doi.org/10.1007/978-3-030-95722-3_5

Onačillová, K., Gallay, M., Paluba, D., Péliová, A., Tokarčík, O., Laubertová, D.: Combining Landsat 8 and Sentinel-2 data in Google earth engine to derive higher resolution land surface temperature maps in urban environment. Remote Sens. **14**(16) (2022). https://doi.org/10.3390/rs14164076

Oral, H.V., et al.: A review of nature-based solutions for urban water management in European circular cities: a critical assessment based on case studies and literature. Blue-Green Syst. **2**(1), 112–136 (2020). https://doi.org/10.2166/bgs.2020.932

Rahman, H.Z.: Urbanization in Bangladesh : Challenges and Priorities, pp. 1–8 (2014). http://www.pri-bd.org/upload/file/bef_paper/1414213688.pdf

Rahman, M.S., Islam, B., Ahmed, B.: An overview on rapid urbanization and induced disaster risk factors in Bangladesh. World Town Planning Day 2012, November, pp. 56–64 (2012). https://doi.org/10.13140/RG.2.1.1418.5206

Rana, M.M.P.: Urbanization and sustainability: challenges and strategies for sustainable urban development in Bangladesh. Environ. Dev. Sustain. **13**(1), 237–256 (2011). https://doi.org/10.1007/s10668-010-9258-4

Rashid, H., Hunt, L.M., Haider, W.: Urban flood problems in Dhaka, Bangladesh: slum residents' choices for relocation to flood-free areas. Environ. Manag. **40**(1), 95–104 (2007). https://doi.org/10.1007/s00267-006-0233-7

Reddy, S.J.: Urban Flooding : [The Role of Urbanization & Climate Change], pp. 3–21 (2019)

Reedevi, P.D.S., Wais, S.O., Han, H.H.K., Hmed, S.A.: Morphometric analysis of a watershed of South India using SRTM data and GIS, **73**(56), 543–552 (2009)

Ciurean, R.L., Schröter, D., Glade, T.: Conceptual frameworks of vulnerability assessments for natural disasters reduction. Intech, pp. 225–240 (2016). https://www.intechopen.com/books/advanced-biometric-technologies/liveness-detection-in-biometrics

Schreider Schreider, S.Y., Smith, D.I., Jakeman, A.J.: Climate change impacts on urban flooding. Res. Urban Sociol. **12**, xiii–xvii (2012). https://doi.org/10.1108/s1047-0042(2012)0000012003

Samad, R.B.: Urbanization and urban growth dynamics: a study on Chittagong City. J. Bangladesh Inst. Plan. **8**(December), 167–174 (2015)

Samela, C., Manfreda, S., Paola, F.D., Giugni, M., Sole, A., Fiorentino, M.: DEM-based approaches for the delineation of flood-prone areas in an ungauged basin in Africa. J. Hydrol. Eng. **21**(2), 1–10 (2016). https://doi.org/10.1061/(asce)he.1943-5584.0001272

Serje, J.: Preliminary extensive risk analysis for the global assessment report on disaster risk reduction GAR 2011. October 2010, vol. 38 (2010)

Sreedevi, P.D., Sreekanth, P.D., Khan, H.H., Ahmed, S.: Drainage morphometry and its influence on hydrology in an semi arid region: using SRTM data and GIS. Environ. Earth Sci. **70**(2), 839–848 (2013). https://doi.org/10.1007/s12665-012-2172-3

Stevens, M.: Cities and flooding: a guide to integrated urban flood risk management for the 21st century. J. Reg. Sci. **52**(5), 885–887 (2012). https://doi.org/10.1111/jors.12006_6

Grimmond, S.: Urbanization and global environmental change: local effects of urban warming, **173**(1), 83–88 (2019)

Sunarta, I.N., Saifulloh, M.: Coastal tourism: impact for built-up area growth and correlation to vegetation and water indices derived from Sentinel-2 remote sensing imagery. Geojournal Tour. Geosites **41**(2), 509–516 (2022). https://doi.org/10.30892/gtg.41223-857

Tanim, A.H., Akter, A.: Estimating urban flood hazard zones using SWMM in Chittagong City, **13**(October 2016), 87–101 (2016)

The World Bank: World Development Report. In Geography (2009)

Tuhin, M., Tajik, M., Faruk, O., Julfikar, M.: Rainwater harvesting to mitigate water logging, a case study on Chittagong, Bangladesh, February 2021

Vojinovic, Z., et al.: Effectiveness of small- and large-scale Nature-Based Solutions for flood mitigation: the case of Ayutthaya, Thailand. Sci. Tot. Environ. **789**, 147725 (2021). https://doi.org/10.1016/j.scitotenv.2021.147725

Xu, H.: Modification of normalised difference water index (NDWI) to enhance open water features in remotely sensed imagery. Int. J. Remote Sens. **27**(14), 3025–3033 (2006). https://doi.org/10.1080/01431160600589179

Wisner, B., Blaikie, P., Cannon, T., Davis, I.: At Risk: Natural Hazards, People's Vulnerability and Disasters, 2nd edn., p. 471. Routledge, London (2004)

Yang, X., Zhao, S., Qin, X., Zhao, N., Liang, L.: Mapping of urban surface water bodies from sentinel-2 MSI imagery at 10 m resolution via NDWI-based image sharpening. Remote Sens. **9**(6), 1–19 (2017). https://doi.org/10.3390/rs9060596

Zevenbergen, C., Cashman, A., Evelpidou, N., Pasche, E., Stephen Garvin, R.A.: Urban flood management. In: Dictionary Geotechnical Engineering/Wörterbuch GeoTechnik (2014). https://doi.org/10.1007/978-3-642-41714-6_211004

Short and Long Term Climate-Related Disaster Risk: Connecting the Dots

Demosthenes Tambakis[1(✉)] and Saite Lu[2]

[1] Pembroke College, University of Cambridge, Cambridge, UK
dnt22@cam.ac.uk

[2] Emmanuel College, University of Cambridge, Cambridge, UK
sl590@cam.ac.uk

Abstract. We employ non-homogeneous Markov chains to study the impact of anthropogenic CO_2 emissions on stationary climate-related disaster frequency (SDF). Annual incidence rates are calibrated with a logit model of global disasters and CO_2 growth estimated from 1960–2022. We find that the baseline SDF ranges from approximately 7 to 14%. Up to one half of the SDF is attributed to variability in CO_2 emissions, which may be influenced by policymakers. Our findings highlight the significant role of CO_2 emissions in driving disaster frequency, aligning with the Sendai Framework's first priority for action: understanding disaster risk. We also conduct sensitivity analysis to compare the long-term risk reduction potential of different IPCC RCP scenarios.

Keywords: Climate change · Disaster risk · Non-homogeneous Markov chain · Stationary frequency

1 Introduction

Pour ce qui est de l'avenir, il ne s'agit pas de le prévoir, mais de le rendre possible. - Antoine de Saint Exupéry, Citadelle, 1948

There is clear evidence that the *short term* (annual) incidence of climate-related extreme events—floods, storms, draughts, heatwaves and forest fires—defined as the ratio of extreme events per country-year has risen steadily since the mid-20th century [1,39]. Upward trending likelihood has grown alongside extreme events' severity, as measured by the number of affected people [5,15]; negative GDP and productivity extremes [3,11]; their correlated, interconnected dimension [25]; and financial market and systemic risk [2,4,30]. There is also overwhelming evidence of a causal anthropogenic influence on compound extreme event risk proxied by greenhouse gas emissions—primarily carbon dioxide (CO_2) [21]. Short-term increases in CO_2 emissions can cause immediate atmospheric changes, potentially leading to extreme weather events. Figure 1 highlights the

H. Tatano et al. (Eds.): GSRIDRR 2023, DRRGADRIBS, pp. 348–369, 2025.
https://doi.org/10.1007/978-981-96-6609-6_24

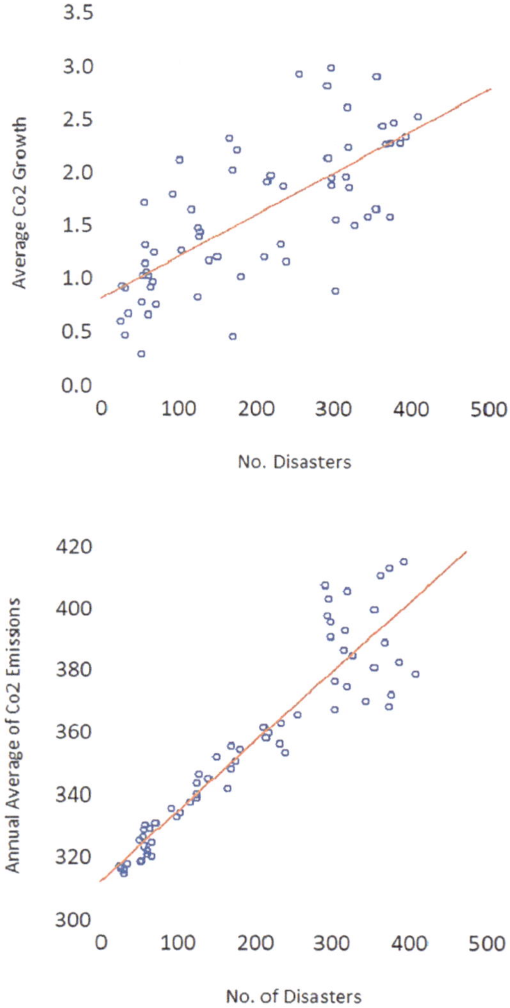

Fig. 1. Number of Global Climate-Related Extreme Events and CO_2 Emissions Levels and Growth Rates: 1960–2022. Source: CRED/EMDAT and NOAA

positive relationship between annual extreme event frequency and CO_2 emissions in the short term.[1]

By comparison, the *long term* (2100 and beyond) impact of curbing carbon emissions or continuing with some form of business-as-usual receives relatively

[1] Both the average growth rate and average level of carbon emissions are positively correlated with the frequency of climate-related extreme weather events. The anthropogenic influence is evident despite the brief decline in 2020 CO_2 emissions attributed to the COVID-19 pandemic [18].

less research attention. That is even though long term projections of climate-related extreme event frequency are a key input to global efforts to address catastrophic climate change and the associated mounting human and economic losses, [7,14].

Here we argue that short and long time horizons are linked because the annual incidence of extreme events predicted by CO_2 emissions combined with average duration jointly determine their long-run frequency—measured by the share of time a country spends in extreme events. The decadal progression of disaster average duration in country-weeks is reported in Table 1:

Average Duration, Impacts and Descriptive Statistics of Incidence Rates[2]

Table 1. Average Duration, Impacts and Descriptive Statistics of Incidence Rates

A. Average Duration per Country-Year and Average Impacts: 1960-2022

Period	*Ave. duration [weeks]*	*Ave. losses [million USD]*	*Ave. affected [million persons]*
1960-1969	0.52	104.71	166.59
1970-1979	1.61	54.68	147.29
1980-1989	2.55	57.61	210.80
1990-1999	3.00	144.93	294.09
2000-2009	5.41	88.59	180.03
2010-2019	5.64	125.56	135.83
2020-2022	6.34	153.23	93.42

B. Descriptive Statistics of Incidence Rates

Period	*1960-2022*	*1980-2022*	*1990-2022*
Obs. $\hat{p}$	63	43	33
Mean	0.062	0.083	0.093
St.Dev.	0.046	0.041	0.035
Skewness	0.294	−0.188	−0.075
Kurtosis	1.858	2.075	1.958
JB	4.333	1.785	1.523
(p-value)	(0.12)	(0.41)	(0.47)

Note: Annual disaster incidence $\hat{p} = \frac{\text{Annual total duration}}{\text{no. of countries} \times 52 \text{ weeks}}$. Kurtosis values above (*below*) 3 indicate the data is platy- (*lepto-*) kurtic. Jarque-Bera (JB) joint Gaussian distribution test p-values given in parentheses. Bootstrap histograms available upon request.

Note that average incidence is trending up since 1960 while its standard deviation is declining.

Specifically, we apply the Markov chain methodology of [36] to study the long-run frequency implications of the 1960–2022 annual global disaster record employing the CRED Emergency Events Database (*EMDAT*). The model is in two stages. First, following [33], we estimate a dynamic panel logistic model of extreme event incidence as a function of CO_2 emission growth, controlling for geographic factors through fixed effects, and demographic factors such as countries' urban population share. Our logit specification is standard. A salient theoretical property of logit model is that, because the implied incidence rates are a strictly convex function of CO_2 emission growth, the total expected annual

2 Proxying extreme weather events' impact by their average duration may cover some of the under-reporting that hampers data collection efforts on economic losses. Even since 2000 only 35% of all extreme events reported any economic losses; in Africa only 13% [5].

incidence exceeds its certainty-equivalent ("no uncertainty") counterpart evaluated at average emission growth.

In the second stage, simulated incidence rates from stage 1 are calibrated to match the 1960–2022 extreme event average likelihood, $\sim$ 6.2% p.a. and fed to a 2-state Markov chain. Long-run stationary disaster event frequencies, *SDF* henceforth, are then computed conditional on a constant exit rate—the inverse of an event's average duration—and stochastic incidence rate. Time-varying incidence controls the short-term frequency of extreme events while the fixed exit rate determines their persistence. Based on EMDAT, we have compiled a novel duration measure of climate-related events' impact with weekly increments. Table 1 indicates our measure generally correlates well with expected loss proxies such as total affected population and economic cost.[3]

Unlike incidence, measuring the annual conditional likelihood of an extreme event, SDF captures the share of time a country spends in such events; it thus proxies for long-term risk. The convex incidence property implies that the total SDF measure also exceeds its certainty-equivalent component. The resulting gap, henceforth the *uncertainty wedge* component of SDF, widens in the mean and persistence of CO_2 growth (fundamental uncertainty) and the elasticity of incidence to the CO_2 emissions process (parameter uncertainty). Conversely, the wedge narrows if either type of uncertainty is scaled down.

There are two main findings. First, for the baseline parameterization with urban population share as the single control variable, the mean certainty-equivalent SDF for extreme events lasting up to 1 year ($T_1 = 1$) is $\sim$ 6.2%. The corresponding total SDF range is $[8.6, 12.2]\%$ of time spent in extreme events depending on estimation method. The uncertainty wedge baseline range then is $[2.4, 6]\%$. Hence, fundamental (CO_2 emissions-driven) uncertainty may contribute up to one half of total long-term disaster risk. These comparisons also relate to the total annual incidence rate which exceeds its corresponding certainty-equivalent component at each average duration. Beyond 1 year, total SDF grows non-linearly with event average duration, underscoring the urgency of rapid decarbonization policies implemented at global scale.

Second, we explore the baseline's sensitivity to CO_2 emission trajectories in the 21st century, modeled based on the IPCC's Representative Concentration Pathways (RCP), [22,37]. The most ambitious mitigation scenario—RCP1.9, assuming successful decarbonization reaches peak radiative forcing CO_2 growth by 2100 and achieves net zero emissions by 2050—would yield a sharp decline in total SDF: $\sim$ $[5, 10]\%$ depending on the estimation method. Moreover, the uncertainty wedge collapses to nearly zero under RCP1.9, but remains $\sim$ 5%

[3] Average duration also depends on countries' geography, infrastructure and other socioeconomic drivers. For example, from 1960–2022 average disaster duration for non-OECD countries was much higher than the OECD counterpart. This not only reflects differences in natural environments but also the lack of risk management capacity in the Global South, reflected in the United Nations' Disaster Risk Reduction progress score [39]. In this paper we abstract from such determinants and focus on the basic model calibration.

under RCP8.5, the worst-case trajectory. Delaying mitigation and relying on adaptation— unchanged average CO_2 growth coupled with a sharp reduction in emissions unconditional variability—would imply SDF $\sim 6\%$, comparable to certainty-equivalent SDF under the baseline. Reducing parameter uncertainty by specifying lower elasticities in the logit transfer function would result in large SDF declines only under the most adverse scenarios, baseline and RCP8.5. A caveat is that these estimates refer to extreme events lasting about 1 year, the finest resolution for the annual frequency EMDAT data.[4]

These findings may inform thinking about disaster risk reduction along UNDRR's Sendai Framework Priorities [*1: Understanding disaster risk*].[5] While in principle higher (lower) uncertainty implies that expected disaster costs—however proxied—will grow (decline), the results suggest the value of differentiating between fundamental and parameter uncertainty. On the one hand, our simulations indicate drastic scaling down of carbon emissions (consistent with RCP1.9 and RCP3.4) could lower SDF by between 6 and 10% (against RCP8.5), with a parallel smaller reduction in the uncertainty wedge. On the other hand, weaker logit transmission—effectively, attributing less extreme event risk to climate change drivers—narrows the uncertainty wedge but the certainty equivalent SDF is roughly unchanged. The tentative policy implication is that lower average CO_2 growth has more long term impact than adaptation policies targeting less variable emissions while maintaining their mean.

A growing number of geophysical and meteorological studies model the time-dependence of climate-related processes using Markov chains.[6] They primarily focus on forecasting the short term likelihood of extreme weather events. [19] shares our econometric approach to predicting incidence, applying cointegration methods, and they also control for socio-economic drivers; see also [23]. However they focus on short term (annual) incidence and limit their data to hydrometeorological subset of the EMDAT database. In a companion paper [20] we calibrate short-term (1-year-ahead) global disaster incidence based on 1960-2022 data using our methodology.

The main innovation of our approach is directly obtaining extreme weather events' long term frequency from their time-varying incidence. Research on

[4] Our results assume accurate logit elasticity estimates and reliable vulnerability data, proxied here by urban population share. Estimating small, or insignificant elasticities may also suggest that employing historical averages as "best guesses" of extreme events' frequency underestimates their true long-run frequency.

[5] See UNDRR's mid-term review of the Sendai Framework for Disaster Risk Reduction from 2015–2030, [38].

[6] [8] employ a non-homogeneous Markov chain similar to ours to forecast U.S. tornado activity but do not model its stationary probabilities. Research has studied Bayesian updating of wind speeds ([10]), temperature and precipitation extremes ([32]) and heatwaves ([34]) with first- and higher-order homogeneous Markov processes to compute short-term co-exceedance probabilities for alternative extreme value distribution functions, but not modeled stationary probabilities. An exception is [40], who compute stationary probabilities for wet and dry day occurrences in South Korea from 1960–2014.

World Weather Attribution (WWA) has facilitated identifying the contribution of climate change to the likelihood of extreme events close to real time [6,27]. By contrast, our non-homogeneous Markov chain model only maps carbon emissions to stationary probabilities via the logit function's parameters estimated over the period between 1960 and 2022. Thus, it may be thought of as a long run complement to WWA. Our approach allows some cautious optimism insofar as a lower long run mean *and/or* variance of CO_2 emissions have the potential to limit SDF. In terms of missing risks [29], our approach aims to connect short-term weather-related impacts to long-term climate ones.

In what follows, disaster events and average duration are defined in Sect. 2 and catalogued in Appendix A. The simulation methods and sensitivity analysis to IPCC RCP are discussed in Appendices B and C. Section 3 estimates the panel logit model of disaster incidence. Countries' urban population share of a country is used to control for vulnerability, although vulnerability to extreme events is complex and endogenous [9,28]. Section 4.1 calibrates the model and produces baseline SDF estimates. Section 4.2 considers the baseline's sensitivity; and Sect. 5 concludes.

We interchangeably refer to disaster and extreme event risk, and short- and long-term disaster risk denote the 1-year-ahead conditional incidence and SDF measures, respectively.

2 Extreme Weather Event Incidence and Duration

We employ the CRED/EMDAT dataset[7] to measure the conditional incidence rate (annual likelihood) and duration of extreme weather events. The analysis is restricted to climatological (draughts and wildfires), hydrological (rainfall-based flooding) and meteorological (extreme heat and coldwaves, storms, hurricanes and tropical cyclones) events occurring from 1959–2022[8]. Geophysical (earthquakes and volcanism) and epidemiological (virus-based) events such as the COVID-19 pandemic are excluded (See Appendix A).

The incidence measure determines the conditional probability of observing an extreme weather event in a country during a given period of time. It therefore records the number of extreme events, while the incidence rate refers to the ratio of the total number of events scaled by the duration in each exposed country. The annual incidence rate is calculated using:

$$\underline{p}^{year} = \frac{Annual\, total\, duration}{Number\, of\, countries \times 52\, weeks} \tag{2.1}$$

[7] The dataset has been widely used to study the socio-economic impacts of natural disasters. See [5,13,16,24].

[8] Continued global warming is certain to increase the frequency of extreme weather, including longer droughts and more powerful storms, and contribute to rising sea levels. In addition, [5,39] document the on-going coincidence of mounting disaster risk and the risk of collapse in biospheres and ecosystems, reinforcing vulnerability and spiralling human losses.

The average incidence rate is 0.062 with a standard deviation of 0.046. Its time series and empirical histogram are shown in Fig. 2:

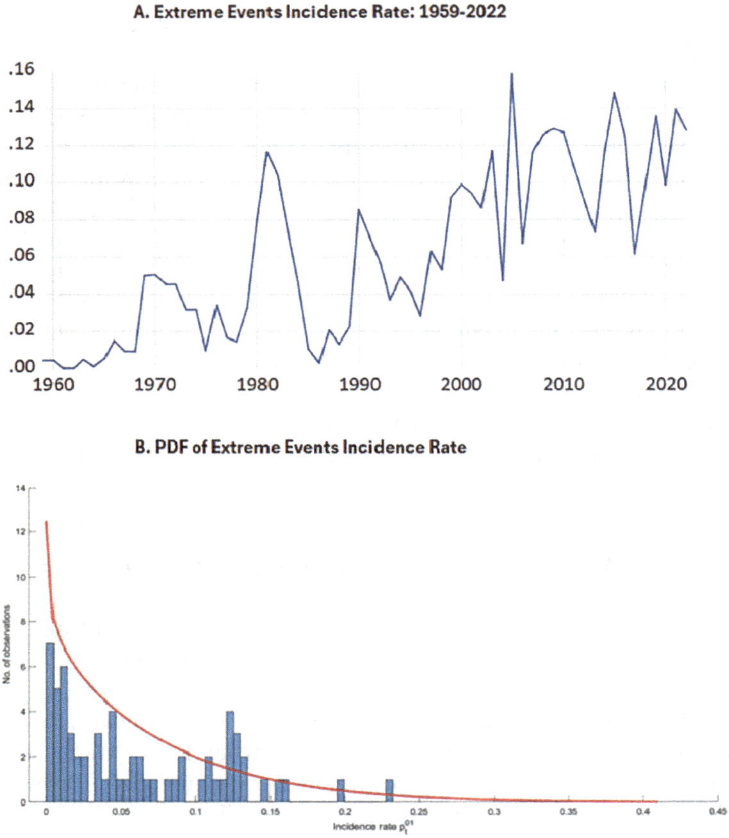

Fig. 2. Extreme Weather Event Incidence Rates and PDF. Source: CRED/EMDAT and author calculation.

Extreme Weather Event Incidence Rates: 1959–2022

Regarding event duration, EMDAT records the starting and ending date for each entry. We have normalised the duration using number of weeks. For example, 1959–0023-CAN refers to a storm that took place in Canada on 16 February 1959. Its starting and ending dates are the same in the database, so it will be recorded as 0.14 weeks in the analysis. Spanning the 1959–2022 sample period, the average extreme event duration per country-year is estimated at 3.12 weeks.

Extreme event duration varies widely across countries. The average duration for OECD countries is 2.68 weeks while for the rest of the world it is 3.54 weeks, consistent with the geography hypothesis in the economic development literature, [31], and the inequity issue that the most vulnerable world regions have contributed least to carbon emissions, [6, 26, 28]. Further, extreme weather events' average duration has risen sharply from 0.52 to over 6 weeks from the 1960s to the 2020s (See Table 1). In Sect. 4.1, we set baseline duration to one year, the shortest duration consistent with EMDAT dataset's annual resolution.

3 Logit Specification for Extreme Event Incidence

The stylized persistence in carbon emissions since 1960 evident from Fig. 1 may be represented by a stationary autoregressive process:

$$c_t = \rho_0 + \sum_{j=1}^{P} \rho_j \; c_{t-j} + \xi_t \ , \ \Sigma_{j=1}^{P} \mid \rho_j \mid < 1 \tag{3.1}$$

where $c_t \stackrel{\text{def}}{=} \log(C_t - C_{t-1})$ is the log of year-on-year CO_2 emission growth; C_t is the emissions level in year t; and $\xi_t \sim N(0, \sigma_\xi)$ is Gaussian iid noise. ARMA-ML estimates of $\{\rho_j\}_{j=0}^{P}$ are reported in Table 2:

Table 2. CO_2 Emission Growth: AR(p) Regressions 1960–2022

Estimation method	Growth rate of mean	Mean growth rate	Mean growth rate
Parameter	AR(2)	AR(2)	AR(3)
ρ_0	0.446***	1.605***	0.6150***
	(0.000)	(0.000)	(0.0000)
ρ_1	0.361**	0.354***	0.2598**
	(0.010)	(0.006)	(0.037)
ρ_2	0.113	0.312**	0.2149
	(0.419)	(0.035)	(0.184)
ρ_3	–	–	0.3212**
	–	–	(0.018)
Adj. obs.	63	64	64
Mean c_t (μ_c)	0.447	1.624	1.624
S.D. c_t (σ_c)	0.163	0.680	0.680
AIC/SBIC	$-0.88/-0.74$	1.78/1.91	1.71/1.88
DW	2.07	2.17	2.12
JB^{res} (p-val)	(0.31)	(0.47)	(0.24)
Adj. R^2	0.14	0.30	0.36
Regr. S.E. (σ_ξ)	0.151	0.568	0.544

Note: The estimated AR(p) model is in Eq. (3.1). p-values given in parentheses. Optimal lag order determined using Akaike (AIC) and Schwartz (SBIC) Information Criteria.

Data: Annual CO_2 growth rates in percent, U.S. National Oceanic and Atmospheric Administration (https://www.esrl.noaa.gov/gmd/ccgg/trends/).

In the preferred AR(3) estimation, both 1- and 3-year lags (the AR(1) and AR(3) coefficients) are statistically significant at the 5% level. The AR(2) coefficient's significance is sensitive to model specification and the CO_2 growth data. The fitted residuals pass the joint Gaussian pdf test (JB^{res}); this will be relevant in Sect. 4.2. The 1960–2022 pdf of CO_2 growth is shown in Fig. 3:

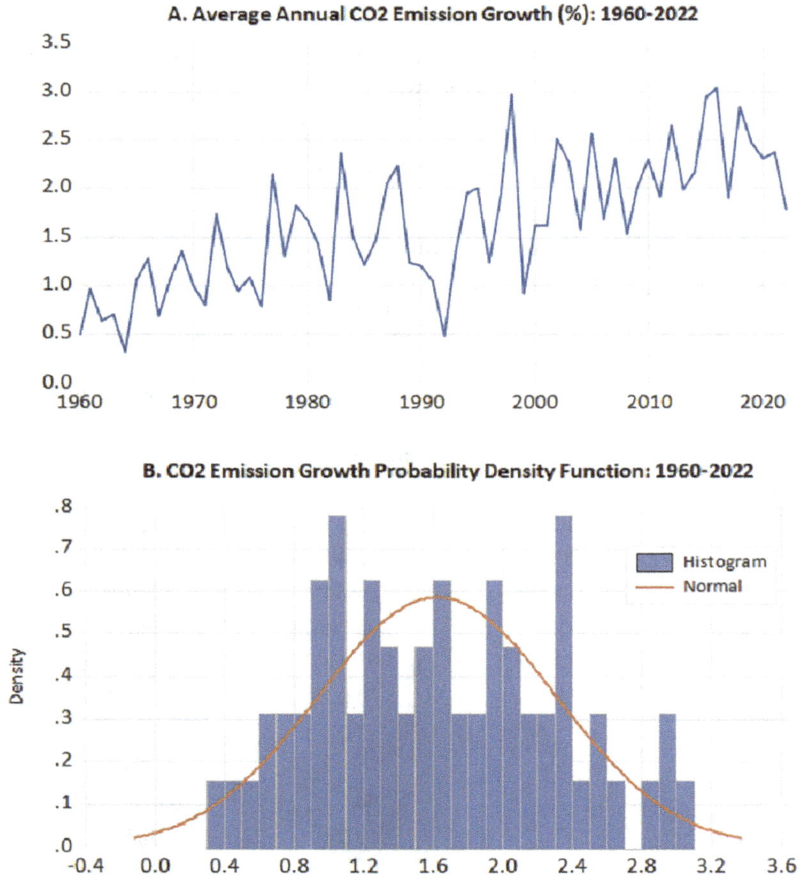

Fig. 3. CO_2 Emission Growth and Probability Density Function: 1960–2022. Source: Created by author. Sample moments: Mean 1.6245 [1.4769, 1.7999]; St.dev. 0.6797, [0.6036, 0.7809]; Skewness 0.1911 [−0.1490, 0.6049]; Kurtosis 2.1821 [1.8120, 2.7048]. 95% bootstrap confidence bands for 1000 replications.

CO_2 Emission Growth Probability Density Function: 1960–2022
Extreme event incidence in year $t+1$, p_{t+1}, is the conditional probability of a disaster event (state $s_{t+1} = 1$) occurring in year $t+1$ given no disaster in year t (state $s_t = 0$). Let p_{t+1} be a logistic function of log CO_2 growth in year $t-j$, c_{t-j}, with $j = 0, ...P$, and constant unconditional moments $E[c_{t-j}] = \mu_c$ and

$var[c_{t-j}] = \sigma_c^2$, all j by stationarity, and $\mathbf{X}_t = [X_{1t}, ..., X_{Kt}]$ be a $K \times 1$ vector of other relevant year-t fundamentals. Specifically:

$$\begin{aligned} p_{t+1} &\stackrel{\text{def}}{=} p(c_t, \mathbf{X}_t) = \Pr\{s_{t+1} = 1 | s_t = 0, X_{kt}\} \\ &= \frac{\exp(\overline{h} + \sum_{j=0}^{P} h_j c_{t-j} + \sum_{k=1}^{K} h_k X_{kt})}{1 + \exp(\overline{h} + \sum_{j=0}^{P} h_j c_{t-j} + \sum_{k=1}^{K} h_k X_{kt})} \end{aligned} \tag{3.2}$$

where $h_j \geq 0$ measures the elasticity of p_{t+1} to CO_2 growth at lag j, h_k is the elasticity of p_{t+1} to fundamental $k = 1, ..., K$, and $\overline{h}$ is a scaling constant such that $0 \leq p_{t+1} \leq 1$. Evaluating p_{t+1} at μ_c and $E[\mathbf{X}_t]$ yields the certainty-equivalent incidence rate, denoted $\underline{p}$:

$$\underline{p} \stackrel{\text{def}}{=} p(\mu_c, E[\mathbf{X}_t]) = \frac{\exp(\overline{h} + \sum_{j=0}^{P} h_j \mu_c + \sum_{k=1}^{K} h_k \mu_k)}{1 + \exp(\overline{h} + \sum_{j=0}^{P} h_j \mu_c + + \sum_{k=1}^{K} h_k \mu_k)} \tag{3.3}$$

Historical average CO_2 growth ($\mu_c = 1.624\%$ p.a. from Table 2) directly impacts $\underline{p}$ while variability ($\sigma_c = 0.68\%$ p.a.) does not. Further, because $p(\cdot)$ is strictly convex left of its unique inflexion point, the expected incidence rate exceeds certainty-equivalent incidence:[9]

$$UW_p \stackrel{\text{def}}{=} E[p_{t+1}] - \underline{p} > 0 \tag{3.4}$$

Taking logs of Eq. (3.1) and introducing a composite error term ϵ yields the multivariate logit transfer function:

$$logit(p_{it+1}) = \overline{h} + \sum_{j=0}^{P} h_j c_{it-j} + h_k \log \mathbf{X_{it}} + \epsilon_{it} \tag{3.5}$$

where $\text{logit}(p) = \log(p/(1-p))$ is the log of the odds ratio and h_j correspond to the elasticities in Eq. (3.3).[10] Our panel dataset covers all countries and territories detailed in the EMDAT database; c_{it} is the CO_2 growth rate reported by the National Ocean Atmospheric Administration (NOAA) between 1960 and 2022; vector $\mathbf{X}$ contains all control variables other than CO_2 growth; and ϵ_{it} is assumed to satisfy the standard assumptions. The logit parameter estimates are reported in Table 3:

In the absence of fixed effects (FE) the logit-implied certainty-equivalent incidence rate, $\underline{p}$ in *row 8* far exceeds the historical disaster likelihoods, $\hat{p}$ from Table 1B. By contrast, when 1- and 2-way FE are included (*columns 4–7*) the model is calibrated accurately: plugging the corresponding $\{h_{j=0}^3, h_{urban}, h_{y_{t-1}}\}$ estimates into Eq. (3.3) implies $\underline{p} \to \hat{p} \approx 0.062$. For the 1-way FE specification,

[9] This property follows from Jensen's inequality, [12,36]. A closed-form approximation for $E[p_{t+1}] - \underline{p} > 0$ is available under a Gaussian pdf.

[10] The intercept and standard error of Eq. (3.5) are calibrated as $\overline{h} = (1-\rho_1-\rho_2-\rho_3)\mu_c$ and $\hat{\sigma}_\epsilon$ employing the AR(3) estimates from Table 2.

Table 3. "Extreme Event Incidence: Logistic Regressions 1960-2022"

Specification	No FE		1-way FE		2-way FE	
Control	None	Urban	None	Urban	None	Urban
$\overline{h}$	-3.607^{***}	-6.005^{***}	-5.2	-9.6	-5.7	-6.4
$\hat{h}_0$ (lag 0)	0.497^{***}	0.485^{***}	0.504^{***}	0.456^{***}	-0.279	-1.693^{*}
$\hat{h}_1$ (lag 1)	0.347^{***}	0.305^{***}	0.360^{***}	0.277^{***}	0.616	1.200^{***}
$\hat{h}_2$ (lag 2)	0.423^{***}	0.358^{***}	0.432^{***}	0.326^{***}	1.242^{***}	1.719^{***}
$\hat{h}_3$ (lag 3)	0.297^{***}	0.272^{***}	0.306^{***}	0.245^{**}	0.603^{**}	1.033^{**}
h_{urban}	–	0.747^{***}	–	1.276^{***}	–	0.204
$h_{y_{t-1}}$	0.282^{***}	0.067^{***}	0.382^{***}	0.296^{***}	0.175^{***}	0.121^{**}
$Cert.Eq.\underline{p}$	26.2	29.8	7.1	7.1	7.0	7.4
$\hat{p}$	6.2	6.2	6.2	6.2	6.2	6.2
Pseudo R^2	0.128	0.156	0.130	0.144	0.170	0.180
AUROC			0.8616	0.8599	0.8750	0.8722
Obs.	13,298	12,000	13,115	11,820	13,115	11,820

Note: Columns 2-3, 4-5, and 6-7 report maximum likelihood estimates for the logit regression in Eq. (3.5) estimated for 1960-2022 without fixed effects and with one- and two-way fixed effects, respectively. ***, **, * denote statistical significance at 1, 5, and 10 percent levels, respectively. The estimated coefficients are displayed to 3 d.p. Country-specific (δ_i^c) and time-fixed (δ_t^g) dummy coefficient estimates are not reported to save space; they are available upon request. Certainty-equivalent incidence $\underline{p}$ is calibrated to $\hat{p} \simeq 6.2\%$ by choice of $\overline{h}$.
Data: EM-DAT is the CRED international hazard and disaster events inventory: www.emdat.be/. The urban population share is available online: data.worldbank.org.

contemporaneous CO_2 growth and its first 3 lags are all significantly positive and the coefficient on urban population share exceeds one (*column 5*). When time-fixed effects are introduced, lags 1 to 3 remain significant at 99% confidence while urban population share is insignificant (*column 7*). The fact that elasticities tend to be highly significant across estimation methods indicates a clear long-run attribution of extreme events to anthropogenic carbon emissions, in line with [6,7,27]. The time-fixed effects can help capture the potential contributions from unspecified, time-specific events such as the Kyoto Protocol of 1997 and the Paris Agreement in 2015. To decide between 1 and 2-way fixed effects, the distribution of predicted incidence rates $\tilde{p}$ against the actual moments is displayed in Fig. 4 and reported in Table 4.[11] The bootstrap confidence intervals

[11] The logit model of [33] had a long time span but only 14 countries, so they made explicit that time dummies (2-way FE) could not be included. They also reported that only lag 2 was statistically significant. In Appendix B we describe simulating stochastic incidence calibrated to historical CO_2 growth; see also [20].

correspond to logistic and Gaussian CO_2 shocks calibrated to 1960–2022 carbon emissions:

Table 4. Simulated Incidence Rates 97.5 Bootstrap C.I.

CO_2 *growth*	Gaussian $\sigma_\xi = 0.54$	Logistic	*Gaussian* $\sigma_\xi = 0.25$
Pseudo $\tilde{p}$	2000	2000	2000
Mean	$[0.107, 0.121]$	$[0.235, 0.266]$	$[0.067, 0.074]$
St.Dev.	$[0.148, 0.166]$	$[0.328, 0.347]$	$[0.0480, 0.0549]$
Skewness	$[2.212, 2.444]$	$[1.055, 1.257]$	$[1.834, 2.317]$
Kurtosis	$[7.418, 9.557]$	$[2.531, 3.052]$	$[7.848, 11.118]$

Data: EM-DAT is the CRED international hazard and disaster events inventory: www.emdat.be/. The urban population share is available online: data.worldbank.org.

4 Computing Stationary Disaster Frequencies

Consistent with Eq. (3.2), let $\{Z_t\}_{t\geq 0}$ be a discrete 2-state Markov chain (MC) such that $Z_t = 0$ in normal times ($s_t = 0$) and $Z_t = 1$ in extreme event state ($s_t = 1$). Homogeneous MC have constant transition probabilities $p^{ij} \stackrel{\text{def}}{=} Pr\{s_{t+1} = j|s_t = i\}$ with $p^{ii} + p^{ij} = 1$ and $0 < p^{ij} < 1$, $i, j \in \{0, 1\}$. Defining 1-year-ahead conditional incidence and exit rates as $p \stackrel{\text{def}}{=} p^{01}$ and $q \stackrel{\text{def}}{=} p^{10}$, respectively, the Markov transition matrix is $\mathbf{M} = \begin{pmatrix} 1-p & p \\ q & 1-q \end{pmatrix}$. The stationary distribution of $\mathbf{M}$ is 1×2 vector $\underline{\pi} = [\pi_0, \pi_1]$ solving $\underline{\pi} = \mathbf{M}\underline{\pi}$ [17]:

$$\lim_{T\to\infty} [\prod_{t=1}^{T} \mathbf{M}] = \begin{pmatrix} \pi_0 & \pi_1 \\ \pi_0 & \pi_1 \end{pmatrix} \tag{4.1}$$

$$\pi_1 = \frac{p}{q+p} \in (0,1) \quad , \pi_0 = 1 - \pi_1 \tag{4.2}$$

Evaluating Eq. (4.2) at $\underline{p}$ from Eq. (3.3) yields the certainty-equivalent stationary frequency of extreme events, henceforth *SDF*:

$$\pi_1^{CE} = \frac{\underline{p}}{q+\underline{p}} \quad , \quad \pi_0^{CE} = 1 - \pi_1^{CE} \tag{4.3}$$

Against this benchmark, following [17,36] in this paper we model non-homogeneous MC with time-varying incidence. The Markov transition matrix then becomes $\boldsymbol{\mathcal{M}}_{t+1} = \begin{pmatrix} 1-p_{t+1} & p_{t+1} \\ q & 1-q \end{pmatrix}$, where p_{t+1} follows Eq. (3.1) and $0 \leq q \leq 1$ is fixed. The stationary distribution of $\mathcal{M}_t$ is given by the ∞-step-ahead transition:

$$\lim_{T\to\infty}[\prod_{t=1}^{T}\boldsymbol{\mathcal{M}}_t] \stackrel{\text{def}}{=} [\pi_0^\infty, \pi_1^\infty] \tag{4.4}$$

Equalizing the expected probabilities of entering and exiting each state, $\pi_1^\infty q = (1-\pi_1^\infty)E[p_{t+1}]$, yields the stochastic SDF analog to Eq. (4.3):

$$\pi_1^\infty = \frac{E[p_{t+1}]}{q+E[p_{t+1}]} \quad , \quad \pi_0^\infty = 1-\pi_1^\infty \tag{4.5}$$

$0 \leq \pi_1^\infty \leq 1$ measures the equilibrium share of time countries spend in extreme weather events. We note that π_1^∞ increases smoothly in average incidence $E[p_{t+1}]$ and widens at a nonlinear rate with average duration ($T_1 = q^{-1}$). Intuitively, with longer-lasting extreme events countries spend more time in state 1. Non-homogeneous MC converge slowly to the equilibrium distribution; in our case we focus on the finite sample properties of the relevant time horizons. The resulting uncertainty wedge, denoted UW_π, is the gap between the SDF measures in eqs. (4.5) and (4.3):

$$UW_\pi \stackrel{\text{def}}{=} \pi_1^\infty - \pi_1^{CE} > 0 \tag{4.6}$$

The wedge is always positive as it is a strict monotone transform of the incidence wedge in Eq. (3.4). We turn to compute π_1^{CE} and sample statistics of π_1^∞ and UW_π for climate-related disaster events.

4.1 SDF Baseline Calibration

Certainty-equivalent incidence $\underline{p}$ is evaluated from Eq. (3.3), and the resulting SDF $\pi_1^{CE}(q)$ follows from Eq. (4.3) at each $q \geq 1$. The stochastic SDF measure $\tilde{\pi}_1^\infty$ is driven by the mean, variance and persistence of CO_2 emission growth, respectively μ_c, σ_c, $\{\rho_j\}_{j=0}^3$ from Table 2; logit parameter estimates $\overline{h}$, $\{h_j\}_{j=0}^3$, h_{urban} and $h_{y_{t-1}}$ from Table 3; control variable mean μ_{urban}; and exit rate q as a free parameter. The methodology for simulating $\tilde{\pi}_1^\infty$ is in Appendix B.

The left, center and right columns in Fig. 4 below display simulated SDF schedules for the $T_{2100} = 87$y, $T_{2300} = 287$y and $T_{2500} = 487$y time horizons based on the 2-way FE parameter estimates.[12] The certainty-equivalent π_1^{CE} (*stochastic* $\tilde{\pi}_1^\infty$) schedules are shown in black (*red*). The top panels zoom in on extreme events lasting about a year ($0.9 \leq q \leq 1$) while the middle and bottom panels display $\tilde{\pi}_1^\infty$ and $UW_{\tilde{\pi}_1^\infty}$ (in blue) for each run over the whole duration range $0.01 \leq q \leq 1$:

[12] The corresponding schedules based on 1-way FE parameter estimates show a slight downward shift in all measures driven by the leftward incidence shift (See Fig. 4). They are available upon request.

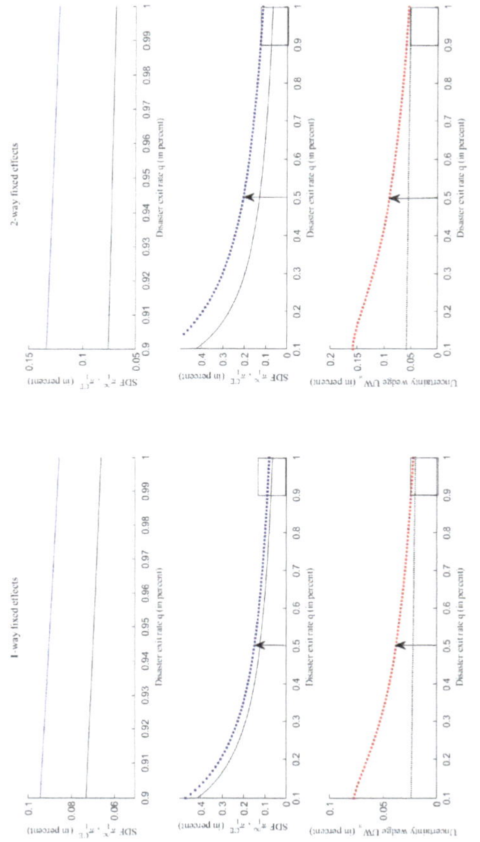

Fig. 4. Baseline Stationary Disaster Frequencies. Source: Created by author

Baseline Stationary Disaster Frequncies

We make three points. First, the range of stochastic $\tilde{\pi}_1^{\infty}$ schedules generate a band pattern across average durations. All else equal, the band narrows as the MC converges in distribution to π_1^{∞}, its unique equilibrium distribution in Eq. (4.5). Hence, the band narrows with the horizon across panels.

Second, from top panels the SDF schedules appear nearly linear in q: at the 2100 horizon, the stochastic SDF band is $\pi_1^{\infty} \sim [10, 12]\%$. Put differently, the long-term disaster frequency for the minimum average duration consistent with post-1960 EMDAT data is $\sim$ 8-10 events per century for the "average country".[13] However, such inference is misleading: the middle panels indicate that SDF rises sharply with average duration, reflecting the hyperbolic dependence of π_1^{CE} and

[13] The quotation marks serve to acknowledge that we have assumed constant urban population share, and that averaging is highly unsatisfactory given the great disparities in countries' climate change exposure, persisting along the North-South divide despite the drive toward inter-connectedness in recent decades; see [6,25].

$\tilde{\pi}_1^\infty$ on q in eqs. (4.3) and (4.5). For example, if disaster events last 2 years on average ($q = 0.5$) then $\tilde{\pi}_1^\infty \simeq 16\%$ of time.

Third, as π_1^{CE} is fixed, the uncertainty wedge inherits the stochastic properties of π_1^∞. Thus, for $0 \leq q \leq 1$ the wedge is $UW_\pi \simeq 6\%$. For extreme events lasting 2 years on average, UW_π grows to $\sim 9\%$.

4.2 Sensitivity to Representative Concentration Pathways

We assess the baseline calibration's robustness to 3 alternative emission scenarios specified in IPCC Representative Concentration Pathways (RCP) [21,27,37] (See Appendix C). A sufficient indicator of less anthropogenic interference with climate-driven extreme events is a less convex logistic function (3.5). This amounts to smaller elasticity estimates $\overline{h}$, $\{h_j\}_{j=0}^3$, indicating less extreme event incidence associated with carbon emissions after controlling for countries' urban population share. The schedules plotted in Fig. 5 are for extreme events lasting between 1-2 years ($0.5 \leq q \leq 1$) and refer to RCP8.5, the most adverse climate scenario for the 2100 horizon:[14]

Average annual CO_2 growth under RCP8.5 is $\mu_c = 1.54$, somewhat below the 1960–2022 baseline, while historical emissions variance is assumed fixed at $\sigma_c = 0.68$ (shock variance $\sigma_\xi = 0.54$). On account of lower μ_c, certainty-equivalent SDF π_1^{CE} is $\sim [5, 10]\%$, below its baseline range $\sim [7, 10]\%$ (Fig. 4). To highlight the uncertainty wedge contribution to SDF we maintain μ_c at RCP8.5 but lower σ_c to 0.16, estimated in Table 2, column 2. Note the certainty-equivalent SDF schedule is unchanged at each q, but $UW_\pi \to 0$ as $\tilde{\pi}_1^\infty$ converges towards π_1^{CE}. Thus, less volatile carbon emissions contributes to lower long-run frequency of extreme events *independently* of decarbonization policies targeting μ_c.[15]

Panel D (*lower right*) contrast the above extreme pathways with RCP3.4 and RCP6, two intermediate trajectories consistent with average annual CO_2 growth $\mu_c = 0.34$ and 0.91%, respectively, in the 21st century. Additionally, we assume global emissions *also* involve less variable CO_2 growth. In particular, we halve historical σ_c to 0.34% p.a.[16] Following RCP3.4 (blue dotted line), UW_π would shrink to $\sim 0.5\%$ and total SDF drop to $\pi_1^\infty \simeq 1\%$ at $T_1 = 1$ year. Following RCP6, the total SDF (red dotted line) would be $\sim 4\%$ at $T_1 = 2$ years, while $\pi_1^\infty(q = 1) \simeq 2\%$ of time. Therefore, similar to Panel C, targeting more stable emissions contributes to lower long term extreme event risk independently of any parallel mitigation policies.

[14] Under this worst-case climate change scenario, global atmospheric CO_2 concentration would continue to rise throughout the 21st century, peaking at $\sim$ 1370ppm by 2100, [37].

[15] This case can be viewed as the reverse of a mean-preserving spread common in financial risk management, [12]. Equivalently, less variable emissions could result from avoiding boom-and-bust business cycles, [33].

[16] Less persistent carbon emissions could equivalently be generated by smaller absolute AR(p) coefficients in Table 2. To be sure, our scenarios obscure the key role that emission pathways play in determining climate outcomes; see for example the debate on early vs. late action for achieving net zero CO_2 emissions, [35].

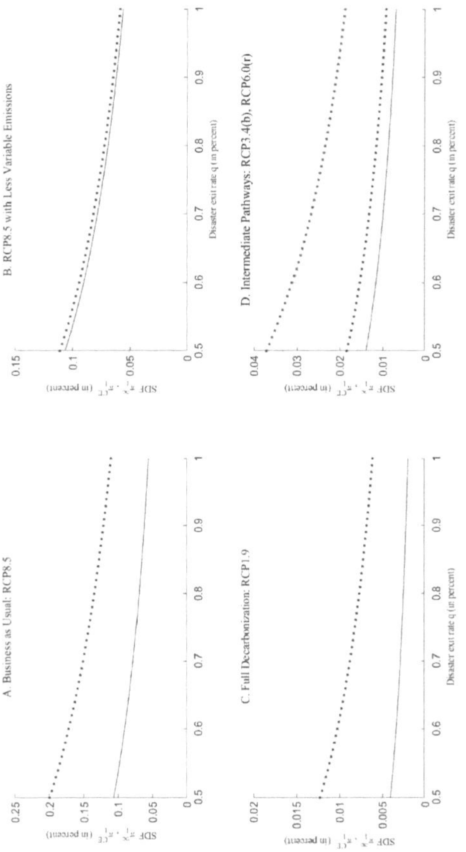

Fig. 5. SDF Sensitivity to Selected IPCC RCP. Source: Created by author

In the previous exercise, each RCP trajectory featured different unconditional mean and variance of CO_2 growth based on their end-horizon CO_2 concentrations. It is also possible that logit elasticity parameters $\{h_j\}_{j=0}^{3}$ are lower—reflecting more effective adaptation policies—or higher, indicating adaptation limits or maladaptation. To illustrate the SDF implications of such parametric—rather than fundamental CO_2— uncertainty, in Table 5 we compare the incidence and SDF magnitudes implied by 1- and 2-way fixed effect parameter estimates for the same 4 RCP scenarios. Average duration is set to 1 year:

For baseline and RCP8.5, the 2-way FE higher elasticity estimates imply ~ 4.5 (~ 3)% higher incidence (SDF) than their lower 1-way FE counterparts. Hence, a relatively more convex logistic transfer appears to generate in excess of 40% more short- and long-term extreme event risk. For the remaining 3 pathways, however, parameter uncertainty does not have a substantial impact ($< 1\%$) on the implied incidence or SDF measures. Therefore, weaker transmission from carbon emissions to incidence impacts SDF less than targeting lower μ_c, *i.e.*

Table 5. Short and Long-Term Disaster Risk ($T_1 = 1$): Parameter Uncertainty

Horizon	Annual				SDF			
Estimation	1-way FE		2-way FE		1-way FE		2-way FE	
Risk measure	$\underline{p}$	$\tilde{p}$	$\underline{p}$	$\tilde{p}$	π_1^{CE}	π_1^{∞}	π_1^{CE}	π_1^{∞}
Baseline	*7.1*	*9.4*	*7.4*	*13.9*	*6.6*	*8.6*	*6.9*	*12.2*
$\Delta[\tilde{p}]$	—	2.3	—	2.5	—	—	—	—
RCP8.5	6.3	8.4	6.1	12.1	6.0	7.7	5.8	10.8
RCP8.5 $\sigma_c = 0.16$	6.3	6.3	6.1	6.4	6.0	6.0	5.8	6.0
RCP1.9	0.9	1.3	0.2	0.7	0.9	1.3	0.2	0.6
RCP3.4	1.9	2.0	0.7	0.9	1.8	1.9	0.7	0.9
RCP6	2.9	3.1	1.5	2.0	2.8	3.0	1.5	2.0

Note: All entries are in percent, rounded to 1 d.p. Disaster average duration is set to 1 year throughout. The uncertainty wedge for extreme event incidence (UW_p in Eq. (3.4)) is the difference between columns 2 and 3 (4 and 5) consistent with 1-way (2-way) fixed effects panel logit estimation controlling for urban population share. The corresponding SDF wedge (UW_π in Eq. (4.6)) is the difference between columns 6 and 7 (8 and 9). SDF π_1^{∞} is computed as the average stochastic disaster frequency over 50 simulation runs of $1,000$ years each. $\Delta[\tilde{p}]$ is the cumulative impact on predicted incidence of 1% higher CO_2 growth in percent p.a. It is computed as $\Delta[\tilde{p}] = \sum_{i=0}^{3}\left(\frac{e^{\hat{h}_i}}{1+e^{\hat{h}_i}}\right)$ where the $\hat{h}_i$, $i = 1, ..., 3$ are in Table 3.

aggressive decarbonization policies. This tentative implication is supported by *row 5*, showing the predicted incidence decline from 1% lower CO_2 growth ($\Delta[\tilde{p}]$) exceeds 2%. Against historical incidence of $\sim$ 6.2%, rapid decarbonization pays short run dividends, as emphasized in [21].

5 Concluding Remarks

Our research aligns closely with the Sendai Framework's first priority for action: understanding disaster risk. We showed that modeling short-term incidence as a convex function of global CO_2 emissions drives a wedge between the stochastic and certainty-equivalent stationary (multi-decadal to $>$ 100 year long) disaster frequencies. This uncertainty wedge widens in the elasticity of extreme events' incidence to underlying anthropogenic drivers. We estimated the model and calibrated a non-homogeneous Markov chain to the global disaster and CO_2 records from 1960–2022. After drawing baseline SDF and UW from time-varying short-term incidence, sensitivity analysis facilitated evaluating different IPCC RCP scenarios' 2100 horizon risk reduction potential. We found that while parameter uncertainty (*UC2*) about logit transfer elasticities contributes to the SDF wedge, the bigger contribution comes from fundamental, or model (*UC3*) uncertainty. The possibility of non-Gaussian CO_2 growth shocks fits within trajectory uncertainty (*UC4*) and is a subject for future research.

In addition, the low-dimensional Markov model may be applied to assess the long-term extreme event prospects of different world regions. The predicted SDF of different RCP may also be an input to evaluating the long run fiscal burden of extreme event risk. Overall, our framework represents a response to policy calls to promote long-termism in order to break the *tragedy of the horizon* [2,4] and avert the prospect of abrupt and irreversible climate change.

Acknowledgments. We thank Xiyu Jiao, Glen Peters, Donald Robertson, and Susan Solomon for helpful comments, and Nigel Stone for excellent research assistance.

Appendix A. Data Definitions and Sources

Disaster Categories
The two main CRED/EMDAT hazard types are hydrometeorological (precipitation events, such as floods and related landslides, storms and heatwaves and extreme cold events) and climatological (draughts and wildfires). We focus on these categories of extreme events by country and year and exclude geophysical (earthquakes and volcanism) and biological events (virus outbreaks and epidemics). EMDAT has recently expanded its coverage to report extra-tropical cyclones and atmospheric river extreme events.

Population and Vulnerability
Population vulnerability to extreme weather events is proxied by the urban share of countries' population (coefficient h_{urban} in logit specification (3.5)). The data is from the World Bank and they are collected and smoothed by United Nations Population Division.

Greenhouse Gas Concentration Levels (in ppm)
Atmospheric carbon dioxide emissions data, including annual mean levels and growth rates, are measured, documented and updated by the US National Oceanic and Atmospheric Administration (NOAA). The CO_2 contribution is an underestimate of the growth of total greenhouse gases; it excludes atmospheric methane (CH_4), nitrous oxide (N_2O) and sulfur hexafluoride (SF_6). NOAA began providing monthly CH_4 and N_2O concentration measurements in 1983 and 2001, respectively.

Appendix B. Model Calibration and Simulation

Logit Specification
The logit intercept $\overline{h}$ is computed for the baseline specification from Table 3. Without fixed effects, $\overline{h}$ is estimated in columns 2–3. With fixed effects, it is calibrated such that the resulting certainty-equivalent disaster incidence in Eq. (3.3) matches historical annual disaster incidence in Table 1B: $\underline{p} \to \hat{p} \approx [0.062 - 0.093]$ per annum. The $\overline{h}$ and matching $\underline{p}$ are shown in italics in Table 3, columns 4–5 (6–7) for the coefficient estimates using 1(2)-way fixed effects.

Disaster Risk Projections
The stochastic SDF measure π_1^∞ in Eq. (4.5) is computed as follows. With 2023 as start year, for time horizon $hor = 2100$, 2300 and 2500, $T_{2100} = 87$, $T_{2300} = 287$ and $T_{2500} = 487$ Gaussian i.i.d. CO_2 growth shocks are drawn from $N(0, \sigma_\xi)$, where σ_ξ is the AR(p) regression std.error reported in Table 2. Fed to the AR(p) coefficients in Eq. (3.1), each shock sequence produces a $T_{hor} - p$ persistent (non-i.i.d.) CO_2 growth path. $S = 100$ such shock sequences (*runs*) are generated and their average moments calibrated to μ_c and σ_c (Sect. 4.1) or specific moments (Sect. 4.2).

The artificial $T_{hor} \times S$ CO_2 years are fed to Eq. (3.2). Assuming the control variables are fixed to their historical means $\{\mu_k\}_{k=1}^K$, this produces S stochastic incidence vectors $\{p_s\}_{s=1}^S$ for each exit rate increment $q \in (0, 1)$. While Eq. (3.3) generates one certainty-equivalent incidence $\underline{p}$ for each parameter constellation. Jointly with fixed exit rate q, they form a $T_{hor} \times S$ array of 2×2 Markov transition matrices, $[\mathcal{M}_{t,s}]_{t=1,s=1}^{T,S}$. Post-multiplying $\mathcal{M}_{T,s}$ for each run s and averaging over all S runs determines π_1^∞ by Eq. (4.4). The resulting SDF uncertainty wedge is $\pi_1^\infty - \pi_1^{CE}$, by Eq. (4.6).

90% confidence intervals around point π_1^∞ are also generated from 50 replications of each $T = 1,000$ sample path based on the logit coefficients' 90% confidence intervals. They are tightly spaced as the point estimates are highly significant, and are available upon request. Subject to iid Gaussian shocks to c_t, the quality of calibration is determined by the convergence of the simulated mean and s.d. of CO_2 growth to within $< 1\%$ of their actual (historical) values μ_c and σ_c.

Appendix C. Mean CO_2 Growth Calibration to IPCC RCP

RCPs were introduced to coordinate climate modeling assumptions ahead of the IPCC Fifth Assessment Report (AR5) in 2014. There are 6 RCPs altogether spanning the range of radiative forcing values (in W/m^2), from 1.9 to 8.5 W/m^2, produced by total GHG atmospheric concentrations in the year 2100 found in the climate modeling literature and converted to CO_2 equivalent concentrations by the IPCC. We have selected all except RCP4.5. The pathways are "representative" in that each represents a wider range of scenarios. Each RCP includes a CO_2 concentration level (in parts per million, ppm) reached in 2100.

A range of 5 Shared Socioeconomic Pathways (SSP) is also available and was employed to help produce the IPCC Sixth Assessment Report (AR6) in 2021. However, SSP are not suited for our purposes as they are conditional on specific projected socio-economic outcomes, including future agriculture yields, population dynamics, energy use etc. As we only condition on future CO_2 (GHG) emission paths, we set control variables to their unconditional mean, so RCP is more appropriate.

The end-horizon concentration allows computing an average growth rate when compared with current global atmospheric concentration levels and the remaining number of years. Given present (2023) CO_2 concentration of $C_{2023} \simeq$ 423ppm, an RCP with terminal CO_2 concentration level C_{2100} is consistent with annual average growth rate x, where $\log(CO2_{2023}) + (2100 - 2023) * \log(1 + x) = \log(C_{2100})$. That implies $x \simeq 1.54\%$ p.a. for RCP8.5; 0.91% p.a. for RCP6; 0.56% p.a. for RCP4; 0.34% p.a. for RCP3.4; and 0.19% p.a. for RCP2.6.

References

1. Asian Development Bank. Global increase in climate-related disasters, working paper, November 2015
2. Bank for International Settlements: The Green Swan. Basel, Switzerland (2020)
3. Cantelmo, A., Melina, G., Papageorgiou, C.: Macroeconomic outcomes in disaster-prone countries. International Monetary Fund WP2019/2017, Washington, D.C (2019)
4. Carney, M.: Breaking the tragedy of the horizon – climate change and financial stability, Speech by the Governor of the Bank of England and Chairman of the Financial Stability Board at Lloyd's of London, London, 29 September 2015 (2015). https://www.bis.org/review/r151009a.pdf. Accessed 15 Jan 2023
5. Centre for Research in the Epidemiology of Disasters (CRED): Human cost of disasters. An Overview of the Last Twenty Years, Louvain-la-Neuve, Belgium and UNDRR, Geneva, Switzerland (2020)
6. Clarke, B., Otto, F., Stuart-Smith, R., Harrington, L.: Extreme weather impacts of climate change: an attribution perspective. Environ Res: Clim **1**, 012001 (2022). https://doi.org/10.1088/2752-5295/ac6e7d
7. Collins, M., Knutti, R.: IPCC Fifth Assessment Report, Chapter 12: Long-term Climate Change: Projections, Commitments and Irreversibility (2014)
8. Drton, M., Marzban, C., Guttorp, P., Schaefer, J.T.: A Markov chain model of tornadic activity. Am. Meteorol. Soc. Monthly Weather Rev. **131**, 2941–2953 (2003)
9. Fan, J., Liu, B., Ming, X., Sun, Y., Qin, L.: The amplification effect of unreasonable human behaviors on natural disasters. Humanit. Soc. Sci. Commun. **9**, 322 (2022). https://doi.org/10.1057/s41599-022-01351-w
10. Fawcett, L., Walshaw, D.: Markov chain models for extreme wind speeds. Environmetrics (2006). https://doi.org/10.1002/env.794
11. Gampe, D., et al.: Increasing impact of warm droughts on northern system productivity over recent decades. Nat. Clim. Chang. **11**, 772–779 (2021). https://doi.org/10.1038/s41558-021-01112-8
12. Grinblatt, M., Linnainmaa, J.: Jensen's inequality, parameter uncertainty, and multi-period investment. Rev. Asset. Pricing Stud. **1**, 1–34 (2011)
13. Guha-Sapir, D., D'Aoust, O., Vos, F., Hoyois, P.: The Frequency and Impact of Natural Disasters. In: Guha-Sapir, D., Santos, I. (eds.) The Economic Impacts of Natural Disasters. Oxford University Press, Oxford (2013)
14. International Monetary Fund: The Effect of Weather Shocks on Economic Activity: How Can Low-Income Countries Cope? World Economic Outlook, October 2017, Chapter 3, Washington, D.C (2017)
15. Japan Meteorological Agency: Climate Change Monitoring Report (2017)
16. Jones, R.L., Guha-Sapir, D., Tubeuf, S.: Human and economic impacts of natural disasters: can we trust the global data? Sci. Data **9**, 572 (2022)

17. Kelbert, M., Suhov, Y.: Markov Chains: A Primer in Random Processes and their Applications. Cambridge University Press, Cambridge (2008)
18. Le Quéré, C., et al.: Temporary reduction in daily global CO2 emissions during the COVID-19 forced confinement. Nat. Clim. Chang. **10**, 647–653 (2020)
19. Lopez, R., Thomas, V., Troncoso, P.: Impacts of carbon dioxide emissions on global intense hydrometeorological disasters. Clim. Disast. Dev. J. **4**, 30–49 (2020). https://doi.org/10.18783/cddj.v004.i01.a03
20. Lu, S., Tambakis, D.: Stabilizing CO2 concentration as a channel for global disaster risk mitigation. Nat. Sci. Rep., 29120 (2024). https://www.nature.com/articles/s41598-024-79437-5
21. Masson-Delmotte, V., et al.: Summary for Policymakers. In: Climate Change 2021: The Physical Science Basis. Contribution of Working Group I to the Sixth Assessment Report of the Intergovernmental Panel on Climate Change. Cambridge University Press (2021)
22. Meinshausen, M., et al.: The RCP greenhouse gas concentrations and their extensions from 1765 to 2300. Clim. Change **109**, 213–241 (2011). https://doi.org/10.1007/s10584-011-0156-z
23. Mora, C., Spirandelli, D., Franklin, E., et al.: Broad threat to humanity from cumulative climate hazards intensified by greenhouse gas emissions. Nat. Clim. Chang. **8**, 1062–1071 (2018). https://doi.org/10.1038/s41558-018-0315-6
24. Noy, I.: The macroeconomic consequences of disasters. J. Dev. Econ. **88**, 221–231 (2009)
25. O'Connor, J., et al.: Interconnected Disaster Risks. United Nations University - Institute for Environment and Human Security (UNU-EHS), Bonn, Germany (2021)
26. Pelling, M., Garschagen, M.: Put equity first in climate adaptation. Nature **569**, 327–329 (2019)
27. Philip, S., et al.: A protocol for probabilistic extreme event attribution analyses. Adv. Stat. Climatol. Meteorol. Oceanogr. **6**, 177–203 (2020). https://doi.org/10.5194/ascmo-6-177-2020
28. Raju, E., Boyd, E., Otto, F.: Stop blaming the climate for disasters. Commun. Earth Environ. **3**, 1–2 (2022)
29. Rising, J., Tedesco, M., Piontek, F., Stainforth, D.A.: The missing risks of climate change. Nature **610**, 643–651 (2022). https://doi.org/10.1038/s41586-022-05243-6
30. Rudebusch, G.: Climate change is a source of financial risk. Federal Reserve Bank of San Francisco Econ Lett 2021-03 (2021). https://www.frbsf.org/wp-content/uploads/sites/4/el2021-03.pdf. Accessed 15 Jan 2023
31. Sachs, J.D.: Tropical underdevelopment. In: NBER Working Paper 8119 (2001)
32. Sedlmeier, K., Mieruch, S., Schadler, G., Kottmeier, C.: Compound extremes in a changing climate - a Markov chain approach. Nonlinear Process. Geophys. **23**, 375–390 (2016). https://doi.org/10.5194/npg-23-375-2016
33. Schularick, M., Taylor, A.: Credit booms gone bust: Monetary policy, leverage cycles, and financial crises, 1870–2008. Am. Econ. Rev. **102**(2), 1029–1061 (2012). https://doi.org/10.1257/aer.102.2.1029
34. Shaby, B.A., Reich, B.J., Cooley, D., Kaufman, C.G.: A Markov-switching model for heat waves. Ann. Appl. Stat. **10**, 74–93 (2016). https://doi.org/10.1214/15-AOAS873
35. Sun, T., Ocko, I., Sturcken, E., Hamburg, S.: The path to net zero is critical to climate outcome. Nat. Sci. Rep. **11**, 22173 (2021). https://doi.org/10.1038/s41598-021-01639-y

36. Tambakis, D.: A Markov chain measure of systemic banking crisis frequency. Appl. Econ. Lett. **28**, 1351–1356 (2020). https://doi.org/10.1080/13504851.2020.1817300
37. van Vuuren, D., et al.: The representative concentration pathways: an overview. Clim. Change **109**, 5–31 (2011). https://doi.org/10.1007/s10584-011-0148-z
38. UNDRR: The Report of the Midterm Review of the Implementation of the Sendai Framework for Disaster Risk Reduction 2015–2030. Geneva, Switzerland, UNDRR (2023)
39. United Nations Office for Disaster Risk Reduction (UNDRR) (2022) Global Assessment Report on Disaster Risk Reduction: Our World at Risk: Transforming Governance for a Resilient Future. Switzerland, Geneva (2022)
40. Yoo, C., Lee, J., Ro, Y.: Markov chain decomposition of monthly rainfall into daily rainfall: Evaluation of climate change impact. Adv. Meteorol. (2016). https://doi.org/10.1155/2016/7957490

Initiatives of Green and Sustainable Development – A Case Study of Sardar Patel University and Anand District Community Science Centre, India and Proposed Comprehensive Climate Action Plan

Vibha S. Vaishnav(✉)

Department of Electronics, Anand District Community Science Centre, Sardar Patel University, Vallabh Vidyanagar, India
vaishnavvibha@gmail.com, vaishnavvibha@spuvvn.edu

Abstract. Sardar Patel University stands as a beacon of rural revitalization in post-independent India, exemplifying sustainable education and environmental stewardship through its innovative green initiatives. Pioneering in the nation, this Green University hosts a district-level Community Science Centre, effectively bridging higher education with the community through various science awareness programs. This model serves as a compelling example for universities worldwide, aligning with UNESCO's vision of education for objective learning. By replicating such endeavors, institutions can contribute significantly to mitigating the impacts of climate change and fostering a sustainable future.

In today's world, where pressing challenge of climate change necessitates collective action from all segments of society, the proposed inclusive climate action programme aims at an integral approach and transform the district into a hub of learning society about the climate related issues and the importance of green practices.

Keywords: Climate Action · Green Practice · SDG · HEI · Eco-Friendly practices · Environmental Initiatives

1 Introduction

1.1 Scope and Purpose

Climate change and global warming have emerged as existential crises. As outlined in a United Nations publication, addressing climate change and achieving the Sustainable Development Goals (SDGs) are inextricably intertwined [1]. Synergizing both propels us more swiftly toward creating a safer and more peaceful world. SDG 4 emphasizes the imperative of ensuring inclusive and equitable quality education, thereby implying that lifelong learning for climate action holds solutions to these existential challenges [2]. Higher Education Institutions (HEIs) play pivotal role in this pursuit.

H. Tatano et al. (Eds.): GSRIDRR 2023, DRRGADRIBS, pp. 370–385, 2025.
https://doi.org/10.1007/978-981-96-6609-6_25

1.2 Literature Review

Numerous HEIs are actively engaged in advancing these goals. Serafini et al. conducted a systematic review of HEIs' efforts toward the SDGs, noting an increasing focus on sustainable development knowledge formation within research agendas [3]. AWGNM Abeyrathna (2021) underscores the significance of rethinking university curriculum design to incorporate green education [4]. A survey conducted at a Brazilian University suggests that sustainable dissemination strategies within universities enhance students' understanding of the importance and relevance of SDGs [5]. Alberto Romolini et al. (2021) in the case study of University of Florence concludes that integrating green issues across all dimensions of university activities is essential [6]. Similarly, a study on green practices in Malaysian higher education institutions highlights the challenge of insufficient information on green indicators [7]. While research on Green Human Resource Management (GHRM) and universities' role in cultivating green human resources exists, case studies or promising results with comprehensive approaches remain scarce [8–13].

Sardar Patel University in India stands out as an exceptional HEI, demonstrating decades-long implementation of green practices and fostering green literacy within the community through its unique Community Science Center.

This paper presents a case study of Sardar Patel University and its Community Science Center at Vallabh Vidyanagar, India along with a proposed inclusive climate action program.

2 Methodology

The research methodology employed in this study relies primarily on a literature review within the field. Additionally, it draws upon the findings of a green audit conducted at Sardar Patel University, along with the associated documentation generated from this audit. This approach ensures a robust foundation for the study, integrating both theoretical insights from existing literature and empirical data derived from the university's specific environmental assessment. By combining these two sources of information, the research aims to provide a nuanced understanding of green practices and sustainability initiatives within the context of Sardar Patel University.

3 Observations and Discussion

3.1 Case Study of Green Practices and Education at Sardar Patel University

Sardar Patel University, situated in Vallabh Vidyanagar achieved the distinction of being accredited as the First Green University of Gujarat with a Platinum ranking by Green Mentors, an organization recognized by the United Nations General Assembly (UNGA) for its nature-driven research initiatives. The accreditation was conferred in 2021 following a comprehensive audit, the details of which can be found in the audit report accessible at spuvvn.edu/orbit-cdn/uploads/green_audit/SPU Green Audit Report.pdf. [14, 15].

Establishment of the University

Known as a Green Town in the district of Anand, Vallabh Vidyanagar was established in 1946 with the primary objective of developing a university township, ultimately giving rise to Sardar Patel University (SPU) in 1955, which became India's First Rural Residential University. This accomplishment stemmed from the post-independence rural resurgence through education project envisioned by Sardar Vallabhbhai Patel, the first Deputy Prime Minister of India, and brought to fruition by Engineer Shri Bhailalbhai Patel, the founding Vice-Chancellor of the University. Vallabh Vidyanagar serves as a remarkable example of the transformation of once dacoit-infested barren land into a lush green township with over 80 diverse tree species lining its roadsides [16, 17]. This meticulously planned township, which set a precedent for town planning across post – independent India, boasts perhaps the highest density of trees in the state within a small area. Additionally, the town hosts the largest colony of rose-ringed parakeets, with the latest count exceeding 65,000 [18].

Here are the highlights of the green and sustainable initiatives of Sardar Patel University:

3.1.1 Creation of a New Green Belt

Continuing the legacy of the university's founders, a new green belt has been established at the satellite campus of the Biosciences Department (Fig. 1). According to reports, faculty and students have actively engaged in cultivating a balanced ecosystem, resulting in the documentation of over 450 species of various life forms across approximately 8 acres of land. This thriving habitat includes 90 tree species, more than 70 species each of butterflies and spiders, approximately 90 bird species, and numerous other life forms.[19, 20] The tree planting initiative transcends mere symbolism; rather, it entails ongoing maintenance efforts, including daily care such as watering, integrated seamlessly into the department's informal teaching practices. Students are not only involved in compiling a comprehensive checklist of species but have also contributed over 1000 photographs to a Google Photos album [21, 22] and are actively engaged in the barcoding of trees [23, 24]. This hands-on approach extends beyond traditional classroom and laboratory settings typically associated with botany and zoology education in universities. Notably, there is a remarkable instance of student and staff collaboration in the successful revival of a fallen tree through re-plantation efforts. Additionally, the department maintains a dedicated botanical garden and prioritizes the preservation of special species within its care.

3.1.2 Air Quality of Vallabh Vidyanagar

As shown in Fig. 2, the air quality in Vallabh Vidyanagar (Gujarat, India) (VVN) surpasses that of the surrounding villages/towns, as indicated by the 2023 data. This is the result of the green cover in Vallabh Vidyanagar. The air temperature is also less by 2–3 degree C.

Fig. 1. Satellite campus of Bioscience Department created into a green cover by the faculty and students in a land where there was not a single blade of grass.

3.1.3 Rain Water Harvesting Systems

According to the University Estate Section's report, the University has 07 Rainwater Harvesting/Recharge Wells [25, 26]. The initiative began in 2003 with the construction of the first underground water harvesting tank at the VC Bungalow, capable of storing 10,000 L of rainwater. Subsequently, another 10,000-L underground rainwater harvesting tank was installed at the Community Science Centre (Fig. 3) In 2004, 02 water recharge wells were added to the infrastructure, designed to collect rainwater throughout the monsoon season. These efforts have significantly contributed to maintaining the water table on the university campus, consequently reducing electricity bills. In 2013, a water harvesting recharge well was established near the University Women's Hostel. This recharge well has effectively harvested 250,000 L of water from the hostel's catchment area of approximately 5000 sq.ft.

3.1.4 Rain Natural Air Conditioning system in Sports Complex

The University has implemented an innovative natural air cooling system powered by solar energy in the Sports Complex. Utilizing 6-inch diameter PVC pipelines, the system runs along the Short Wall and is buried beneath the ground at a depth of 10 feet, with the outlet positioned in the Basketball Court area, effectively cooling the space. To facilitate air circulation, turbo ventilators have been installed in the ceiling to draw out hot air, further enhancing the cooling effect.

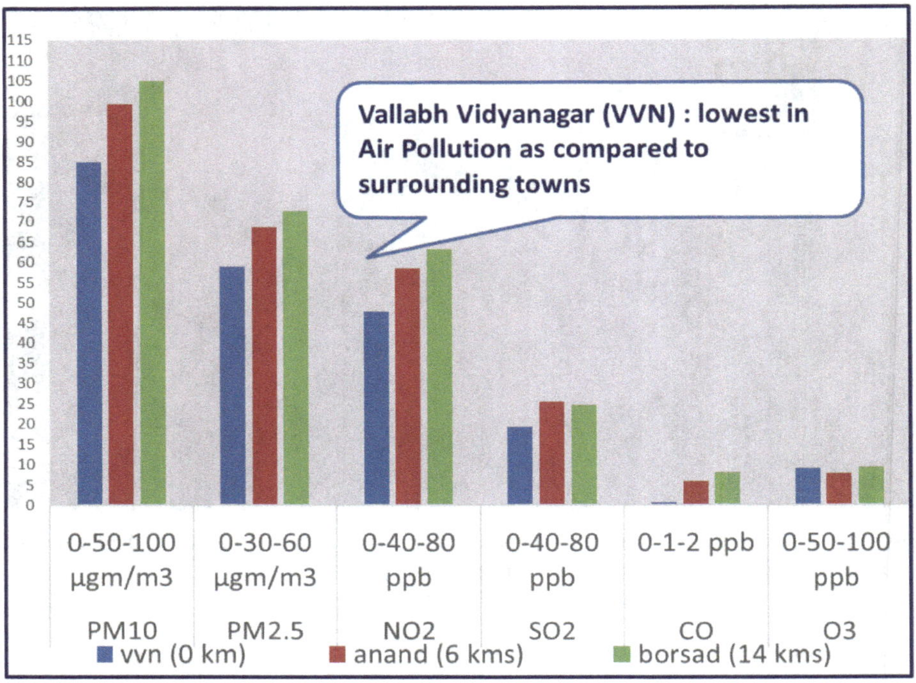

Fig. 2. PM 10, PM 2.5, NO_2, SO_2, CO and O_3 levels in atmosphere in Vallabh Vidyanagar (VVN), Anand and Borsad.

Fig. 3. Construction of Rain water harvesting well of 10,000 L capacity at Community Science Centre.

3.1.5 Wind-Solar Power Generator and Utilization of Energy-Efficient Devices

In a groundbreaking initiative, the University developed a pioneering Small Wind generator in 2004, marking a first-of-its-kind achievement in the country. Additionally, the University led the way in adopting Wind Solar hybrid Inverter systems, showcasing its commitment to sustainable energy solutions. Another notable milestone was the installation of Dual Axis Solar Tracking systems across four buildings, further solidifying the institution's status as an innovator in renewable energy technologies. Other sustainable practices include the use of eco-friendly refrigerants and energy-efficient electrical devices. Furthermore, Roof-Top Solar Power Plants have been successfully integrated into 11 buildings, significantly reducing the University's carbon footprint [24]. Additionally, the adoption of Solar water heating systems in hostels and the use of CNG gas as a fuel source in laboratories underscore the institution's dedication to environmentally conscious practices.

3.1.6 Eco Friendly Buildings

The majority of the buildings adhere to the principles of Open to Sky spaces (OTS) and Universal Design, incorporating anthropometrically appropriate dimensions for learning spaces, ample natural sunlight, and ventilation. Furthermore, some departments have embraced the practice of conducting outdoor classes when weather conditions permit, further enhancing the environmental sustainability of our educational facilities.

3.1.7 Paper Recycling

The University prioritizes digitization for issuing marksheets and other documents, significantly reducing paper consumption. Additionally, SPU diligently ensure the recycling all answer books from exams, a remarkable feat considering the volume of over 60,000 students per semester.

3.1.8 Waste Management

Stringent waste management practices are enforced for chemical, biological, and clinical waste. Moreover, electronic waste (e-waste) is disposed of in accordance with government guidelines to ensure proper handling and environmental protection.

3.1.9 Research, Education, Innovation, and Extension

Sardar Patel University has been at the forefront of addressing environmental crises, making significant contributions through over 300 research publications, numerous research projects, Ph.D. dissertations, and several patents and books over the past decade focused on environmental issues. [27–40] Various departments, including basic and applied sciences, management, economics, business studies, home science, humanities, and social sciences, are actively engaged in research endeavors related to green technology, pollution detection sensors, energy sector studies, rural development, environmental economics, environmental accounting, natural dyes, and other eco-friendly practices. The university offers a range of undergraduate, postgraduate, and Ph.D. programs with courses dedicated to environmental studies. One noteworthy project funded under the Student Startup and Innovation Policy (SSIP) involves the production of compost from bio-waste generated on the university campus, contributing to a zero-leaf-burning policy. Additionally, the agro-economic research center, established with support from the Government of India, promotes water and soil conservation practices and encourages the adoption of renewable energy in agriculture. The museum and the community radio station play active roles in disseminating messages advocating for the preservation, care, and respect of Mother Nature, reaching approximately 70,000 individuals annually.

3.2 Case Study of Community Science Centre

The C C Patel Community Science Centre serves a District Level Community Science Centre (CSC) in Anand District, Gujarat, sponsored by the Gujarat Council on Science and Technology (GUJCOST) and managed by Sardar Patel University. Notably, Sardar Patel University is the FIRST and ONLY public university in the country to oversee a Community Science Centre. Since its inception in 2004, the CSC has been guided by the principle of "Science to People and People towards Science".

Benefitting from the ample resources of its Post Graduate Departments and a cadre of highly qualified university faculty, including both active and retired members, the CSC has been able to conduct its activities effectively. The center's approach to scientific awareness generation and popularization is comprehensive and inclusive, as illustrated in Fig. 4.

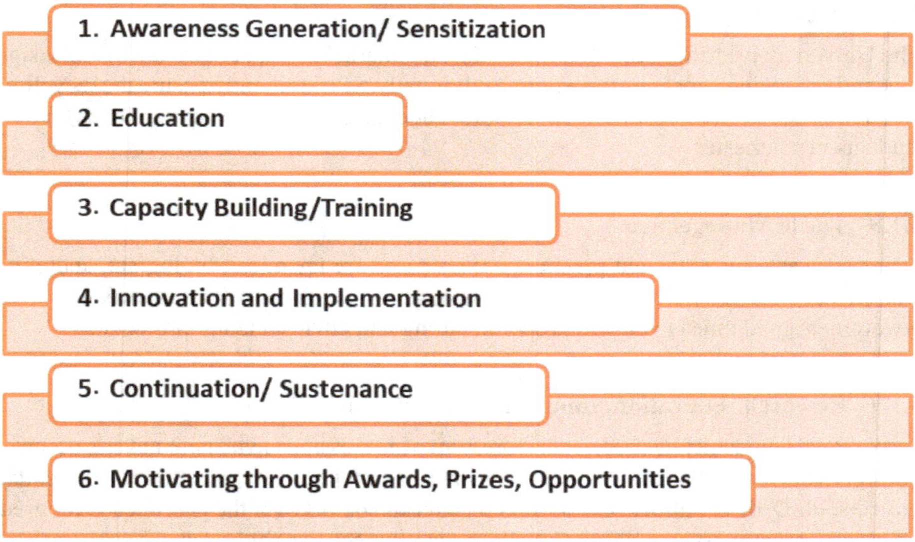

Fig. 4. Comprehensive approach of CSC.

CSC caters to individuals of all ages and backgrounds, offering programs involving all age groups ranging from 10 to 85 years old (Fig. 4). All programmes have participation from these 04 categories of age-groups (Fig. 5).

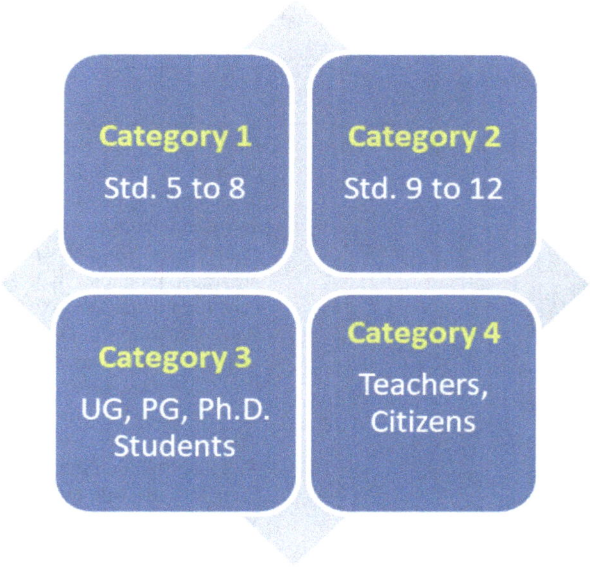

Fig. 5. Categories of participation in CSC activities by different age-groups.

A sample of outreach programmes organised by CSC to tackle environmental issues is depicted in Table 1.

Figure 6 showcases the number of beneficiaries involved in CSC's Climate Action initiatives over the past 12 years. The increasing response and participation serve as indicators of the activities' impact.

The significant contributions of CSCs are depicted in the Table 2.

The CSC's commitment to inclusivity is reflected in its growing participation rates, heightened community awareness, increased student involvement in projects, and enhanced interaction with the local community. Notably, observations indicate that no major environment-related health hazards have been identified among the individuals engaged with the center [42].

3.3 Proposed Comprehensive Action Plan

Both Sardar Patel University and the Community Science Centre (CSC) have actively participated in environmental initiatives, yet a more concentrated effort is needed to educate the entire society.

The proposal is to establish a Center of Excellence for Inclusive Climate Action within the district, as shown in Fig. 6. The objective is to install meteorological units in every sub-district cell with support from National/International Agencies, aiming to achieve "net-zero" emissions district-wide. It is in lines with the Priorities of "Sendai Framework for Disaster Risk Reduction" P1, P2, P3, P4 [43].

Table 1. Details of the activities organized by CSC to tackle environmental issues.

Sr No.	CSC's Outreach initiatives	Details
1.	Celebration of special Internation/ National Days	For over a decade, the CSC Anand has upheld the tradition of commemorating international or national days as a means to raise public awareness about environmental and climate change issues through various educational programs. Examples include World Water Day, Earth Hour Day, World Health Day, World Environment Day, World Ozone Day, Climate Change Day, National Bird Day, Lion Day, Tiger Day, Biodiversity Day, Zoo Fest, Sparrow Day, Wildlife Week, Energy Conservation Day, World Meteorological Day, National Pollution Prevention Day, Mother Earth Day, Bicycle Day, National Nutrition Week, World Meteorological Day, Renewable Energy Day, and more
2.	Talks	by national/state level experts
3.	Seminars	Individuals from all walks of life and age groups are encouraged to participate. The seminars include self-directed learning sessions and interactive events
4.	Training	Some of the topics include (1) butterfly taxonomy (2) hydroponics and vertical farming (3) compost making (4) growing Indoor plants- rich in oxygen (5) use of smokeless stove in villages (6) bird counting (7) nurturing trees (8) Making of bird feeders (9) making of cloth bags to replace plastic bags from unused/discarded cloths (10) Re-using plastic
5.	Exhibits on Biogas, Green House, Solar Water Heater	Approximately 50 schools from Anand District send students to visit the Community Science Centre annually. The Science Hall within the center showcases a range of exhibitive models illustrating how technology can contribute to environmental protection and climate change prevention
6.	Community Meetings	The staff and volunteers of the Community Science Centre regularly visit various villages in the Anand district to conduct demonstrations of energy-saving kits and raise awareness by providing information to the villagers

(continued)

Table 1. (*continued*)

Sr No.	CSC's Outreach initiatives	Details
7.	Demonstration of Working Models	Like use of Solar Energy, Wind Mill, Energy Conservation, Rain Water Harvesting
8.	Radio Programmes	Talks/ Interviews are arranged on University's community radio station
9.	Posters on Social Media	Public awareness programs are conducted by displaying posters on social media
10.	e- publications	brought out under the Enhancing Scientific Awareness program
11.	Festival Celebration	An array of online and offline events are arranged to commemorate festivals in an eco-friendly manner, such as eco-friendly Ganesha-making, Christmas, Diwali, kite-flying, and bird-related activities
12.	Nature Strolling	Nature and Environment Club, CSC Anand organizes various events for Nature strolling during summer camps
13.	Ban on Single use plastic	Throughout the year, a variety of events are arranged. Periodically, prizes for plastic plogging and incentives are introduced. Workshops teaching how to repurpose sarees into bags as eco-friendly alternatives are conducted. Additionally, a Memorandum of Understanding (MoUs) with an industry focusing on plastic reuse is established
14.	Bio-compost	People are guided through a workshop aimed at creating compost from kitchen waste
15.	Awareness generation on Biodiversity	CSC at times hosts activities in collaboration with a prominent local NGO, Nature Conservancy Services (formerly known as Vidyanagar Nature Club) as well as Forestry Division, Anand. The Nature and Environment organizes no. of event events including bird watching, distributing bird houses, decorating, crafting birdfeeders, installing bird water feeders, and at times conducting bird censuses
16.	Nature/Wildlife photography	a regular event under Wildlife Week celebration

(*continued*)

Table 1. (*continued*)

Sr No.	CSC's Outreach initiatives	Details
17.	Research based projects by School Children	Children and teachers are offered support and guidance to engage in the National Children's Science Congress, particularly focusing on research projects related to sustainable development
18.	Film Show/Library Week	Film screenings and library events centered around environmental themes are organized
19.	Disaster Risk Prevention and Diminution	A range of programs addressing common hazards such as floods, droughts, earthquakes, cyclones, fires, pandemics, epidemics, tobacco-related health issues including cancer, pollution, increased usage of gasoline vehicles, plastics, electronic waste disposal, bio-waste management, heatwaves, and climate change are coordinated
20.	Solutions to the problems of community	CSC has developed an innovative solution to provide comfort to animals during heatwaves, which has been officially recognized with a design patent by the Indian Patent Office. [41]

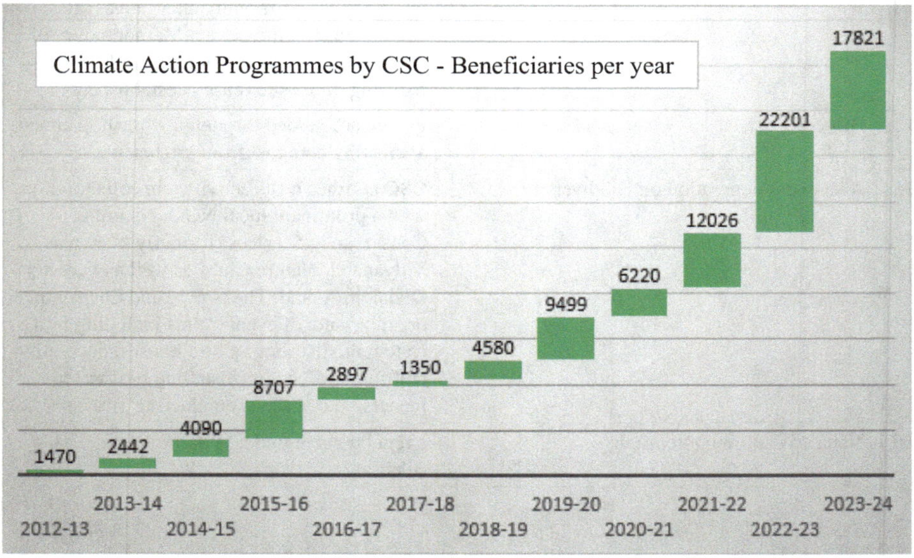

Fig.6. No. of Beneficiaries of Climate Action Initiatives organized by CSC.

Table 2. Significant contributions of CSCs

1.	Bridging the silos of age, social, economic, educational, regional backgrounds, yet providing Science Education to entire Community equitably
2.	Encouraging and Involving University students and faculty to find out and give solutions to the problems of community through science and technology
3.	Educating through demonstration and example the importance of scientific temper, the applications of Mathematics and Green Technology
4.	Providing cognitive, conative & experiential learning when anyone visits the Science centre or the Staff visits schools/villages/other centres/social clubs
5.	Training the Trainers: Education to the B.Ed. Students and young Teachers
6.	Trying to develop interest in Science among children by programmes on Learning Science with Fun
7.	Cultivating Science centre as a Hobby Centre
8.	Accessibility to one and all: Science to Community, Community to Science
9.	Arranging to have an exposure of the top most intellectuals, social workers and outstanding people of the society to the young children
10.	Providing education on the contribution of ancient Indian sages and modern scientists
11.	Developing a sense that a Science Centre is a Universal temple of Learning, showing Spirituality and Science are ONE
12.	A sense of fulfilment and satisfaction among one and all who get associated

The goals include:

(i) adopting a multidisciplinary approach involving all fields of study,
(ii) raising awareness about the crisis to educate the entire community,
(iii) establishing observatories and monitoring systems at various locations,
(iv) assembling a proficient data analysis team to aid researchers,
(v) fostering innovative solutions through research projects,
(vi) promoting the implementation of green practices in the University and the entire district, ultimately fostering behavioral change through Green literacy among all citizens in an inclusive manner.

The action plan involves:

(1) installing Continuous Air Quality Monitoring Systems and Weather Stations, particularly in secondary and higher secondary schools,
(2) establishing a Data Monitoring, Analysis, and Control Room,
(3) establishing a network with the community and schools through the Community Science Centre and Community Radio Station for information dissemination,
(4) setting up a HAM Radio Station for service purposes in case of disasters,
(5) training research students and organizing extension activities for awareness generation programs,
(6) incorporating innovations and implementing modern green technology.

The expected outcome is a halt to further environmental exploitation, encouragement of young minds to innovate suitable solutions, promotion of a healthier way of living, and advancement towards achieving the mission of net-zero emissions.

This approach has the potential to facilitate behavioral change. Additionally, as advocated by the Government of India, adopting the LiFE (Lifestyle for Environment) movement can further solidify behavioral change [44].

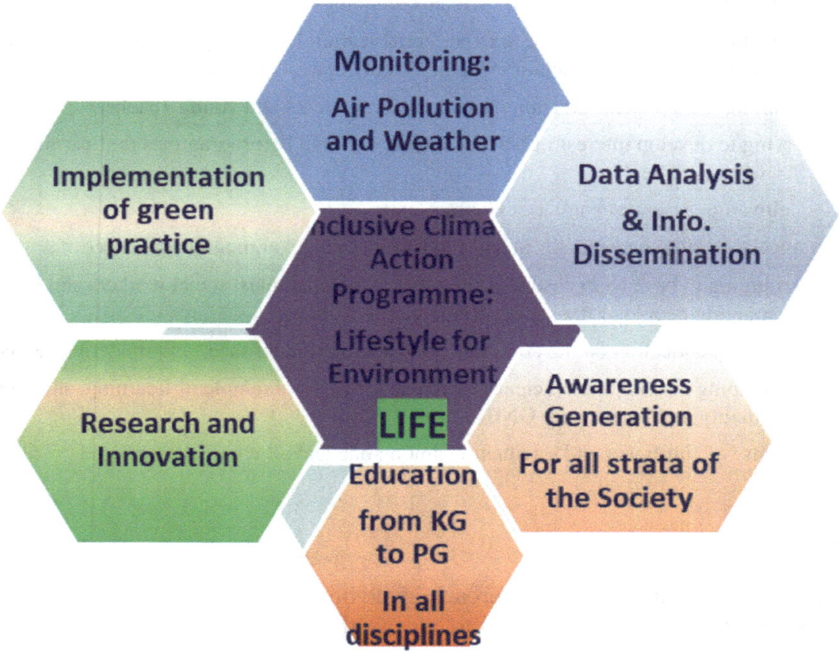

Fig. 7. Comprehensive Action plan for a mass movement in the entire district.

4 Conclusion

Sardar Patel University remains steadfast in its dedication to societal transformation by cultivating a holistic ecosystem and Global Human Resource Management. Furthermore, the university actively promotes awareness and delivers Green Literacy programs and training to fulfil Sustainable Development Goals (SDGs) and mitigate disaster risks through its innovative initiative of Community Science Centre. The proposed comprehensive action plan seeks to encourage broad participation, acknowledging that tackling the climate change crisis demands collective action, widespread awareness, and behavioural shifts nurtured through education.

Message: Climate change is an urgent global challenge that cannot be left solely to specialists or treated merely as a research problem. While innovations in climate monitoring, analysis, technology, and policy are essential, they alone are insufficient. What's crucial is the widespread adoption of UNESCO's "Learning to be" ethos, with a heightened emphasis on "Learning to be eco-friendly." This imperative must be embraced by all Higher Education Institutions (HEIs) and implemented as a matter of urgency. Furthermore, according to the Government of India's recommendations, the LiFE (Lifestyle for Environment) movement could prove to be a significant catalyst for change. This initiative seeks to catalyze a change in consciousness, and promote the adoption of appropriate Green Practices and emerging Sustainable Technologies.

Acknowledgements. The author is thankful to Sardar Patel University for partial financial assistance to present the work at 6th GADRI, Kyoto, Japan. The data taken from the Green Audit report and SSR, NAAC from the SPU's website is acknowledged. Thanks are due to Sardar Patel University Sustainable and Innovative Education and Development Cell and the Department of Electronics. Special acknowledgements are to the Advisor, GUJCOST and members of the Community Science Centre for all support.

References

1. UN Climate SDG Synergies Report-091223B_1.pdf
2. https://sdgs.un.org/goals/goal4
3. Serafini, P.G., de Moura, J.M., de Almeida, M.R., Rezende, J.F.D.: Sustainable development goals in higher education institutions: a systematic review. J. Clean. Prod. **370**, 133473 (2022). https://doi.org/10.1016/j.jclepro.2022.133473
4. Abeyrathna, A.W.G.N.M.: Green Education in a University Classroom: Benefits and Challenges. In: Proceedings of the 7th International Research Conference on Humanities & Social Sciences (IRCHSS) 2021, University of Sri Jayewardenpura, Sri Lanka (2021). https://doi.org/10.2139/ssrn.3809215
5. Ribeiro, J.M.P., et al.: Green campus initiatives as sustainable development dissemination at higher education institutions: students' perceptions. J. Clean. Prod. **312**, 127671 (2021). https://doi.org/10.1016/j.jclepro.2021.127671
6. Fissi, S., Romolini, A., Gori, E., Contri, M.: The path toward a sustainable green university: the case of the University of Florence. J. Clean. Prod. **279**, 123655 (2021). https://doi.org/10.1016/j.jclepro.2020.123655
7. J Anthony Jnr, B.: Green campus paradigms for sustainability attainment in higher education institutions – a comparative study. J. Sci. Technol. Policy Manag. **12**(1), 117–148 (2020). https://doi.org/10.1108/JSTPM-02-2019-0008
8. Zaleniene, I., Pereira, P.: Higher education for sustainability: a global perspective. Geogr. Sustain. **2**, 99–106 (2021). https://doi.org/10.1016/j.geosus.2021.05.001
9. Tairu, Z.O.: Green human resource management—delivering high-performance human resource systems at divine word university papua new guinea. In: Leal Filho, W. (eds.) Handbook of Sustainability Science and Research. World Sustainability Series, pp. 489–511. Springer, Cham (2018). https://doi.org/10.1007/978-3-319-63007-6_30
10. Ali, Q.M., Nisar, Q.A., Abidin, R.Z.u., et al.: Greening the workforce in higher educational institutions: The pursuance of environmental performance. Environ. Sci. Pollut. Res. **30**, 124474–124487 (2023). https://doi.org/10.1007/s11356-022-19888-3
11. Tiyarattanachai, R., Hollmann, N.M.: Green Campus Initiative and Its Impacts on Quality of Life of Stakeholders in Green and Non-Green Campus Universities. Springerplus **5**, 84 (2016). https://doi.org/10.1186/s40064-016-1697-4

12. Abbas, Z., Sarwar, S., Rehman, M.A., Zamecnik, R., Shoaib, M.: Green HRM promotes higher education sustainability: a mediated-moderated analysis. Int. J. Manpower. **43**, 827–843 (2021). https://www.emerald.com/insight/0143-7720.htm
13. Binacardi, A., Colasante, A., D'Adamo, I., Daraio, C., Gastaldi, M., Uricchio, A.F.: Strategies for developing sustainable communities in higher education institutions. Sci. Rep. **13**, 20596 (2023). https://doi.org/10.1038/s41598-023-48021-8
14. https://www.spuvvn.edu/orbit-cdn/uploads/green_audit/SPU%20Green%20Audit%20Report.pdf
15. https://www.youtube.com/watch?v=OYN1Mqcbiq8&t=2s
16. Trivedi, R.M.: Saga of Seven Decades: Journey of Charutar Vidyamandal in Education (1945–2014). Vallabh Vidyanagar: Charutar Vidyamandal (2015)
17. Reddy, A.S.: Trees in and around Vallabh Vidyanagar, in Chai, Buddies and Parrots, Photo-Essay on Vallabh Vidyanagar, Adesara, S. (Ed.) (2020). https://www.youtube.com/watch?v=imBMIN5dvpk
18. https://vncindia.org/project/digital-tree-mapping/
19. Vaishnav, V. : Sardar Patel University contributing to creating Green and Sustainable world through Education, Research, Training and Extension, Activties during the period 2016–2021. https://www.spuvvn.edu/ssr/Criterion-7/7.1/7.1.6/7.1.6-Any_other_relevant_information/1_SPU_Green_initiatives_in_last_5_years.pdf; https://drive.google.com/file/d/1gF0fIXoZEgcwaIahhUbjEfczW0ZosqT5/view?usp=sharing
20. https://www.spuvvn.edu/academics/departments/bio-sciences/campus-diversity/
21. https://www.spuvvn.edu/academics/departments/bio-sciences/campus-diversity/#1626948762132-f3236b25-1964
22. https://photos.google.com/share/AF1QipPiWuyRP8GvIM0VBnK5Hamwg1peRocR81HYQgZKOgNlwlXlpMew_9RaCPcLZiLyVQ?pli=1&key=TXJLV0FxaVJjTjdSZGNyaXBTZ0ZiOWE0S2VwVzZn
23. Patel, J., Trivedi, U., Vasant, R.: Spider Diversity of Garden Ecosystem in Department of Biosciences, Sardar Patel University, Gujarat, India. Munis Entomology & Zoology, 18(Supplement), ISSN 1306–3022. https://www.munisentzool.org/
24. Variya, M.H., Trivedi, U.B.: Observations on Nectar Food Plants of Zizina otis (Fabricius, 1787) from Gujarat, India (Lepidoptera: Lycaenidae). SHILAP Revista de Lepidopterología. **52**(205), 5. eISSN: 2340-4078. ISSN:0300-5267(2024). https://shilap.org/revista/article/view/849
25. https://www.spuvvn.edu/ssr/Criterion-7/7.2/Best_Practice_2/08_Rain_Water_Harvesting_at_SPU.pdf
26. https://www.spuvvn.edu/ssr/Criterion-7/7.2/Best_Practice_1/Best_Practice_Inclusiveness_Freeships,_Scholarship_and_Other_Financial_Aids__to_the_Students,_Academic_and_Administrative.pdf
27. Patel, H., Chapla, D., Shah, A.: Bioconversion of pre-treated sugarcane bagasse using enzymatic and acid followed by enzymatic hydrolysis approaches for bioethanol production. Renew. Energy **109**, 323–331 (2017)
28. Bhatt, B., Prajapati, V., Patel, K., Trivedi, U.: Kitchen waste for economical amylase production using Bacillus amyloliquefaciens KCP2. Biocatal. Agric. Biotechnol. **26**, 101654 (2020)
29. Vasava, D., Kher, M.M., Nataraj, M., da Silva, J.A.T.: Bael Tree (Aegle marmelos (L.) Corrêa): importance, biology, propagation, and future perspectives. Trees **32**, 1165–1198 (2018)
30. Yadav, S., Kalamkar, S.S., Chandna, V., Kaur, U., Rani, U.: Climate change and Indian agriculture in the present scenario. The Horizon – J. Soc. Sci. **1**, 43–58 (2021)
31. Vadwala, Y., Kola, N.: Natural dyes extracted from waste leaves of Terminalia catappa locally known as tropical almond and its application on silk fabrics pretreated with eco-friendly and

non-eco-friendly mordants. Int. J. Res. - Granthaalayah **5**(5), 125–137. https://doi.org/10.29121/granthaalayah.v5.i5.2017.1845
32. Gandhi, J.R., Jha, H., Jha, S.N., Patel, D.S.: Renewable energy-based floating power generator (rivers and canals). Int. J. Eng. Res. Appl. **6**, 49–52 (2016)
33. Bhatt, S., Kalamkar, S.S.: Implications of Solarized Irrigation: Evidence from Gujarat. In Water Governance in India: Issues and Challenges. Allied Publishers Private Limited, New Delhi (2018)
34. Patel, D.M., Vala, R.M., Sharma, M.G., Rajani, D.P., Patel, H.M.: A practical green visit to the functionalized [1,2,4] Triazolo [5,1- B] Quinazolin-8(4 H)one scaffolds using the group-assisted purification (GAP) chemistry and their pharmacological testing. ChemistrySelect **4**, 1031–1041 (2019)
35. Sharma, M.G., Rajani, D.P., Patel, H.M.: Green approach for synthesis of bioactive hantzsch 1,4-dihydropyridine derivatives based on Thiophene moiety via multicomponent reaction. R. Soc. Open Sci. **4**, 170006 (2017)
36. Patel, R.H., Mishra, V.K., Kapatel, P., Kachhia, P., Pokiya, M.: Studies on biodegradable polyether ester urethanes from natural resource and their application in coatings. Mater. Today Proc. **24**, 1376–1382 (2020)
37. Dalvadi, Y., Rao, M.: A study on green baking's contribution in digitalization. Int. J. Appl. Financ. Manage. Perspect. **6**, 35–38 (2017)
38. Gandhi, T.: Green accounting and reporting practices as a part of CSR: a comparative study of selected index-based Indian and Australian companies. In: Ajmeri, S., Vohra, S.M. (Eds.) Emerging Trends in International Business and Commerce. Allied Publication Pvt. Ltd., New Delhi. ISBN 978-81-942903-5-3
39. Talati, J.S., Mishra, M.K.: Natural Mosquitoes Repellent Formulation. Indian Patent Application No. 201821001819, 2 February 2018 (2018)
40. Vaishnav, V.S., Patel, P.D., Patel, N.G.: Indium tin oxide thin film gas sensors for detection of ethanol vapours. Thin Solid Films **490**(1), 94–100 (2005)
41. Sardar Patel University, Marwadi, C., Solanki, S., Vaishnav, V.: Timer-Based Automatic Cooling System for Livestock. Indian Patent No. 404391-001. 7 March 2024 (2024)
42. https://www.youtube.com/watch?v=k2mxPg2z7p4&t=23s
43. https://www.undrr.org/implementing-sendai-framework/what-sendai-framework
44. https://missionlife-moefcc.nic.in/

The Importance of System Thinking in Achieving Desirable Housing

A. Shahmohammadian[1(✉)] and M. Ghafory-Ashtiany[2]

[1] Department of Civil Engineering, Science and Research Branch, Islamic Azad University, Tehran, Iran
a.shahmohammadian@srbiau.ac.ir
[2] International Institute of Earthquake Engineering and Seismology (IIEES), Tehran, Iran
ashtiany@iiees.ac.ir

Abstract. Housing is one of the most basic needs of communities. Residential buildings constitute the highest percentage of all structures in communities, evident at state, country, and local scales. Economically, a home is generally a household's largest financial investment. From a resilience perspective, housing damage and collapse account for a significant portion of disaster losses. Reports indicate that access to desirable housing has been one of the biggest challenges for governments in many developing countries. These governments have struggled to solve housing problems due to their one-dimensional, linear, and traditional view of the issue. Desirable housing, in general, is defined by its safety against natural and human-made hazards (such as earthquakes, floods, and hurricanes), affordability, and acceptable quality. These criteria provide comfort and convenience for tenants and ensure the safety, resilience, and sustainable growth and development of society. Achieving these goals involves many parameters and variables. Desirable housing encompasses social, political, economic, legal, and technical and engineering aspects, all of which are interconnected. A change in one aspect can affect others. Furthermore, desirable housing involves numerous stakeholders, and differing goals among these stakeholders can hinder the main objective of the system (achieving desirable housing). Given the variable parameters influencing desirable housing and the existence of stakeholders with diverse goals, desirable housing is a complex phenomenon. Therefore, this article aims to take a comprehensive look at housing issues in developing countries using a system-thinking approach. It examines the importance of system thinking in achieving desirable housing and considers complexity theory within the housing system.

Keywords: Housing · System Thinking · Complexity · Interconnection

1 Introduction

Housing is one of the fundamental needs of people, households, and human societies. Adequate housing (affordable, safe, and sustainable) enables families to prosper and contributes to sustainable community development. Nowadays, providing appropriate and desirable housing is one of the major challenges for both people and governments

H. Tatano et al. (Eds.): GSRIDRR 2023, DRRGADRIBS, pp. 386–400, 2025.
https://doi.org/10.1007/978-981-96-6609-6_26

in many developing countries. Many proposed policies for achieving desirable housing have not been successful, leaving everyone dissatisfied with the current conditions. One of the main reasons for the failure of housing policies in developing countries can be attributed to the one-dimensional, traditional, and segmented approach to housing issues. A simplistic consideration to the complex issue of housing has led many cities to face increased urbanization rates, rising housing prices, environmental and human health damages, unsuitable social conditions, Unsafe housing against natural disasters, reduced access to natural resources and urban services, and ultimately decreased all kinds of citizen welfare and sustainable development. A deep and systemic view of housing issues can provide solutions to address these problems in such countries. This perspective offers deep and fundamental solutions instead of insincere and temporary painkillers to achieve desirable housing and sustainable urban development. This article attempts to highlight the complexities of the housing system and the motives for using system thinking in housing governance, demonstrating the advantages of using system thinking concepts for framing the understanding of the housing area.

System thinking offers a set of beneficial concepts, mental models, and ways of thinking that can help structure housing policy and develop strategies. To track this chain of reasoning, the article includes: a) Presentation of the main ideas related to systemic thinking with an emphasis on complexity-based perspectives, b) Explanation of the reasons for utilizing system thinking in housing, and c) How to apply the systemic model in the housing area.

There are some key Points which has been mentioned as follows:

1. System thinking focuses on the interactions and links between elements of a system that lead to changes in the general behavior of the system. Systems analysis has grown into various and numerous approaches and is applied in many policy areas. This set of tools facilitates scientific policymaking in more rational ways.
2. One of the most effective methods for housing policy is using a systemic approach. When policymaking in the housing area is accompanied by this way of thinking, it improves. This approach leads to a deeper understanding of the housing system's behavior and characteristics such as interdependency, feedback loops, emergence, initial conditions, path dependency, and evolutionary and adaptive learning.
3. System thinking principles should not be adopted uncritically. Mental models are simplified, and causal loop diagrams are only initial approximations. Their value can be increased with proper and continuous use of data, indicators, and even simulation models.

2 System Thinking

System thinking is a method that enhances thinking to understand more about the world's issues and complexities and find new opportunities to solve problems. A system is a set or combination of objects or components that form a complex and unified entity, representing its structure and organization. The term organized implies that the components of the system interact with each other through specific relationships, creating unique behavior. The system is coherently organized to move towards achieving a goal. The multiple interactions between system components create features and behaviors that are greater than all of the system's components put together, potentially exhibiting adaptive, dynamic, goal-seeking, and sometimes evolutionary behaviors.

A system consists of three parts: components, interactions, and goals. The components of a system are the easiest part to identify as many are visible and tangible. However, the interactions and links between components (due to their vague and ambiguous nature) are the least considered aspect in understanding the system, leading to incomplete comprehension of issues. The system's goal represents the behavior the system exhibits, not what policymakers claim or promote. The goal of subsystems and their behavior can lead to unpredicted results that are not desirable for the whole system.

The behavior of a system comes from its structure, often surprising policymakers. Every element or component within the system can work both correctly and logically, yet these seemingly generous and logical actions can result in undesirable consequences.

System thinking shifts from partial or limiting views towards a more comprehensive and interconnected understanding of the integrated nature of society. It also modifies the recognition of multiple perspectives on reality. Different stakeholders view a real-world system from various points of view, with different priorities and goals (based on their subjective models). Therefore, a comprehensive and systemic basis for analysis is required.

In system thinking, understanding interconnections, commitment to multiple perspectives, and awareness of boundaries are particularly important. Interconnections are always at the heart of system thinking. Complex patterns of links and interdependence fundamentally form a system's behavior. These interconnections show why linear simplification models can be misleading, making policy based on such models ineffective or reverse. Systemic perspectives emphasize the importance of understanding the diverse perspectives of actors and institutions that form critical elements or nodes of a system. Understanding this diversity and divergence in views can be key to applying system thinking for social variations. Encouraging different stakeholders to adopt systemic thinking can help overcome change barriers, enabling each stakeholder to better understand others' perspectives and collectively support what is needed for change. It is also necessary (and inevitable) to define boundaries and distinguish interactions within a system from external shocks. However, attention must also be paid to where boundaries are drawn, as this is an analytical decision that can shape the nature of the analysis in a continuous system.

2.1 Systems and Complexity

Systems create complexities due to the interactions between their components, which may be ignored in policymaking, leading to policy failure. The main ideas of systems and complexity can be expressed as follows:

Interdependence: Systems may significantly differ, but they all fundamentally involve the interdependence of various elements. This directs decision-makers to focus on the system's interrelationships. Complexity begins when a system cannot be explained simply by breaking it down into its components. In other words, a policy problem should be analyzed not only by its relevant components but also by considering how different components interact with each other. Therefore, multi-disciplinary and interdisciplinary teams are often needed to address the complex nature of systems.

Emergence: A common feature of a complex system is that it displays behavior at the system level that cannot be derived from the knowledge and functioning of its components. Interactions between system components indicate that the system's observed outcomes are the product of complex causality, often beyond the comprehension of policymakers. The rules governing how local nodes and actors interact are crucial in shaping what happens within the system. These processes may produce unpredictable periods of stability and rapid short-term changes, making it challenging for policymakers to control system behavior predictably.

Dynamics: System behavior and relationships change over time, sometimes rapidly and sometimes slowly, often unpredictably. The existence of temporal changes, sometimes accompanied by delays, can produce complex and unpredictable effects on system behavior. Dynamics are not only about short-term and long-term changes but also about recognizing time as a real constraint or obstacle to progress. Implementing policies and strategies to change a system requires time and patience. Dynamic considerations change how systems are analyzed and approached, indicating the inadequacy of short-term static models.

Temporal and Spatial Delays Between Cause and Effect: Another crucial complexity is the temporal and spatial delays between cause and effect in systems, often invisible, leading to incorrect root causes identification by policymakers. The cause of a change in system behavior might have occurred long before the observed change or could result from changes in other subsystems of the overall system.

Initial Conditions: System behavior significantly depends on what has been done before, such as the built environment or the economic history of an area. Thus, systems are path-dependent, and controlled by resources, institutions, and initial conditions. Initial conditions limit and shape the development of a system, also constraining the ability to change the system. Initial conditions indicate the feasibility of adopted policies in terms of their capacity.

Evolution, Learning, and Adaptive Behavior: Dynamics and dependence on initial conditions lead to changes in stakeholder behavior. Actors and institutions adjust their assumptions and behavior based on feedback from processes, results, and performance, exhibiting evolutionary and self-adaptive behavior. This behavior leads to evolving relationships within the system, changing its structure and behavior over time. Evolutionary, learning, and adaptive behaviors add to the complexity of systems.

Tipping Points: Humans often perceive processes as simple and predictable, unaware of many fundamental processes occurring before they appear as problems. When systems experience problems, the occurrence is usually neither simple nor predictable. Like a wall built of blocks, removing one block might initially leave the wall stable, but instability slowly increases until removing one more block causes the wall to collapse. Similarly, when a system reaches a certain instability tipping point, it might collapse or fundamentally change. Identifying tipping points in a system and predicting when they will be reached is challenging, increasing system complexity.

Feedback Loops (Reinforcement - Balancing): Identifying and controlling feedback loops is crucial. Positive feedback loops can amplify initial impacts, leading to nonlinear changes. Such feedback might be socially or economically negative (e.g., a negative shock is reinforced, like when decreased GDP reduces employment, further lowering economic activities, which again reduces GDP). Negative feedback loops reduce the impact of initial stimuli, potentially balancing a system and resisting change.

2.2 Iceberg Model

The Iceberg Model is a method designed to enhance understanding and identify the root causes of issues. The root causes of problems often occur at deeper levels and are less visible and less considered. Typically, when an event happens, everyone focuses on that event, and the media also amplifies this focus. When addressing issues and phenomena, it is always necessary to examine events over time and identify trends from these events, then recognize the structure that has formed these trends. Focusing solely on events prevents attention to the main root causes, which weakens the response to the problem. Identifying the structure behind the trends increases decision-makers' understanding and leads to solutions with minimal unexpected results. This model concerns unintended consequences where a quick solution with good intentions is chosen but targets the events rather than addressing the root causes. Such solutions usually work in the short term but worsen the situation in the long term.

Among policymakers, there is often debate between short-term solutions targeting symptoms and long-term root solutions. Sometimes, even when the problems and root solutions are known, short-term solutions are adopted due to a lack of motivation and resources, while fundamental solutions, despite being more effective in resolving the issue, are avoided due to delayed results. Short-term solutions have a side effect of reducing the motivation for implementing fundamental solutions. Not applying fundamental solutions, in turn, exacerbates the initial roots of problems and worsens conditions in the long run. An example can be seen in government budget deficits and the creation of money.

3 Housing System (Housing as a System)

Housing is more than just a shelter; it should be affordable, safe, sustainable, and provide comfort and well-being for households. Using a system thinking approach offers a new perspective on the housing system. This interdisciplinary approach encourages data-driven decision-making and a focus on long-term sustainability. By adopting policies based on a systemic approach, the housing system can become more inclusive, sustainable, and resilient. This approach provides a powerful framework for understanding the complex dynamics of housing and its impact on communities, emphasizing that housing is not an isolated entity. Adhering to a comprehensive systemic approach integrates economic, social, environmental, cultural, natural resource, engineering, and technical considerations, ensuring a holistic and multi-level perspective on housing.

In developing countries, the term "housing system" is often used interchangeably with the housing sector or even the housing market. Housing is highly segmented, indicating a lack of comprehensive and multi-level understanding of housing's root problems. Housing comprises various sub-sectors that differ not only in quality, safety, economy, and sustainability but also geographically and regionally. A simplistic view of the complex housing system causes problems within this system. This complexity makes describing and understanding issues difficult and complicates responses, mobilizing knowledge and subsequent actions. One reason for this complexity is the interdependencies, connections, and links between housing system components and its dependence on the environment, education, social security, economy, and other subsystems.

3.1 Why Housing is Considered as a Complex System

1. **Multiple Dimensions and Subsystems**: The housing system includes various economic, social, technical, cultural, environmental, natural resource, governance, and urban planning dimensions.
2. **Dynamic Parameters and Indicators**: Each dimension or subsystem has dynamic parameters and indicators whose characteristics and behaviors change over time.
3. **Interdependencies**: Each influencing parameter or factor has direct and indirect interactions and interdependencies, meaning changes in one parameter or behavior can lead to changes in other dimensions, resulting in unpredictable conditions.
4. **Diverse Stakeholders**: Numerous stakeholders with different viewpoints, perspectives, and goals play roles in the housing system. The diversity of stakeholder goals and the lack of coordination in their views contribute to the complexity of providing desirable housing solutions.

Communities should develop a comprehensive, systematic, and evidence-based approach to understand the housing system's performance, focusing on the complex and dynamic linkages and dependencies among its different subsystems (Fig. 1).

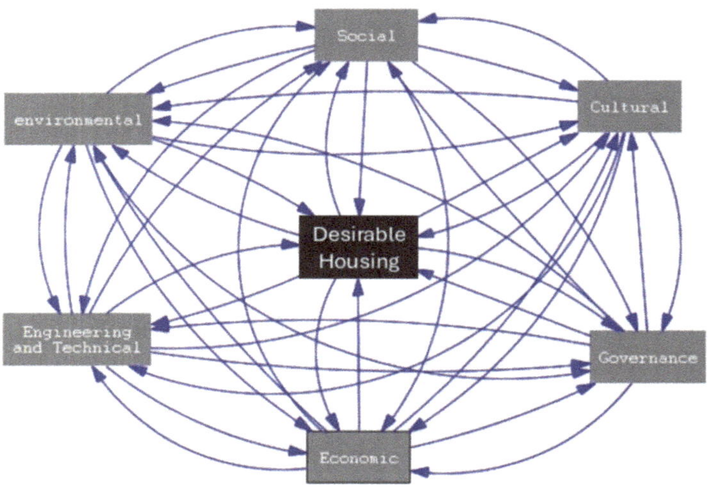

Fig. 1. Desirable Housing Subsystems and Interactions

All interventions and policies must be appropriate and derived from the housing system to be effective and efficient. Figure 2 illustrates a simplified model depicting the parameters and interactions that influence the desirable housing system. System thinking improves decision-makers' understanding of housing processes by:

- **Clarifying interdependencies**: Highlighting the interactions and dependencies between system nodes, components, and elements.
- **Unanticipated Policy Effects**: Identifying unexpected outcomes due to the complex relationships between multiple aspects and subsystems, making the consequences of interventions complicated to predict.
- **Temporal Effects and System Dynamics**: Policies operate in real-time and incorporate delays, meaning that ignoring delays can result in different behaviors than anticipated.
- **Evolution, Learning, and Adaptive Behavior**: Recognizing that the dynamic housing system evolves, learns from feedback, revises plans, and adapts, which is essential for decision-makers' behavior.
- **Feedback Loops**: Emphasizing the importance of feedback loops to provide a clear picture of future outcomes based on policies before implementation.

3.2 Aspects to Consider in Housing Policy

1. Economic Aspect

Housing is a significant driver of the economy and economic growth in many developing countries, contributing substantially to gross domestic product (GDP) and related industries. Housing production fosters economic growth and reduces unemployment. The economic dimension of housing includes affordability, market dynamics, and housing finance. A systemic approach recognizes the impact of macroeconomics, the interplay between supply and demand, housing's broader economic impact, and the need for sustainable financing mechanisms. Studies on demographic changes and income distribution are essential to ensure the supply and demand alignment and housing provision for all income deciles.

2. Social Aspect

Housing is closely linked to social aspects like community cohesion, inclusiveness, and social equity. A system thinking approach emphasizes the need for housing policies that address social equity, strengthen social networks, and enhance social well-being. Considering social dimensions, plans such as housing for all income deciles or social development programs can create inclusive and dynamic communities. This involves examining social demographics, community needs, and access to amenities to promote cohesion and social equity. The social dimension of housing can significantly impact health issues, educational inequalities, social isolation, security and crime, stigma and discrimination, economic damage, mental health effects, and intergenerational impacts. Therefore, housing is vital for households, providing security, stability, and a sense of belonging, affecting physical and mental well-being, education, economic stability, and community participation. Beyond mere shelter, housing plays a crucial role in overall family happiness, quality of life, and social growth.

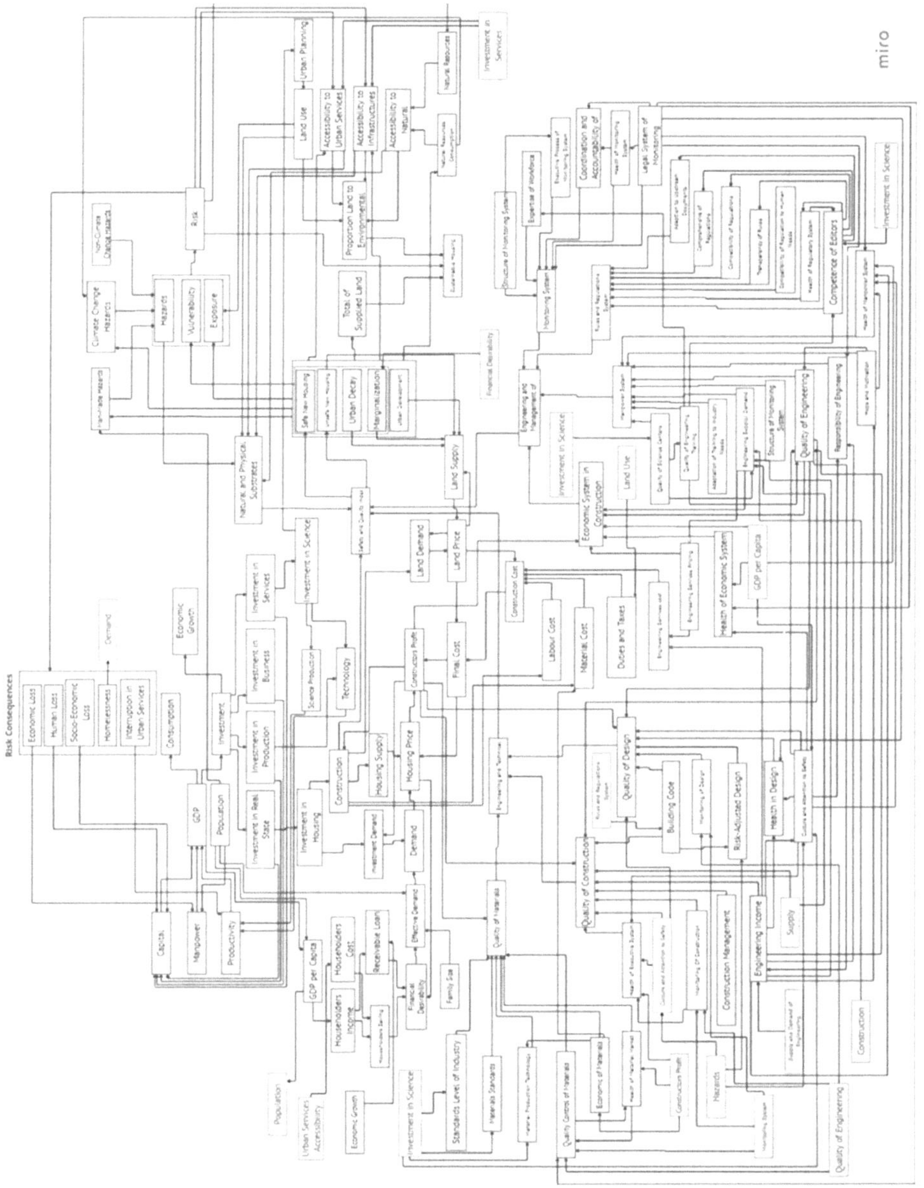

Fig. 2. A simplified model depicting the parameters and interactions that influence the housing system

3. Environmental Aspect

Housing has significant environmental implications, including energy consumption and greenhouse gas emissions. Increased greenhouse gas emissions lead to global warming and climate change, which intensifies the frequency and severity of natural disasters, affecting the housing sector. Additionally, housing development should align with energy supply system capacities to avoid housing and energy access unsustainability for households. Attention to sustainable design, energy-efficient technologies, energy optimization, zero carbon, and integrating renewable resources is crucial to minimize housing's environmental impact.

4. Safety Aspect

The housing system is vulnerable to increased frequency and intensity of natural (floods, earthquakes, etc.) and human-made (fires, subsidence, etc.) disasters. Housing comprises a significant portion of urban capital, exposed to disaster risks. Major economic and human losses in developing countries post-disasters are often due to housing damage. Therefore, incorporating policies and measures to enhance safety, resilience, and disaster risk reduction in housing design, construction, and location considerations is necessary to ensure resident safety and well-being. The cascading effects of economic and human losses from disasters on other housing subsystems, such as the economy and society, are also significant.

5. Natural Resources Aspect

Appropriate housing development decision-making must consider natural resources. Development should be compatible with natural resource capacities, as disproportionate development can cause severe stress on natural resources, leading to a lack of potable water, air pollution, soil security issues, subsidence, urban unsustainability, and increased natural disaster risks. Effects of ignoring natural resources in housing development decisions include deforestation, water scarcity, pollution, waste production in densely populated urban areas, habitat destruction, urban sprawl, loss of agricultural land, and threatened food security.

6. Urban Planning Aspect

Focusing on the residential environment rather than just shelter plays a crucial role in urban planning and development. Housing development planning should integrate factors like transportation, infrastructure, urban services, and public spaces to create dynamic and sustainable communities. Appropriate housing development brings urban services closer to residents, increases walkability, household health, social network formation, and social dynamics. Conversely, disproportionate housing development with urban services increases distance to such services, higher vehicle usage, energy consumption, traffic, pollution, reduced social dynamics, and increased corruption and disorder. Urban development planning emphasizes managing city administration costs, municipalities, and financing methods, where neglect can lead to reduced quality of life, and increased urban corruption and speculative attack.

7. Housing Governance Aspect

Effective governance improves the production and provision processes of desirable housing. Good governance practices, including competence, transparency, accountability, stakeholder participation, participatory decision-making, and alignment of housing policies with social goals, play a significant role in achieving desirable housing. Involving stakeholders in decision-making and ensuring governance transparency makes housing policies responsive to community needs. This includes studying governance structures, policy frameworks, and stakeholder engagement processes to enhance housing governance effectiveness and inclusivity.

To achieve desirable housing, all housing aspects and their complexities and interconnections must be considered. Figure 3 illustrates a simplified schematic of the graph network within the housing system.

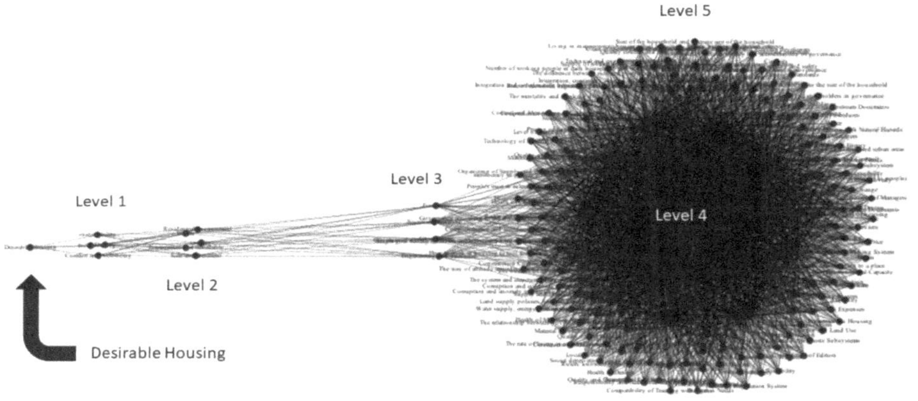

Fig. 3. Graph network of the desirable housing system

A deep understanding of all housing aspects and their interactions, facilitated by approaches to increase understanding and clarify housing issues' complexities, is essential. A System thinking approach enables a comprehensive examination of these dimensions, enhancing the integration of diverse perspectives and solutions. It also emphasizes interdisciplinary collaboration, data-driven decision-making, and a focus on long-term sustainability to achieve desirable housing outcomes. Systemic complexity thinking improves our understanding of housing processes explained as follows:

Interdependence: System thinking focuses on the flows and relationships between the nodes of a system. In addition to identifying influential parameters in different parts of housing, attention should be paid to the connections and prevailing flows between these parameters. Often, implementing a policy to improve one parameter can worsen other parameters in the housing system, ultimately exacerbating the situation. For instance, government-provided housing for low-income groups increases the supply of affordable housing. However, if financing for its production is neglected, it can lead to budget deficits, money creation, and inflation. Inflation disproportionately affects low-income individuals, meaning a policy intended to improve their welfare might end up putting more pressure on them. Another example is the relationship between elements affecting housing prices. An increase in housing prices leads to more housing supply, which subsequently reduces prices. However, lower prices boost demand, driving prices up again.

Emergence: The effects of policies and interventions can be greater than expected due to the emergent properties arising from the system's interconnected components, leading to behaviors and outcomes that are hard to predict. For example, measures to reduce housing prices in cities with high immigration can increase urbanization and migration rates to these cities. This urbanization surge increases housing demand, raising prices again and causing the policy to fail. Additionally, rising housing prices can lead to increased slum development, unconventional housing, and poor living conditions. Such shelters reduce urban safety, resiliency, and security, escalating economic and human losses in developing countries. These economic losses and uneven development affect GDP, and slower economic growth results in reduced investment in housing, lower supply, and thus higher housing prices.

Dynamics: Identifying the roots of housing changes and problems over time is crucial. Addressing housing issues requires examining their historical timeline and pinpointing when the symptoms accelerated. For example, corruption and negligence by engineers concerning safety (especially against natural disasters) and the production of unsafe housing can be traced back to reduced engineers' incomes and economic pressures. One policy could be to adjust engineering service fees according to inflation. However, since the main cause of low engineering income is the mismatch between supply and demand for engineering services, pricing engineering services will fail. Policies must target root causes to resolve issues sustainably and effectively. Root-cause solutions usually require more actions, budget, and time, which is why policymakers often choose superficial measures that only delay and exacerbate problems. Policies must account for real-time delays and dynamics, as changes in housing conditions require patience. Ignoring delays and housing sector dynamics during decision-making processes can lead to unexpected behaviors and policy failures. For example, building housing units involves delays, including investors' reaction times to market analysis (informational delay) and the time required to construct the housing units (physical delay).

Initial Conditions: Housing policies should align with the country's initial conditions, capabilities, and resources. Policies that disregard a country's context will not be implemented effectively and might worsen the housing situation. Unrealized policies weaken public trust in the government and decision-makers, reducing their credibility, power, and influence, which diminishes the success of future policies, even if they are well-suited. Housing decisions must consider actual housing supply and demand, public and private investment capacity, government budget, housing production capability, land availability, and access to natural resources, energy, and urban services.

Evolution, Learning, and Adaptive Behavior: The housing market is dynamic, learning from feedback, revising plans, and adapting behavior. Policymakers must recognize that housing market actors behave rationally according to economic theory, seeking to maximize their benefits. As conditions change, actors learn and adjust their behavior to maximize benefits, complicating policy-making. Often, housing market actors do not follow prescribed policies but act in ways that maximize their benefits (rational behavior). Policymakers should create conditions that improve housing situations and better align policies with the collective goals of housing actors to increase the likelihood of policy success. For example, price controls in certain housing segments do not balance and stabilize prices; they create black markets and speculative activities in those segments.

Reinforcing and Balancing Feedback: Feedback means that a change in one part of the system eventually affects the same part through a process. Recognizing feedback loops in the housing system clarifies the system and anticipates policy outcomes before implementation. Many examples mentioned above indicate numerous feedback loops in the housing system. Often, a policy aimed at improving part of the housing system may be effective in the short term, but due to other feedback loops, it can worsen conditions in the long term. For example, distributing land and horizontally expanding cities with low-rise buildings might reduce land prices, housing prices, and urban density in the short term. However, lower housing prices increase the attractiveness of migrating to the targeted city, raising housing demand and prices again. Moreover, horizontal city expansion increases urban management costs. Without proper financial solutions to cover municipal expenses, this can lead to increased sale of building density, corruption, and rising land prices in the long term. Hence, horizontal city expansion policies intended to reduce urban density and housing prices ultimately worsen housing and urban conditions in the long run.

3.3 Few Points Worth Mentioning

- Housing is path-dependent and requires long-term policies. Housing policy faces short-term parliamentary, electoral, and political changes. It is crucial to engage all parties and groups in creating cross-party consensus, identifying housing challenges, and then outlining and implementing long-term political strategies.
- In areas where policy intervention involves multiple public and private organizations, problems arise regarding goals, resources, and time scales, leading to adopted rational strategies for individual implementation. However, they fail to provide the necessary collaboration and alignment to create real change. These types of problems hinder and limit the capacity to take effective actions in the housing sector.
- Research indicates that the economic growth of leading cities and metropolitan areas in various countries crucial for future national economic growth is increasingly hampered by a lack of affordable housing. As land becomes scarcer, rents and prices rise, undermining productivity over time. Without housing policy reforms across the board, growth will be severely limited.
- Most current solutions delay rather than fundamentally solve housing issues, focusing on short-term fixes. While addressing immediate concerns may temporarily improve conditions, systemic and transformative solutions have the potential to drive housing towards a more sustainable future. Coordinated and coherent solutions derived from a comprehensive approach can shift focus from delaying problems to root-cause changes.

4 Final Insight

A housing system and system thinking illustrate a range of approaches that can help organize individual perspectives, identifying system housing structures, characteristics, and behaviors. A systemic approach encourages better analysis by considering and identifying emerging interactions and features, preventing errors stemming from linear thinking and excessive focus on individual elements or nodes.

In system thinking, responses to problems arising from complex system behaviors must be compatible and proportional to the system's complexity; efforts to simplify complex issues will face challenges. A structured solution requires a sufficient understanding of system complexity Policy failure occurs when there is an imbalance between system problems and policy responses. Complex problems require multi-faceted and intricate solutions.

5 Conclusion

This article emphasizes the necessity of adopting systemic thinking to discover more effective solutions for achieving desirable housing and resolving housing issues in developing countries. Viewing housing as a complex and interconnected system enhances understanding of various factors and dynamics affecting housing quality, safety, affordability, and sustainability.

Through a systemic lens, housing is revealed not as an isolated system but as a product of complex interactions among different stakeholders including policymakers, producers, households, etc. Recognizing these interdependencies and adopting a holistic perspective enables the identification of leverage points and potential interventions that can lead to positive and sustainable changes in the housing sector.

A systemic approach stresses the importance of collaboration and coordination among diverse sectors and disciplines involved in housing, such as urban planning, architecture, engineering, governance, economics, and social sciences. This integration allows for the development of comprehensive strategies capable of addressing multifaceted housing challenges.

System thinking highlights the need to consider long-term consequences and unintended effects of policy interventions. By anticipating potential impacts of policy decisions, inappropriate short-term solutions that may exacerbate existing housing problems or introduce new problems can be avoided. Moreover, it can contribute to ensuring sustainable and resilient housing that meets the needs of current and future generations.

Ultimately, adopting a system thinking approach is crucial for achieving desirable outcomes. It facilitates understanding of complex dynamics, identifies effective interventions, and develops sustainable solutions. Embracing this approach can engage all housing stakeholders in creating a comprehensive, affordable, resilient, sustainable, and environmentally and socially compatible housing system.

References

Alexander, D.E.: Resilience and disaster risk reduction: an etymological journey. Nat. Hazard. **13**, 2707–2716 (2013)

Arigoni, D.: Affordable Housing and Smart Growth, Making the Connection, National Neighborhood Coalition, Washington DC (2001)

Cavallo, A.: Complex systems thinking in preparation for unexpected risks: building general resilience in South Australia. In: Handbook of Disaster Risk Reduction & Management (2017)

Cavallo, A., Ireland, V.: Preparing for complex interdependent risks: a system of systems approach to building disaster resilience. Int. J. Dis. Risk Reduct. **9**, 181–193 (2014). https://doi.org/10.1016/j.ijdrr.2014.05.001

Clarvis, M., Bohensky, E., Yarime, M.: Can resilience thinking inform resilience investments? Learning from resilience principles for disaster risk reduction. Sustainability **7**(7), 9048–9066 (2015). https://doi.org/10.3390/su7079048

Deegan, M.A.: Defining the Policy Space for Disaster Management: A System Dynamics Approach to U.S. Flood Policy Analysis, Rockefeller College University at Albany, State University of New York (2010)

Gallent, S., Mart, N., Tewdwr-Jones, M.: Housing in the European Countryside. Rural Pressure and Policy in Western Europe, Rutledge (2003)

Gan, Q., Hill, R.J.: Measuring housing affordability: looking beyond the median. J. Hous. Econ. **18**(2), 115–125 (2009)

Gawande, A.: The Checklist Manifesto. Metropolitan Books: New York For the Practical And Life-Changing Adoption of Checklists in Different Public Spheres (2010)

Ghafory-Ashtiany, M.: System approach for safe and affordable (desirable) housing. In: IIASA Asia Conference (2021)

Ghafory-Ashtiany, M., Shahmohammadian, A.: Integrated system modelling for desirable housing. In: IIASA Asia Conference (2021)

Lane, D.: Till the muddle in my mind have cleared awa: can we help shape policy using systems modelling? Syst. Res. Behav. Sci. **33**, 633–650 (2016)

Maclennan, D, Pawson, H, Gibb, K, Hulchanski, D., Chisholm, S.: Shaping Futures: Changing the Housing Story – Full Report. University of Glasgow, Policy Scotland (2019). https://shapingfutures.gla.ac.uk/index.php/changing-the-housing-story

Meen, G., Gibb, K., Leishman, C., Nygaard, C.: Housing Economics: A Historical Approach. Macmillan, Basingstoke (2016)

Sanderson, I.: Intelligent policy making for a complex world: pragmatism, evidence and learning. Polit. Stud. **57**, 699–719 (2009)

Shahmohammadian, A., Ghafory-Ashtiany, M.: Game theory applications in managing stakeholder conflicts for building safety and resilience against natural disasters. Prog. Disaster Sci. **26**, 100409 (2025). https://doi.org/10.1016/j.pdisas.2025.100409

Shahmohammadian, A., Ghafory-Ashtiany, M., Mansouri, B., Motamed, H., Nasserasad, K.: FiCat-EQ: An integrated system-based tool for assessing financial vulnerability and economic impact of seismic risk, case study: housing sector of Iran. Int. J. Disaster Risk Reduct. **123**, 105474 (2025). https://doi.org/10.1016/j.ijdrr.2025.105474

Shahmohammadian, A., Ghafory-Ashtiany, M.: The Need of Using System Thinking Approach to Achieve Safe and Affordable Housing, Integrated Disaster Risk Management Conference, IDRiM2021 Conference (2021)

Sterman, J.D.: Business Dynamic. Systems Thinking and Modelling for a Complex World. McGraw-Hill, Boston (2000)

UNHABITAT: Housing 2030, Effective Policies for Affordable Housing in the UNECE Region (2023)

Van der Heidjen, H.: West European Housing Systems in a Comparative Perspective. IOS Press, Amsterdam (2013)

Analyzing Thermodynamic Condition of Different Atmospheric Hazards that Affect Lives and Livelihoods of the People in Bangladesh

Fatima Akter(✉) and Saurav Dey Shuvo

Department of Meteorology, Faculty of Earth and Environmental Sciences, University of Dhaka, Dhaka 1000, Bangladesh
{fatima.geoenv,sauravshuvo}@du.ac.bd

Abstract. This research aims to enhance our understanding of atmospheric thermodynamics in severe convection. In this research total 11 cases of lightning, thunderstorm, and hailstorm have been studied. For lightning events, the atmosphere undergoes significant changes before and during the occurrences. The temperature profile from the Skew-T diagram illustrates this. Notable variations are seen in the dewpoint temperature for each event. Additionally, the Level of Free Convection (LFC) shifts before and during lightning events. In contrast, the thermodynamics of the atmosphere show fewer changes in thunderstorm cases. Therefore, instability indices have been used to analyze these cases in this research. The distribution of Convective Available Potential Energy (CAPE) values across all events demonstrates that a comprehensive assessment of temperature dynamics is essential for detecting thunderstorm occurrences. The hailstorm events provided a significant opportunity to examine the atmospheric thermodynamic conditions. Prior to these storms, the atmosphere was highly turbulent, and the thermal changes occurred rapidly during the hail events. As a result, the vorticity conditions were notably chaotic during these storms. The primary goal of this study is to provide valuable insights that could enhance forecasting methods. The results offer crucial details that can refine the prediction of convective events, which frequently affect lives and livelihoods in the country.

Keywords: Thermodynamics · Severe events · Hailstorm · Lightning · Skew-T diagram

1 Introduction

Geographically, the location of Bangladesh is such that – the country is easily susceptible to any sort of natural disasters, and more so – the atmospheric phenomena and hazards (Shuvo & Sultana, 2022.b). The country has experienced devastating floods and tropical cyclones; fatal lightning events; catastrophic thunderstorms and hailstorms; severe spells of heat waves and cold waves (Islam et al., 2021; Shuvo & Sultana, 2022.a; Sultana &

H. Tatano et al. (Eds.): GSRIDRR 2023, DRRGADRIBS, pp. 401–425, 2025.
https://doi.org/10.1007/978-981-96-6609-6_27

Shuvo, 2022). All of them took a huge toll on people's lives and livelihoods (Shuvo, 2018). Incidentally, not all of them have formed within the territory of the country (Nahar et al., 2022; Sarker et al., 2021). In recent years, there is a growing tendency of increased such major events within the country. But, the reasoning behind them is mostly unknown (Shuvo, 2021; Shuvo & Awal, 2021). So, research is required in this regard to identify the causes behind increasing levels of severe convective events. Severe convective events could be characterized from several perspectives. But, no other characteristic of the events is more influential than the thermodynamics of those events (Islam et al., 2024). Thermodynamic profiles of the atmosphere are always given importance in meteorology (Akter et al., 2023). So, deciphering the reasons behind increased convective activities in recent years will be aptly guided by the delineation of their thermodynamic profiles.

A series of favorable weather conditions have to come together for severe storm formation. Instability and moisture (CAPE, CINE) are used to determine the strength of a thunderstorm's precipitation (rate of rain release, hail generation) and wind shear determines the storm's movement and life span (Weisman and Rotunno, 2000, 2004; Weisman et al., 1988; Chaudhari et al., 2010). Strong wind shear can produce supercells and make storms as well as keep continued updraft and downdraft separation and by rotating updraft falls into the storm with the help of directional shear. (Das et al., 2015). The maximum updraft speed measured by radars for a typical cumulonimbus over Europe by Schmidt et al. (2012) was about 10.5 m s-1. Stratiform rain areas are found behind the initial line while squall lines bow out because of the formation of the high-pressure system (Davis et al., 2004; Cotton and Anthes, 1989; Houze, 1993). The results of the simulated values are lower than the TRMM hydrometeor profiles. The value of wind shear containing 6–8 ms-1 is situated in the lower toddle troposphere. Solenoidal field and warm advection are found where geopotential contours cut across isothermal flow over Bangladesh at 300 hPa (Das et al., 2015).

The initiation and growth of thermals are strongly influenced by intense low-level heating and cooperative positive vorticity in the wind field (Das et al., 2015). Deep convective systems, particularly cumulonimbus clouds characterized by robust convective core echo structures, are predominantly observed over land due to extreme buoyancy driven by sensible heat flux from heated land surfaces. Field experiments such as JASMINE have provided insights into convective regimes over the northern Indian Ocean, adjacent to Sri Lanka and the Maldives (Webster et al., 2002). Comparable convective systems, including super convective structures, are observed in the western Pacific and other regions (Chen et al., 1996; Tao et al., 1991; Tao et al., 2003).

Using MST radar, Uma and Narayana Rao (2009) identified intense updrafts of 15–20 m/s associated with deep convection near Gadanki on India's southeast coast. Thunderstorms (TS) pose significant threats to aviation due to the strong winds they produce, which account for the highest aviation hazards during such events (Litta et al., 2012). Approximately 72% of tornadoes are linked to Nor'westers (Chaudhuri, 2008; Ghosh et al., 2008). Rajeevan et al. (2010) examined the sensitivity of microphysical schemes in simulating thunderstorms. The occurrence of TS depends on three primary factors: intense instability, a sufficiently deep humid layer in the lower and middle troposphere, and an updraft to initiate convection (Johns and Doswell, 1992; McNulty, 1995). Increasing Convective Available Potential Energy (CAPE) is associated with a higher likelihood of significant severe weather (Brooks et al., 2003). Over Bangladesh, 94% of Nor'westers occur when the Showalter Stability Index (SWI) exceeds 100 (Karmakar and Alam, 2006).

Surface temperature and relative humidity are valuable parameters for forecasting thunderstorm likelihood (Lopez et al., 2007). Vertical motion plays a critical role in initiating convection, as indicated by May and Rajopadhyaya (1999). Higher convective updraft speeds are linked to more intense storms, which transport greater amounts of ice into anvils and produce increased lightning activity (Zipser et al., 2006). Severe weather development is strongly influenced by atmospheric instability, and instability indices can effectively indicate the potential for convective weather (Tajbakhsh et al., 2012). Tornado occurrence is more probable when the Severe Weather Threat Index (SWEAT) exceeds 400 (Henz, 2009). CAPE values below 300 J/kg are insufficient for significant convective potential, while a Convective Inhibition (CIN) value exceeding 200 J/kg may suppress convection (Knutsvig, 2009). Thunderstorm probability is largely determined by latent instability (especially near the surface), potential instability, and conditional instability (Haklander and Delden, 2003). The role of orography in thunderstorm development has been comprehensively addressed by Whiteman (2003). In regions like Iran, thunderstorms are most prevalent during spring and summer (Alijani, 2009).

Convection, driven by warm air rising and cold air sinking, plays a central role in severe thunderstorm formation. Variables such as atmospheric humidity and vertical velocity, derived from numerical weather prediction (NWP) models, are critical for diagnosing atmospheric instability and supporting thunderstorm forecasting. Parameters including vertical motion and low-level moisture instabilities are essential for the development of thunderstorms.

2 Data and Methodology

The Global Data Assimilation System (GDAS), employed by the National Center for Environmental Prediction (NCEP) within the Global Forecast System (GFS) model, integrates observational data into a gridded model framework to initialize weather forecasts. GDAS data are accessible through the NOAA National Operational Model Archive and Distribution System (NOMADS), hosted by the National Centers for Environmental Information (NCEI). These data include both the input observations assimilated into GDAS and the gridded output fields it produces. NCEP's FNL (Final) operational global analysis and forecast datasets, derived from GDAS, are provided on a $0.25° \times 0.25°$ grid and are updated operationally every six hours (Shuvo et al., 2021).

The GDAS system continuously collects observational data from the Global Telecommunications System (GTS) and other sources to conduct multiple analyses. The FNL product, generated using the same model as GFS, is prepared approximately one hour after GFS initialization to allow the incorporation of additional observational data. In contrast, the GFS runs earlier to meet time-sensitive forecasting requirements and utilizes the FNL from the prior six-hour cycle during its initialization.

The GDAS analyses are produced across various levels, including the surface, 26 mandatory pressure levels (from 1000 mb to 10 mb), the surface boundary layer, sigma layers, the tropopause, and additional specialized levels. These analyses encompass a wide array of meteorological and atmospheric parameters, such as surface pressure, sea level pressure, geopotential height, temperature, sea surface temperature, soil properties, ice cover, relative humidity, u- and v- wind components, vertical motion, vorticity, and ozone.

The selected events have been simulated with assimilated data through the WRF-ARW model. Therefore, the observed conditions can be visualized from these simulations. Since the objective of this research is to understand the thermodynamic condition of the events, assessing the temperature conditions has been prioritized during the analyses. The zonal and meridional distribution of temperature as well as composite characteristics of the atmosphere has been prevalently studied. These are thoroughly done with some indices; like - CAPE. The selection of these indices is based on the review of works of literature. Validation with observed data (be it gridded or station data) has been done wherever possible. Sometimes it is not possible to check the thermodynamics of the atmosphere due to the unavailability of data, especially for upper levels of the atmosphere. The following table (Table 1) contains all the information of the experimental setup for this research.

Table 1. Experimental setup of WRF-ARW for this research.

Model Properties	Specification
Vertical levels	28
Dynamics	Non-hydrostatic
Map projection	Mercator
Initial and lateral boundary condition	0.25° X 0.25° six hourly global data assimilation system (GDAS) data
Physics option	**Parameterization schemes**
Microphysics	1) Morrison double-moment scheme with CESM aerosol (8) [for Lightning] 2) Milbrandt-Yau Double-Moment 7-class scheme (9) [for Thunderstorms] 3) NSSL 2-moment with Cloud Condensation Nuclei for hail and graupel number concentration (17) [for Hailstorms]
Cumulus Parameterization	Kain-Fritsch scheme (1)
PBL option	Yonsei University scheme (1)
Longwave Radiation	Rapid Radiative Transfer Model (1)
Shortwave Radiation	Dudhia scheme (1)

3 Events Studied in This Research

In this research – lightning, thunderstorm, and hailstorm events have been studied. A brief description of the events that have been studied is given below (Table 2). All of them have been simulated with the ARW model, and then further analyzed from the simulated results.

Table 2. A brief about the events that are being studied in this research.

(a) Lightning Events			
Index	Date	Time (UTC)	Location
Case 1	24.03.2017	2230	90.8ºE, 24.8ºN
Case 2	23.04.2017	1100	90.6ºE, 22.6ºN
Case 3	29.04.2018	2145	90ºE, 25.2ºN
Case 4	31.03.2019	1730	90.5ºE, 25ºN
Case 5	02.04.2019	0745	89.2ºE, 25.3ºN
Case 6	01.04.2019	1645	92.1ºE, 24.8ºN
(b) Thunderstorm Events			
Index	Date	Time (UTC)	Location
Case 1	17.05.2019	1250	Dhaka
Case 2	31.03.2019	1222	Dhaka
Case 3	03.03.2020	0300	Comilla
Case 4	23.04.2017	2300	Rangpur
(c) Hailstorm Event			
Index	Date	Time BST(+6 = UTC)	Location
Case 1	27.02.2019	0800, 1800, 1900, and 2000	Narail

4 Assessing the Thermodynamics of the Atmosphere During Lightning Events

The Skew-T diagram shows the temperature, dew point, and wind speed/direction. From these basic values, a wealth of information can be obtained concerning the meteorological condition of the upper air. There are six basic sets of fixed lines that comprise the Skew-T diagram. They are temperature, pressure, dry adiabatic line, moist (or saturated) adiabatic line, mixing ratio, and wind direction. Skew-T diagram shows the thermodynamic condition of a severe weather system. It also shows the upper atmospheric instability conditions. Unstable weather triggers different weather phenomena like Lightning, thunderstorms, tornado, cyclones, hailstorms, heavy rain, etc. In this study, for this section, Skew-T diagram is given more importance to assess the thermodynamic profiles of the

Table 3. A brief description of the Skew-T diagram of the individual cases derived from the WRF model (15 min before intense storm occurrence).

Event Date	Time UTC	Location	LFC (hPa)	Air Temp (° C)	Dewpoint Temp (° C)	LI	CAPE (J/Kg)
Case 1 24.03.2017	2230	90.8°E, 24.8°N	950	23.1	18.3	−5	1618
Case 2 23.04.2017	1100	90.6°E, 22.6°N	1000	29.2	25.3	−9	3275
Case 3 29.04.2018	2145	90°E, 25.2°N	1000	23.5	21.4	−3	1023
Case 4 31.03.2019	1730	90.5°E, 25°N	800	22.2	15.9	−5	1239
Case 5 02.04.2019	0745	89.2°E, 25.3°N	1000	28.2	19.3	−9	2526
Case 6 01.04.2019	1645	92.1°E, 24.8°N	1000	24.7	16.6	−2	334

atmosphere during an intense storm situation at the point where the maximum lightning flashes have occurred (Table 4).

In this study, for the first case, the maximum flash count (156 flashes) was occurred (90.8˚E & 24.8˚N) at 2245 UTC. For the event on 24th March 2017, Skew-T diagram represents that the lifted condensation level was at 950 hPa level and the temperature at that level was 22.9 ˚C (dew point temperature was 18 ˚C). According to the CAPE Value (1633 J/Kg), it represents a moderately unstable atmosphere. According to the LI value of −5, the atmosphere is unstable (thunderstorms and lightning are likely, and some could be severe with a lifting mechanism).

Figure 2 (a), Shows the thermodynamic condition on 24th March 2017 before 15 min of the total lightning flash occurrence. Before 15 min of that time (2245 UTC), at the location of 24.8° N & 90.8° E, according to CAPE (1618 J/Kg) and KI (−5) index, thermodynamic condition was at a moderately unstable atmosphere (Table 5).

In the second case on 23rd April 2017, the lifting condensation level was found to be at level 1000 hPa, and the temperature at this level was 29 ˚C (dew point temperature was 25.3 ˚C). On this day at 1115 UTC maximum lightning flashes counts (64) were observed at 90.6˚E & 22.6˚N. According to Skew-T diagram, CAPE (3150 J/Kg) and LI index (-9) suggests an extremely unstable atmospheric conditions. Before 15 min of that time (1100 UTC), at the location of 22.6° N & 90.6° E, according to CAPE (3275 J/Kg) and KI (−9) index, the thermodynamic condition preempts that it was an extremely unstable atmosphere.

In the third case, on 29th April 2018, maximum lightning flashes count (14) were counted at 90˚E & 25.2˚N at 2157 UTC. Figure 2 (c) represents the Skew-T diagram at 2200 UTC, where the lifting condensation level is at 950 hPa and the air temperature at this level was 23.4 ˚C (dew point temperature was 21.5 ˚C). CAPE (1221 J/Kg) value

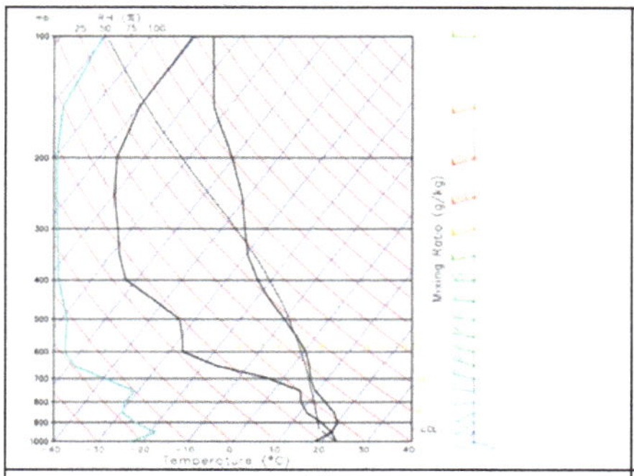

(a)

K=30; TT=49; Surface Temp = 23.1°C; dewP = 18.3°C; Most unstable Press = 950 mb; LI= -5; CAPE= 1618 J/kg

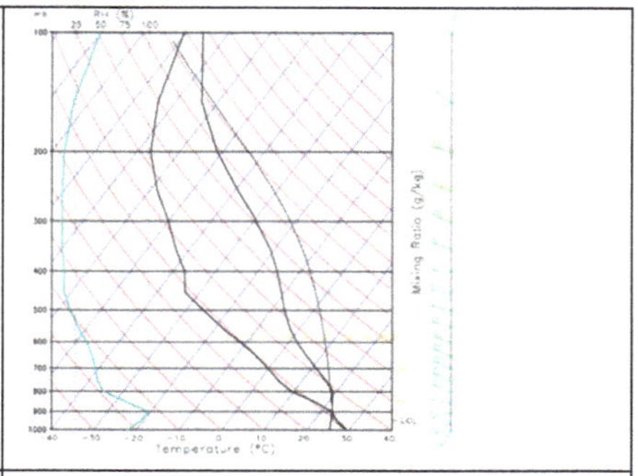

(b)

K=33; TT=51; Surface Temp = 29.2°C; dewP = 25.3°C; Most unstable Press = 1000 mb; LI= -9; CAPE= 3275 J/kg

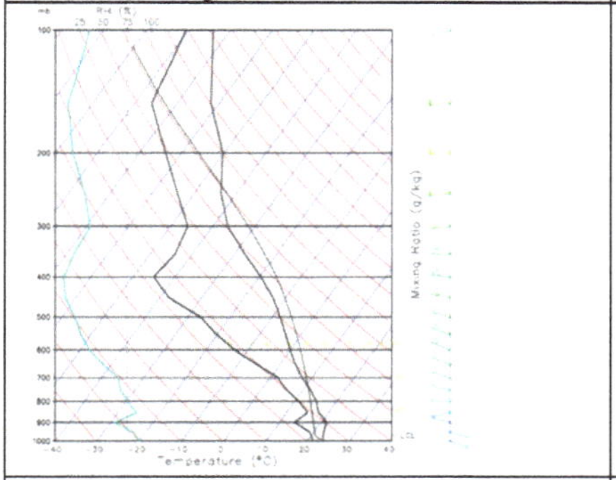

(c)

K=34; TT=48; Surface Temp = 23.5°C; dewP = 21.4°C; Most unstable Press = 1000 mb; LI= -3; CAPE=1023 J/kg

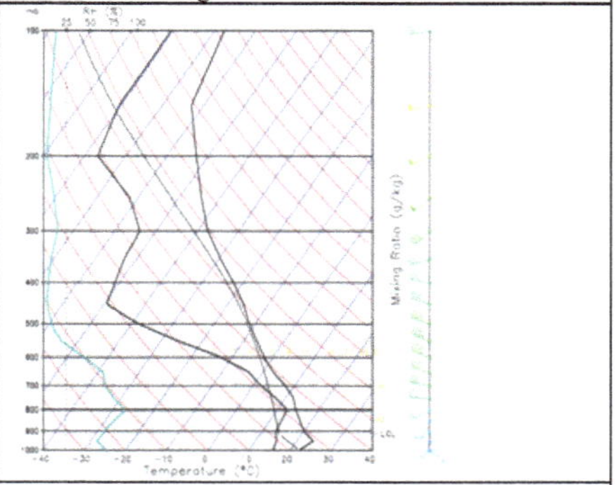

(d)

K=35; TT=53; Surface Temp = 22.2°C; dewP = 15.9°C; Most unstable Press = 800 mb; LI= -5; CAPE=1239 J/kg

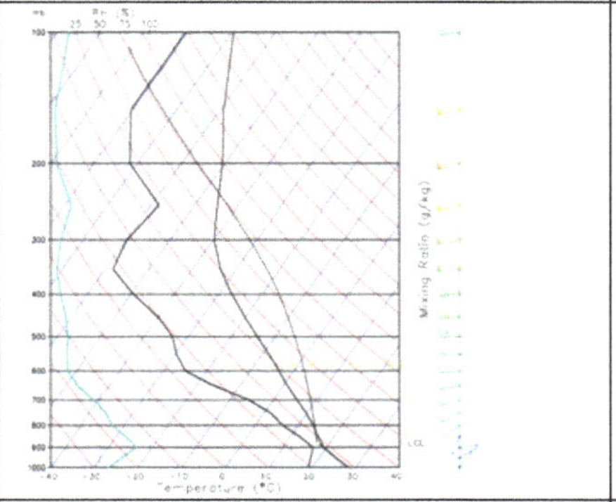

(e)

K=31; TT=56; Surface Temp = 28.2°C; dewP = 19.3°CMost unstable Press = 1000 mb;LI= -9; CAPE=2526 J/Kg

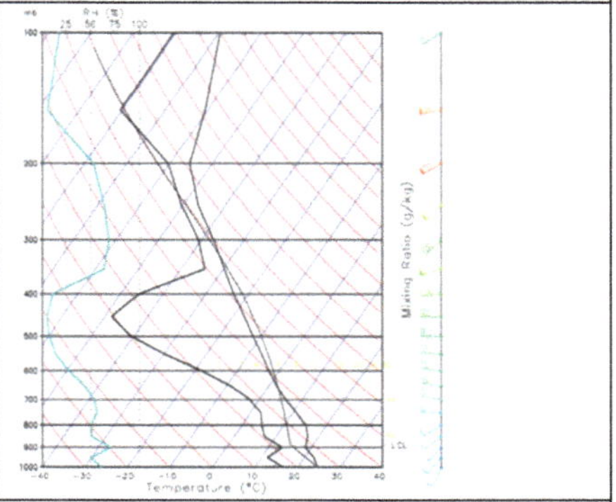

(f)

K=28; TT=48; Surface Temp = 24.7°C; dewP = 16.6°C; Most unstable Press = 1000 mb; LI= -2; CAPE=334 J/kg

Fig. 1. Skew-T diagram (before 15 min) at the maxima points of the lightning flashes count during the intense storm condition driven from the WRF model simulations for the selected cases - (a) Case 1, (b) Case 2, (c) Case 3, (d) Case 4, (e) Case 5; and (f) Case 6.

Table 4. The brief description of the Skew-T diagram of the individual cases derived from the WRF model around the time of intense storm occurrence.

Event Date	Time UTC	Location	LFC (hPa)	Air Temp (° C)	Dew Point Temp (° C)	LI Index	CAPE (J/Kg)
Case 1 24.03.2017	2245	90.8°E, 24.8°N	950	22.9	18	−5	1633
Case 2 23.04.2017	1115	90.6°E, 22.6°N	1000	29	25.3	−9	3150
Case 3 29.04.2018	2200	90°E, 25.2°N	950	23.5	21.5	−3	1221
Case 4 31.03.2019	1745	90.5°E, 25°N	800	21.6	16	−5	1259
Case 5 02.04.2019	0800	89.2°E, 25.3°N	1000	28.5	19.2	−9	2482
Case 6 01.04.2019	1700	92.1°E, 24.8°N	1000	24.7	15.5	−1	133

indicates a moderately unstable atmosphere but LI value of −3 indicates a marginally unstable atmosphere. Figure 2 (c) shows the thermodynamic condition on 29th April 2018 before 15 min of the total lightning flash occurrence. Before 15 min of that time (2145 UTC), at the location of 25.2° N & 90° E, according to CAPE (1023 J/Kg) and KI (−3) index, the thermodynamic condition was that of a moderately unstable atmosphere.

In the fourth case, on 31st march 2019 CAPE value is only 1259 J and the LI index (−5) represents the moderately unstable atmosphere but the LIS ISS satellite base counts total lightning flashes (retrieved from https://ghrc.nsstc.nasa.gov/lightning/data/data_lis_iss.html) were 151. Figure 2 (d) shows the Skew-T diagram and the lifting condensation level was 800 hPa and the temperature was 21.6 °C (dew point temperature was 16 °C). Figure 1 (c), shows the thermodynamic condition on 31st March 2019 before 15 min of the total lightning flash occurrence. Before 15 min of that time (1730 UTC), at the location of 25° N & 90.5° E, according to CAPE (1259 J/Kg) and KI (−5) index, thermodynamic condition was representative of a moderately unstable atmosphere.

In the fifth Case, on 2nd April 2019, the CAPE value (2482 J/Kg) indicates a moderately unstable atmosphere but the LI index (−9) indicates an extremely unstable atmosphere (severe thunderstorms and tornadoes likely). LIS ISS satellite-based maximum total lightning flashes were counted (72 flashes) at 0754 UTC. Figure 2 (e) represents the Skew-T diagram of 2nd April 2019, when the the lifting condensation level was at 1000 hPa and the temperature at this level was 28.5 °C (dew point temperature was 19.2 °C). Figure 1 (e) shows the thermodynamic condition on 2nd April 2019 before 15 min of the total lightning flash occurrence. Before 15 min of that time (0730 UTC), at the location of 25.3° N & 89.2° E, according to CAPE (2526 J/Kg) and KI (−9) index the thermodynamic condition of the atmosphere was extremely unstable.

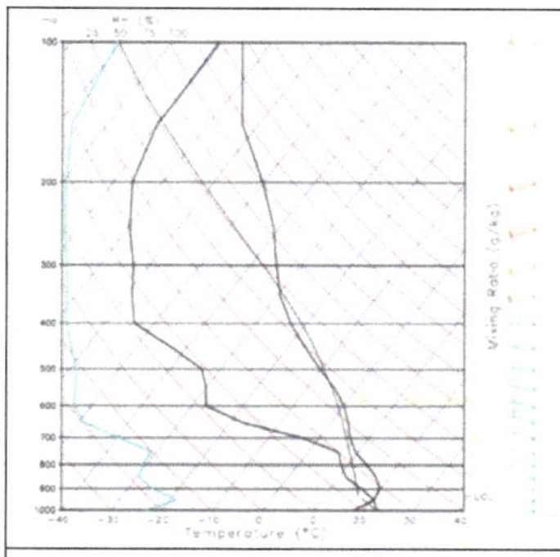	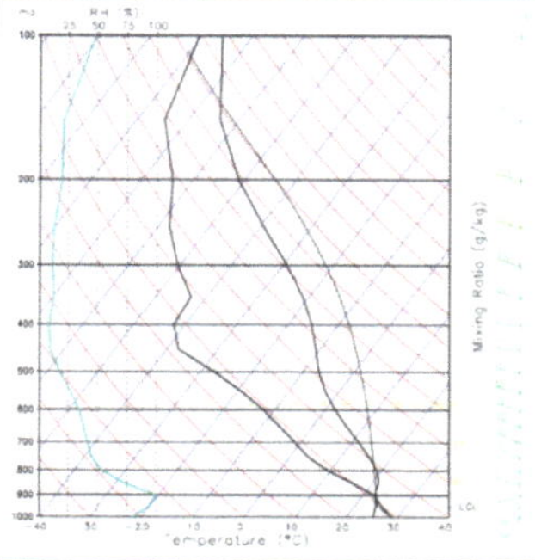
(a)	(b)
K=29; TT=49; Surface Temp = 22.9°C; dewP = 18°C; Most unstable Press = 950 mb; LI= -5; CAPE= 1633 J/kg	K=33; TT=51; Surface Temp = 29°C; dewP = 25.3°C Most unstable Press = 1000 mb; LI= -9; CAPE= 3150 J/kg
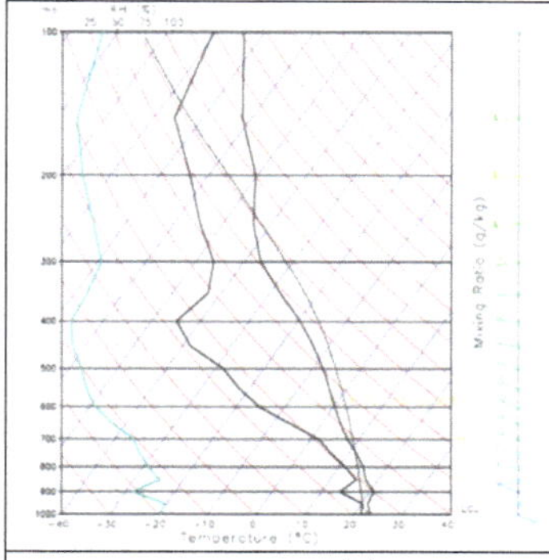	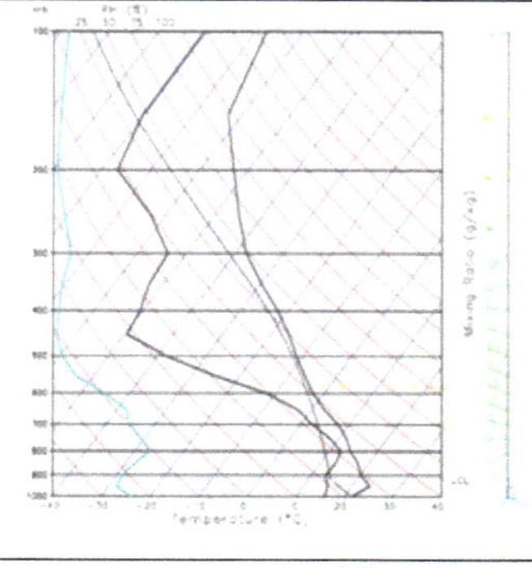
(c)	(d)
K=34; TT=48; Surface Temp = 23.4°C; dewP = 21.5°C; Most unstable Press = 950 mb; LI= -3; CAPE=1221 J/kg	K=37; TT=53; Surface Temp = 21.6°C; dewP = 16°C; Most unstable Press = 800 mb; LI= -5; CAPE=1259 J/kg
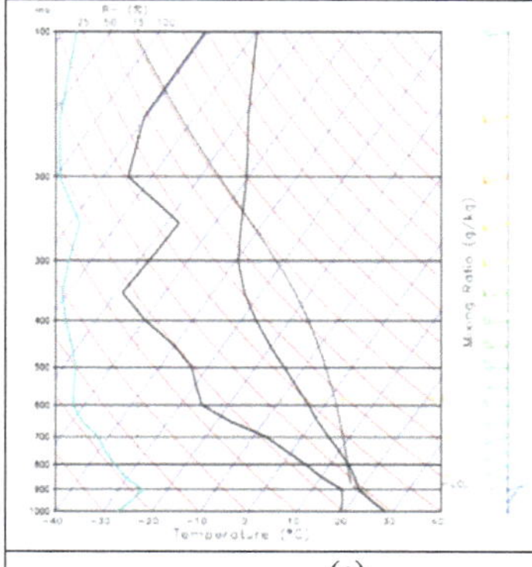	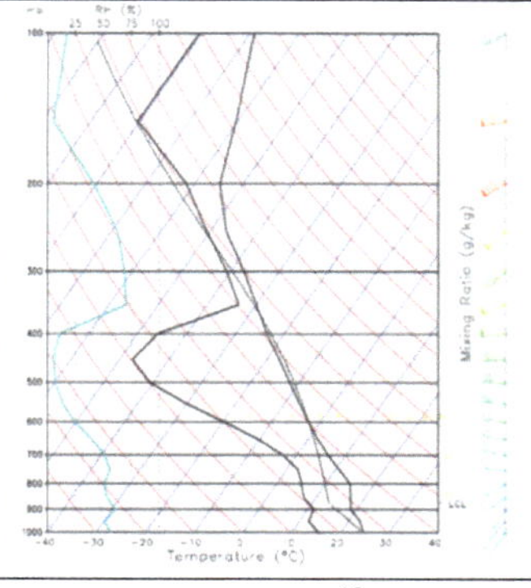
(e)	(f)
K=28; TT=55; Surface Temp = 28.5°C; dewP = 19.2°CMost unstable Press = 1000 mb;LI= -9; CAPE=2482 J/kg	K=27; TT=48; Surface Temp = 24.7°C; dewP = 15.5°C; Most unstable Press = 1000 mb; LI= -1; CAPE=133 J/kg

Fig. 2. .

Table 5. LIS (ISS) satellite-based maximum flashes count at occurrence place according to individual cases.

Case	Date	Time (ISS) UTC	latitude	longitude	Maximum Flash	Place
Case 1	24.03.2017	22:47:55	90.8	24.8	156	Atpara, Netrokona
Case 2	23.04.2017	11:10:02	90.6	22.6	64	Daulatkhan, Barishal
Case 3	29.04.2018	21:57:13	90	25.2	14	Jhenaigati, Sherpur
Case 4	31.03.2019	17:47:09	90.5	25	151	Gautola, Netrokona
Case 5	02.04.2019	07:54:57	89.2	25.3	72	Rasik Nagor, Rangpur
Case 6	01.04.2019	16:56:32	92.1	24.8	26	Tilpara, Sylhet

In the last case of this study, on 1st April 2019, the LIS ISS satellite-based maximum flash count at 1656 UTC was 26 flash. Figure 2 (f), Skew-T diagram shows the lifting condensation level was at 1000 hPa and temperature was 24.7 °C (dew point temperature was 15.5 °C). CAPE value of 133 J/Kg indicates a stable atmosphere but LI index (−1) indicates a marginally unstable atmosphere condition. Figure 1 (f) shows the thermodynamic condition on 24th March 2017 before 15 min of the total lightning flash occurrence. Before 15 min of that time (1645 UTC), at the location of 24.8° N & 92.1° E, according to CAPE (334 J/Kg) and KI (−2) index thermodynamic condition was that of a moderately unstable atmosphere (Table 6).

Table 6. The brief description of the Skew-T diagram of the observed cases near the time of intense storm occurrence over Dhaka.

Case	Date	Time (UTC)	Location	LFC hPa	Air Temp. °C	Showalter Index	CAPE (J/Kg)
Case 1	25.03.2017	0000	90.38° E 23.77° N	900	26.5	−0.10	1299
Case 2	23.04.2017	1200	90.38° E 23.77° N	925	24.5	−5.68	483.3

To verify the model-derived Skew-T diagram with the observation two cases, the available observational Skew-T are presented in this section. Based on the availability of the radiosonde data, as well as nearest location from the maximum lightning flashes occurrence point and time of intense storm occurrence, these two (24th march 2017 and 23rd April 2017) Skew-T diagrams are collected from the website for the University of Wyoming (downloaded from https://weather.uwyo.edu/upperair/sounding.html). However, the observational data (Skew-T) for the third case is not available on this website.

From the observation on 24th March 2017, the LCL was present at 900 hPa level and the temperature at that level was 26.5 °C. CAPE value was 1299 J/Kg which indicates moderately unstable atmosphere, but LI index (−0.10) indicates a stable atmosphere at 2500 UTC. Comparison to model-driven values of CAPE and LI index using Skew-T diagram shows CAPE (1633 J/Kg) and LI index (−5) both indicate moderately unstable atmosphere. As there are some spatial (0.42° E & 1.04° N) and temporal (75 min) variations, so it could be due to that reason.

On 23rd April 2017, the model and observation shows dissimilarities in terms of the thermodynamic condition of the atmosphere. Observation shows CAPE of 483.3 J/Kg, which indicates a marginally unstable atmosphere and LI index (−5.68) indicates a very unstable atmosphere condition over Dhaka at the time of 12 UTC but the model shows the extremely unstable condition of the atmosphere according to CAPE value (3150 J/Kg) and LI value (−9). Spatial and temporal variation between model and observation, in this case, is 0.22° E and 1.16° N, and 45 min respectively. So, they could be responsible for this dissimilarity.

5 Assessing the Thermodynamics of the Atmosphere During Thunderstorm Events

For this study, 04 cases have been simulated. Among them, two cases are TS active (Case 1 and Case 2) days and another two are TS inactive (Case 3 and Case 4) days. The mean sea level pressure (MSLP) is the atmospheric pressure at the mean sea level (PMSL). The average sea-level pressure is 1013.25 hPa. The model-derived sea level pressure (hPa) at 1000 hPa level during the occurrence of storm for all selected cases is represented in the figure (Fig. 4) below. For Case 1 (12:50 UTC; 17th May 2019; Location: Dhaka), the model derived sea level pressure at 1000 hPa level is 1003.32 hPa which is valid at 1200 UTC (a) over Dhaka. On the western side, pressure is lower and on the southern side, it is high.

For Case 2 (12:22 UTC; 31st March 2019; Location: Dhaka), the model derived MSLP at 1000 hPa level is 1003.98 hPa which is valid for 1200 UTC (b) over Dhaka. Pressure is lowest on the western side.

For Case 3 (3:00 UTC; 3rd March 2020; Location: Comilla), MSLP from the model is 1012 hPa which is valid for 0300 UTC over Comilla (c). On this day, high pressure is observed throughout the day.

For Case 4 (23:00 UTC; 23rd April 2017; Location: Rangpur), the model derived MSLP at 1000 hPa is 1003 hPa and it is valid for 2300 UTC over Rangpur. Pressure is lowest on the north-western side whereas the highest value is on the opposite side.

The model-derived dew point temperature (in °C) at 950 hPa level during the storm for all selected cases is represented in the following figure (Fig. 5). For Case 1 (12:50 UTC; 17th May 2019; Location: Dhaka), the model-derived temperature at 950 hPa level is 26°C which is valid for 1200 UTC over Dhaka.

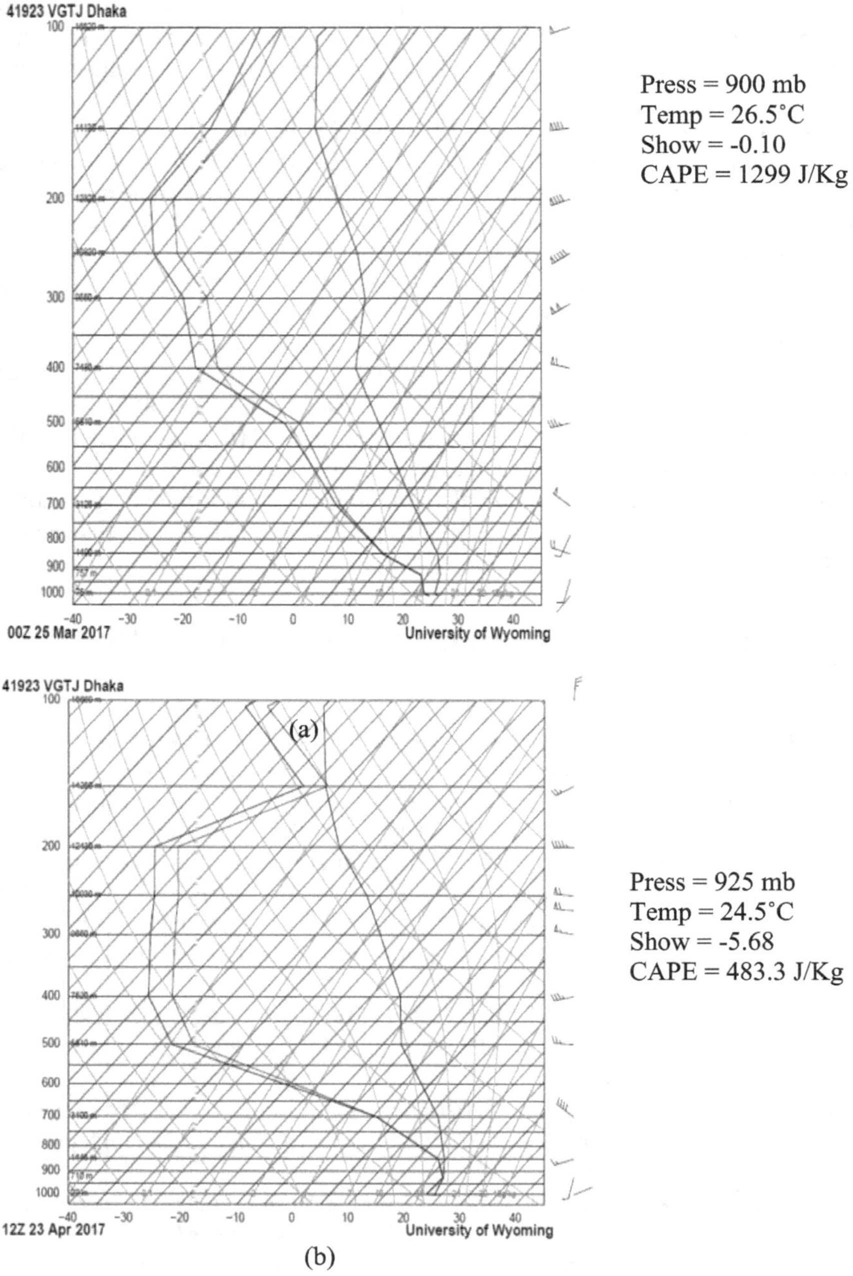

Fig. 3. Skew-T diagram from observed data over Dhaka during the nearest place and nearest time to the maximum lightning frequency - (a) Case 1; and (b) Case 2.

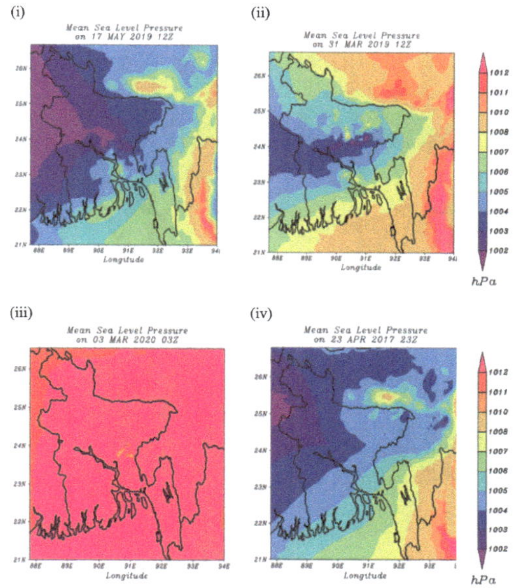

Fig. 4. Mean Sea-Level Pressure at 1000 hPa level - (i) Case 1 over Dhaka, (ii) Case 2 over Dhaka, (iii) Case 3 over Comilla; and (iv) Case 4 over Rangpur.

For Case 2 (12:22 UTC; 31st March 2019; Location: Dhaka), the model-derived temperature at 950 hPa level is 26°C which is valid for 1200 UTC over Dhaka.

For Case 3 (3:00 UTC; 3rd March 2020; Location: Comilla), the model-derived temperature at 950 hPa level is 20°C which is valid for 0300 UTC over Comilla.

For case-4 (23:00 UTC; 23rd April 2017; Location: Rangpur), the model-derived temperature at 950 hPa level is 23°C which is valid for 2300 UTC over Rangpur.

Relative Humidity (RH) is the ratio of the partial pressure of water vapor to the equilibrium vapor pressure of water at a given temperature. Relative Humidity (RH) depends on temperature and the pressure of the system of interest. Relative Humidity (RH) of over 80% is one of the conditions for TS formation.

For the first case, the distribution of model simulated relative humidity (%) valid at 1200 UTC of 17th May 2019 (over Dhaka) is 70%. The vertical cross-section of relative humidity displays that the RH condition in different heights of the cumulonimbus clouds indicates the presence of TS. In the North-South (NS) vertical cross-section of relative humidity, Longitude is fixed at 90.417 and for East-West (EW) vertical cross-section, Latitude is fixed at 23.78. The updraft and downdraft are seen in the vertical profile of moisture. The anvil shape of the thunderstorm at the mature stage is also clearly seen in the moisture profile.

For the second case, the distribution of model simulated relative humidity (%) valid at 1200 UTC on 31st March 2019 is 50%. For NS vertical cross-section, the Longitude is fixed at 90.417 and for EW vertical cross-section, the Latitude is fixed at 23.78.

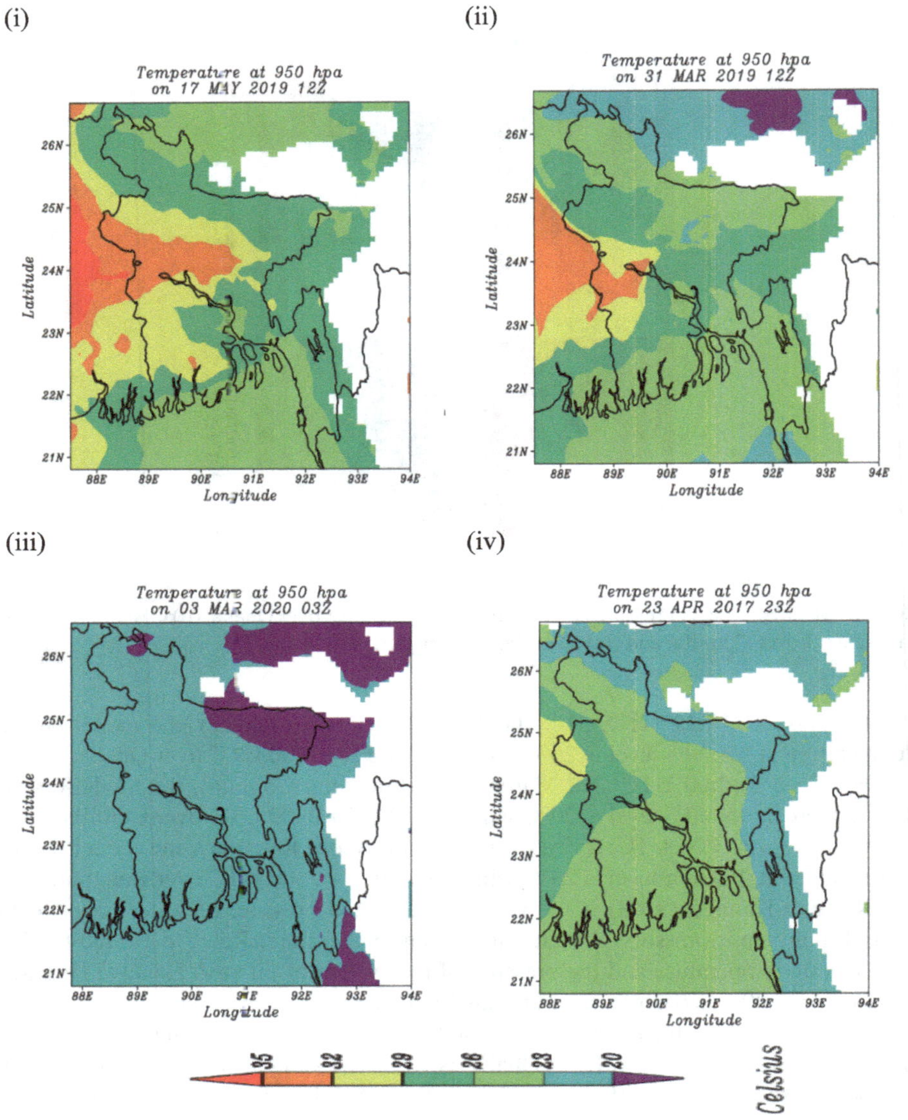

Fig. 5. Dew point temperature at 950 hPa level for - (i) Case 1 over Dhaka, (ii) Case 2 over Dhaka, (iii) Case 3 over Comilla; and (iv) Case 4 over Rangpur.

During the third case, the model simulated relative humidity (%) valid at 0300 UTC of 3rd March 2020 is 60%. Latitude is fixed at 23.4607 for EW vertical cross-section & Longitude is fixed at 91.1809 for NS vertical cross-section.

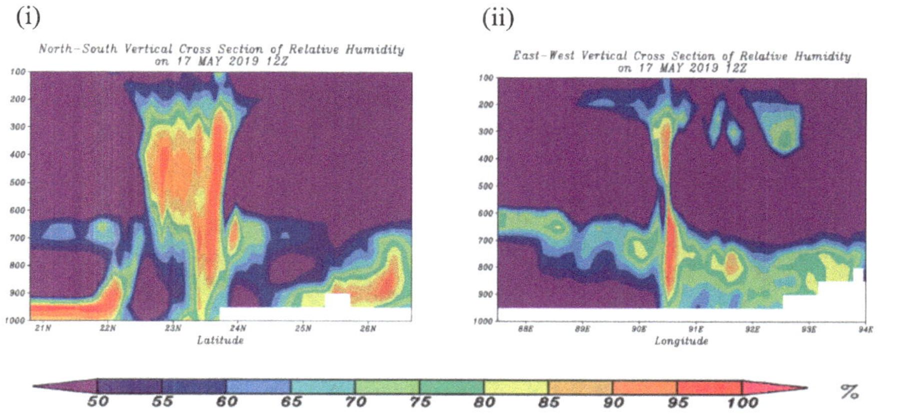

Fig. 6. Vertical profile of relative humidity (%) (i) NS and (iii) EW valid at 1200 UTC of Case 1 (17th May 2019; over Dhaka).

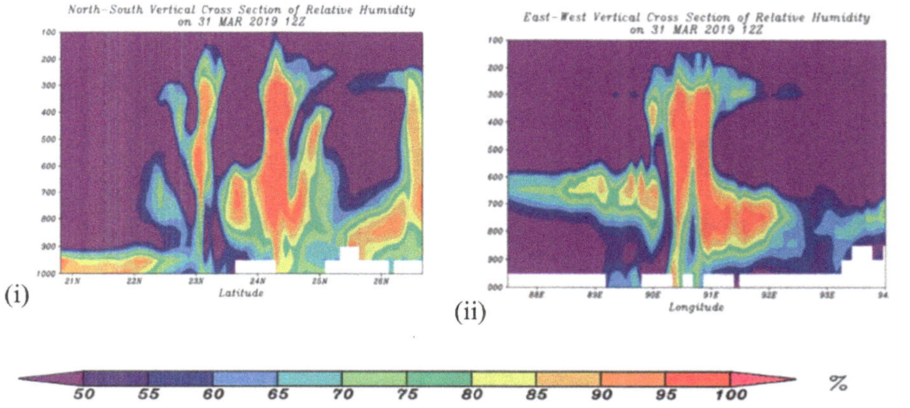

Fig. 7. Vertical profile of relative humidity (%) (i) NS and (iii) EW valid at 1200 UTC of Case 2 (31st March 2019).

For the fourth case, the model simulated relative humidity (%) valid at 2300 UTC on 23rd April 2017 is 95%. For NS vertical cross-section, the Longitude is fixed at 89.2752 and for EW vertical cross-section, the Latitude is fixed at 25.7439.

6 Assessing the Thermodynamics of the Atmosphere During Hailstorm Events

The model derived dew point temperature (°C) at 1000 (hPa) level valid for BST (Bangladesh Standard Time) 08:00 am; 06:00 pm; 07:00 pm; and 08:00 pm of 27th February 2019 are represented in Fig. 4.10. The simulated dew-point ahead of the trough was high reaching a temperature of more than 19 °C throughout Dhaka. These dew points

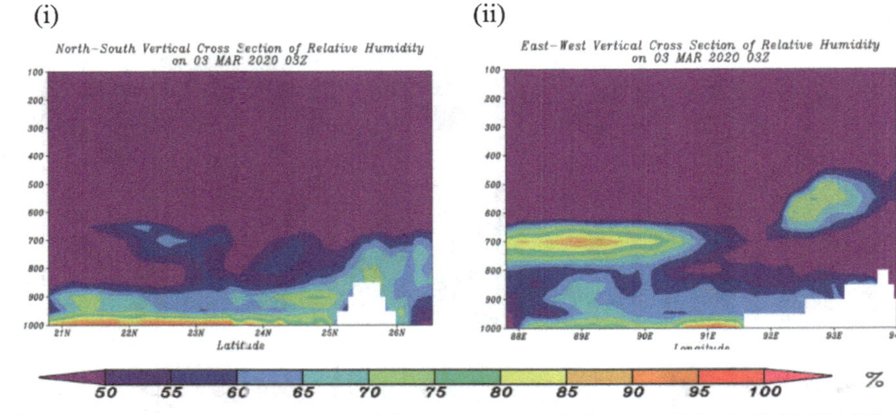

Fig. 8. Vertical profile of relative humidity (%) (i) NS and (ii) EW valid at 0300 UTC of Case 3 (3rd March 2020).

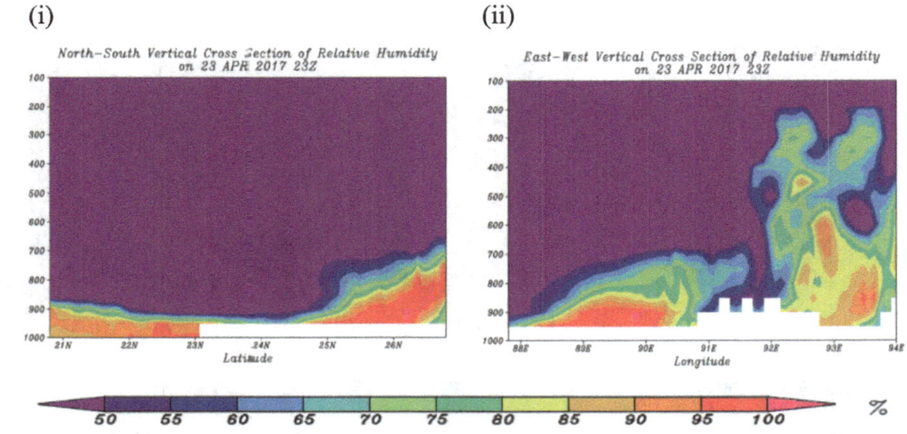

Fig. 9. Vertical profile of relative humidity (%) (i) NS and (ii) EW valid at 2300 UTC for Case 4 (23rd April 2017).

were high due to the southeasterly wind with hail at 08:00 am. And as per the wind analysis at 08:00 am of 27th February for 950 hPa level, a circulation was laid over Dhaka with a strong vertical wind shear supporting lower level convergence. And 06:00BST dew-point dropped at 12–17 °C and at BST 07:00 pm, the dew-point temperature increased above 19 °C over Rajshahi, Naogaon, Jaypurhat, and Jhenaidah. At BST 08:00 pm, dewpoint also increased over Gaibandha, Rangpur, and Kurigram and beside the highest dew point there was a dry airline. Along the boundary, there was a noticeable boundary of dry line with the dew point dropping to around 7 °C at the north-western part and central and the south-western part of Bangladesh with a westerly wind.

The model derived CAPE (J/kg) at 1000 (hPa) level valid for one hour before the event (Fig. 11) and during the time of the event (Fig. 12) at BST 07:00 am, 05:00 pm,

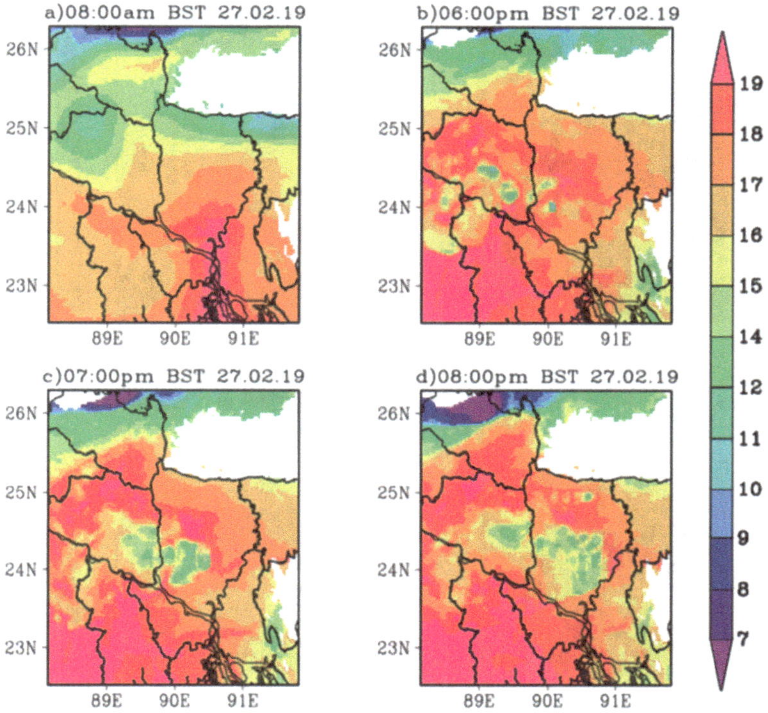

Fig. 10. Simulated dew point temperature (°C) at 1000 (hPa) level valid for BST 07:00 am; 06:00 pm; 07:00 pm; and 08:00 pm of 27th February 2019.

06:00 pm, 07:00 pm, 08:00 am, 06:00 pm, 07:00 pm, and 08:00 pm of 27th February 2019 are represented respectively.

Before an hour of the event, the CAPE value was more over Dhaka - about 1100 to 1200J/kg at 07 am BST, and about 1100 to more than 1300J/kg at 05 pm BST over the southwestern part of Kushtia, Jhenaidaha, Meherpur, Chuadanga and besides the highest CAPE there was a noticeable low CAPE value less than 100 J/kg over southwestern part. As the highest CAPE value was moving toward the northeast the boundary line of the lowest CAPE was more noticeable at BST 06:00 pm and 07:00 pm before an hour of the event.

During the time of the event CAPE value was less over Dhaka was about 0 to 100J/kg at 08 am BST and also at 06 pm BST over southwestern part of Kusthia, Jhenaidaha, Meherpur, Chuadanga; and at BST 07:00 pm - the lowest CAPE was noticeable beside the highest CAPE value about 100–300 J/kg over Magura, Faridpur and highest CAPE was the central and north part about 800–1000 J/kg. Around the lowest CAPE value, there was a noticeable high CAPE value moving toward the north over Gaibandha, Rangpur, and Kurigram at BST 08:00 pm more than 1100 J/kg. The difference in CAPE value between the time of occurrence and before the time of occurrence was more noticeable over Dhaka at BST 07:00 am and 08:00 am (Figs. 11 and 12).

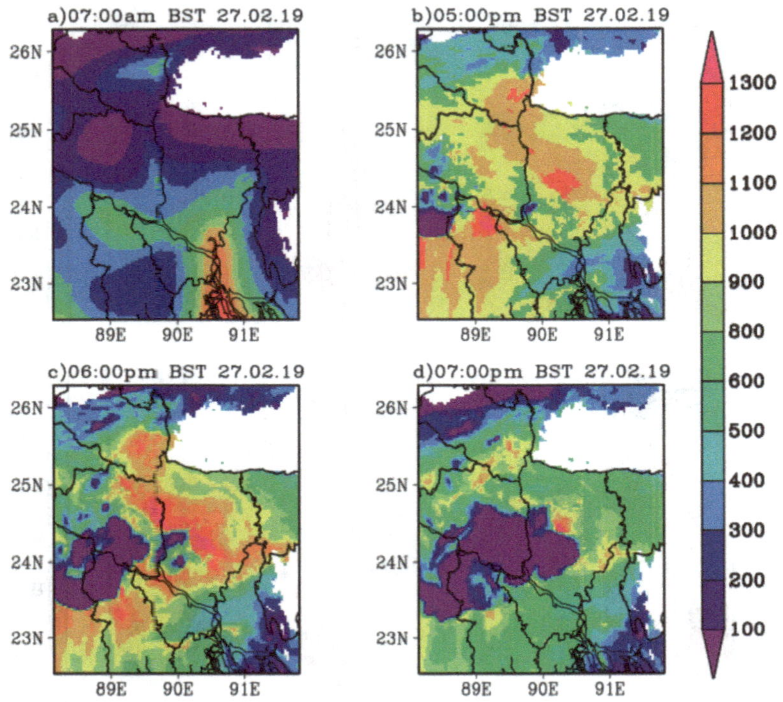

Fig. 11. Simulated CAPE (J/kg) at 1000 (hPa) level valid for BST 07:00 am, 06:00 pm, 07:00 pm, and 08:00 pm of 27th February 2019.

The model simulated Skew-T (J/kg) diagrams at a) BST 07:00 am and b) BST 08:00 am on the 27th of February at latitude 23.53°N and longitude 90.19°E before an hour of the event and during the time of the event are represented in Fig. 13 and 14 respectively. The model simulated result shows that at BST 07:00 am, CAPE was 449J/kg with 5-10kt southerly winds and as deep as surface level, before turning southwesterly with 25kt winds at 850 hPa and west-westerly 120kt at 300 hPa level, temperature and dew point was 20.4 °C and 19.4 °C, most unstable layer was 1000 (hPa) level which indicates that before the event there was convection. The deeper layer of moisture at 800 hPa to 450 hPa level whilst dry air intrusion started from 400 hPa level and temperature inversion occurred at 350–300 hPa level. And Lifted Index (LI) was negative (−3) means marginally unstable in the middle troposphere.

At 08:00 am on 27.02.19 (Fig. 4.13) temperature and dew-point temperature both dropped to 18.7 °C and 17.4 °C respectively with the south-easterly wind at the upper level, the most unstable layer was the 850 (hPa) level, during the time of event convection was less than before. CAPE was 115J/kg with 5-10kt south-easterly winds and as deep as surface level, before turning south-westerly 15kt at 900 hPa and west-westerly 115kt at 300 hPa level. The deeper layer of moisture at 700 hPa to 450 hPa level whilst dry air intrusion started from 400 hPa level and temperature inversion occurred at 950-850hPa

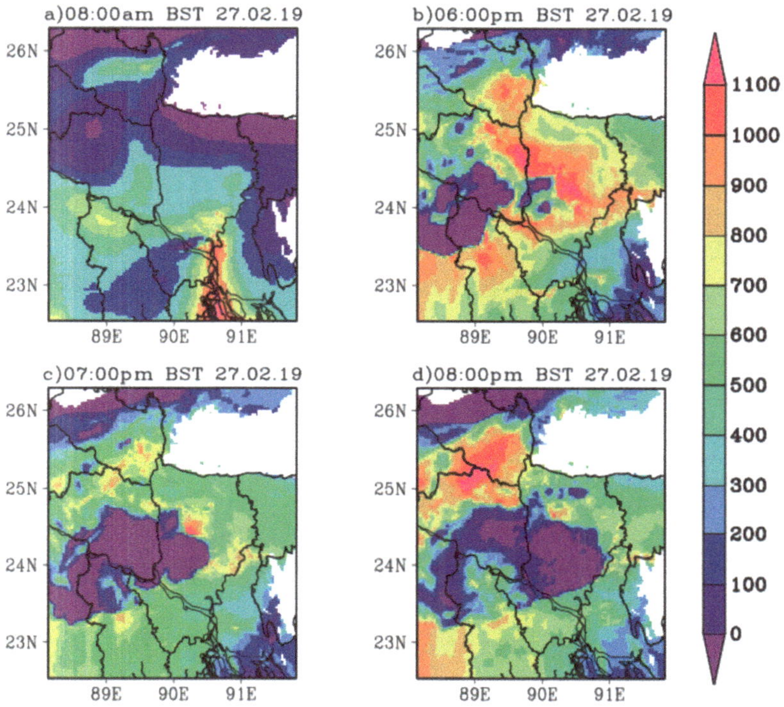

Fig. 12. Simulated CAPE (J/kg) at 1000 (hPa) level during the time of event valid for BST 08:00 am, 06:00 pm, 07:00 pm, and 08:00 pm of 27th February 2019.

level. And Lifted Index (LI) was positive 1 at the surface and non-negative (0) at 850 hPa means marginally unstable in the middle troposphere.

To further illustrate the simulated storm development within the trough, potential vorticity (PVU) and absolute vorticity ($10^{-5}s^{-1}$) were investigated (Fig. 15). Across the domain, a greater tendency to develop more isolated patches of potential vorticity of more than 5 PVU at 300hPa level and absolute vorticity of more than 100 $10^{-5}s^{-1}$ at 500 hPa level.

The model simulated storm-relative helicity ($m^{-2}s^{-2}$) is shown near the storm - an isolated region of intense lowest helicity persisting between 07:00 am to 09:00 am over Dhaka; and then peaking around 08:00 am. In the preceding environment, values were up to 700 $m^{-2}s^{-2}$ at highest hail accumulation. At 07:00 am and 08:00 am, there were two isolated areas of lowest helicity which were no longer than around 50 to 240 $m^{-2}s^{-2}$. A persistent region of intense positive helicity is noticeable throughout this period, particularly from 07:00 am, as it approached and moved further towards east.

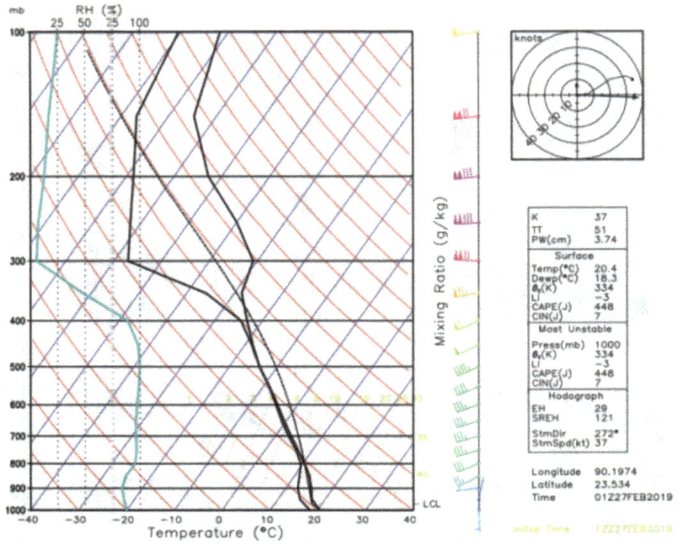

Fig. 13. Model simulated Skew-T diagram at 07:00 am on the 27th of February at latitude 23.53°N and longitude 90.19°E (1 h before the event).

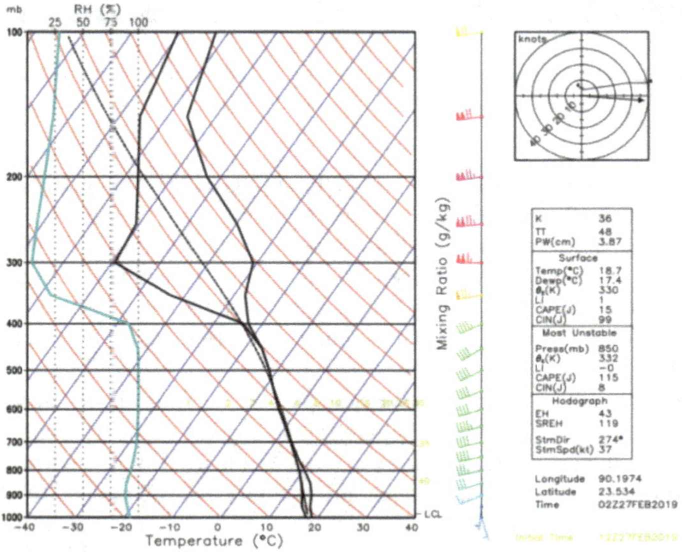

Fig. 14. Model simulated Skew T-log p diagram at 08:00 am on the 27th of February at latitude 23.53°N and longitude 90.19°E (during the time of the event).

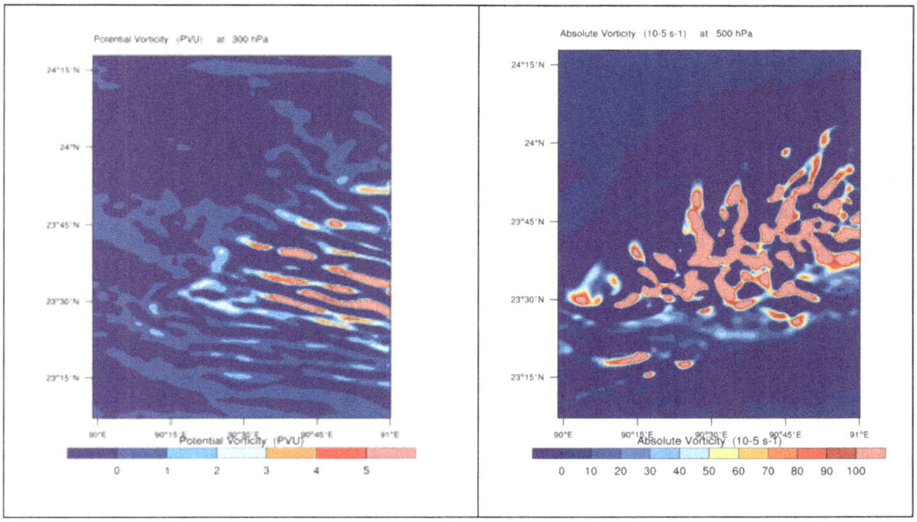

Fig. 15. Simulated potential vorticity (PVU) and absolute vorticity ($10^{-5}s^{-1}$) at 300 hPa and 500 hPa level at BST 08:00 am of 27th February 2019.

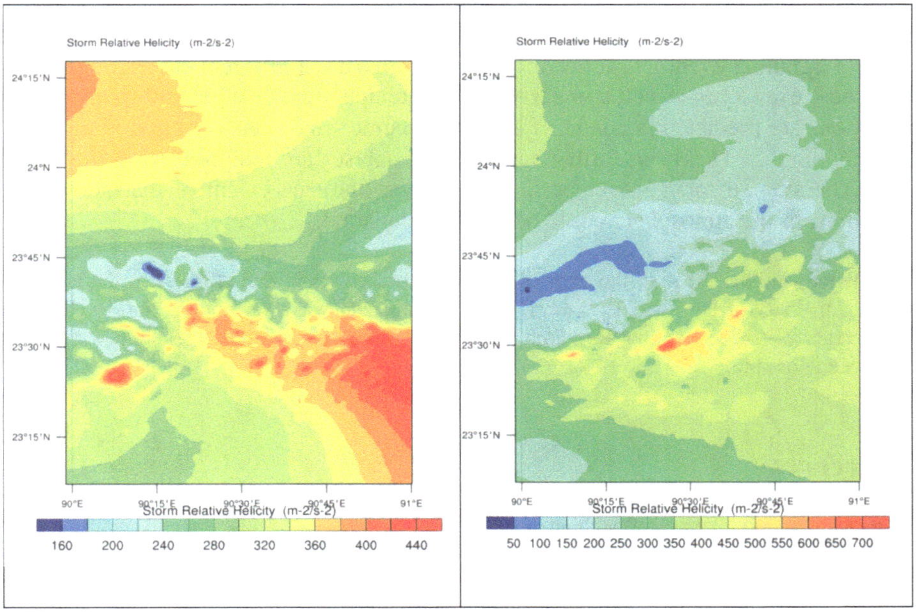

Fig. 16. Simulated storm-relative helicity ($m^{-2}s^{-2}$) at BST 07:00 am and 08:00 am of 27th February 2019.

7 Conclusions

The objective of this research was to explore the thermodynamic conditions of the atmosphere during severe convective events; like – as thunderstorms, lightning, or hailstorms. So, the experimental setup and simulations were set up in a manner to envisage the thermodynamics of the atmosphere for those events. The major findings from the simulated outputs are listed below –

For lighting, the atmosphere has considerable changes before and during the events. The temperature profile from the Skew-T diagram proves this point. Marked changes are observed in the dew-point temperature for each of the events. The LFC of the atmosphere changes before and during lightning events. Thermodynamics of the atmosphere has fewer changes for thunderstorm cases. So, thunderstorm cases have been analyzed with instability indices for this research. The distribution of CAPE values for all the events proves that the dynamics of temperature changes should be thoroughly assessed for detecting any thunderstorm event. The hailstorm events presented a great cause for studying the thermodynamic conditions of the atmosphere. The atmosphere was very turbulent before the hail events. The thermal changes were very rapid for the hailstorm events. So, the vorticity conditions were discernably tumultuous for the hailstorm event.

The results from this research can be apprehended to improve the forecasting methods for these convective events. For example, the vorticity conditions before a hailstorm event should be given due importance if the location of the hail downburst is to be realized. Similarly, thermodynamics of the atmosphere posits a good inference about when and where a lightning event might occur. Thunderstorms are a different case though. Even then, these events can be better understood if the temperature changes and the instability conditions are predicted in advance. Thermodynamic conditions could provide a good lead in doing so. However, further studies and analyses are required. Also, the events should be studied for a longer period. Only then, the true extent of thermodynamic conditions of the atmosphere could be better understood.

Acknowledgement. At first, the authors assert deep respect and thankfulness to the authority of the University of Dhaka. The authority of the University was enormously kind to allow the research grant – the Centennial Research Grant (CRG), without which this research would have never been possible.

References

Albrecht, R., Goodman, S., Buechler, D., Chronis, T.: Tropical frequency and distribution of lightning based on 10 years of observations from space by the lightning imaging sensor (LIS). In: Prepr, pp. 2–12 Phoenix, AZ, American Meteorological Society, Fourth Conf. Meteorol. Appl. Light. Data (2009)

Ahasan, M.N., Islam, M.N., Ferdousi, N.: A diagnostic study on synoptic and environmental conditions associated with Nor'westers in and around Bangladesh using MM5 model. In: Proceedings of the SAARC Seminar on Nor'westers and Tornadoes over the SAARC Region and their Forecasting & Preparedness, pp. 20–21 (2008)

Ahasan, M.N., Quadir D.A., Khan, K.A., Haque, M.S.: Simulation of a thunderstorm event over Bangladesh using WRF-ARW model. J. Mech. Eng. **44**(2), 124–131 (2014)

Battan, L.J.: Some factors governing precipitation and lightning from convective clouds. J. Atmos. Sci. **22**(1), 79–84 (1965)
Brooks, C.E.P.: The distribution of thunderstorms over the globe. Geophys. Mem. **3**(24), 147–164 (1925)
Brooks, H.E., Lee, J.W., Craven, J.P.: The spatial distribution of severe thunderstorm and tornado environments from global reanalysis data. Atmos. Res. **67**, 73–94 (2003)
Chaudhari, H.S., Sawaisarje, G.K., Ranalkar, M.R., Sen, P.N.: Thunderstorms over a tropical Indian station, Minicoy: role of vertical wind shear. J. Earth Syst. Sci. **119**(5), 603–615 (2010)
Chowdhury, M.H.K., Karmakar, S.: Pre-monsoon nor'westers in Bangladesh with case studies. In: Proceedings of the SAARC Seminar on Local Severe Storms, pp. 147–166 (1986)
Das, R.C., Munim, A.A., Begum, Q.N., Karmaker, S.: A diagnostic study on some local severe storms over Bangladesh. J. Bangladesh Acad. Sci. **18**, 81 (1994)
Das, S.: Composite characteristics of nor'westers observed by TRMM and simulated by WRF model. SAARC Meteorological Research Centre (SMRC) (2009)
Das, S.: Climatology of thunderstorms over the SAARC region. SAARC Meteorological Research Centre (2010)
Das, S.: Severe thunderstorm observation and modeling–a review. Vayu Mandal **43**(2), 1–29 (2017)
Davis, C., et al.: The bow Echo and MCV experiment: observations and opportunities: observations and opportunities. Bull. Am. Meteor. Soc. **85**(8), 1075–1094 (2004)
Dewan, A., Hossain, M.F., Rahman, M.M., Yamane, Y., Holle, R.L.: Recent lightning-related fatalities and injuries in Bangladesh. Weather Climate Soc. **9**(3), 575–589 (2017)
Gomes, C., Ahmed, M., Abeysinghe, K.R., Hussain, F.: Lightning accidents and awareness in South Asia: experience in Sri Lanka and Bangladesh. In: Proceedings of the 28th International Conference on Lightning Protection (ICLP). pp. 18–22 (2006)
Goodman, S.J., Christian, H.J.: Global observations of lightning. Atlas of Satellite Observations Related to Global Change, RJ Gurney, JL Foster, and CL Parkinson, Eds. Cambridge University Press (1993)
Goodman, S.J., Buechler, D.E., Knupp, K., Driscoll, K., McCaul, E.W., Jr.: The 1997–98 El Nino event and related wintertime lightning variations in the southeastern United States. Geophys. Res. Lett. **27**(4), 541–544 (2000)
Gubenko, I.M., Rubinshtein, K.G.: Analysis of the results of thunderstorm forecasting based on atmospheric instability indices using the WRF-ARW numerical model data. Russ. Meteorol. Hydrol. **40**(1), 16–24 (2015)
Hamid, E.Y., Kawasaki, Z., Mardiana, R.: Impact of the 1997–98 El Niño event on lightning activity over Indonesia. Geophys. Res. Lett. **28**(1), 147–150 (2001)
Holle, R.L.: Lightning-caused deaths and injuries related to agriculture. In: Preprints of the 33rd International Conference on Lightning Protection, Estoril, pp. 25–30 (2016)
Holle, R.L., López, R.E., Zimmermann, C.: Updated recommendations for lightning safety–1998. Bull. Am. Meteor. Soc. **80**(10), 2035–2042 (1999)
Johns, R.H., Doswell, C.A.: Severe local storms forecasting. Weather Forecast. **7**(4), 588–612 (1992)
Kandalgaonkar, S.S., et al.: Spatio-temporal variability of lightning activity over the Indian region. J. Geophys. Res. Atmos. **110**(D11) (2005)
Kar, S.K., Liou, Y.-A., Ha, K.-J.: Aerosol effects on the enhancement of cloud-to-ground lightning over major urban areas of South Korea. Atmos. Res. **92**(1), 80–87 (2009)
Karmakar, S.: Climatology of thunderstorm days over Bangladesh during the pre-monsoon season. Bangladesh J. Sci. Technol. **3**(1), 103–112 (2001)
Mansell, E.R., Ziegler, C.L.: Aerosol effects on simulated storm electrification and precipitation in a two-moment bulk microphysics model. J. Atmos. Sci. **70**(7), 2032–2050 (2013)

Murugavel, P., Pawar, S.D., Gopalakrishan, V.: Climatology of lightning over Indian region and its relationship with convective available potential energy. Int. J. Climatol. **34**(11), 3179–3187 (2014)

Nath, A., Manohar, G.K., Dani, K.K., Devara, P.C.S.: A study of lightning activity over land and oceanic regions of India. J. Earth Syst. Sci. **118**(5), 467 (2009)

Piepgrass, M.v., Krider, E.P., Moore, C.B.: Lightning and surface rainfall during Florida thunderstorms. J. Geophys. Res. Oceans **87**(C13), 11193–11201 (1982)

Rajeevan, M., Kesarkar, A., Thampi, S.B., Rao, T.N., Radhakrishna, B. and Rajasekhar, M.: Sensitivity of WRF cloud microphysics to simulations of a severe thunderstorm event over Southeast India. In: Annales Geophysicae (Vol. 28, No. 2, pp. 603–619). Copernicus GmbH (2010)

Ranalkar, M.R., Chaudhari, H.S.: Seasonal variation of lightning activity over the Indian subcontinent. Meteorol. Atmos. Phys. **104**(1–2), 125–134 (2009)

Reeve, N., Toumi, R.: Lightning activity as an indicator of climate change. Q. J. R. Meteorol. Soc. **125**(555), 893–903 (1999)

SAARC-STORM. SAARC STORM Pilot Field Experiment 2011, SAARC Meteorological Research Centre (SMRC). SMRC Sci. Rep. **46**, 51 (2013)

Singh, D., Buchunde, P.S., Singh, R.P., Nath, A., Kumar, S., Ghodpage, R.N.: Lightning and convective rain study in different parts of India. Atmos. Res. **137**, 35–48 (2014)

Stolz, D.C., Rutledge, S.A., Pierce, J.R.: Simultaneous influences of thermodynamics and aerosols on deep convection and lightning in the tropics. J. Geophys. Res. Atmos. **120**(12), 6207–6231 (2015)

Tajbakhsh, S., Ghafarian, P., Sahraian, F.: Instability indices and forecasting thunderstorms: the case of 30 April 2009. Nat. Hazards Earth Syst. Sci. **12**(2) (2012)

Thompson, C., Lin, Y.: Pre-storm stability conditions of the meso-scale thunderstorm environment during AVE. SESAME–V (20–21 May 1979). In: Proceedings for the 14th Conference on Severe Local storms Indianapolis, Indiana, vol. 29, pp. 131–134 (1985)

Tinmaker, I.R., Chate, D.M.: Lightning activity over India: a study of east–west contrast. Int. J. Remote Sens. **34**(16), 5641–5650 (2013)

Tyagi, A., Sikka, D.R., Goyal, S., Bhowmick, M.: A satellite based study of pre-monsoon thunderstorms (Nor'westers) over eastern India and their organization into mesoscale convective complexes. Mausam **63**(1), 29–54 (2012)

Vujović, D., Paskota, M., Todorović, N., Vučković, V.: Evaluation of the stability indices for the thunderstorm forecasting in the region of Belgrade, Serbia. Atmos. Res. **161**, 143–152 (2015)

Webster, P.J., et al.: The JASMINE pilot study. Bull. Am. Meteor. Soc. **83**(11), 1603–1630 (2002)

Williams, E.R.: Lightning and climate: a review. Atmos. Res. **76**(1–4), 272–287 (2005)

Williams, E., Rosenfeld, D., Madden, N., Gerlach, J., Gears, N., Atkinson, L., et al.: Contrasting convective regimes over the Amazon: Implications for cloud electrification. In: Journal of Geophysical Research: Atmospheres, vol. 107, no. D20, pp. LBA-50 (2002)

Williams, E.R.: The Schumann resonance: a global tropical thermometer. Science **256**(5060), 1184–1187 (1992)

Williams, E.R.: Global circuit response to seasonal variations in global surface air temperature. Mon. Weather Rev. **122**(8), 1917–1929 (1994)

Williams, E.R.: Meteorological aspects of thunderstorms. Handbook Atmos. Electrodynamics **1**, 27–60 (1995)

Williams, Earle R.: The electrification of severe storms. In: Severe Convective Storms, pp. 527–561. Springer (2001)

Williams, E.R., et al.: A radar and electrical study of tropical "hot towers." J. Atmos. Sci. **49**(15), 1386–1395 (1992)

Yamane, Y., Hayashi, T.: Evaluation of environmental conditions for the formation of severe local storms across the Indian subcontinent. Geophys. Res. Lett. **33**(17) (2006)

Yamane, Y., Hayashi, T., Dewan, A.M., Akter, F.: Severe local convective storms in Bangladesh: Part I. Climatology. Atmos. Res. **95**(4), 400–406 (2010)

Yuan, T., Remer, L.A., Pickering, K.E., Yu, H.: Observational evidence of aerosol enhancement of lightning activity and convective invigoration. Geophys. Res. Lett. **38**(4) (2011)

Zheng, Y., Alapaty, K., Herwehe, J.A., Del Genio, A.D., Niyogi, D.: Improving high resolution weather forecasts using the weather research and forecasting (WRF) model with an updated kain-fritsch scheme. Mon. Wea. Rev. **144**, 833–860 (2016)

Zipser, E.J., Cecil, D.J., Liu, C., Nesbitt, S.W., Yorty, D.P.: Where are the most intense thunderstorms on Earth? Bull. Am. Meteor. Soc. **87**(8), 1057–1072 (2006)

Akter, S., Rashid, T., Quadir, D.A., Ahasan, M.N., Shuvo, S.D., Hasan, M.S.: Study of arrival and withdrawal of southwest summer monsoon over Bangladesh and analysis of extreme early and late arrival events using RegCM. Dhaka Univ. J. Earth Environ. Sci. **11**(2), 101–111 (2023). https://doi.org/10.3329/dujees.v11i2.68868

Islam, M.A., Meandad, J., Shuvo, S.D., Kabir, A.: Modeling of lightning events using WRF-derived Microphysical Parameters. Dhaka Univ. J. Earth Environ. Sci. **8**(2), 41–50 (2021). https://doi.org/10.3329/dujees.v8i2.54838

Islam, M.M., Paul, P., Shuvo, S.D., Akter, F., Khan, S.M.: Exploring the use of WRF-ARW model for analyzing heatwaves in Bangladesh. Dhaka Univ. J. Earth Environ. Sci. **12**(1), 9–28 (2024). https://doi.org/10.3329/dujees.v12i1.70461

Nahar, S., Quadir, D.A., Mannan, M.A., Shuvo, S.D.: Prediction of track of super cyclone Amphan using WRF-ARW model. Dhaka Univ. J. Earth Environ. Sci. **10**(2), 35–42 (2022). https://doi.org/10.3329/dujees.v10i2.57513

Sarker, M., et al.: Simulation of structure, intensity and track of super cyclone Amphan using high resolution WRF-ARW model. Dhaka Univ. J. Earth Environ. Sci. **8**(2), 17–23 (2021). https://doi.org/10.3329/dujees.v8i2.54835

Shuvo, S.D.: Climate change in Sundarbans after Cyclone Sidr and Aila: An interpretation of Meteorological Data. In: 5th International Conference on Natural Sciences and Technology (ICNST'18), Chittagong, Bangladesh (2018). https://www.researchgate.net/publication/340983001_Climate_change_in_Sundarbans_after_Cyclone_Sidr_and_Aila_An_interpretation_of_Meteorological_Data

Shuvo, S.D.: Climatology of frequency, life period, energy and speed for tropical disturbances and cyclones over the bay of Bengal. Dhaka Univ. J. Earth Environ. Sci. **10**(1), 23–31 (2021). https://doi.org/10.3329/dujees.v10i1.56277

Shuvo, S.D., Awal, M.R.: Assessing atmospheric instability over the bay of Bengal during October and November months between 2007–2018. Dhaka Univ. J. Earth Environ. Sci. **9**(2), 45–54 (2021). https://doi.org/10.3329/dujees.v9i2.55089

Shuvo, S.D., Rashid, T., Panda, S.K., Das, S., Quadir, D.A.: Forecasting of pre-monsoon flash flood events in the northeastern Bangladesh using coupled hydrometeorological NWP modelling system. Meteorol. Atmos. Phys. **133**(6), 1603–1625 (2021). https://doi.org/10.1007/s00703-021-00831-z

Shuvo, S.D., Sultana, S.S.: Are Tropical Cyclones in Bay of Bengal Getting Stronger? 2022, A52M-1134. AGU Fall Meeting Abstracts (2022a). https://ui.adsabs.harvard.edu/abs/2022AGUFM.A52M1134S

Shuvo, S.D., Sultana, S.S.: Assessing the climatology and synoptic conditions of tropical cyclone recurvature over the bay of Bengal, Bangladesh. Dhaka Univ. J. Earth Environ. Sci. 131–141 (2022b). https://doi.org/10.3329/dujees.v10i3.59079

Sultana, S.S., Shuvo, S.D.: Fog Stability Index: A Combination of NWP Model and Machine Learning Algorithm for Forecasting Fog over Bangladesh during Winter Season. 2022, A55P-1336. AGU Fall Meeting Abstracts (2022). https://ui.adsabs.harvard.edu/abs/2022AGUFM.A55P1336S

Strengthening Health System Resilience: Addressing Vaccine Hesitancy Using Behavioral Theory-Based Social Media Interventions: A Framework and Case Study

Lan Li(✉) and Patty Kostkova

Centre for Digital Public Health in Emergencies, Institute for Disaster and Risk Reduction, University College London, London, UK
lan.li.19@ucl.ac.uk

Abstract. Leveraging social media to enhance vaccination behaviors faces challenges such as a lack of informed strategies, low user engagement, and difficulties in maintaining engagement. To tackle these issues, we propose a systematic framework that integrates behavioral theory, human-centered design, and rigorous evaluation. This paper introduces the framework and assesses its feasibility through a case study, aiming to elucidate lessons learned and potential implications. The case study focuses on developing and evaluating a WeChat intervention to improve influenza vaccine uptake among Chinese university students in the UK. A three-course natural experimental design using mixed methods has been implemented among Chinese university students studying in the UK as the case study. In phase I, a cross-sectional survey was conducted using the Theoretical Domain Framework with mapping to the Capability, Opportunity, Motivation-Behaviour Model. In Phase II, three focus groups were held to collect feedback from users as evidence for a user-centered intervention design. In Phase III, pre- and post-survey was conducted in a quasi-experimental design to evaluate the effectiveness of the intervention. The results showed that social media is a promising tool for enhancing vaccination knowledge and attitudes among Chinese university students and the proposed framework facilitates the intervention design process and provides insights into how it could be tailored to other contexts. Further evaluation is warranted to refine its usability and effectiveness across diverse intervention contexts.

Keywords: Vaccine Hesitancy · Health Intervention · Human-centered Design · Social Media

1 Introduction

The advent of social media has revolutionized human interaction with the world, offering unprecedented opportunities for designing, evaluating, and disseminating impactful health behavior change interventions [1, 2]. Despite its recognized potential, achieving

H. Tatano et al. (Eds.): GSRIDRR 2023, DRRGADRIBS, pp. 426–443, 2025.
https://doi.org/10.1007/978-981-96-6609-6_28

meaningful improvements in vaccination behaviors through social media remains a significant challenge [3, 4]. Research indicates that while social media is extensively used for accessing health information and seeking peer interaction, its effectiveness in influencing vaccination behaviors is limited by various factors [5]. Large-scale social media campaigns often lack theoretically informed strategies and rigorous evaluation, resulting in low user engagement and diminishing participation over time [6]. Conversely, smaller-scale interventions, while more likely to be theoretically grounded and empirically validated, struggle with sustaining user engagement and miss out on the advantages of iterative refinement from user feedback [7].

Addressing these challenges requires enhancements in several key areas. Firstly, interventions should be firmly grounded in behavioral theory to ensure they are underpinned by established psychological principles that drive behavior change [5]. Secondly, they should be supported by quantitative results from population data, providing a solid empirical basis to inform and refine the intervention strategies [8]. Thirdly, a deep qualitative understanding of the target population is crucial to tailor interventions that resonate with the specific attitudes, beliefs, and contexts of the audience [9]. Moreover, the development process should be rapid and iterative, incorporating multiple stages of user feedback to continuously improve the intervention, making it more relevant and effective. Additionally, these interventions should undergo rigorous evaluation to scientifically validate their effectiveness, ensuring they achieve the intended health behavior outcomes [3, 10]. Finally, enhancing user engagement is essential, involving creating interventions that are not only informative but also compelling and interactive, maintaining the audience's interest and encouraging sustained engagement [11].

Existing frameworks for social media interventions lack comprehensive integration of behavioral theory, human-centered design, and evaluation methods tailored for vaccine behavior change. To address this gap, this study aims to develop a systematic framework that synthesizes these elements and applies them in a case study to test its feasibility and applicability. By providing a structured approach, this framework aims to guide the development of effective social media interventions, ultimately contributing to improved vaccine behavior and population health.

2 Proposed Framework

The existing frameworks for designing interventions often fall short of fully integrating behavioral theory, design thinking, and evaluation strategies, particularly for social media interventions targeting health behavior modification. While widely used frameworks like the Behavior Change Wheel offer comprehensive views, they may not address the nuanced requirements of social media interventions [12, 13]. Similarly, frameworks specifically focused on social media delivery may lack the depth needed for designing interventions from scratch, neglecting crucial elements like behavioral theory and design thinking [11]. Thus, there is a clear gap in frameworks that provide detailed, step-by-step guidelines tailored for developing social media interventions aimed at improving vaccine uptake, highlighting the need for a comprehensive framework that integrates these essential elements.

We propose a novel framework integrating behavior change theory, human-centered design, and rigorous evaluation to address intervention development challenges. Utilizing a multi-theoretical approach, such as the COM-B model, allows for a comprehensive understanding of behavior change determinants, particularly valuable for tackling vaccine hesitancy (Liu & Liu, 2021; Garbutt et al., 2018; Williams et al., 2020). Human-centered design (HCD) offers an iterative problem-solving approach rooted in interdisciplinary collaboration, aiming to create empathetic, user-tailored solutions [16]. Rigorous evaluation, guided by frameworks such as MRC guidelines, ensures methodological rigor and valid outcomes, shedding light on intervention effectiveness and challenges [10].

Building on these, we present a comprehensive framework (see Fig. 1) integrating the three elements, specifically tailored for social media interventions targeting vaccine uptake. This structured approach consists of ten steps categorized into three phases: background analysis and conceptualization, human-centered design and development,

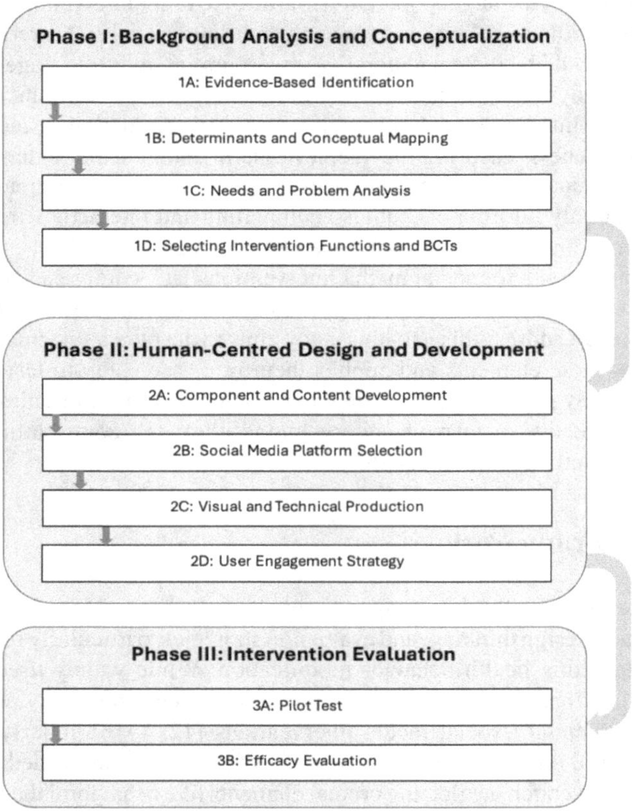

Fig. 1. The proposed framework for designing evidence-based, theory-informed, human-centered social media intervention for improving vaccine behavior.

and evaluation. Each phase is crucial for guiding the intervention development process, ensuring its effectiveness in promoting vaccine behavior change on social media platforms.

The details of the framework are as below:

Phase I: Background Analysis and Conceptualization:

1A: Evidence-Based Identification: Conduct a systematic literature review to identify effective social media strategies for vaccine hesitancy.

1B: Determinants and Conceptual Mapping: Align identified determinants with the COM-B model and TDF domains, facilitating comprehensive behavioral analysis.

1C: Needs and Problem Analysis: Gather feedback from the target population through surveys or interviews to understand barriers and enablers to vaccine acceptance.

1D: Selecting Intervention Functions and BCTs: Translate behavioral insights into actionable intervention strategies by aligning with intervention functions and Behavioral Change Techniques (BCTs).

Phase II: Human-Centered Design and Development:

2A: Component and Content Development: Develop intervention components and content that resonate with the target audience, incorporating evidence-based information and creative formats.

2B: Social Media Platform Selection: Choose appropriate social media platforms based on alignment with intervention goals, target demographics, and content delivery capabilities.

2C: Visual and Technical Production: Create visually appealing and technically robust materials, applying user interface design principles and iterative refinement.

2D: User Engagement Strategy: Develop a detailed plan for engaging users, considering factors like group size, posting frequency, and interaction strategies.

Phase III: Intervention Evaluation:

3A: Pilot Test: Conduct a comprehensive pilot test to assess intervention usability, feasibility, and initial effectiveness, incorporating usability assessment, user feedback, feasibility evaluation, and iterative refinement.

3B: Efficacy Evaluation: Evaluate intervention efficacy through appropriate study designs, outcome assessments, and analytic methods, considering factors like research questions, study design, outcome selection, and statistical analysis plans.

This framework integrates the Behavior Change Wheel (BCW) and Human-Centered Design (HCD) approaches, addressing gaps in existing frameworks for social media health interventions. It emphasizes iterative design, user empathy, theoretical grounding, and rigorous evaluation, aiming to develop effective, user-friendly interventions that promote vaccine behavior change on social media platforms. To evaluate its feasibility, a case study is conducted in the following section.

3 Case Study

3.1 Background

Studies suggest that influenza vaccination rates among university students, both in the UK and globally, are suboptimal, falling short of establishing herd immunity [17]. For instance, vaccination coverage among university students in countries like the USA and Japan remains low [18–20]. Cultural factors and low awareness contribute to these low vaccination rates, with Chinese students likely exhibiting even lower rates, given concerns about vaccine safety prevalent in Chinese society [21, 22]. However, Chinese university students in the UK present an intriguing case for potential intervention studies. They exhibit high levels of eHealth literacy and engage in frequent social media usage, spending significantly more time on social media compared to their American counterparts [23–25]. China boasts the most active social media environment globally, with a vast majority of the population, particularly young people, actively using social media platforms [26].

Harnessing social media for health interventions targeting Chinese university students in the UK holds promise due to the wide reach and influence of the platform [27]. Social media can serve as a vital tool for disseminating health information, addressing vaccine hesitancy, and promoting behavior change [5]. Despite the potential of social media, its utilization for vaccine-related interventions among Chinese students remains largely untapped, presenting an opportunity for further research and intervention development.

Therefore, while previous studies underscore the importance of influenza vaccination for university students, there remains a notable gap in understanding and addressing the needs of international students, particularly those from China. Targeting Chinese university students in the UK for influenza vaccine interventions could yield valuable insights and contribute to broader public health efforts. It is therefore selected as the case study to test the feasibility of the proposed framework for designing and evaluating the intervention for improving vaccination behavior.

3.2 Methods and Results

Phase I

Procedure

The study focused on understanding the barriers and enablers of receiving influenza vaccination among Chinese university students in the UK, where students do not qualify for free vaccines. They can access vaccines for approximately GBP 14.5 from local centers or pharmacies like Boots or Superdrugs [28]. The study was conducted between May and June 2022, deliberately outside the influenza season, and amidst lifted COVID-19 restrictions in the UK. Ethical approval was obtained from the UCL Ethics Committee before initiation (ref no. 21647/001), and a cross-sectional design was adopted. Participants consented online via a survey platform, acknowledging their rights and risks. The study detail was published in 2023 [29].

Chinese university students aged 18 and above residing in the UK were targeted, with inclusion criteria specifying Chinese nationality for participants and their parents. Convenience and purposive sampling were used, with initial dissemination through university WeChat accounts and subsequent snowball sampling within student group chats. They were screened for eligibility including university name and year of study.

Measurement

The study employed the COM-B model and TDF domain framework for vaccination behavior, localized based on existing literature [30]. Questionnaire refinement followed pilot testing, resulting in a final version comprising 36 items covering 7 TDF domains. Likert scale responses assessed agreement levels, with sociodemographic questions preceding. Vaccine acceptance history and intention were measured, with intention questions placed at the beginning and end to mitigate order effects. Descriptive statistics calculated frequencies and percentages, with normality verified using Shapiro-Wilk tests. The paired Wilcoxon Sign Rank test assessed differences in vaccine intention scores. Mean scores for TDF domains were derived, and binary logistic regressions identified factors associated with vaccine acceptance and intention, adjusting for demographics.

Outputs

Participants' influenza vaccination decisions were categorized within the TDF framework. Notable findings include uncertainties about vaccine necessity, availability, and side effects, with students relying heavily on online information. Factors enabling vaccination included belief in vaccine efficacy and motivation to protect vulnerable populations. Logistic regression analysis revealed significant associations between knowledge, reinforcement, and emotion domains with both previous vaccine acceptance and future vaccination intention. Higher knowledge and emotion scores, along with lower belief about consequences scores, correlated with greater intention to receive the vaccine. However, belief about consequences and intention negatively affected vaccine intention among C-UK students and were negatively correlated with previous vaccine acceptance.

The main barrier to influenza vaccination among C-UK students is a lack of knowledge and belief in its consequences, suggesting a need for targeted educational campaigns emphasizing vaccine safety and benefits. Tailoring communication strategies to match the preferences of the target group, such as leveraging social media and influential community members, could enhance vaccination promotion. Limited knowledge about accessing vaccines and related information, particularly among foreign-born Chinese students, highlights the importance of providing clear guidance on vaccine services in the UK. Incorporating the Theoretical Domains Framework (TDF) and COM-B model can aid in understanding vaccination behavior and identifying areas for intervention. Four intervention functions aligned with relevant Behavior Change Techniques (BCTs) were identified, informed by previous studies and survey findings, to guide the development of effective intervention strategies [30]. Table 1 presents the details.

Phase II

Procedure

As a continuation of Phase 1, this phase aims to design and develop a social media intervention to bridge the gaps, informed by the proposed framework. WeChat has been selected as the main mode of delivery due to its popularity. Focus groups were utilized to develop a WeChat-based intervention using human-centered design principles. Convenience sampling via WeChat recruited 11 Chinese university students in the UK, meeting inclusion criteria through pre-chats. Three focus groups, each lasting over two hours, were conducted in a university meeting room, with participants aged 18 or above, native Chinese speakers, and enrolled in a UK university. Consent was obtained verbally and on paper.

Table 1. Intervention functions identified by applying the TDF and BCTTv1 to the study found

COM-B Element	Behavior Targeted	TDF Domain	BCTs	Intervention Function	Potential Intervention Strategies
Capability	Increase knowledge	Knowledge	Provide information about health consequences	Education	Design educational materials emphasizing the safety, necessity, and benefits of the influenza vaccine tailored to the preferences of Chinese university students studying in the UK
			Provide information on where and how to access the vaccine		Offer clear information on accessing vaccines and related information in the UK, addressing language and cultural barriers, and providing guidance on navigating the healthcare system effectively
Opportunity	Enhance accessibility	Environmental	Improve access to vaccination services on campuses and nearby areas	Environmental Restructuring	Enhance the convenience and accessibility of vaccination services to overcome barriers related to accessibility
		Social influences	Leverage social networks and influential community members	Social Support	Engage influential community members and social networks to promote vaccination and disseminate accurate information about the influenza vaccine
		Environmental	Restructure the overall context involving stakeholders to reduce barriers to vaccine and information access	Environmental Restructuring	Collaborate with public sector, universities, and media to create an enabling environment that facilitates vaccine access and information dissemination

(*continued*)

Table 1. (*continued*)

COM-B Element	Behavior Targeted	TDF Domain	BCTs	Intervention Function	Potential Intervention Strategies
Motivation	Address misconceptions	Beliefs about consequences	Challenge misconceptions about vaccine safety and effectiveness	Persuasion	Develop targeted educational campaigns to address misconceptions and enhance confidence in the vaccine's effectiveness
	Increase intention	Emotion	Address emotional factors such as fear of living alone, COVID-19 concerns, and fear of spreading the virus	Persuasion	Tailor interventions to address emotional factors influencing vaccine intention, such as fear of living alone and concerns about COVID-19, to increase vaccination intention among Chinese students
		Reinforcement	Provide reminders and reinforcement about the importance of vaccination	Enablement	Implement reminders and reinforcement strategies to maintain awareness and motivation regarding influenza vaccination, potentially leveraging online platforms frequently used by Chinese students

Focus groups and intervention sessions were held in a university setting, scheduled based on participants' convenience. Ground rules were established, emphasizing privacy and respect, with sessions audio-recorded and notes taken. Data collection included demographic information and discussions on influenza vaccine perceptions and intervention preferences. Iterative feedback guided intervention development, validated through member fact-checking.

Analysis
R Studio analyzed demographic data, while transcribed audio recordings underwent independent verification. Content analysis, utilizing NVivo, involved independent coding reconciled collaboratively. Consensus agreement on themes was reached through participant validation, ensuring alignment with views expressed during focus group sessions.

Output
The focus group discussions highlighted six key themes outlining barriers to receiving the influenza vaccine among C-UK students. These barriers included a lack of awareness, access challenges, concerns about vaccine effectiveness, motivation issues, financial barriers, and potential drawbacks of certain intervention components. For each identified barrier, corresponding solutions were discussed and proposed by the co-creators. These solutions collectively formed a set of intervention principles aimed at addressing the specific challenges faced by the target population. Table 2 presents the details.

Based on the themes identified, the intervention focused on a multi-component approach to support vaccine decision-making among C-UK students. This approach aimed to combat the lack of awareness by providing diverse, scientifically validated information on the importance of vaccination. Access challenges were mitigated by offering clear details on vaccination locations and enabling easy appointment booking. Concerns about vaccine effectiveness were addressed by sharing success stories and scientific evidence. Motivation was enhanced through interventions highlighting the broader societal benefits of vaccination and providing reminders. Financial barriers were addressed through the use of discount incentives and leveraging social platforms for promotion.

The co-creators identified three crucial components to be included in the intervention (See Fig. 2): educational posts, an auto-reply function, and reminders. Educational posts were designed to provide key information related to vaccine access, safety, and necessity, while also addressing access and financial issues. The auto-reply function aimed to deliver personalized instructions and support decision-making regarding vaccination behavior. Reminders played a critical role in promoting timely vaccine uptake by nudging individuals to take proactive steps towards protecting their health.

The intervention delivery mode was selected as Tuisong (creating an official account in WeChat) due to its popularity and high feasibility. Articles emerged as the primary form of information to be shared through the official account. Participants participated in choosing topics, layout details, and the preferred time of day for delivery. They also suggested avoiding article dominance and incorporating interactive features such as Q&A sections to enhance engagement.

Fig. 2. Overview of the final co-created three-component intervention

Following the focus group discussions, a content pool containing 12 articles distributed across three categories was established (See Fig. 2). These categories focused on factual information, sharing experiences, and providing guidance on accessing vaccines. The articles were carefully integrated into the final intervention, which aimed to prompt behavior change effectively by maximizing the synergy between its components.

The intervention was scheduled to send articles three times a week, incorporating educational content, reminders, and interactive features. Each article followed a standard layout and was designed to capture the audience's attention and encourage the uptake of influenza vaccination among C-UK students.

Phase III

Procedure

With the final intervention completed, this phase aims to assess the effectiveness of the intervention in improving the knowledge, attitude, and behavior of the targeted population towards influenza vaccination. A mixed-methods approach with intervention and comparison groups was employed due to resource constraints and ethical concerns regarding a new intervention. A randomized selection within the baseline survey reduced selection bias. The intervention was conducted in the UK from December 19, 2022, to January 16, 2023. A control group, blinded to the intervention, ensured comparative effectiveness. Baseline and follow-up assessments were conducted for both groups.

Recruitment through Chinese social media platforms ensured participant diversity. Consent was obtained digitally, emphasizing voluntariness. Participants were incentivized monetarily for completing surveys and referrals. Inclusion criteria encompassed

Table 2. Themes of barriers and solutions, corresponding intervention principles

Excerpts		Intervention principles
Barriers	Solutions	
Theme 1: Lack of awareness and understanding		
"Not everyone seems to know how important it is to get vaccinated, which means not many people are doing it…"	"Providing a variety of information is important; we prefer making decisions based on judging the information by ourselves."	Provide diverse, scientifically validated information about the importance of vaccination
Theme 2: Access challenges		
"Getting to the places where you can get the shot can be a bit of a mission with limited access and weird hours."	"I need to know where I can get vaccinated, preferably with the option for someone to book it for me so I just need to show up."	Providing clear information on vaccination locations, booking appointments in engaging and digestible formats
Theme 3. Perceived effectiveness		
"Some of my friends got vaccinated, but still got the flu."	"…any data available to demonstrate its effectiveness…", "Is there anyone else's experience I can refer to?"	Providing evidence of scientific validation and possibly sharing success stories
Theme 4. Lack of motivation		
"I always think it (flu vaccine) is not necessary, so I don't have the motivation to take the risk." "Living alone here is tough…" "I always delay receiving it as I think it is not urgent"	"…does this vaccine have any other benefits? like herd immunity…." "If getting vaccinated can help prevent illness, I would still want to get vaccinated." "A reminder would be helpful, especially with timeline showing the deadline"	Emphasize the broader benefits of vaccination to instil a sense of social responsibility and motivation Using the scenario of sick alone may help increase motivation Personal reminders would be helpful
Theme 5. Concerns about side effects		
"I'm not sure about the potential side effects, and that's something I'm concerned about."	"Explaining all the side effects clearly would be more helpful…"	Increase transparency by explaining potential side effects clearly
Theme 6. Financial barriers		
"Money is another thing – it's not always easy for me to shell out for healthcare, including vaccines."	"…discount information might be quite useful…" "Use the discount as the headline, so that people are more willing to click through."	Providing discount information and strategically using discounts to attract individuals, leveraging social platforms for promotion

Chinese international students aged 18 or above studying in the UK, proficient in Mandarin, and familiar with WeChat. The details of the study are available online in pre-print format [31].

Analysis
Questionnaires assessed socio-demographic characteristics and Knowledge, Attitude, and Behavior (KAB) scores at baseline and post-intervention. Control group questions minimized data contamination. Primary outcomes focused on KAB scores, while secondary outcomes included user engagement data and satisfaction with the intervention. Descriptive statistics and correlation tests were employed, with paired Wilcoxon Signed Rank Tests assessing within-group differences in KAB scores and clustered independent Wilcoxon Signed Rank Tests evaluating between-group differences. User satisfaction was analyzed descriptively, and qualitative data underwent thematic analysis.

Output
The primary outcome analysis of the study showcased substantial improvements in knowledge, attitude, and intended behavior scores within the intervention group compared to the control group. Specifically, the intervention group exhibited significant increases in knowledge and attitude scores, while the control group showed only minimal improvements in intended behavior scores. While there was no significant difference in intended behavior scores between the two groups, the vaccination rate in the intervention group was slightly higher than that in the control group.

Furthermore, regression analysis identified gender and self-reported health status as significant factors influencing knowledge scores within the intervention group. In contrast, several factors including gender, health status, past flu vaccination, annual flu vaccination habits, and COVID-19 vaccine acceptance influenced changes in intended behavior scores within the intervention group. These findings highlighted the complex interplay of sociodemographic and health-related factors in shaping knowledge, attitude, and behavior related to influenza vaccination.

The secondary outcome analysis provided insights into participant engagement with the WeChat official account, indicating consistent usage patterns and high user satisfaction. Participants appreciated the detailed information, professional presentation, and news updates provided by the account. However, challenges such as users' numerous subscriptions to various public accounts and the perceived limited appeal of medical information were also noted, suggesting areas for improvement in future interventions.

4 Discussion

4.1 Overview

This paper presents a novel framework aimed at guiding the design, development, and evaluation of social media interventions geared toward enhancing vaccination behavior. It offers a systematic, step-by-step approach designed to assist researchers in leveraging evidence and behavior change theories to create impactful interventions. Building upon the foundation of the Behavior Change Wheel, this framework introduces additional phases for conceptualization and background analysis [30]. It then seamlessly links the outcomes of these phases to identify intervention functions and behavior change techniques that can enhance intervention effectiveness. Moreover, it provides clear guidance on leveraging behavioral strategies to inform the development of more effective technologies. Drawing upon the strengths of human-centered design principles [32], the framework emphasizes co-creation and incorporates Pagoto's social media design principles to tailor interventions for optimal delivery via social media platforms [11]. Notably, while previous literature has highlighted the importance of integrating behavioral theory and design thinking [33], this framework stands out as the first to offer detailed, step-by-step guidance on seamlessly integrating these approaches.

Informed by the framework, the intervention developed for the case study, effectively enhanced students' knowledge and attitudes toward seasonal influenza vaccination. The study highlights social media's vital role in disseminating vaccine-related information and influencing attitudes. However, it suggests the intervention lacked evidence to induce behavioral change, possibly due to external factors like concurrent NHS campaigns promoting flu vaccination during a severe flu season. Demographic factors like gender and health status influenced knowledge enhancement, emphasizing the need for tailored interventions. User engagement with the intervention platform was consistent, with high satisfaction regarding content credibility and relevance. Feedback suggests improvements in emotional content integration, vaccine comparison, and platform usability. Future interventions may benefit from streamlined functionalities and the exploration of alternative platforms.

4.2 Lessons Learned on How to Use the Framework

Enhancing vaccination behavior via social media interventions requires a comprehensive approach that integrates various factors crucial for success. Factors such as behavioral theory, evidence-based practices, human-centered design, and appropriate evaluation methods all play pivotal roles in the effectiveness of such interventions [33]. Addressing these factors collectively demands the involvement of a multidisciplinary team that transcends disciplinary boundaries. This team comprises human-centered designers, programmers, behavioral researchers specialized in behavioral intervention development, and industry partners. A multidisciplinary team possesses the collective knowledge and expertise essential for navigating the complexity of factors involved in vaccination behavior interventions on social media platforms [34]. The engaged team can progress iteratively with agility, fostering collaboration and garnering buy-in at every stage. This inclusive practice ensures alignment with user needs and a firm grasp of the theoretical underpinnings of the intervention. Furthermore, harnessing diverse perspectives within the team fosters the generation of various ideas and prototypes, enhancing creativity and efficacy in the final intervention designs.

Similar to design thinking, the framework for improving vaccination behavior via social media interventions thrives on flexibility [35]. Teams should adapt and combine stages in various ways to suit their needs, enabling exploration of new directions and iterative refinement. High-performing teams often seamlessly blend ideation, prototyping, and feedback-gathering phases, transitioning between them as necessary. For example, they may shift from brainstorming sessions directly into creating rudimentary prototypes to convey ideas, spark discussions, and gather user feedback. This iterative process maximizes creativity, collaboration, and innovation, allowing teams to dynamically navigate the framework.

Balancing divergent and convergent thinking processes is critical in the framework application. Divergent thinking, essential in the early stages, ensures a thorough exploration of diverse ideas and user experiences. As the process evolves, convergent thinking becomes equally important, narrowing down ideas to refine potential solutions and avoid continual revision. Successful teams exhibit a nimble approach, shifting between divergence and convergence as needed throughout the design process [35]. During concept generation, vigorous debate and divergence foster creativity, while during concept selection, focused attention facilitates the refinement of promising solutions. This dynamic approach enhances the likelihood of uncovering potent solutions in the ever-evolving landscape of social media platforms.

4.3 Strengths and Limitations of the Proposed Framework

Strengths

The framework boasts several strengths that enhance intervention design, making it a valuable toolkit for developers seeking to create effective solutions. One notable advantage is its integration of behavioral strategies rooted in the design process. These strategies offer a nuanced approach not fully captured by existing taxonomies of behavior change techniques [36], thereby enriching the repertoire available to intervention designers. Theory, while rich in insights, can often seem abstract and challenging to translate into practical intervention features [37]. It addresses this challenge by advocating for iterative use of ideation, prototyping, and user feedback. This iterative process facilitates the incremental translation of theoretical insights into highly relevant and actionable intervention components. Additionally, The framework sheds light on the importance of creativity, an aspect often overlooked in mHealth and eHealth design literature [33]. Moreover, the framework emphasizes the integration of end-users involvement into the design process. This approach guards against potential mismatches between intervention goals and user preferences, which can undermine user satisfaction and adherence.

Limitations

While the proposed framework offers valuable guidance for intervention development, several limitations warrant consideration. Firstly, the complexity of the framework may pose challenges for less-experienced users, hindering their ability to navigate and implement its strategies proficiently. Practical challenges also arise when working with multidisciplinary teams, a key aspect of the framework [33]. Managing divergent perspectives and fostering productive collaboration within such teams can prove challenging, impacting the efficacy of the intervention development process. Additionally, while the behavioral strategies presented within the framework offer valuable insights, they are not exhaustive. This limitation may restrict the range of approaches available to intervention developers, potentially limiting the effectiveness of interventions developed using the framework.

Another notable limitation is the absence of a suggestion for randomized evaluations of each component and function separately [10, 38]. While it is commonly argued that combining behavioral strategies yields better results, evaluating individual intervention components in isolation may not provide a comprehensive understanding of their efficacy within the broader intervention context [39]. It is important to acknowledge that the framework is not the only approach to developing interventions. Numerous alternative approaches exist, each offering unique insights and methodologies that may complement or advance the principles outlined in this framework. Therefore, it is imperative to consider the diversity of approaches available and adapt methodologies accordingly to suit specific intervention contexts.

Given these limitations, future research should prioritize systematic evaluation of the framework. A rigorous assessment of its effectiveness, usability, and impact on intervention outcomes will provide valuable insights for refining and optimizing the framework for widespread application in intervention development contexts.

5 Conclusion

We proposed a novel framework aimed at guiding the design and evaluation of social media interventions geared toward enhancing vaccine behavior. This framework integrates principles from behavioral theory, design thinking, user-centered design, and evaluation. The primary objective of this framework is to expedite the translation of behavioral theory and evidence into practical interventions, particularly within the realm of addressing vaccination hesitancy. To demonstrate its efficacy, we applied the framework within a specific case study context: improving influenza vaccine uptake among Chinese university students in the UK. Our findings indicate that the final intervention indeed led to significant improvements in vaccine knowledge and attitudes among the target population. Furthermore, we have distilled key insights and lessons learned from our application of the framework, along with identifying potential limitations for future studies. We recommend other researchers adopt and adapt this framework in their work, emphasizing the importance of transparently reporting on their processes and outcomes. By sharing experiences and outcomes, we can collectively enhance the quality, efficacy, and effectiveness of social media interventions aimed at addressing vaccine behavior and beyond.

References

1. Shaw, J.M., Mitchell, C.A., Welch, A.J., Williamson, M.J.: Social media used as a health intervention in adolescent health: a systematic review of the literature. DIGITAL HEALTH SAGE Publications Ltd; 1, 2055207615588395 (2015). https://doi.org/10.1177/2055207615588395
2. Simeon, R., et al.: Behavior change techniques included in reports of social media interventions for promoting health behaviors in adults: content analysis within a systematic review. J. Med. Internet Res. **22**(6), e16002 (2020). https://doi.org/10.2196/16002
3. Guo, Y., Bian, J.: Social media-based health interventions: where are we now? In: Bian, J., Guo, Y., He, Z., Hu, X. (eds.) Social Web and Health Research: Benefits, Limitations, and Best Practices Cham: Springer International Publishing, p. 15–30 (2019). https://doi.org/10.1007/978-3-030-14714-3_2

4. Welch, V., Petkovic, J., Pardo, J.P., Rader, T., Tugwell, P.: Interactive social media interventions to promote health equity: an overview of reviews. Health Promot. Chronic Dis. Prev. Can. **36**(4), 63–75 (2016). PMID:27077792
5. Li, L., Wood, C.E., Kostkova, P.: Vaccine hesitancy and behavior change theory-based social media interventions: a systematic review. Translational Behavioral Medicine ibab148 (2021). https://doi.org/10.1093/tbm/ibab148
6. Winandy, M., Kostkova, P., Quincey, E.D., Louis, C.S., Szomszor, M.: Follow #eHealth2011: measuring the role and effectiveness of online and social media in increasing the outreach of a scientific conference. J. Med. Internet Res. **18**(7), e191 (2016). https://www.jmir.org/2016/7/e191 Journal of Medical Internet Research; 2016 Jul;18(7):e4480. PMID:27436012
7. Escobar-Viera, C.G., et al.: A systematic review of the engagement with social media–delivered interventions for improving health outcomes among sexual and gender minorities. Internet Interv. **1**(25), 100428 (2021). https://doi.org/10.1016/j.invent.2021.100428
8. Elaheebocus, S.M.R.A., Weal, M., Morrison, L., Yardley, L.: Peer-based social media features in behavior change interventions: systematic review. J. Med. Internet Res. **20**(2), e8342 (2018). https://doi.org/10.2196/jmir.8342
9. O'Connor, S., Hanlon, P., O'Donnell, C.A., Garcia, S., Glanville, J., Mair, F.S.: Understanding factors affecting patient and public engagement and recruitment to digital health interventions: a systematic review of qualitative studies. BMC Med. Inform. Decis. Mak. **16**(1), 120 (2016). https://doi.org/10.1186/s12911-016-0359-3
10. Skivington, K., et al.: A new framework for developing and evaluating complex interventions: update of Medical Research Council guidance. BMJ Br. Med. J. Publishing Group **374**, n2061 (2021). PMID:34593508
11. Pagoto, S., et al.: Adapting behavioral interventions for social media delivery. J. Med. Internet Res. **18**(1), e5086 (2016). https://doi.org/10.2196/jmir.5086
12. Michie, S., van Stralen, M.M., West, R.: The behaviour change wheel: a new method for characterising and designing behaviour change interventions. Implementation Sci. **6**(1), 42 (2011). https://doi.org/10.1186/1748-5908-6-42
13. Michie, S., West, R., Brown, J., Gainforth, H.: ABC of Behaviour Change Theories. London: Silverback (2014). https://www.silverbackpublishing.org/products/abc-of-behaviour-change-theories. Accessed 15 Feb 2022
14. Garbutt, J.M., Dodd, S., Walling, E., Lee, A.A., Kulka, K., Lobb, R.: Theory-based development of an implementation intervention to increase HPV vaccination in pediatric primary care practices. Implement. Sci. **13**(1), 45 (2018). https://doi.org/10.1186/s13012-018-0729-6
15. Williams, L., et al.: Towards intervention development to increase the uptake of COVID-19 vaccination among those at high risk: outlining evidence-based and theoretically informed future intervention content. Br. J. Health Psychol. **25**(4), 1039–1054 (2020). https://doi.org/10.1111/bjhp.12468
16. Jacobson, R.: Information Design. MIT Press (2000). ISBN:978–0–262–60035–4
17. Shon, E.-J., Choe, S., Lee, L., Ki, Y.: Influenza vaccination among U.S. college or university students: a systematic review. Am. J. Health Promot. SAGE Publications Inc **35**(5), 708–719 (2021). https://doi.org/10.1177/0890117120985833
18. Kawahara, Y., Nishiura, H.: Exploring influenza vaccine uptake and its determinants among university students: a cross-sectional study. Vaccines Multi. Digital Publishing Inst. **8**(1), 52 (2020). https://doi.org/10.3390/vaccines8010052
19. Matsui, D., Shigeta, M., Ozasa, K., Kuriyama, N., Watanabe, I., Watanabe, Y.: Factors associated with influenza vaccination status of residents of a rural community in Japan. BMC Public Health **11**(1), 149 (2011). https://doi.org/10.1186/1471-2458-11-149
20. Shropshire, A.M., Brent-Hotchkiss, R., Andrews, U.K.: Mass media campaign impacts influenza vaccine obtainment of university students. J. Am. Coll. Health Taylor Francis **61**(8), 435–443 (2013). PMID:24152021

21. Wang, Q., et al.: Influenza vaccination coverage of population and the factors influencing influenza vaccination in mainland China: a meta-analysis. Vaccine **36**(48), 7262–7269 (2018). PMID:30340886
22. Yang, X., et al.: Influenza vaccine effectiveness in mainland china: a systematic review and meta-analysis. Vaccines Multi. Digital Publishing Inst. **9**(2), 79 (2021). https://doi.org/10.3390/vaccines9020079
23. iimedia.cn. 2020–2021 China Mobile Social Industry Report. iimedia (2021). https://www.iimedia.cn/c400/76205.html. Accessed 9 Nov 2021
24. Jamie, S.: 2022 Chinese Social Media Statistics And Trends Infographic. makeawebsitehub.com (2020). https://makeawebsitehub.com/chinese-social-media-statistics/. Accessed 9 Jan 2022
25. Li, X., Liu, Q.: Social media use, ehealth literacy, disease knowledge, and preventive behaviors in the COVID-19 pandemic: cross-sectional study on Chinese netizens. J. Med. Internet Res. **22**(10), e19684 (2020). https://doi.org/10.2196/19684
26. Chiu, C., Ip, C., Silverman, A.: Understanding social media in China. McKinsey Quarterly, p. 4 (2012). https://www.mckinsey.com/~/media/McKinsey/Business%20Functions/Marketing%20and%20Sales/Our%20Insights/Understanding%20social%20media%20in%20China/Understanding%20social%20media%20in%20China.pdf
27. Embassy of China in the UK. Basic information of overseas students, Chinese-funded enterprises, overseas Chinese and other groups in the UK (2021). https://www.mfa.gov.cn/ce/ceuk//chn/lsfw/yingredianfangzuifangfan2021/t1877042.htm. Accessed 8 May 2022
28. NHS. Flu vaccine (2019). https://www.nhs.uk/conditions/vaccinations/flu-influenza-vaccine/. Accessed 8 May 2022
29. Li, L., Yang, L., Wang, Q., Wood, C.E., Kostkova, P.: Comparing factors influencing seasonal influenza vaccine acceptance and intentions among Chinese university students residing in China and UK: A cross-sectional study. Human Vaccines Immunotherapeutics Taylor Francis **19**(3), 2290798 (2023). PMID:38111087
30. Michie, S., van Stralen, M.M., West, R.: The behaviour change wheel: a new method for characterising and designing behaviour change interventions. Implement. Sci. **6**(1), 42 (2011). https://doi.org/10.1186/1748-5908-6-42
31. Li, L., Wood, C.E., Kostkova, P.: Evaluating a WeChat-based Intervention to Enhance Influenza Vaccination Knowledge, Attitude and Behavior among Chinese University Students Residing in the UK: A Controlled, Quasi-experimental, Mixed-methods Study [Pre-print]. JMIR Preprints (2024). https://preprints.jmir.org/preprint/55706. Accessed 28 Mar 2024
32. Vial, S., Boudhraâ, S., Dumont, M.: Human-centered design approaches in digital mental health interventions: exploratory mapping review. JMIR Mental Health **9**(6), e35591 (2022). https://doi.org/10.2196/35591
33. Mummah, S.A., Robinson, T.N., King, A.C., Gardner, C.D., Sutton, S.: IDEAS (Integrate, Design, Assess, and Share): a framework and toolkit of strategies for the development of more effective digital interventions to change health behavior. J. Med. Internet Res. **18**(12), e5927 (2016). https://doi.org/10.2196/jmir.5927
34. Hermawati, S., Lawson, G.: Managing obesity through mobile phone applications: a state-of-the-art review from a user-centred design perspective. Pers. Ubiquit. Comput. **18**(8), 2003–2023 (2014). https://doi.org/10.1007/s00779-014-0757-4
35. Seidel, V.P., Fixson, S.K.: Adopting design thinking in novice multidisciplinary teams: the application and limits of design methods and reflexive practices. J. Prod. Innov. Manag. **30**(S1), 19–33 (2013). https://doi.org/10.1111/jpim.12061
36. Michie, S., et al.: The behavior change technique taxonomy (v1) of 93 hierarchically clustered techniques: building an international consensus for the reporting of behavior change interventions. Ann. Behav. Med. **46**(1), 81–95 (2013). PMID:23512568

37. Glanz, K., Bishop, D.B.: The role of behavioral science theory in development and implementation of public health interventions. Ann. Rev. Public Health Ann. Rev. **31**, 399–418 (2010). https://doi.org/10.1146/annurev.publhealth.012809.103604
38. Collins, L.M., Murphy, S.A., Strecher, V.: The multiphase optimization strategy (MOST) and the sequential multiple assignment randomized trial (SMART): new methods for more potent ehealth interventions. Am. J. Prev. Med. **32**(5), S112–S118 (2007). Elsevier, https://doi.org/10.1016/j.amepre.2007.01.022
39. Fjeldsoe, B.S., Miller, Y.D., O'Brien, J.L., Marshall, A.L.: Iterative development of MobileMums: a physical activity intervention for women with young children. Int. J. Behav. Nutr. Phys. Act. **9**(1), 151 (2012). https://doi.org/10.1186/1479-5868-9-151

Temporal Variation of Large Outer Rise Earthquakes at the Circum-Pacific Subduction Zones

Ling Bai[1,2(✉)]

[1] State Key Laboratory of Tibetan Plateau Earth System, Resources and Environment, Institute of Tibetan Plateau Research, Chinese Academy of Sciences, Beijing 100101, China
bailing@itpcas.ac.cn

[2] University of Chinese Academy of Sciences, Beijing 100049, China

Abstract. Due to the subduction of oceanic plates beneath overriding continental plates, large coupling areas can form in the shallow region of a subduction zone. The rapid rupture of the coupling zone is directly related to the occurrence of large earthquakes at the plate boundaries. The earthquakes within the subducting plate near the trench axis (i.e., outer rise earthquakes) can be used to infer large-scale stress regimes around the interplate coupled region. Over one hundred $M_W \geq$ 6.0 outer rise earthquakes are listed in the Harvard University centroid moment tensor catalog dating back to 1990. We discuss individual correlations between large outer rise seismicity and interplate seismicity for three regions: the Kurile Islands, northeastern Japan, and the northern Tonga. We find that both tensional and compressional outer rise earthquakes have occurred in strong coupled subduction zones. The reverse faulting outer rise events tend to occur before, and the normal faulting outer rise events tend to occur after a large nearby interplate earthquake. The stress regime around the plate subduction zone varies both temporally and spatially due to the cyclic influences of megathrust earthquakes.

Keyword: Subduction zones · Interplate earthquakes · Outer rise earthquakes · Stress regimes

1 Introduction

The subduction of an oceanic plate beneath an overriding plate can create a trench as a result of plate movement. As the subducting plate moves toward the trench, it bends and forms a topographic rise outside of the trench. Outer rise earthquakes occur within the subducting plate on the outside of the trench axis. These events fall into two main categories: compressional (thrust) events and tensional (normal) events. Large normal-faulting outer rise earthquakes are though to be an important way that subducting plates transport water and other volatiles into the deep mantle. They have also been observed to correlate with the seismic cycle of large interplate earthquakes. The study of outer rise earthquakes is therefore useful for assessing the chemical reactions such as serpentinization in the lower crust and the seismic potential of a megathrust earthquake on an intermediate time scales.

H. Tatano et al. (Eds.): GSRIDRR 2023, DRRGADRIBS, pp. 444–453, 2025.
https://doi.org/10.1007/978-981-96-6609-6_29

Outer rise earthquakes were first reported by Stauder and Bollinger (1966), who has studied a normal faulting event in the Aleutian trench and suggested that the cause of this event was correlated to the bending of the subducting oceanic plate. The largest instrumentally recorded outer rise earthquake is the 1933 Sanriku-Oki earthquake ($M_W = 8.6$) in northeastern Japan (Kanamori, 1971). This normal faulting event has generated huge tsunamis of up to 25 m high along the coast. Large amplitude short-period seismic waves are typical for outer rise faulting (Lay et al., 2010). The reverse faulting outer rise events tend to occur before, and normal faulting outer rise events tend to occur after a large nearby interplate earthquake (e.g., Christensen and Ruff, 1983; Christensen and Ruff, 1988; Lay et al., 1989).

In this study, we examine the distribution of outer rise earthquakes based on individual publications or the Harvard University centroid moment tensor. We discuss individual correlations between large outer rise earthquakes and interplate earthquakes for three regions: the Kurile Islands, northeastern Japan, and the northern Tonga. This analysis provides new evidence for the temporal link between large interplate thrust activity and outer rise earthquakes.

2 Distribution of Large Earthquakes and Outer Rise Earthquakes

According to the USGS PDE catalog, there were 34 earthquakes of $M_W \geq 8.0$ in the time period of 1962–2022. All of them are located at the circum-Pacific subduction zones. Among them, 25 earthquakes are of thrust type, 5 are of strike slip type, and 4 are of normal fault type. Most of them are shallow earthquakes in the depth range of 0–75 km, except for 2 deep normal fault located at the Tonga and south America area.

Outer rise earthquakes are distinguished from other subduction zone seismicity by their proximity to the trench. Events occurring seaward of the trench are easily separated from the interplate seismicity. However, events which are located on the subducting side of the trench require additional constrains to separate them from possible understusting events or seismicity in the overriding plate. Focal mechanisms and focal depths can be used to distinguish them from non-outer rise earthquakes.

Considering the uncertainty in the focal depth of earthquakes, the location of the earthquake epicenter, and the location of the trench, we select earthquakes with the following characteristics as outer rise events: 1) They are located within 200 km seaward of the trench; 2) They are located within 50 km landward of the trench with a focal depth between 0 and 50 km.

The selected outer rise earthquakes are shown at the Fig. 1 and Table 1. Among the 63 outer rise earthquakes of $M_W \geq 6.5$, 12 events are of compressional type and 50 are tensional type. For the thrust earthquakes, the focal mechanisms are clearly distinguished from the nearby interplate seismicity. Most of these compressional outer rise earthquakes are located at the western Pacific subduction zones.

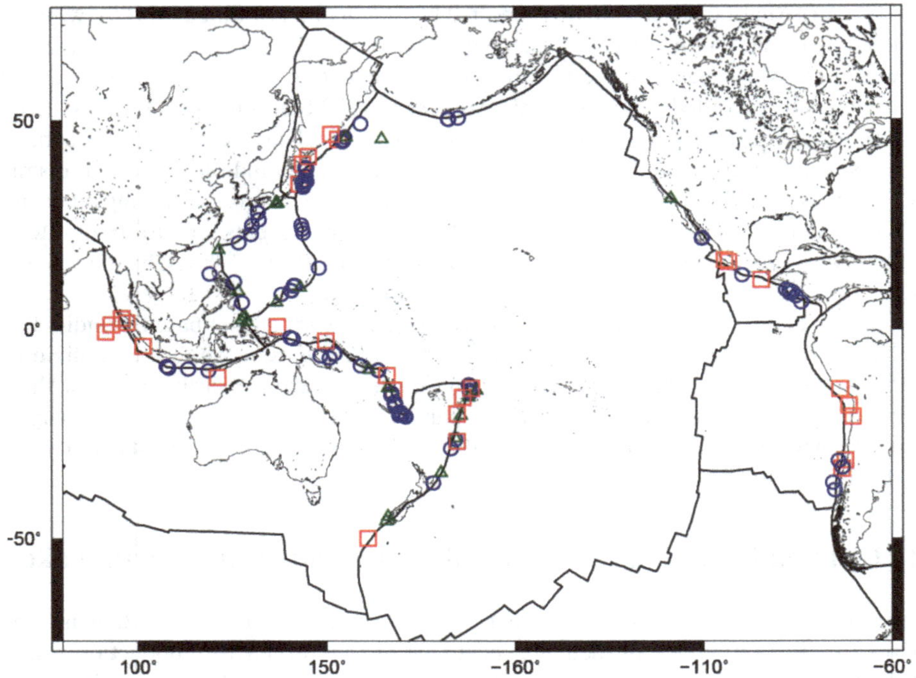

Fig. 1. Distribution of $M_W \geq 8.0$ interplate earthquakes (red squares), and $M_W \geq 6.0$ tensional (blue circles) and compressional (green triangles) outer rise earthquakes. The earthquakes are taken from individual studies and the Harvard CMT catalog. Black thick lines are the plate boundary.

3 Results

We discuss individual correlations between large compressional outer rise seismicity and subduction seismicity for three regions: the Kurile Islands, northeastern Japan, and the northern Tonga. These regions are known to have interplate megathrust earthquakes.

3.1 Kurile Islands

The Kamchatka-Kurile region is centered on a 500-km long segment, where the Pacific plate is subducting beneath the Okhotsk plate at a convergence rate of 7 ~ 8 cm/yr (Fig. 2). The trench segment is flanked by the aftershock areas of the January 4, 1952 ($M_W = 9.0$) Kamchatka event to the northeast and the October 13, 1963 ($M_W = 8.5$) Kurile Islands event to the southwest (Kelleher et al., 1973; Schwartz and Ruff, 1987). A large compressional outer rise event (Mb = 7.7) occurred on March 16, 1963 in the center of this region (Stauder and Bollinger, 1966). An underthrusting earthquake of $M_W = 8.3$ occurred on November 15, 2006 near the trench (Ammon et al., 2008), which is further seaward than typical great earthquakes along the Kurile island arc.

Before this event, the segment has no known historical occurrence of large underthrusting events. An event in 1915 may have ruptured part of this region, but it is not known whether this event represents interplate thrusting (McCann et al., 1979; Lay et al., 1982). A large tensional outer rise event ($M_W = 8.2$) occurred on January 13, 2007 southwest of the 1963 event (Ammon et al., 2008). A compressional outer rise earthquake of $M_W = 7.4$ occurred on January 15, 2009 northeast of the 1963 event.

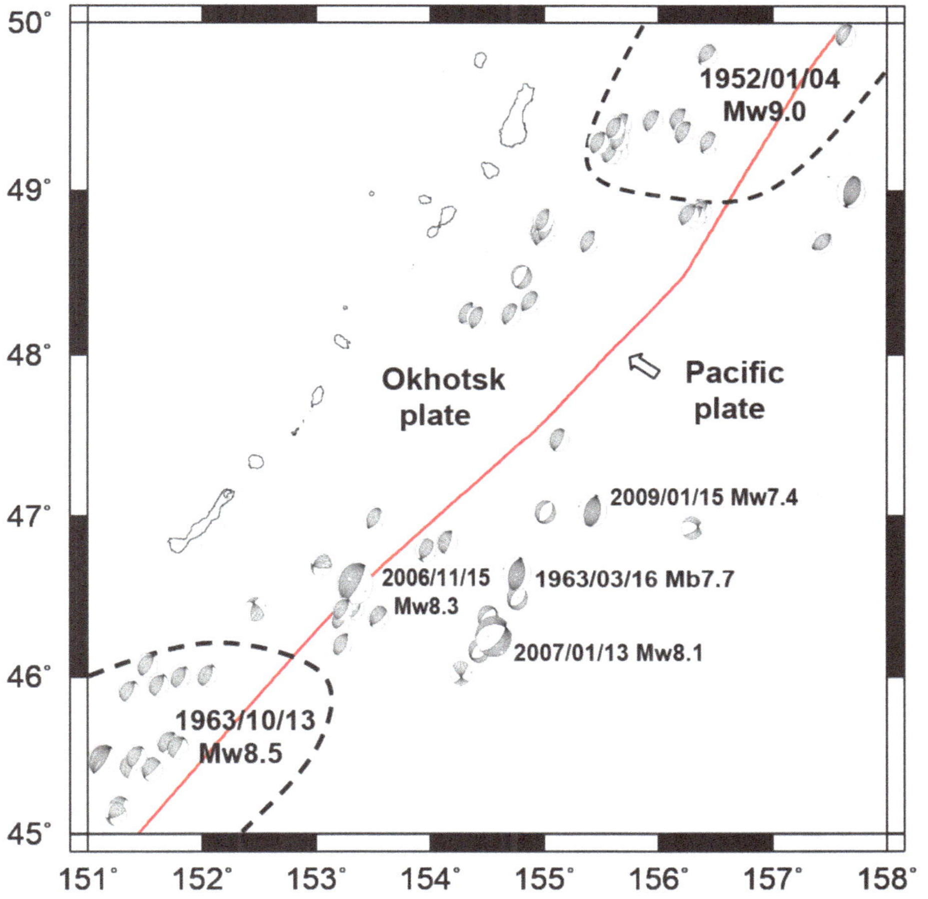

Fig. 2. Interplate earthquakes and outer rise earthquakes in the Kurile region. The red line shows the plate boundary between the Pacific plate and the the Okhotsk plate.

3.2 Northeastern Japan

The Japan trench off the densely populated Tohoku region on the island of Honshu is one of the major earthquake generating regions on the Earth. Here the Pacific plate is subducting beneath the overriding Eurasian plate in an average direction of N62°W at a rate of ~ 8 cm/yr (e.g., DeMets et al., 2000) (Fig. 3). Due to the high convergence rate and strong interplate coupling, the Japan trench has hosted dozens of large to great ($M_W = 7.0$–8.5) earthquakes since last century.

On March 11, 2011, an Mw 9.1 great earthquake occurred at the central part of the Japan trench, causing the largest slip more than 50 m in the recorded history and caused devastating damage from the earthquakes and subsequent tsunamis (e.g., Ide et al., 2011; Simons et al., 2011). The aftershocks of the great earthquake have a wide range of focal mechanisms. Two compressional outer rise earthquakes have been recorded in the region. The July 23, 1982 (Mw7.0) and December 20, 2018 (Mw 6.3) earthquakes occurred before the 2011 Mw9.1 earthquake. The December 7, 2012 earthquake (Mw7.2) occurred after the 2011 Mw9.1 earthquake. There is a subset of tensional aftershocks at the outer rise region, including the March 11, 2011 earthquake (M_W7.6) event and the October 25, 2013 earthquake (M_W7.1).

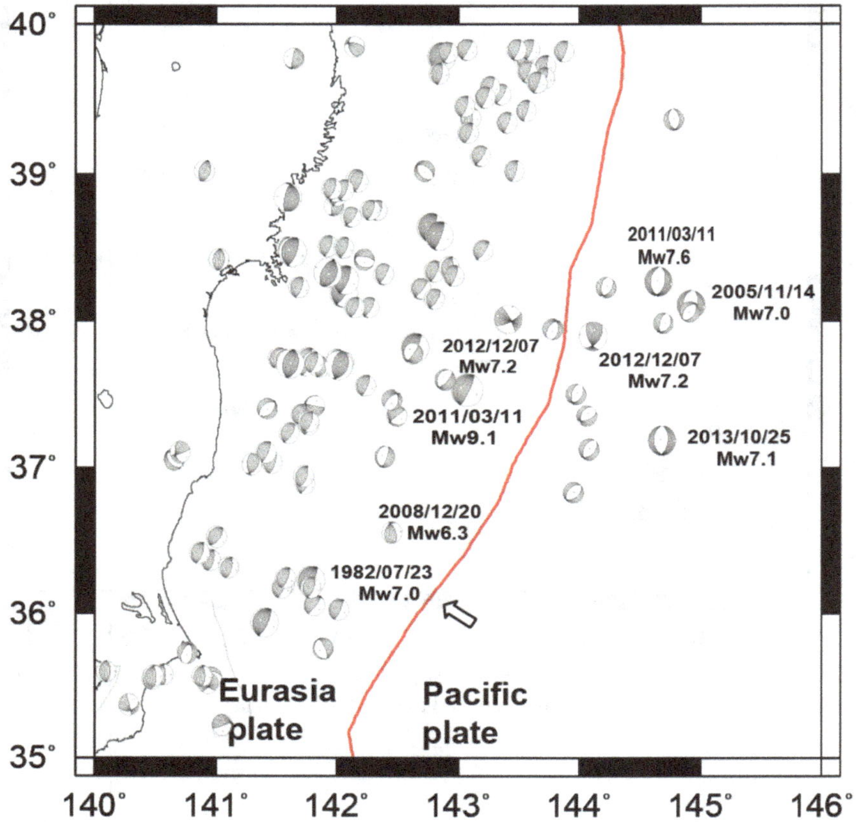

Fig. 3. Interplate earthquakes and outer rise earthquakes in the northeastern Japan region. The red line shows the plate boundary between the Pacific plate and the Euasian plate.

3.3 Northern Tonga

The northern Tonga region is unusual because it is controlled by an unusual corner geometry. The Pacific plate has been subducting beneath the Indo-Australian Plate at a convergence rate of 10–20 cm/yr (Bevis et al., 1995) (Fig. 4). It is a typical deep subduction zone with active seismicity extending to depths of 700 km. Since 1962, there have been approximately 17,200 earthquakes of Mw $\geq$ 5.0, including 11,200 shallow earthquakes (0–70 km), 3,000 intermediate depth (70–300 km) earthquakes and 3,000 deep (300–700 km) earthquakes.

On September 29, 2009, a tensional outer rise earthquake of $M_W = 8.1$ occurred very close to the corner. An underthrust earthquake of $M_W = 8.0$ occurred close to the 2009 event at almost the same time. It is difficult to know which event occurs earlier (Lay et al., 2010; Beavan et al., 2010) because the stresses in this region are very complicated (Christensen and Ruff, 1988). On April 2, 1977 ($M_W = 7.3$), an outer rise event occurred

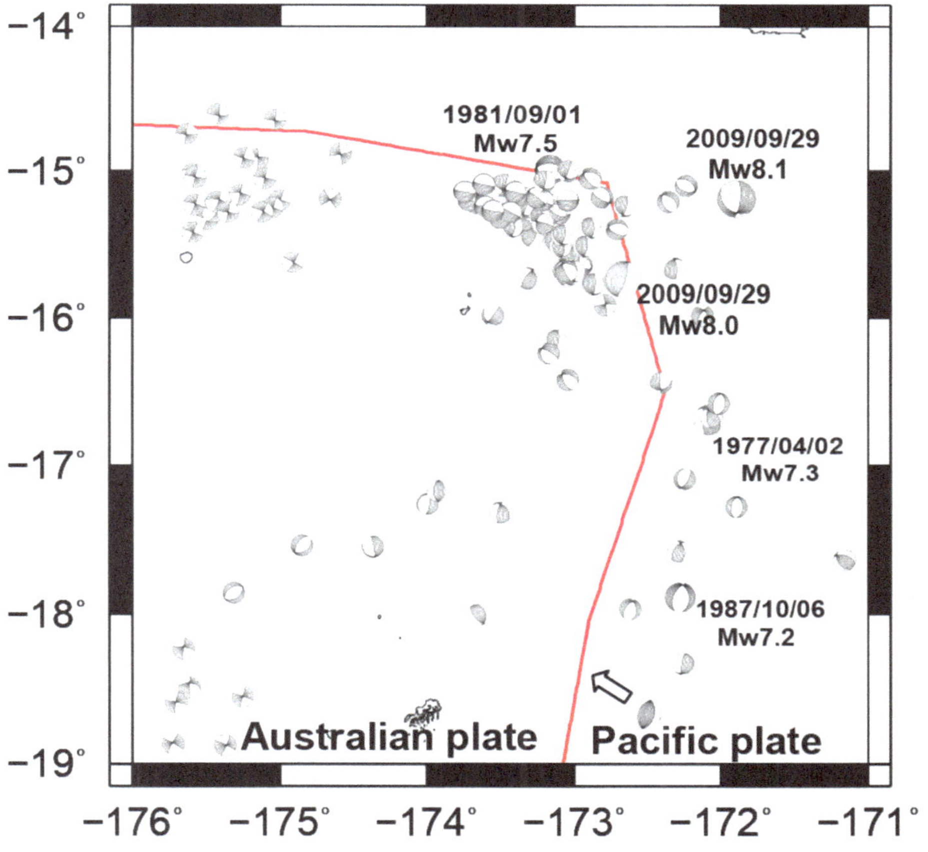

Fig. 4. Interplate earthquakes and outer rise earthquakes in the northern Tonga region. The red line shows the plate boundary between the Pacific plate and the Indo-Australian plate.

Table 1. List of large shallow outer rise earthquakes with Mw ≥ 6.5. T, C and S are tensional (normal), compressional (reverse) and strike-slip faults, respectively. Source parameters are taken from the gCMT catalog.

Type	Data yyyy/mm/dd	Time hh:mm	Lat °N	Lon °E	H km	M_W
South America (−80 ~ −70, −45 ~ −5)						
T	1998/04/01	22:42	−40.32	−74.87	9	6.7
T	2001/04/09	09:00	−32.67	−73.11	11	6.6
T	2010/03/11	14:39	−34.53	−72.09	15	6.9
T	2020/12/27	21:39	−39.34	−74.99	10	6.7
North America (−120 ~ −85, 5 ~ 50)						
T	1978/11/29	19:52	16.01	−96.59	18	7.7
T	1982/06/19	06:21	13.25	−89.40	83	7.3
T	1998/08/23	13:57	11.66	−88.04	55	6.7
T	2001/01/13	17:33	13.05	−88.66	60	7.7
T	2014/07/07	11:23	14.74	−92.41	60	6.9
T	2016/11/24	18:43	11.96	−88.84	10	6.9
T	2019/05/30	09:03	13.24	−89.27	65	6.6
Kurile Islands, Aleutian Islands (151–158, 45–50)						
T	2006/11/15	11:40	46.48	154.73	10	6.7
T	2007/01/13	04:23	46.24	154.52	10	8.1
C	2009/01/15	17:49	47.02	155.40	45	7.4
C	2020/03/25	02:49	48.99	157.69	57	7.5
Japan (140–146, 35–40)						
C	1982/07/23	14:23	36.21	141.76	34	7.0
T	2005/11/14	21:38	38.11	144.90	11	7.0
T	2011/03/11	06:25	38.27	144.63	21	7.6
T	2011/03/12	01:47	37.60	142.87	13	6.5
S	2011/07/10	00:57	38.01	143.39	25	7.0
C	2012/12/07	08:18	37.89	144.09	36	7.2
T	2013/10/25	17:10	37.17	144.66	10	7.1
Ryukyu, Philippines, Izu, Mariana, Solomon, Java Islands						
T	1992/10/31	14.34	−02.39	141.22	33	6.5
C	1993/06/04	10:49	03.74	128.46	28	6.0

(*continued*)

Table 1. *(continued)*

Type	Data yyyy/mm/dd	Time hh:mm	Lat °N	Lon °E	H km	Mw
T	1996/11/06	20:00	28.00	143.54	9	6.5
T	1997/03/11	19:22	07.74	127.65	10	6.8
T	1995/10/18	10:37	27.92	130.34	27	7.1
C	1996/09/05	23:42	21.90	121.50	20	6.8
T	2000/07/16	03:57	−07.75	150.92	10	6.6
C	2003/05/26	19:23	02.35	128.85	31	6.9
T	2005/01/16	20.17	10.93	140.84	25	6.6
T	2005/12/11	14:20	−06.57	152.20	10	6.6
C	2007/07/26	05:40	02.87	127.46	25	6.9
T	2008/03/03	14:11	13.35	125.63	24	6.9
T	2010/12/21	17:19	27.06	143.73	19	7.4
T	2010/08/13	21:19	12.50	141.61	12	6.9
T	2011/04/03	20.06	−10.06	107.72	19	6.7
C	2012/08/31	12:47	10.84	126.70	35	7.6
New Hebrides (155 ~ 175, −25 ~ −8)						
T	1991/02/09	16:18	−09.86	159.09	10	6.9
C	2010/04/11	09:40	−11.02	161.24	36	6.8
T	2004/11/09	23:58	−11.15	163.71	13	6.9
C	1992/02/13	01:29	−15.93	166.30	24	6.8
T	2012/02/02	13:34	−17.81	167.15	10	7.0
T	2010/12/25	13:16	−19.64	168.01	14	7.3
T	2011/05/10	08:55	−20.22	168.15	14	6.7
T	2011/01/13	16:16	−20.60	168.33	13	6.9
T	2004/01/03	16:23	−22.25	169.68	22	7.1
T	1995/05/16	20:12	−23.01	169.89	33	7.7
T	1991/12/11	20.39	−23.39	171.07	34	6.5
T	2003/12/26	21.26	−22.27	169.31	10	6.7
Tonga (−174 ~ −171, −18.5 ~ −14.5)						
C	1977/04/02	07:15	−16.70	−172.10	33	7.3
T	1978/06/17	15:11	−17.10	−172.26	33	6.9
T	1981/09/01	09:29	−14.99	−173.17	15	7.5
T	1987/10/06	14:19	−17.90	−172.29	33	7.2

(continued)

Table 1. *(continued)*

Type	Data yyyy/mm/dd	Time hh:mm	Lat °N	Lon °E	H km	M_W
T	1990/01/04	05:32	−15.05	−172.90	83	6.5
T	1992/05/12	18.05	−16.45	−172.42	33	6.7
T	2006/09/28	06:22	−16.59	−172.03	28	6.9
T	2009/09/29	17:48	−15.19	−171.91	12	8.1
T	2009/09/29	17:48	−15.19	−171.91	12	8.1

close to this earthquake doublet. The normal faulting earthquake of October 06, 1987 ($M_W = 7.2$) occurred further south of the corner.

4 Discussion and Conclusions

We have used over 100 outer rise earthquakes in the circum-Pacific subduction zone with known focal mechanisms to investigate the relationship between stresses in the outer rise and interplate seismicity. These events occur within the downgoing oceanic plate near the trench axes and generally fall into the tensional and compressional categories. The stress regime around the plate subduction zone varies both temporally and spatially due to the cyclic influences of megathrust earthquakes.

In the three example areas, $M_W \geq 7.0$ compressional outer rise earthquakes were followed by megathrust events that occurred 43, 29, and 32 years after for the Kurile, Japan and north Tanga areas, respectively. The occurrence of compressional outer rise earthquakes suggests that compressional stress is accumulating in the adjacent interplate region and that there is the potential for a future large underthrusting event in the region. Compressional outer rise events have nearly the same compressional axes as the nearby megathrust event. Tensional outer rise earthquakes of $M_W \geq 7.0$ tend to occur within a very short time period after the megathrust events, within two month for the Kurile area, and on the same day for the Japan and north Tonga area. Tensional stress axes are also oriented approximately horizontal to the trench.

The stress regime around the plate subduction zone varies both temporally and spatially due to the cyclic influences of megathrust earthquakes. Thus there are three stages of the cycle from underthrusting to tensional outer rise regime to compressional outer rise regime, with only the occurrence of the next underthrusting event required to complete the cycle. The occurrence of compressional outer rise events is useful for assessing the seismic potential of a region on an intermediate time scale.

Acknowledgements. We thank the global CMT for providing the catalogs of earthquake and focal mechanism. The Generic Mapping Tools (Wessel and Smith, 1995) were used to prepare the figures. This research is funded by the grants of the National Nature Science Foundation of China (No. 42130312).

References

Ammon, C.J., Kanamori, H., Lay, T.: Great earthquake doublet and seismic stress transfer cycle in the central Kuril islands. Nature **451**, 561–565 (2008). https://doi.org/10.1038/nature06521

Beavan, J., et al.: Near-simultaneous great earthquakes at tongan megathrust and outer rise in September 2009. Nature **466**, 959–963 (2010)

Christensen, D.H., Ruff, L.J.: Outer-rise earthquakes and seismic coupling. Geophys. Res. Lett. **10**, 697–700 (1983)

Christensen, D.H., Ruff, L.J.: Seismic coupling and outer rise earthquakes. J. Geophys. Res. **93**, 13421–13444 (1988)

DeMets, C., Gordon, R.G., Argus, D.F.: Geologically current plate motions. Geophys. J. Int. **181**, 1–80 (2000). https://doi.org/10.1111/j.1365-246X.2009.04491.x

Ide, S., Baltay, A., Beroza, G.C.: Shallow dynamic overshoot and energetic deep rupture in the 2011 mw 9.0 tohoku-oki earthquake. Science **332**(2011). https://doi.org/10.1126/science.120 7020

Kanamori, H.: Seismological evidence for a lithospheric normal faulting; the Sanriku earthquake of 1933. Phys. Earth Planet. Inter. **4**, 289–300 (1971)

Kelleher, J., Sykes, L., Oliver, J.: Possible criteria for predicting earthqaukes locations and their application to major plate boundaries of the Pacific and the Caribbean. J. Geophys. Res. **78**, 2547–2585 (1973)

Lay, T., Kanamori, H., Ruff, L.: The asperity model and the nature of large subduction zone earthquakes, Earthquake Predict. Res. **1**, 3–71 (1982)

Lay, T., Christensen, D.H., Astiz, L., Kanamori, H.: Temporal variation of large intraplate earthquakes in coupled subduction zones. Phys. Earth Planet. Inter. **54**, 258–312 (1989)

Lay, T., Ammon, C.J., Kanamori, H., Rivera, L., Koper, K.D., Hutko, A.R.: The 2009 Samoa-Tonga great earthquake triggered doublet. Nature **466**(7309), 964–968 (2010)

Lay, T., Ammon, C.J., Kanamori, H., Rivera, L., Koper, K., Hutko, A.R.: The 29 September 2009 Great Samoa earthquake. Nature **466**, 964–968 (2010). https://doi.org/10.1038/nature09214

McCann, W.R., Nishenko, S.P., Sykes, L.R., Krause, J.: Seismic gaps plate tectonics: Seimic potential for majir boundaries, Pure Appl. Geophys. **117**, 1082–1147 (1979)

Schwartz, S.Y., Ruff, L.J.: Asperity distribution and earthquake occurrence in the southern Kurile Islands arc. Phys. Earth Planet. Inter. **49**, 54–77 (1987)

Simons, M., et al.: The 2011 magnitude 9.0 tohoku-oki earthquake: mosaicking the megathrust from seconds to centuries. Science **332**(2011). https://doi.org/10.1126/science.1206731

Stauder, W., Bollinger, G.A.: The S-wave project for focal mechanism studies: earthquakes of 1963. Bull. Seism. Soc. Am. **56**, 1363–1371 (1966)

Wessel, P.S., W.H.F.: New version of the Generic Mapping Tools releases. EOS Trans AGU **76**(33), 329 (1995)

Effect of Pool Fire Toxicity in a Naturally Ventilated Compartment

Aishwarya Narang[1(✉)], Ravi Kumar[2], and Amit Dhiman[3]

[1] Centre of Excellence in Disaster Mitigation and Management, Indian Institute of Technology Roorkee, Roorkee, India
anarang@dm.iitr.ac.in

[2] Mechanical and Industrial Engineering Department, Indian Institute of Technology Roorkee, Roorkee, India
ravi@me.iitr.ac.in

[3] Department of Chemical Engineering, Indian Institute of Technology Roorkee, Roorkee, India
amit.dhiman@ch.iitr.ac.in

Abstract. Toxic gases produced during a fire must be understood while assessing the fire threat. In reality, carbon monoxide poisoning brought on by inhaling the gas accounts for the majority of fire-related fatalities. Smoke is regarded as the greatest threat to human life in the event of unintentional building fires. Additionally, most countries' building fire safety norms and regulations do not include smoke toxicity requirements. Commercial enterprises like chemical, nuclear, and thermal power plants have fuel placements inside enclosures elevated far above the floor.

Additionally, the cabin position of the ship for storing marine diesel fuel is above the base, suggesting enhanced circumstances for a pool fire. From the fire safety perspective, assessing the toxicity of the gases produced under various pool fire circumstances inside the compartment is necessary. Therefore, this study investigates the effect of diesel pool fire in a full-scale compartment on different elevation levels. Full-scale compartment simulations were performed with 0.8 m diameter diesel fuel pans of varying heights in the central location. The analyzed parameters were the flame temperature, door temperature and smoke layer height. The effect of various crucial modeling Smagorinsky constants was studied on the elevated pool fires.

Understanding the complexity of a large-scale test environment with changing combustion conditions both momentarily and geographically is crucial. Therefore, the Fire Dynamic Simulator software results were validated with the experiments, which agrees with the data. The results from the study are significantly valuable for understanding the behavior of hazardous gases in commercial enterprises, promoting disaster risk reduction in case of fire accidents.

Keywords: compartment fire · fire dynamic simulator · flame temperature · toxicity · carbon monoxide

H. Tatano et al. (Eds.): GSRIDRR 2023, DRRGADRIBS, pp. 454–473, 2025.
https://doi.org/10.1007/978-981-96-6609-6_30

1 Introduction

Modern structures are common and highly complex; examples include residential complexes, office buildings, educational institutions, and industrial facilities. Several fire incidents in the past have resulted in the loss of life and expensive assets, raising concerns about building fire safety regulations. A thorough understanding of various fuels under various ventilation settings and the interaction between fire and its surroundings are necessary when designing a fire protection system. Extensive experimentation and better measurement methods can improve the complexity of understanding fire behavior.

Many experiments with solid, liquid, and gaseous fuels have been carried out in full-scale and reduced-scale compartments throughout the past few decades. Numerous studies have examined the characteristics of compartment fires and connections between flame height, plume centerline temperature, velocity, and hot gas layer temperature (Narang et al., 2024). Understanding of turbulent theory, heat transmission, fluid dynamics, and combustion kinetics are all involved in the pool fire inside the compartment. Fuel-burning rate and combustion heat affect how much energy fuel releases (Chaudhary et al., 2018). Liu et al. (2016) investigated burning intensity, combustion yields, and combustion efficiency in compartment fires by conducting pool fire tests with varying ceiling heights. Their experimental findings suggest that a decrease in ceiling height may reduce the mass burning rate. Previous research has demonstrated that direct radiation through flame on the fuel surface boosted the fuel's burning rate by approximately seven times. Many experiments with enclosure fires have been carried out to investigate the thermal environment that arises from burning different fuels (Narang et al., 2022). These experimental studies are expensive and require a large amount of resources. Mathematical modeling proves useful in reducing these costs.

Fire Dynamics Simulator (FDS) is more often used in fire modelling as it focuses on the movement of smoke and heat for low-speed flows in a large eddy simulation (LES) in enclosed fires. The application of FDS is intended for building fire simulation. Steckler et al. (1982) conducted some pioneering work in the field of fire. Numerous validation studies have been performed, considering the work done by Steckler et al. (1982) as a test study for validation purposes. The ability of computational fluid dynamics (CFD) fire field models to forecast the fire environment was also shown by Mahalingam et al. (2012).

The LES modeling equations state that the eddy viscosity is directly proportional to the square of the sub-grid scale coefficient, C_s, and that the eddy viscosity and the turbulent Prandtl number, P_r, on the sub-grid scale are related to the fluid's material diffusivity and thermal conductivity (Smagorinsky, 1963). The sub-grid scale turbulent Prandtl number, P_r, and the sub-grid scale coefficient, C_s, are empirical constants. Yaga et al. (2002) examined that the Sub-grid scale coefficient C_s is dependent on flow and can be optimized within a range of 0.1 to 0.25 through different flow applications. The range of P_r, which is determined by empirical correlation, is between 0.2–0.9.

The Smagorinsky constant effects on the compartment fire simulation have been studied. Smagorinsky (1963) and Deardorff (1970) further investigated the Smagorinsky sub-grid scale model, which is frequently used in LES, assuming that turbulent motion can be separated into large and small eddies. Assuming the small scales in equilibrium, energy production and dissipation are balanced to obtain the eddy viscosity. However, big eddies are simulated, while small eddies are modeled. Smagorinsky (1963) first suggested an empirical constant (C_S) of 0.23 for the eddy viscosity model. The LES applies to Navier-Stokes and uses the low pass filtering principle to remove the solution's small scales. LES conducts the filtering process both implicitly and explicitly. On the other hand, explicit filtering requires finer resolution than implicit filtering. LES techniques yield greater precision than the RANS method because they can resolve larger scales of the flow field solution. The current simulation makes use of the LES model. Sub-grid scale coefficient (C_S), sub-grid scale turbulent Prandtl number (P_r), and sub-grid scale turbulent Schmidt number (S_c) are the three model coefficients in the SGS model. Different researchers have studied the Sub-grid scale coefficient (C_S), which is dependent on flow and optimized over a range of 0.1 to 0.25 by different flow applications (Yaga et al., 2002; Murakami, 1998 and Murakami et al., 1995). The values of 0.16 and 0.2 were discovered to be in better agreement with the experimental results when simulating indoor airflow under various conditions (i.e., natural, forced, and mixed convection) (Zhang and Chen, 2000). In a three-dimensional ventilated room simulation for $C_s = 0.18$, Emmerich and McGrattan (1998) discovered that the fine grid size of 0.14 m shows better agreement with experimental values, while the predicted values by coarse grid were closer to the experimental value. In an experiment with a gasoline pool fire in an ISO 9705 room, Zou et al. (2007) discovered that the simulated results, with C_s values of 0.2 and 0.21, were comparable to the experimental values.

The sub-grid scale turbulent Prandtl number (P_r) is obtained through empirical correlation, ranging from 0.2 to 0.9. Zhang et al. (2002) investigated the effects of turbulent Prandtl number, gird size, and sub-grid scale coefficient while simulating compartment fires. They discovered no discernible difference in the model coefficients C_S and P_r for weak buoyant plumes. However, the Prandtl number has a negligible impact on the mean temperature of the robust and buoyant plume. Yang et al. (2010) examined the impact of the Prandtl number in the context of enclosure fires. They found that while the Pr has no discernible influence on temperature, heat release rate, or CO yield compared to other coefficients, a higher CO yield value was predicted from $S_c = 0.1$ with a relative difference of 59% compared to the default value $S_c = 0.5$. Yoshizawa (2009) obtained $P_r = 0.58$ by applying turbulent statistical theory. According to Nieuwstadt et al. (1991), the turbulent Prandtl number has been optimized to 0.33 for open boundary conditions. The FDS Technical Reference Guide (McGrattan et al., 2013) recommends C_S, P_r, and S_c to have default values of 0.2, 0.5, and 0.5, respectively.

In this study, the FDS is used for the diesel pool fire that was conducted to ascertain the temperature conditions inside the compartment at various elevated fuel positions by Tiwari et al. (2021). The effect of various Smagorinsky constants were studied on the elevated pool fires. However, the position of the fire source in an enclosure is frequently much above the floor, as seen in the case of ship cabin platforms and fuel pipes in chemical, nuclear, and thermal power plants. The burning behavior of diesel fuel was studied in the experiment and simulation, which is a non-standard fuel with blends of several components that burn using a distinct procedure.

2 Description of CFD Model

2.1 Computational Domain

Experiments by Tiwari et al. (2021) were considered for the fire modelling. The diesel pool fire experiment was conducted at various heights with a round mild steel pan in the middle of the compartment. The pan had a 0.8 m diameter and a 0.12 m lip height. The fuel pan height was measured during the experiments and was kept at 0.30, 0.60, 0.90, 1.20, 1.5, and 1.8 m above floor level, or 15%, 30%, 45%, 60%, 75%, and 90% of the door height (H_d) of 2 m. Figure 1 shows the experimental setup used by Tiwari et al. (2021).

The fire test compartment in experimental studies measures 4 m (L) × 4 m (W) × 4 m (H), and the doorway opening measures 1 m (W) × 2 m (H). The computational domain used for the simulation is 6 m (L) × 6 m (W) × 6 m (H), which means it is larger than the actual compartment size in order to maintain the free boundary condition. Figure 1 shows the compartment geometry model. According to the researchers, the extended domain outperformed the non-extended domain, producing better agreement with experiment results. Subsequently, He et al. (2008) presented a study examining the method of choosing the right computational domain and its effect on compartment fire simulation. It was discovered that predicted results agreed with experimental values when the computational domain was extended by half or more of the hydraulic diameter compared to the non-extended domain.

Hydraulic diameter is given by:

$$D_H = \frac{4A_0}{P_0} \tag{1}$$

where D_H is the hydraulic diameter of opening (m), A is the area of the ventilation opening (m2), and P_0 is the perimeter of the ventilation opening (m).

Since all experiments were conducted with the door fully open, the computational domain for all simulations extended to a distance of 2 m from the door opening, as the calculated hydraulic diameter is 1.33. Figure 2 shows the compartment model that have been used for the numerical simulation.

2.2 Material Properties and Boundary Condition

The walls and floor were made of bricks with a thickness of 25 cm. The ceiling was made of concrete with a thickness of 15 mm, and the door was made up of steel with a thickness of 4 mm. Table 1 displays the material properties that are incorporated into the FDS model. The thermocouple sensors were installed in the CFD model at the vent opening and the ceiling of the compartment in four layers. The temperature and heat flux sensors were installed as suggested in the experiments by Tiwari et al. (2021).

In the computational domain, the boundary conditions were configured to be open at the door opening's side, and other sides were set as adiabatic. It was assumed there was no slip as the tangential velocity boundary condition at the solid surfaces. The environment's temperature was fixed at 23 °C.

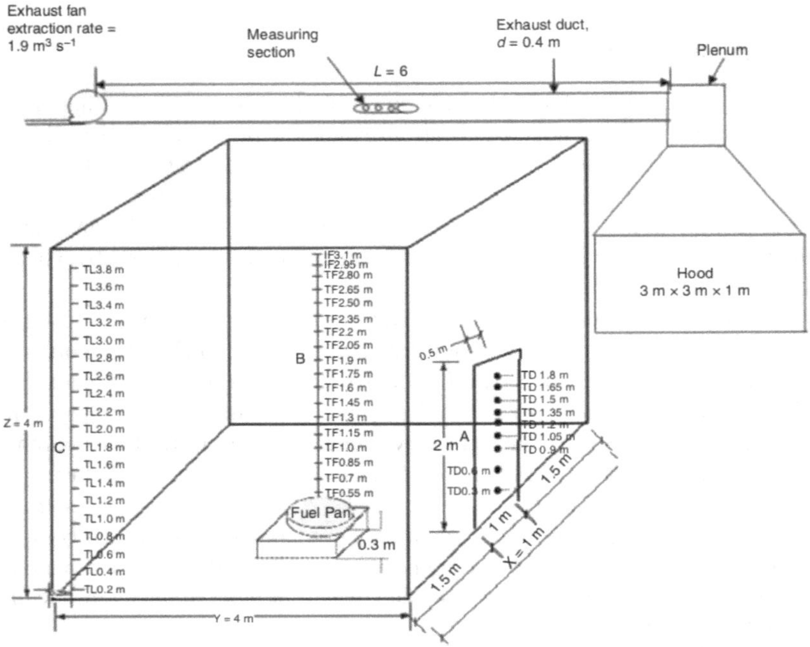

Fig. 1. Schematic of fire test compartment by Tiwari et al. (2021)

2.3 Fire Source and Flow Conditions

According to Babrauskas (1983), the heat release rate in compartment fires indicates how the fire progresses. The precision of the heat release rate forecast determines the accuracy of the numerically derived results from the FDS simulation. Several processes occur during fuel and air reactions, releasing heat and producing product gases. Reaction kinetics and fuel characteristics affect the output of responses. To simulate the combustion of enclosure fires, it is necessary to have accurate values for the fuel properties, rate kinetics, and enclosure surfaces. Because diesel is not standardized, the chemical kinetics of this fuel are not documented in any literature. Additionally, the uncertainty of combustion modeling is mitigated by using the heat release rate as an input. The FDS provides information on the heat release rate measured by the oxygen depletion calorimeter, including the HRRPUA (heat release rate per unit area) and RAMP_Q (heat release rate time history).

2.4 Grid Size Calculations

Grid independence directly affects truncation error or even the rationality of numerical results. Therefore, it is crucial to validate the numerical simulations. Theoretically, a dense grid can guarantee calculation accuracy, but the computation time will increase significantly. Enhancing the grid size can help increase the numerical accuracy. Essential

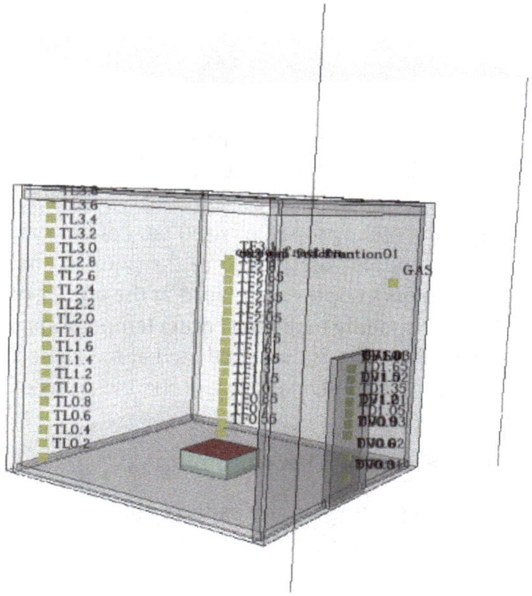

Fig. 2. Computational domain for FDS simulation

Table 1. Material properties adopted for CFD simulation

Materials	Properties				
	Density (Kg/m^3)	Specific heat (kJ/kg.K)	Thermal conductivity (W/m.K)	Absorption (1/m)	Emissivity
Bricks	750	1.04	0.36	5×10^4	0.8
Concrete	2280	1.04	1.8	5×10^4	0.9
Steel	7850	1.46	45.8	5×10^4	0.95

parameters to solve the flow field in buoyant plumes are the characteristic fire diameter D^*, which is derived from the Froude number, and the mesh cell size across the fire Δx.

$$D^* = \left(\frac{\dot{Q}}{\sqrt{g}\rho_0 T_0 C_0}\right)^{\frac{2}{5}} \quad (2)$$

where D^* is the characteristic fire diameter, $\dot{Q}$ is the fire's total heat release rate (kW), T_0 is the ambient temperature (K), ρ_0 is its ambient air density (kg/m^3), and C_0 is its specific heat capacity (J/kg°C). It can be presumed that fire can be effectively resolved if the non-dimensional ratio $D^*/\Delta x$ is significantly large. The USA Nuclear Regulatory Commission conducted a study on grid sensitivity (Stroup and Lindeman, 2013), wherein the non-dimensional ratio $D^*/\Delta x$ was varied for different fire scenarios between 4 and 16.

2.5 Measurements

The thermocouples are placed at different locations, as given in Fig. 1 from the experimental setup of Tiwari et al. (2021). The bare bead thermocouples are used to predict temperature. The emissivity of bare bead thermocouples is 0.85 (default value), and the bead diameter is 1 mm, and other properties used are default values of FDS.

The bare bead thermocouples are placed at the pool's center to predict the flame temperature. Similarly, thermocouples are placed at the corner of the compartment and at the centerline of the door opening to predict corner temperature and doorway centerline temperature, respectively. Velocity is also predicted at the centerline of door opening at different locations, matching with the velocity probe locations of the experiment.

2.6 Temperature Correction of Thermocouples

Thermocouple temperature readings in fire experiments might not accurately reflect the gas temperature. Because of the influence of the remote cool layer, a thermocouple placed in the hot gas layer receives radiation lower than that suggested by the local gas temperature, which results in a temperature reading that is somewhat lower than the actual gas temperature (Luo, 1997; Blevins, & Pitts, 1999). The recorded temperature is significantly higher than the gas temperature in the lower cooler layer. The cause is radiation error; according to Luo (1997), Blevins & Pitts (1999), Francis & Yau (2004), and others, radiation from far-off flames and hot gas layers causes the most significant error in temperature readings from thermocouples. Conduction effects could exist along the length of a thermocouple. The conditions along the thermocouple wire's length must be precisely known to estimate conduction error, but these conditions are unknown. Therefore, correction due to conduction is neglected.

Welch et al. (2007) examined radiation error in temperature measurement using bare bead thermocouples. Equation 3 shows the results of performing energy balance on the thermocouple:

$$T_g = \frac{\epsilon_t \sigma \left(T_t^4 - T_\infty^4\right)}{h_t} + T_t \tag{3}$$

where T_g and T_t represent the true gas temperature and temperature measured by thermocouple respectively. ϵ_t is emissivity of bare bead thermocouple (0.9), h_t is convective heat transfer coefficient and T_∞ is effective surrounding temperature.

The Nusselt number correlation can be used to calculate the convective heat transfer coefficient, which is as follows:

$$h_t = \frac{k_g N_u}{D_t} \tag{4}$$

where N_u is Nusselt number, k_g is gas thermal conductivity (W/mK) and D_t is diameter of thermocouple bead (2 mm).

Equation (5), which comes from McAdams (1954), can be used to calculate the Nusselt number if the thermocouple bead is assumed to be spherical.

$$N_u = 0.37Re^{0.6} \tag{5}$$

where Reynolds number, $Re = \frac{\rho_g V D_t}{\mu}$

In the lower zone, the velocity approaches the 1 m/s doorway velocity. As a result, Re is calculated using this value.

Equation (6) provides the ideal gas law relationship, which can be used to calculate gas density.

$$\rho_g = \rho_\infty \left(\frac{T_\infty}{T_g} \right) \tag{6}$$

where ρ_∞ is the density of ambient air (1.2 kg/m^3) and T_∞ is the ambient air temperature.

Gas thermal conductivity is given as:

$$k_g = \frac{\mu C_p}{P_r} \tag{7}$$

The standard air value of 0.7 is utilized for the Prandtl number P (Jones & Whitelaw, 1982), C_p represents the gas specific heat, and Eq. (8) yields the dynamic viscosity μ derived from Sutherland's law (Anderson et al., 1994).

$$\mu = \frac{C_1 T_g^{1.5}}{T_g + C_2} \tag{8}$$

where constants C_1 and C_2 are 1.458×10^{-6} kg/m/s/K$^{0.5}$and 110.4 K respectively.

The hot gas layer is reached after one meter in height, so the average value of the thermocouples in the hot gas layer is used to approximate the radiation field that is visible from the point of interest when determining the effective surrounding temperature. Equation (3) changes when a relaxation factor r is added for the traditional convergence control method.

$$T_g = \left(\frac{\epsilon_t \sigma \left(T_t^4 - T_s^4 \right)}{h_t} + T_t \right) r + T_{g-1}(1 - r) \tag{9}$$

The local gas temperatures are updated using current temperature estimates in Eq. (9), and as a result, they are all dependent on one another. The measured thermocouple temperature is utilized as the initial approximation for the gas temperature when the solution method starts. Using the estimated temperature that is generated for the following iteration, the process is repeated until the solution converges to a single solution for the temperature field. MATLAB (Release 2022a) was utilized for Eq. 9 and a relaxation factor of 0.3 was taken to calculate the calculations. For experiments 1, 2, 3, 4, 5, and 6, radiation correction was applied to the lower zone temperature; the maximum radiation corrections obtained are, correspondingly, 35, 37, 39, 40, 33, and 32 °C. It is emphasized here that the radiation corrections have solely considered the radiation effect caused by the upper hot gas layer. However, the effects of radiation from adjacent walls and flames cannot be taken into account because of the complex geometry.

3 Simulation Result and Discussion

Simulations were carried out for a pool with a diameter of 0.8 m located in the center of the compartment at various heights. Table 2 contains the details of the simulated experiments. Different mesh sizes, ranging from 2.74 to 18.15 lac cells, are used for simulations. Mesh sizes are chosen based on characteristics fire diameter. Three distinct mesh sizes, from coarse to fine, were used for the simulations. All three mesh sizes were performed in the simulations for experiment number 1, and medium mesh sizes were used for experiments 2–5 simulations.

Table 2. Details of experiments for simulation

Exp. No.	Fuel pan diameter (m)	Fuel pan elevation (m)	Door opening Height (m) × Width (m)	Dimensionless fuel pan height (h/H_d) in %	Fuel pan location
1	0.8	0.3	2 × 1	15	Center
2		0.6		30	
3		0.9		45	
4		1.2		60	
5		1.5		75	
6		1.8		90	

3.1 Simulation of Experiment 1 for Grid Analysis

Three distinct values of $D^*/\Delta x$ have been taken in this study. Table 3 displays the specific mesh sizes. Table 4 shows the boundary conditions and simulation input parameters. Heat release rates from experiments have been used as input for simulations. The HRRPUA has been set for the three simulation sessions at 1472.34 kW/m^2.

Table 3. Mesh size details for the simulation

S. No.	Cell size (m)	Number of cell	Total cells	Characteristics fire diameter, D*
Simulation 1	0.093-0.093-0.093	65-65-65	274625	0.84
Simulation 2	0.06-0.06-0.06	92-92-92	778688	
Simulation 3	0.039-0.039-0.039	122-122-122	1815848	

Table 4. Details of simulation parameter and boundary conditions

Parameter	Values
Initial temperature (°C)	30
Relative humidity (%)	58
Initial Pressure (Pa)	10^5
Simulation type	Large eddy simulation, Transient
Computational domain (m^3)	6.0 (L) × 6.0 (W) × 6.0 (H)
Total simulation time (s)	900
Turbulence model	Deardorff model (Default)
Surface property	
Geometry	Flat
Tangential boundary condition	Slip (Default)
Backing	Air gap (Default)
Initial internal temperature	Ambient temperature
Open property	
Exterior temperature	Ambient temperature
Dynamic pressure	0.0 Pa

Figure 3 shows the comparison of experimental and predicted centerline flame temperature for experiment 1 for different grid sizes. The temperature was measured at 800 s. With refinement in mesh sizes, simulated results are found to be much better, as seen in the graph. The predicted and experimental temperature-height curve follows a similar trend. As the height increases, the flame temperature tends to decrease. This may be because of the reduced oxygen availability at the fuel surface, which results in incomplete combustion. The peak temperatures recorded in the experiment and simulation are 946 °C and 942 °C, respectively, with a relative difference of 4 °C. The average temperature difference between experimental and predicted results at 800 s are 15, 9, and 8 °C, corresponding to simulations 1, 2, and 3, respectively.

Figure 4 shows the comparison of experimental and predicted door temperature distribution for experiment 1 for different grid sizes. The temperature was measured at 800 s. Predicted doorway results are close to the experimental values. The average temperature difference in the doorway ranges from 5 to 12 °C. However, it is the least for simulation 3, which has shown good agreement compared to the other grid sizes. In all simulation runs, the temperature distribution is found to be similar.

3.2 Effect of Key Modeling Constants on Compartment Fire Simulation

Simulations of 0.8 m pool diameter elevated at the center of the enclosure are performed with medium mesh size (0.065), which is at $D^*/dx = 13$. The coarse grid predicted temperature observed the most notable deviations from the experimental data. The medium

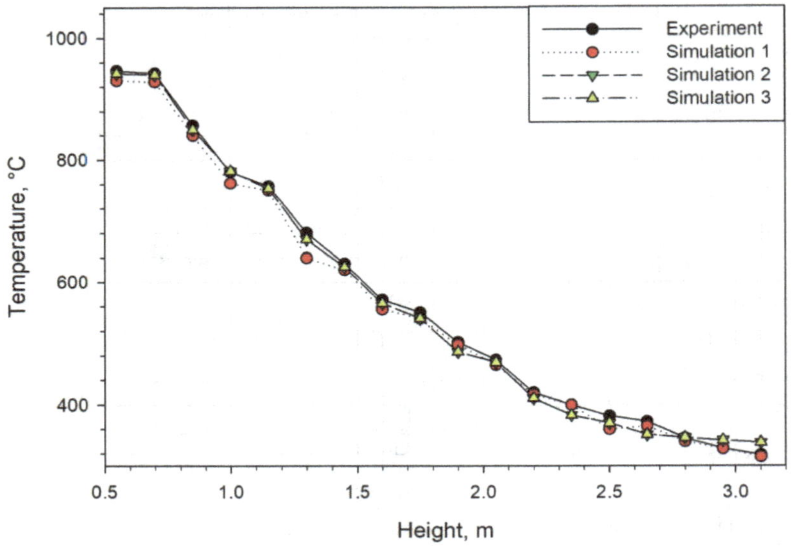

Fig. 3. Comparison of predicted and measured centerline flame temperature variation with compartment height at 800 s

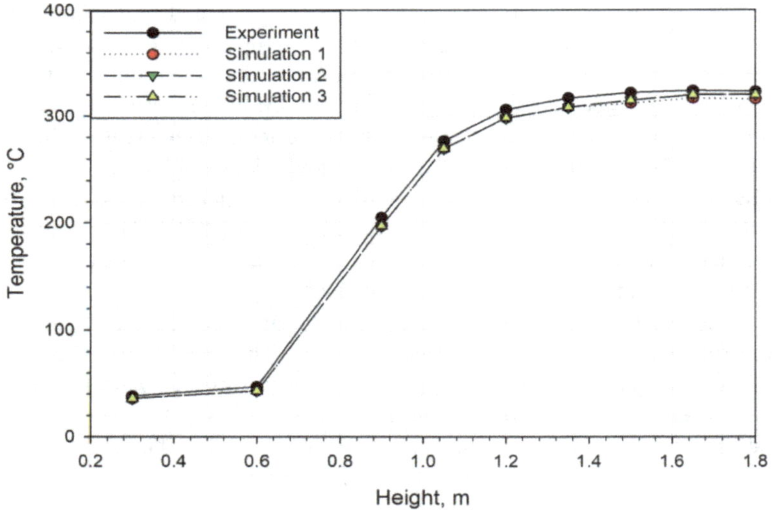

Fig. 4. Comparison of predicted and measured door temperature variation with door height at 800 s

mesh size was found to be reasonably close to both the fine grid and the experiment, but the prediction using the fine grid, at 0.049 m, was the closest to the experimental data. It is observed that the medium mesh size produces favorable results with moderate computational time. Therefore, a medium-mesh size of 0.065 m is used to study the effect of Smagorinsky sub-grid scale (SGS) coefficients. To examine the impact of changing the SGS model's empirical coefficients, 12 numerical simulations were run. The matrix of the numerical simulations run during the current investigation is displayed in Table 5.

Table 5. Matrix of numerical simulation

Grid Size	Pan Elevation	P_r/S_c	Cs
$\Delta x = 0.065$	0.3	$P_r = 0.3$, $S_c = 0.3$, $C_s = 0.2$	$P_r = 0.3$, $S_c = 0.3$, $C_s = 0.18$
	0.6	$P_r = 0.5$, $S_c = 0.5$, $C_s = 0.2$	$P_r = 0.3$, $S_c = 0.3$, $C_s = 0.2$
	0.9	$P_r = 0.7$, $S_c = 0.7$, $C_s = 0.2$	$P_r = 0.3$, $S_c = 0.3$, $C_s = 0.23$
	1.2		
	1.5		
	1.8		

Figure 5 illustrates the variation of centerline flame temperature with the height for different Prandtl and Schmidt numbers for experiments 1 to 6 at 800 s. It can be seen that as the thermocouple height increases, the temperature of the flame tends to decrease. The temperature is over-predicted for higher P_r and S_c values at lower heights and underpredicted as the height increases. For lower values of P_r and S_c, the temperature curve follows the same path as the experimental values. It can be said that heat energy is more concentrated and less diffuse at higher P_r and S_c. The total enthalpy energy aggregate within the compartment is comparatively larger since less heat energy is released through the opening vents. Overall, it is evident that increasing the values of P_r and S_c results in an increase in the temperature of the compartment. This observation aligns with the temperature findings of Zhang et al. (2002). The temperature trends for P_r and S_c values 0.3 are the most accurate of all the situations. At the same time, $P_r = S_c = 0.7$ generally overpredicts the temperature while $P_r = S_c = 0.5$ under predicts it, according to comparisons with experimental observations.

A sensitivity test was also performed on the Smagorinsky constant in this numerical simulation. P_r and S_c were set to 0.3 based on the analysis in the preceding section, and simulation cases were run by altering the Smagorinsky constant from the standard value, which ranges from 0.18 to 0.23. Figure 6 presents the temperature profiles of the centerline flame temperature variation with the height for different pan elevations. The overall temperature profile for $C_s = 0.2$ is found to have the best agreement with the experimental values out of all the simulated findings. The heat and mass flow field gets increasingly diffusive with increasing Cs values. Consequently, the fire's intensity increases when the Smagorinsky constant rises. In conclusion, it is observed that the conventional value of the Smagorinsky constant ($C_s = 0.2$) is better appropriate for the fire simulation scenario of elevated fuel pans.

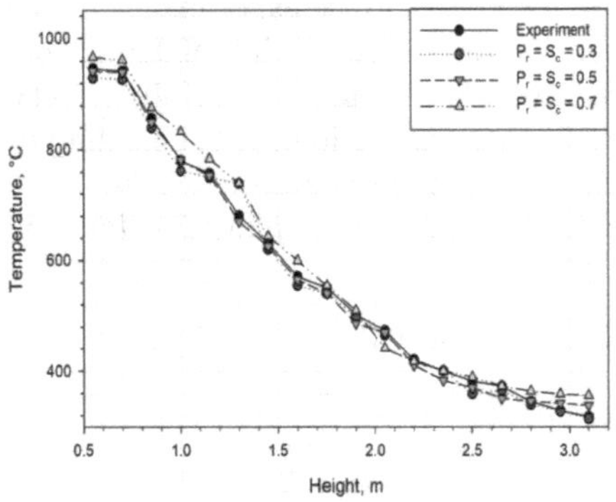

(a) Pan elevation 0.3 m

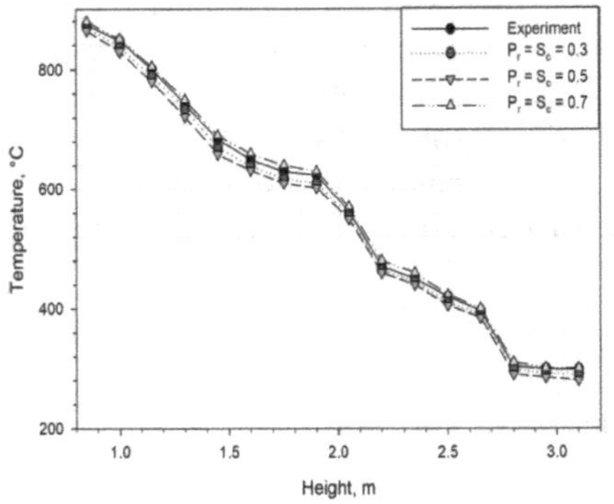

(b) Pan elevation 0.6 m

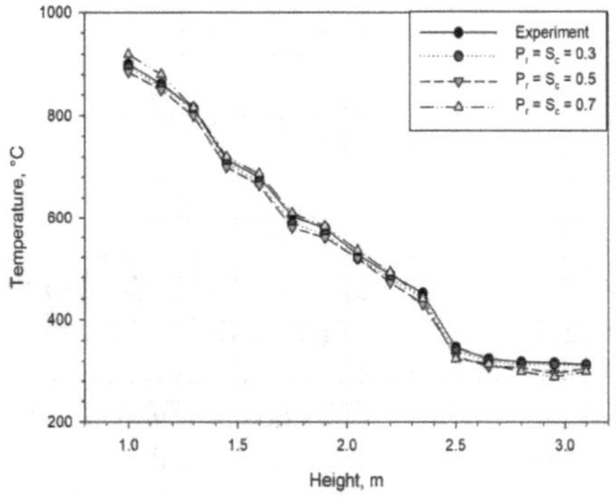

(c) Pan elevation 0.9 m

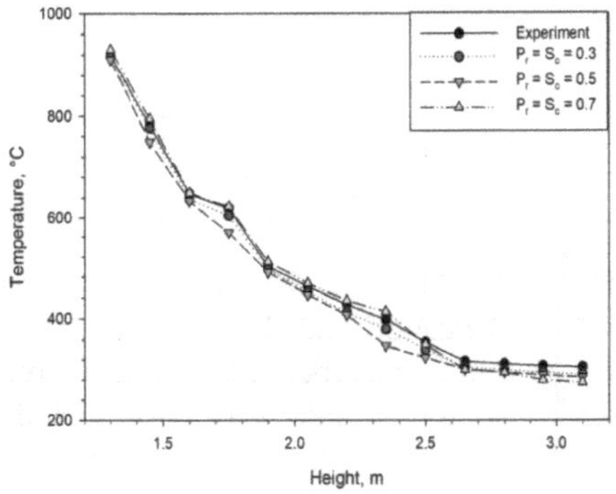

(d) Pan elevation 1.2 m

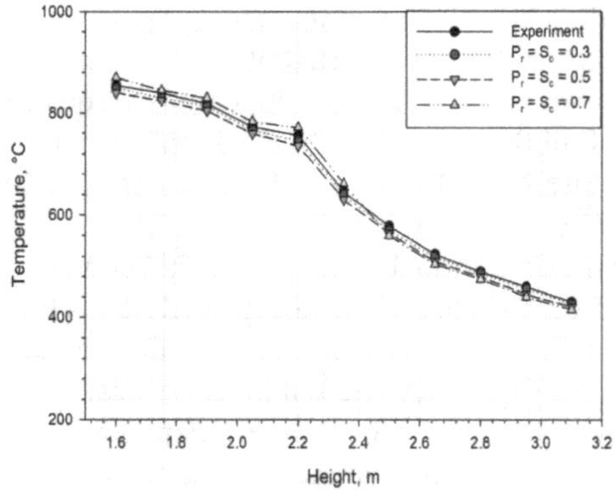

(e) Pan elevation 1.5 m

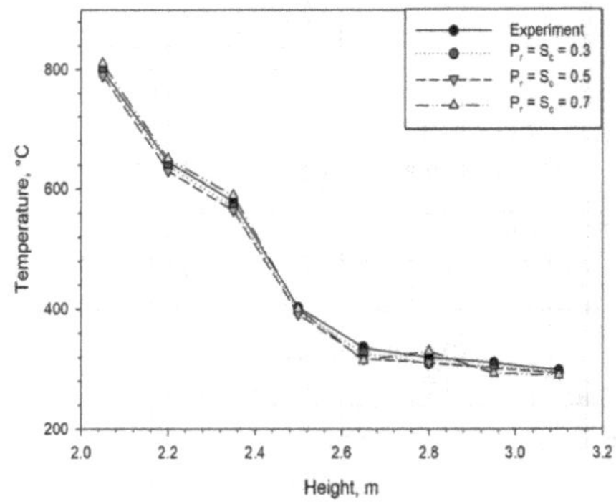

(f) Pan elevation 1.8 m

Fig. 5. Variation of P_r and S_c for the centerline flame temperature for different pan elevations

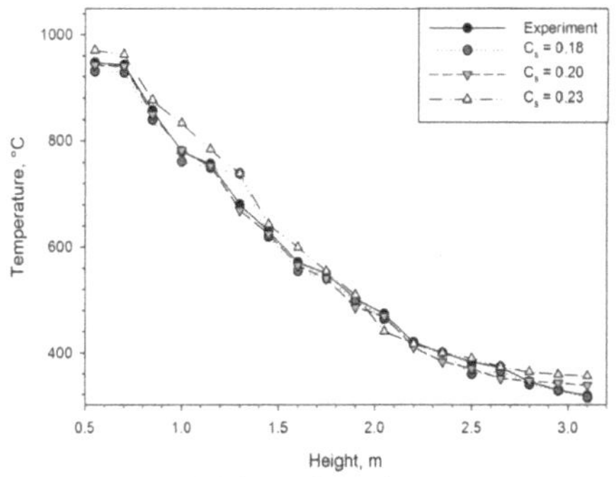

(a) Pan elevation 0.3 m

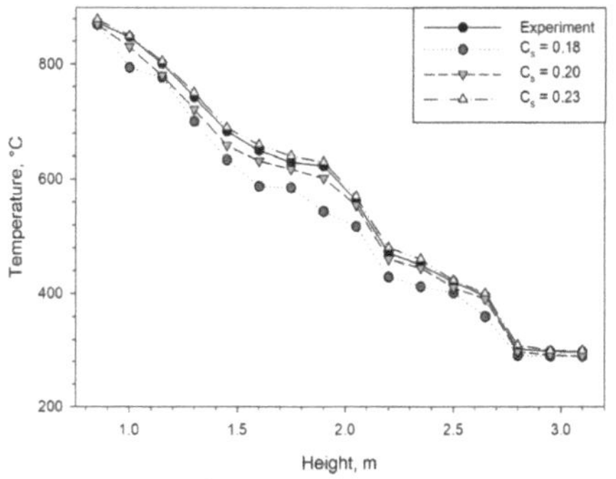

(b) Pan elevation 0.6 m

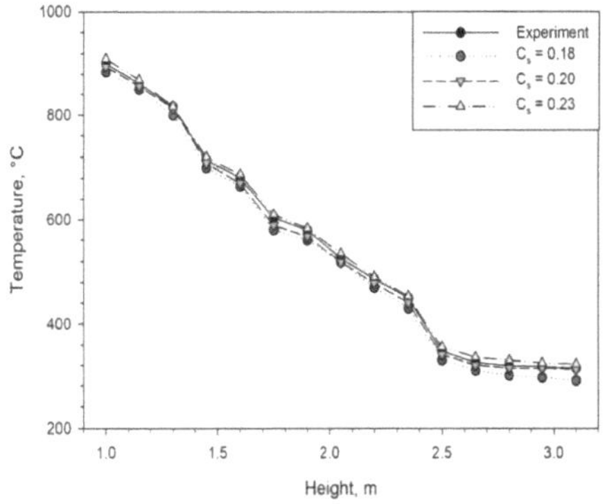

(c) Pan elevation 0.9 m

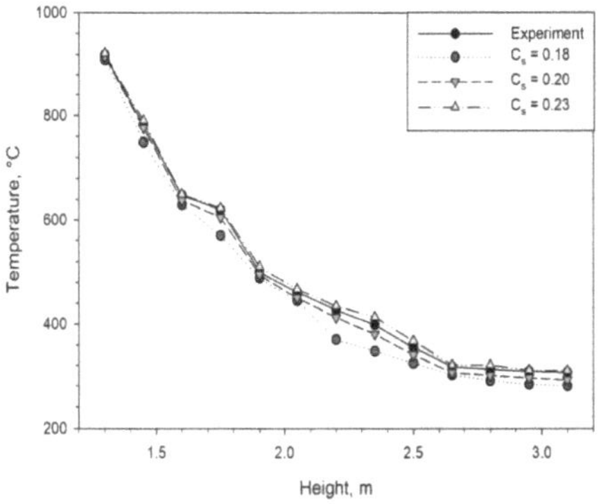

(d) Pan elevation 1.2 m

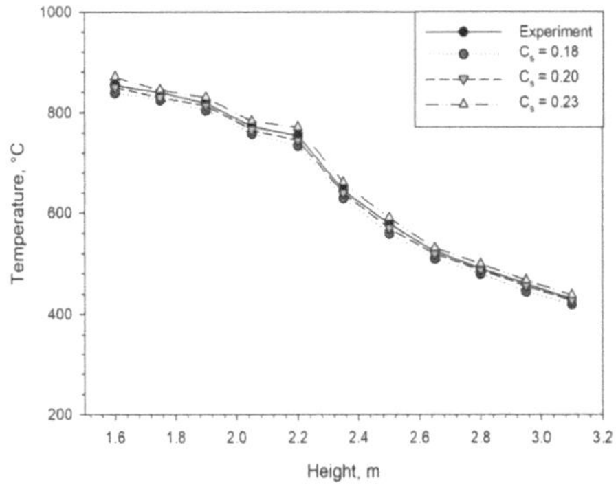

(e) Pan elevation 1.5 m

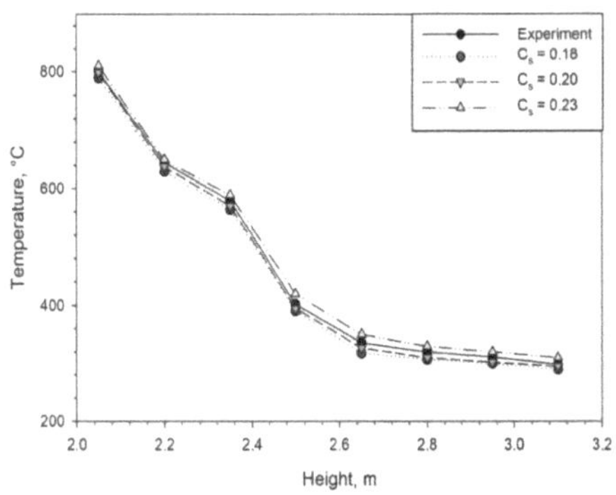

(f) Pan elevation 1.8 m

Fig. 6. Variation of C_S for the centerline flame temperature for different pan elevations

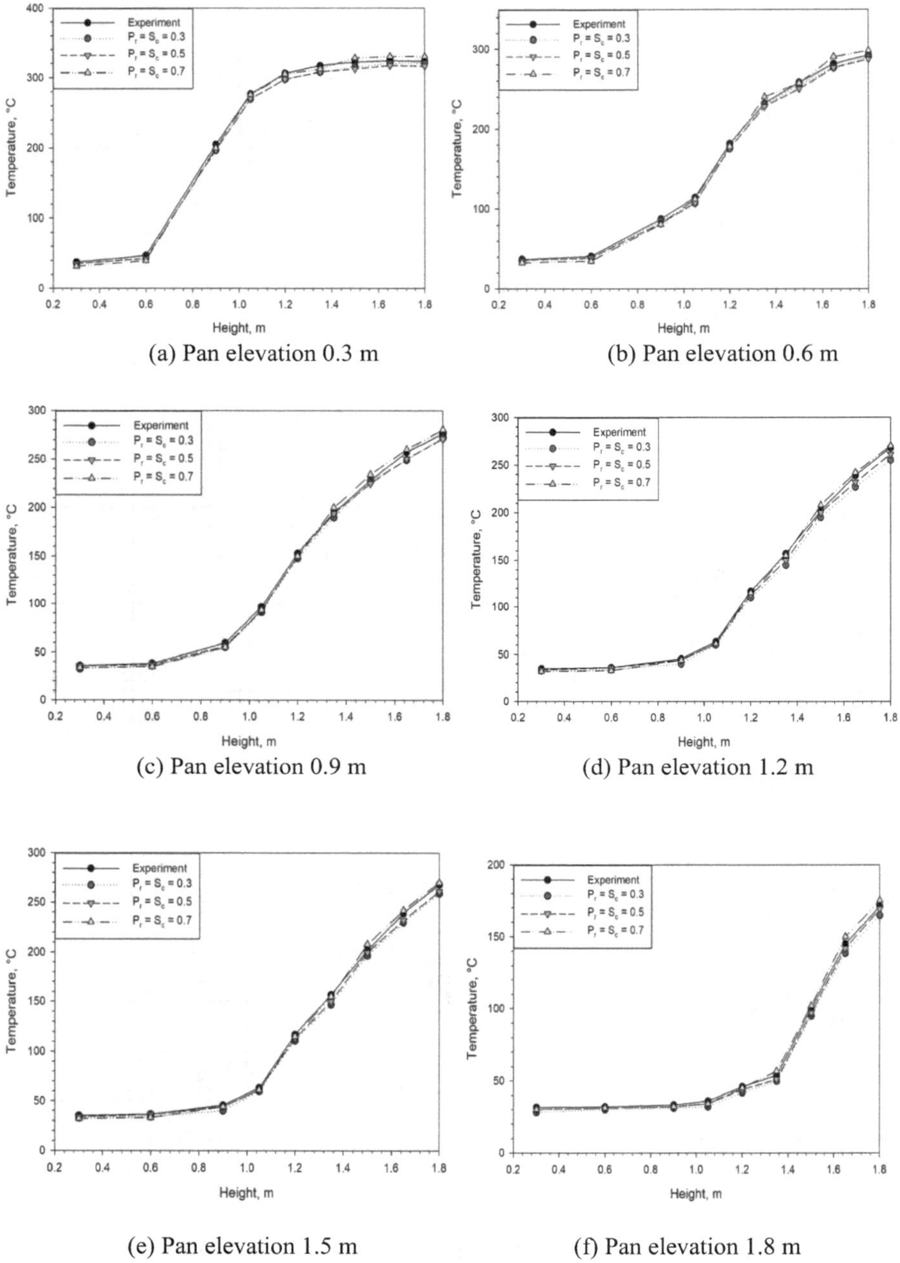

Fig. 7. Variation of P_r and S_c for the door temperature for different pan elevations

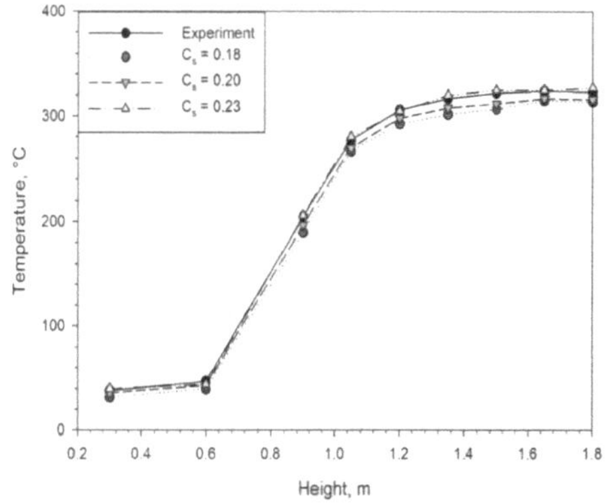

(a) Pan elevation 0.3 m

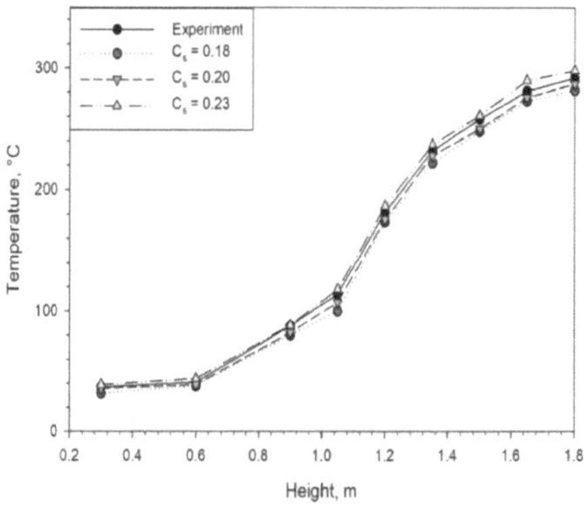

(b) Pan elevation 0.6 m

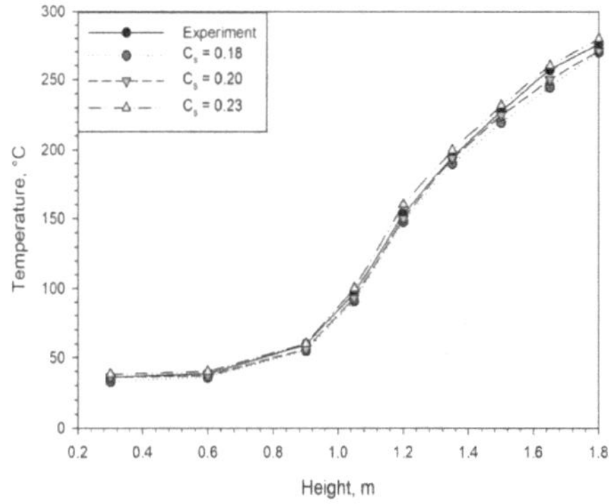

(c) Pan elevation 0.9 m

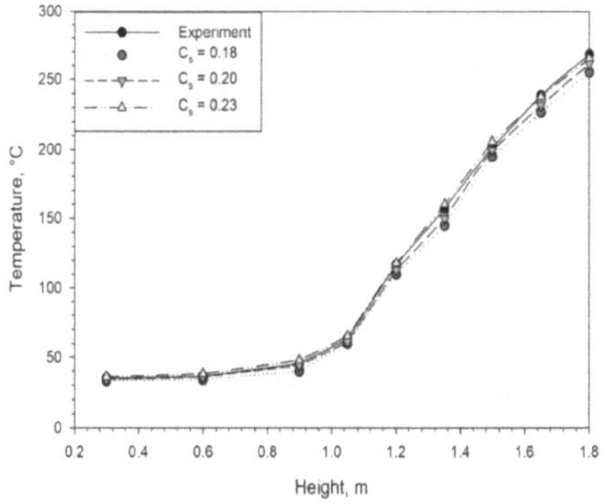

(d) Pan elevation 1.2 m

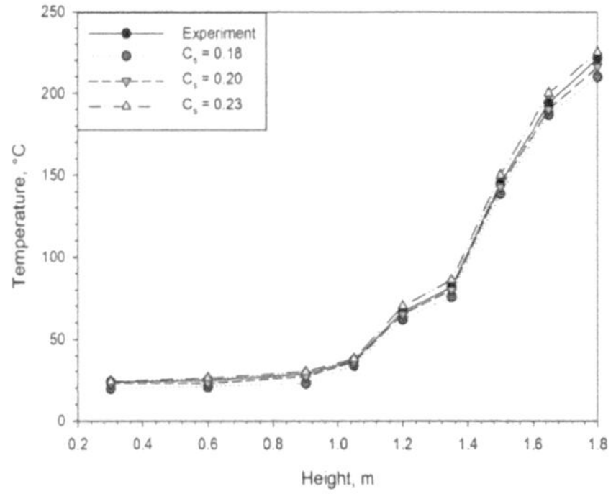

(e) Pan elevation 1.5 m

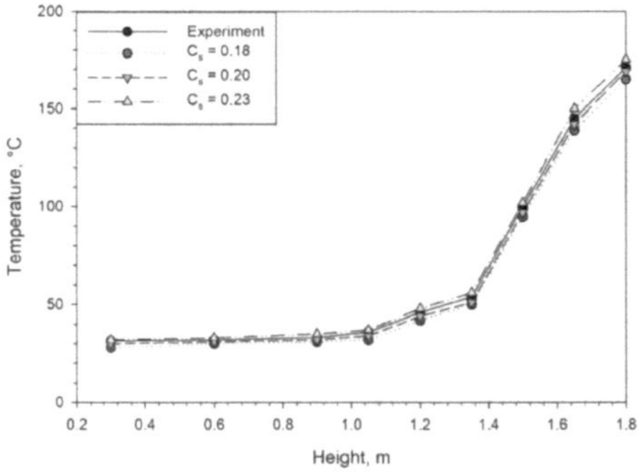

(f) Pan elevation 1.8 m

Fig. 8. Variation of C_S for the door temperature for different pan elevations

(a) Pan elevation 0.3 m

(b) Pan elevation 0.6 m

(c) Pan elevation 0.9 m

(d) Pan elevation 1.2 m

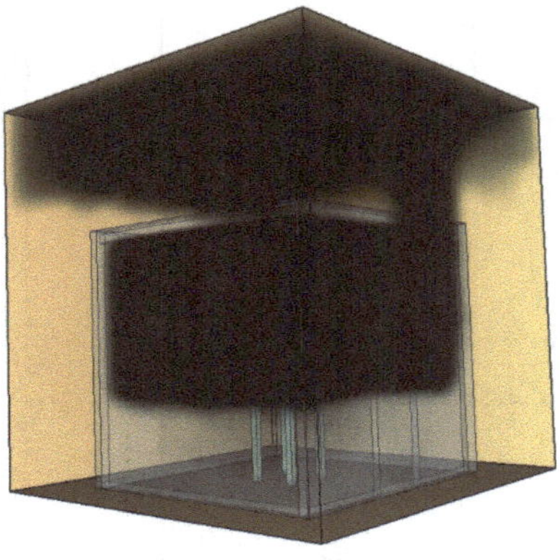

(e) Pan elevation 1.5 m

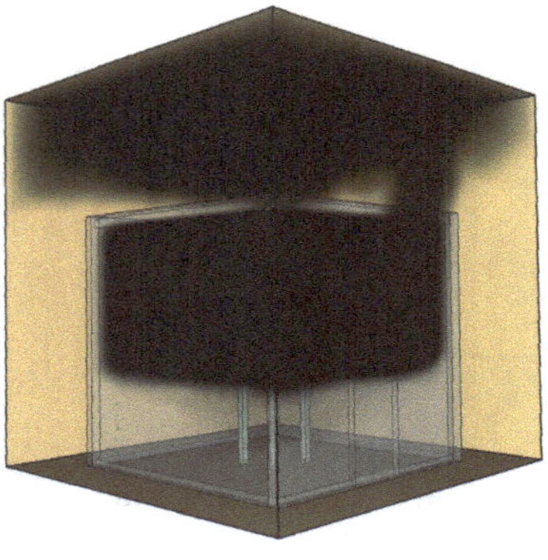

(f) Pan elevation 1.8 m

Fig. 9. Smoke layer height interface for experiment 1- 6 at 800 s

Similar observations were recorded for the doorway temperature. Figures 7 and 8 represent the variation in doorway temperature at 800 s with height for different P_r/S_c and C_S constants, respectively. It can be seen for all pan elevations that temperature shows higher accuracy for lower P_r and S_C values. For C_S, the default value, 0.20, is proven to have better accuracy.

3.3 Smoke Layer Height for Pan Elevations

The thermal zone interfaces for Experiments 1, 2, 3, 4, 5, and 6 were observed to be 0.8 m, 0.9 m, 1.2 m, 1.4 m, 1.7 m, and 1.9 m from the floor, respectively, by Tiwari et al. (2021). Similar observations were noticed while doing the simulations. In Fig. 9, the smoke layer interface in the compartment is shown for all 6 pan elevations. For the simulation, P_r and S_C values were taken as 0.3, while C_S was taken as 0.2. The smoke height layer from the floor after the simulation is observed as 0.7, 0.8, 1.2, 1.4, 1.7, and 1.8 for experiments 1–6 at 800 s, respectively.

4 Conclusions

Numerical simulations for a 0.8 m diesel pool fire were performed under six fire source elevations placed at the center of the compartment. Simulations have been performed to investigate the effect of the various Smagorinsky constants. The main conclusions are as follows:

i. From the grid sensitivity analysis, medium mesh size results are found to be in close agreement with the experimental results. In contrast, course mesh size results are underpredicted in all thermocouple locations. It can be said that the value of D^*/dx can be taken in the range of 10 to 15 to resolve the events of compartment fire accurately.
ii. The effect of different values of SGS constants on 6 pan elevations has been observed. Lesser P_r/S_C ($P_r = S_C = 0.3$) values gave better accuracy than the default values of these constants, and higher values overpredict the centerline flame temperature and doorway temperature.
iii. The default value C_S ($C_S = 0.5$) in FDS gave better accuracy for all pan elevations for centerline flame temperature and doorway temperature.
iv. Smoke layer height is generated in the FDS for all the simulations. The results are in good agreement with the experimental data. After the simulation, the smoke height layer from the floor is observed as 0.7, 0.8, 1.2, 1.4, 1.7, and 1.8 for experiments 1–6 at 800 s, respectively. This shows that FDS can accurately simulate pool fires with blended fuels like diesel.

References

Anderson, D., Tannehill, J.C., Pletcher, R.H.: Computational fluid mechanics and heat transfer. McGraw-Hill, New York (1994)

Babrauskas, V.: Estimating large pool fire burning rates. Fire Technol. **19**(4), 251–261 (1983)

Blevins, L.G., Pitts, W.M.: Modeling of hare and aspirated thermocouples in compartment fires. Fire Saf. J. **33**(4), 239–259 (1999)
Chaudhary, A., Gupta, A., Kumar, S.: Studies on jatropha oil pool fire. Therm. Sci. Eng. Progr. **6**, 104–127 (2018)
Deardorff, J.W.: A Three-dimensional numerical study of turbulent channel flow at large Reynolds numbers. J. Fluid Mech. **41**, 453 (1970)
Emmerich, S.J., McGrattan, K.B.: Application of a large eddy simulation model to study room airflow. ASHRAE Trans. **104**, 1–9 (1998)
Francis, J., Yau, T.M.: On radiant network models of thermocouple error in pre and post flashover compartment fires. Fire Technol. **40**(3), 277–294 (2004)
He, Y., Jamieson, C., Jeary, A., Wang, J.: Effect of computation domain on simulation of small compartment fires. Fire Saf. Sci. **9**, 1365–1376 (2008)
Jones, W.P., Whitelaw, J.H.: Calculation methods for reacting turbulent, flows: a review. Combust. Flame 48, 1–26 (1982)
Liu, J., Chen, M., Lin, X., et al.: Impacts of ceiling height on the combustion behaviors of pool fires beneath a ceiling. J. Therm. Anal. Calorim. **126**(2), 881–889 (2016)
Luo, M.: Effects of radiation on temperature measurement in a fire environment. J. Fire Flammability **15**(6), 443–461 (1997)
Mahalingam, A., Jia, F., Wang, Z., Patel, M.K., Galea, E.R.: Numerical investigation of tube furnace toxicity measurement method (ISO 19700). Fire Mater. **36**(1), 17–30 (2012)
McAdams, W.H.: Heat transmission (No. 660.28427 M32) (1954)
MeGrattan, K.D., Hostikka, S., Floyd, J., et al.: Pire Dynamics Simulator (Version 5): Technical Reference Guide. NIST SP 1018, NIST Special Publication 1018–5, National Institute of Standards and Technology, Gaithersburg, Md, USA (2009)
McGrattan, K., Hostikka, S., McDermott, R., Flyod, J., Weinschenk, C., Overholt, K.: Fire dynamics simulator (version 5) technical reference guide volume 1: mathematical model. Natl. Inst. Stand. Technol. Spec. Publ. **1018**, 147 (2013)
Murakami, S.: Overview of turbulence models applied in CWE-1997. J. Wind Eng. Ind. Aerodyn. **74**, 1–24 (1998)
Murakami, S., Mochida, A., Matsui, K.: Large eddy simulation of non-isothermal room airflow, comparison between standard and dynamic type of Smagorinsky model. J. Inst. Ind. Sci. 47–7–17 (1995)
Narang, A., Pandey, R.S., Kumar, R., Dhiman, A.: An experimental study on compartment crib fires with open ventilation. In ASPS Conf. Proc. **1**(4), 1247–2125 (2022)
Narang, A., Kumar, R., Dhiman, A.K., Pandey, R.S., Sharma, P.K.: Study on the influence of vent area on porosity-controlled wood crib compartment fires prior to flashover. J. Struct. Fire Eng. **15**(2), 177–191 (2024)
Nieuwstadt, F.T.M., Mason, M., Moeng, J.P., Schumann, U.: Large eddy simulation of the convection boundary layer: a comparison of four computer codes. In: Proceedings of the Eighth Symposium on Turbulent Shear Flows, pp. 343–367 (1991)
Steckler, K.D., Quintiere, J.G., Rinkinen, W.J.: Flow induced by fire in a compartment. NBSIR 82-2520, National Bureau of Standards, Washington (1982)
Stroup, D., Lindeman, A.: Verification and validation of selected fire models for nuclear power plant applications. NUREG-1824, supplement 1, United States Nuclear Regulatory Commission, Washington, DC, EPRI 1011999. 37 (2013)
Smagorinsky, J.: General circulation experiments with primitive equations-I the basic experiment. Mon. Weather Rev. **91**, 99–105 (1963)
Tiwari, M.K., Gupta, A., Kumar, R., Sharma, P.K.: Effects of elevated pool fire in a naturally ventilated compartment. J. Therm. Anal. Calorim. **146**, 341–353 (2021)
Welch, S., Jowsey, A., Deeny, S., Morgan, R., Torero, J.L.: BRE large compartment fire tests-Characterising post-flashover fires for model validation. Fire Saf. J. **42**(8), 548–567 (2007)

Yaga, M., Endo, H., Yamamoto, T., Aoki, H., Miura, T.: Modeling of eddy characteristic time in LES for calculating turbulent diffusion flame. Int. J. Heat Mass Transfer **45**, 2343–2349 (2002)

Yang, D., Hu, L.H., Jiang, Y.Q., Huo, R., Zhu, S., Zhao, X.Y.: Comparison of FDS prediction by different combustion models with measured data for enclosure fires. Fire Saf. J. **45**, 298–313 (2010)

Yoshizawa, A.: Eddy-viscosity-type sub-grid-scale model with a variable Smagorinsky coefficient and its relationship with the one equation model in large eddy simulation. Phys. Fluids A **3**, 2007–2009 (2009)

Zhang, W., Chen, Q.: Large eddy simulation of natural and mixed convection airflow indoors with two simple filtered dynamic sub-grid scale models. Numer. Heat Transfer (Part A) **37**, 447–463 (2000)

Zhang, W., Chen, Q.Y.: Large eddy simulation of indoor airflow with a filtered dynamic subgrid scale model. Int. J. Heat Mass Transfer **43**, 3219–3231 (2000)

Zhang, W., Hamer, A., Klassen, M., Carpenter, D., Roby, R.: Turbulence statistics in a fire room model by large eddy simulation. Fire Saf. J. **37**, 721–752 (2002)

Zou, G.W., Tan, H.P., Chow, W.K., Gao, Y.: Effects of varying Smagorinsky constant on simulating post-flashover fires. Int. J. Comput. Fluid Dyn. **21**, 107–119 (2007)

Increase Disaster Data Linkages by GLIDE

Yumi Shiomi(✉)

Asian Disaster Reduction Center, Kobe, Japan
ys-shiomi@adrc.asia

Abstract. In dealing with disaster data and databases, several problems have long been recognized such as inconsistent definitions and classifications of data, data collection methods across databases, and accessibility to data. This presentation depicts the current situations and challenges of disaster data and introduce GLIDE, Global unique disaster IDEntifier, as a solution to these issues to facilitate disaster information management and increase data linkages.

Asian Disaster Reduction Center (ADRC) proposed a concept of GLIDE, which gives common but unique numbers to disasters all over the world in 2001, The GLIDE system, which aims to promote the disaster information sharing among databases developed by a number of relevant stakeholders, has been maintained and operated by GLIDE partner organizations since its establishment.

A GLIDE number is composed of disaster classification (24 types), year of occurrence, serial number and country code. Once a disaster occurs, an operator of GLIDE partner organizations issues a GLIDE number through GLIDE website, the GLIDE number appears on GLIDE website and also is automatically emailed to some 2,000 subscribers. The users apply the GLIDE number to databases and products, linking to other relevant data with the same GLIDE number.

Facing the global initiatives and implementation of the targets like SDGs and SFDRR, there have been increasing needs in establishing national disaster databases for reporting in recent years. With the development of GLIDE API in 2021, GLIDE can further facilitate disaster data integration among various databases of all disaster phases. Thus, it can contribute not only to solving the above-mentioned challenges but also to assisting disaster database linkages, resulting in better disaster data management.

While data integration through GLIDE are active among international and regional organizations at this moment, the concept of GLIDE can also be applicable to national and local database linkages.

Keywords: Disaster Data · Disaster Database · Disaster Information Management · Data Linkage · GLIDE

1 Introduction

The accurate data and its speedy acquisition is a basis not only for disaster information management but also for all phases of disaster risk reduction and management. Furthermore, such data gives more significance when interlinked with other datasets.

H. Tatano et al. (Eds.): GSRIDRR 2023, DRRGADRIBS, pp. 474–480, 2025.
https://doi.org/10.1007/978-981-96-6609-6_31

However, problems and challenges have long been pointed out in dealing with disaster data such as limited access to primary data, languages, verification process of data. Furthermore, such issues become bore apparent in comparing different datasets due to inconsistent definitions and classifications of data, different data collection methods across databases, and limited accesses to definition of hazards, disaster loss and damage and so on. For example, several countries use an international name for a typhoon which is officially agreed in Typhoon Committee for a typhoon, while some others use their own local names to the same typhoon. Also, there were issues of identification in multi-national disasters and multi-hazard disasters such as Indian Ocean Earthquake and Tsunami.

Fakhrudding et al. (2022) [1] list ten challenges of collecting and using disaster data; (1) complex data management mechanisms, (2) limited capacity to manage data, (3) underutilised and unexplored data, (4) lack of useable, clear, and complete data, (5) lack of integration between climate change adaptation and disaster risk reduction, (6) lack of data sharing mechanisms, (7) lack of data standardization, (8) lack of data on human decisions at local state, (9) lack of data availability and (10) lack of data for assessing systemic risks.

In the context of international data, Below (2016) points out five key problems in the context of international disaster data as (intent behind reporting data, (2) dates reported, (3) definition, (4) methods of evaluating damages and (5) geographical location [2].

This presentation aims to depict the current situations and the issues of disaster databases in terms of international initiatives and suggest GLIDE be utilized as a solution to such issues by facilitating disaster information management and increasing in disaster data linkages.

2 What is GLIDE

2.1 Establishment

As seen in the previous section, many organizations relevant to disaster data have disseminated their own disaster information in their own tools that they have designed, developed for various purposes, which have long caused several issues.

In disaster information management, several problems have been pointed out such as disaster identification in/across the countries and technical problems in collection, sharing and publication of DRR information.

The concept of GLIDE was born a solution to these issues. Issuing a GLIDE number to each disaster case facilitates disaster-related information identification, management and integration with other related information. ADRC proposed to develop a standardized code system for managing information on disasters around the world at the Global Disaster Information Network (GDIN) Conference held in Canberra, Australia in March 2001.This proposal was accepted as a pilot project of GDIN. In 2004, glidenumber.net, prototype of GLIDE was jointly developed by ADRC and UNOCHA ReliefWeb with technical assistance from LaRED in order to issue new GLIDE numbers to disasters immediately after their occurrence. Moreover, ADRC, CRED, European Commission/ Joint Research Center (EC/JRC), IRI/Columbia University, the USAID/OFDA, the WMO, IFRC, UNDP, and UNISDR Secretariat (current UNDRR) agreed to use the GLIDE number format as the standard format for disaster identification numbers.

At present, not only these organizations but also more and more organizations utilize GLIDE in their systems including international cooperative project utilizing space technology in DRR, Sentinel Asia, JAXA,, AHA Centre and so on (Fig. 1).

Fig. 1. GLIDE Participating Institutions (https://glidenumber.net/glide/public/institutions.jsp)

2.2 Mechanism of GLIDE

As seen in the previous section, GLIDE was proposed by ADRC and promoted by key DRR stakeholders, as a tool to allow anyone to easily share disaster information provided by various organizations around the world.

Once a disaster occurs, an operator of GLIDE partner organizations issues a GLIDE number through GLIDE website. Instantly the GLIDE number is created on GLIDE website and also is automatically emailed to some 2,000 subscribers. The users apply the GLIDE number to databases and products, linking to other relevant data with the same GLIDE number (Fig. 2).

A GLIDE number is composed of disaster classification (24 types), year of occurrence, serial number and country code. For example, 2023 Turkey earthquake is expressed as EQ-2023-000015-TUR. So far over 7,700 GLIDE numbers have been issued as of March 2023 (Fig. 3).

Since its establishment, GLIDE has been managed and operated by key organizations. With emerging needs identified with the times, organizational and functional changes have been made. First of all, the Steering Committee (SC) was established in 2021, to strengthen its management system. It is composed of ADRC as Secretariat, and SC members, EU/JRC, IDMC, IFRC, Tohoku University, UNDP, UNDRR, UNOCHA and WMO, Chair nominated among the SC members. Also, three Subcommittees were established to deal with thematic issues of API (Application Programming Interface),

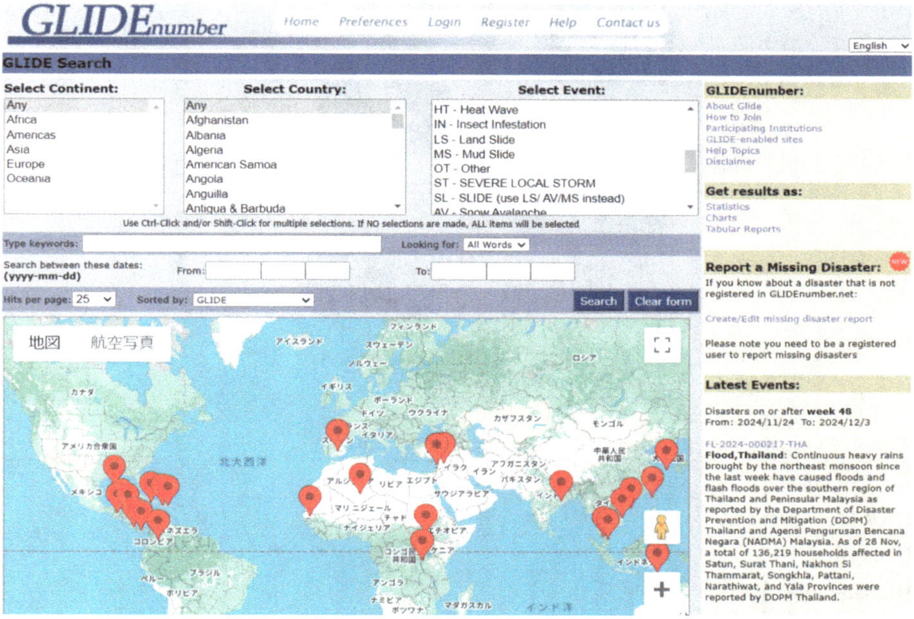

Fig. 2. GLIDE Website (https://glidenumber.net/glide/public/search/search.jsp)

Hazard Code (24)

- CW - Cold Wave
- CE - Complex Emergency
- DR - Drought
- EQ - Earthquake
- EP - Epidemic
- EC - Extra-tropical Cyclone
- FR - Fire
- FF - Flash Flood
- FL - Flood
- HT - Heat Wave
- IN - Insect Infestation
- LS - Land Slide
- MS - Mud Slide
- OT - Other
- ST - Severe Local Storm
- AV - Snow Avalanche
- SS - Storm Surge
- AC - Tech. Disaster
- TO - Tornadoes
- TC - Tropical Cyclone
- TS - Tsunami
- VW - Violent Wind
- VO - Volcano
- WF - Wild fire

Fig. 3. GLIDE Hazard Code

SOP (Standard Operating Procedures) and New Product Development, resulting in the development of the GLIDE API.

The GLIDE API which was developed in 2021 has facilitated the linkages of data across organizations. This drastically facilitated the interlinkages of databases. For example, ADRC's disaster database, Latest Disaster Information, is linked with UNOCHA ReliefWeb and EU/JRC GDACS by GLIDE numbers. Once GLIDE number is issued, data is referred automatically. As GLIDE application increases, various information has become available from different phases like Early Warning, relief operation, loss and damage, response to thematic information like displacement, satellite imageries and so on.

By analyzing the search results of GLIDE numbers on the Internet by Google, which was also a new function added in 2021, the expansion of GLIDE utilization is confirmed. Figure 4 shows the types of organizations that use GLIDE numbers on the Internet. While the number of the search results vary among the numbers (Fig. 5),

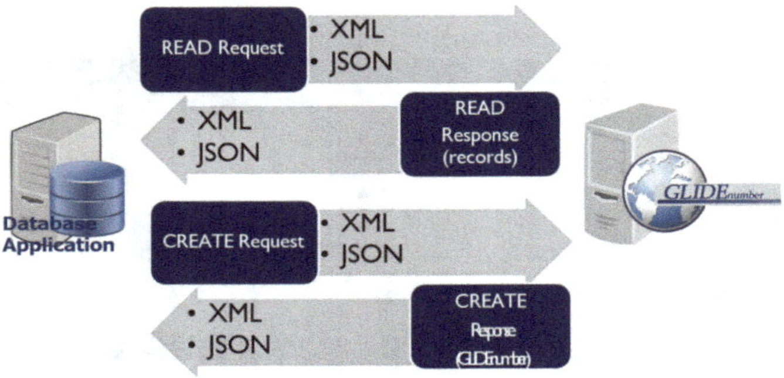

Fig. 4. GLIDE API Concept

Types of Organization that use GLIDE numbers on the Internet
DRR (member)
DRR (non-member)
Geospatial/ GIS
Humanitarian
Academia
Media
Others

Types of Organization that use GLIDE numbers on the Internet
DRR (member)
DRR (non-member)
Geospatial/ GIS
Humanitarian
Academia
Media
Others

Fig. 5. GLIDE search results on the Internet

3 International Initiatives

In implementing the global targets like SDGs and SFDRR, there have been increasing needs in establishing national disaster information management system and database for reporting. The following are some of the initiatives focusing on disaster information.

Sendai Framework Indicators (https://sendaimonitor.undrr.org/), with 7 Global Targets and 38 Indicators, require "optional" disaggregated data in reporting. Global target A: Substantially reduce global disaster mortality by 2030, aiming to lower average per 100,000 global mortality between 2020–2030 compared with 2005–2015, is composed of the number of deaths and missing persons attributed to disasters, per 100,000 population and disaggregation of by hazard, sex, age, disability and income [3]. Disaster data integration is required in collecting such disaggregated data from local level or line organizations.

DesInventar Sendai (https://www.desinventar.net/index.html) is a web-based disaster data inventory system initiative by UNDRR (then UNISDR) and UNDP in 1994. It aims to facilitate disaster information collection, documentation and analysis by filling in predefined fields. Now 110 countries utilize the system with loss and damage databases, which provide a comprehensive picture of human, economic, and infrastructural losses at subnational level.

Disaster Loss and Damage Tracking System (DLDT) a successor of DesInventar Sendai, is now being developed UNDRR, UNDP and WMO. According to UNDRR (2023), there is a "strong need to further understand the triggering factors and causes of each recorded event, to be able to link the impact of events with weather-related and other hazardous events" and it is an "application of data standards related to common terminologies, hazard classification, etc." [4]. DLDT designed as the form of comprehensive and multisectoral database and it is composed of (1) Cataloguing of hazardous events (CHE) data by National Hydro Met Agencies, Geological Surveys, (2) recording of event effects (loss and damages) by National Disaster Management Offices and DRR agencies and (3) Statistical baselines by statistics offices.

Hazard Information Profiles (HIPs) was compiled by UNDRR and International Science Council in 2021. It contains a description of each of the 302 hazard information profiles (HIPs), developed using a consultative process by scientists and experts across the globe and "aligned with the Sendai Framework for Disaster Risk Reduction 2015–2030, but also the Sustainable Development Goals of Agenda 2030, the Paris Agreement on Climate Change and the Addis Ababa Action Agenda on Sustainable Financing" [5]. HIPS include 8 hazard types, 47 clusters and 302 specific hazards. One of the eight hazard types, Meteorological and Hydrological hazard type has 9 clusters and 51 specific hazards. Among them a cluster Flood has 10 specific hazards, namely Coastal Flood, Estuarine (Coastal) Flood, Flash Flood, Fluvial (Riverine) Flood, Groundwater Flood, Ice-Jam Flood Including Debris, Ponding (Drainage) Flood, Snowmelt Flood, Snowmelt Flood, Surface Water Flooding, Glacial Lake Outburst Flood.

EM-DAT, Emergency Events Database is a widely acknowledged global disaster database which the Centre for Research on the Epidemiology of Disasters (CRED) launched in 1988. EM-DAT dataset is often utilized in various analyses. Their disaster classification is 2 Groups, 9 Sub Groups and 77 Subtypes. In Hydrological Sub Group, 12 subtypes like flood (general) and riverine flood.

As seen, these initiatives require reporting of various types of disaster-related data, which result in coordinating across different methodologies, could bring about enormous tasks for persons in charge of disaster information. By using GLIDE, it is possible to fetch required data from linked databases, such reporting could be facilitated as far as GLIDE API is installed in the systems let alone GLIDE could facilitate strengthening their disaster data information management capacity.

4 The Way Forward

Global initiatives referring to disaster-related indicators are on the rise. National Disaster Management Organizations (NDMOs) are increasingly required to report their progress and develop disaster information and data sets and manage them efficiently and effectively.

To support such tasks, interlinking disaster information could facilitate disaster information management and increase the value of data. With the development of GLIDE API, GLIDE can facilitate disaster data interlink and integration among various databases, which can contribute not only to solving the above-mentioned challenges but also to assisting disaster database linkages, resulting better disaster data management.

Regarding to 2023 Turkey earthquake (EQ-2023-000015-TUR), there are 478 hits in the Internet as of 10 March 2023, with the glide number, including not only GLIDE SC member organizations, UNOCHA ReliefWeb, IFRC and EC/JRC, but also Humanitarian OpenStreetMap and media reports. GLIDE is a glue that can integrate and link with a wealth of different information about a disaster, both at national and international level.

References

1. Fakhrudding, B., Kirsch-Wood, J., Niyogi, D., Guoqing, L., Murray, V., Frolova, N.: Harnessing risk-informed data for disaster and climate resilience. Progress Disaster Sci. **16**, 100254 (2022)
2. Below, R.: 30 Years experience in disaster data collection. In: JRC, 7th EU Loss Data Workshop, ISPRA, 10-11 March 2016 (2016)
3. UNDRR: Technical guidance for monitoring and reporting on progress in achieving the global targets of the Sendai Framework for Disaster Risk Reduction (New edition) (2017). https://www.desinventar.net/documentation/Technical_guidance.pdf
4. UNDRR: Progress and advances in monitoring of Sendai Framework for disaster risk reduction. In: Asian Conference on Disaster Reduction, March 2022 (2023)
5. UNDRR and international science council: "Hazard Information Profiles" (2023). (https://www.undrr.org/publication/hazard-information-profiles-hips)

Risk Assessment and Social Application of Local Observation in Communities: Toward the Construction of IoT-COS

Kensuke Takenouchi[1](✉) and Keisuke Hinoda[2]

[1] Faculty of Engineering and Design, Kagawa University, Takamatsu, Japan
takenouchi.kensuke@kagawa-u.ac.jp
[2] Graduate School of Science for Creative Emergence, Kagawa University, Takamatsu, Japan
s24d202@kagawa-u.ac.jp

Abstract. In recent years, the development of Internet of Things (IoT) technology has been remarkable. To utilize IoT technology to prepare for disasters, it is important to link the natural science approach, which focuses on observation and prediction, with the social science approach, which focuses on information utilization and systems. The authors proposed the construction of the IoT–Community Observation System (IoT–COS) as a framework for studying the operation and utilization of local observations with the participation of local residents. In this study, local observation, which is expanding along with the improvement of IoT technology, was analyzed from the perspectives of natural science (risk assessment) and social science (social application) using six case studies on community disaster prevention in Japan. Although risk assessment depends on the type of disaster, social applications also differ from risk assessment. In addition, while the purpose of risk assessment is to assess risk, it is also important to build relationships among the parties involved. An even greater variety of local observation efforts has been identified worldwide. It is necessary to more carefully and actively discuss the social positioning of observation networks in next-generation disaster management.

Keywords: Local observation · Residents' participation · Citizen science · Community disaster prevention · IoT technology

1 Background

In recent years, observations of heavy rainfall disasters have entered a new dimension. This can be discussed from two major perspectives. First, technological improvements in observation methods have enabled highly accurate and high-resolution observations. In recent years, phased-array radar has been developed as a weather radar with a horizontal resolution of 100 m and a temporal resolution of several 10-s intervals. Risk information with a horizontal resolution of 1 km has been available in Japan since 2017. The second is the improved locality of observation locations, enabling more familiar and relevant observations (hereafter referred to as "Local Observations"), which can be attributed to the development of several inexpensive and low-cost observation devices that utilize Internet of Things (IoT) technology.

H. Tatano et al. (Eds.): GSRIDRR 2023, DRRGADRIBS, pp. 481–496, 2025.
https://doi.org/10.1007/978-981-96-6609-6_32

This study focuses on the latter, but the development of IoT technology has been remarkable; according to The Business Research and Company (2024), the global market for IoT technology is worth $46.445 billion, expanding into a very wide range of fields. In Japan, discussions on introducing IoT technology are also underway, and the ICT Regional Revitalization Portal published by the Ministry of Internal Affairs and Communications (MIC 2018) lists 103 advanced cases nationwide, including the "ICT Regional Revitalization Award" winners and support measures. IoT technology is strongly expected to contribute to revitalizing regional economies, solving regional issues, and improving regional attractiveness. Some of these technologies are also expected to be used in the field of disaster prevention.

To utilize IoT technology to prepare for disasters, it is important to link the natural science approach, which focuses on observation and prediction, with the social science approach, which focuses on information utilization and systems. The authors proposed the construction of the IoT–Community Observation System (IoT–COS) as a framework for studying the operation and utilization of local observations with the participation of local residents (Takenouchi and Takeuchi 2023). IoT-COS means "Systems and technologies necessary for its utilization." The purpose of this study is to further the discussion of IoT–COS by organizing examples of IoT observations in Japan from both the aspects of risk assessment from the perspective of natural science and social utilization from that of social science.

2 Related Studies

The Sendai Framework for Disaster Reduction 2015–2030 (source: United Nations (2015)) highlights the importance of local people's participation in disaster reduction and the need to consider the characteristics of each community. The International Federation of Red Cross (IFRC) (2012) also provides guidelines for such efforts and based on various examples from around the world, identifies 13 guiding principles and 4 key areas of importance. Such discussions are increasingly being discussed in a variety of fields, and Follett and Strezov (2015) analyzed a wide range of peer-reviewed articles on citizen science (CS), including not only CS projects but also the theories and methods developed to support the research, to demonstrate that the CS approach highlights its breadth and depth and points out that it encourages cross-fertilization between different disciplines. Guidebooks, such as Pocock et al. (2014), are now available to examine what discussions, such as those in CS, can be appropriate in the first place. From the perspective of observation, the advent of a sensor society, as argued by Andrejevic and Burdon (2014), is being discussed. Discussions on disaster mitigation based on local observations using IoT technology are also advancing. In IFRC (2012), monitoring is also considered important. There are also various discussions on IoT technologies and the relationship between observations and communities; however, the approaches vary widely from case to case.

Rainfall observation is a common example of community participation in observation; Chen and Wu (2014) and Sasada et al. (2015) mention the "Sediment Flow Disaster Prevention Specialist" program in Taiwan, which aims to improve rainfall observation and local disaster preparedness by residents. In Taiwan, the local government has taken the initiative in establishing volunteer "sediment and runoff disaster prevention specialists" among local residents. Such specialists are responsible for increasing the amount of rainfall data in mountainous areas and improving the accuracy of disaster forecasting by observing rainfall using simple rain gauges and reporting the results. Abon et al. (2012) reported a case study in the central Bicol River Basin in the Philippines, where rainfall measurement results from 20 volunteers were obtained via SMS messages and used for early flood warnings by linking the results to a watershed model. The report also points out that the awareness of participation in the system has a significant impact on residents' understanding of floods. Similarly, Starkey et al. (2017) conducted a study in the 42 km^2 Haltwhistle Burn watershed, a tributary of River Tyne in northeast England, collecting and using quantitative and qualitative observations from the public to effectively understand spatial and temporal river responses. Successful collection and quality checking of community-based rainfall, river level, and flood observations, and their use in the construction and implementation of the physics-based spatially distributed catchment model SHETRAN, have shown that local networks of community-based observations, in addition to traditional hydrologic sources, can be used to provide a comprehensive understanding of the spatial and temporal scales to more accurately characterize watershed responses than traditional observations alone. They point to the benefits of such an approach as increasing the credibility of characterization and management actions, encouraging broader community participation, and providing important information for fishery management. Savitri et al. (2022), in a study of the upper Serayu watershed in Indonesia aimed at obtaining accurate rainfall data by utilizing appropriate rain gauge locations and community participation, have found that accurate rainfall data and the participation of younger people (elementary school students) in the local community are important for flood management. The study points out that this could have a positive impact on developing early warning systems (EWS) for landslide hazards.

In addition to rainfall, there are cases in which residents are involved in observing river levels. Smith et al. (2017), in a case study in Nepal, note that in community-based EWS (CBEWS), communities are not passive recipients of early warnings but are engaged in monitoring and analysis and are active owners and promoters of the EWS (IFRC, 2012). Initially, the CBEWS trained community water-level gauge readers to monitor river levels and communicate warnings to downstream communities when river levels exceeded the established warning and hazard thresholds (Mercy Corps and Practical Action, 2010). In addition to rising water levels, water-level gauge readers send out warnings when the observed rainfall in upstream catchments exceeds a certain threshold that could cause flooding downstream. Schmidt-Thome et al. (2018) discussed the case study of a landslide observation network in Thailand involving residents. In this case study, upon receiving an early warning from the GOC, trained villagers are deployed to observe local precipitation and river runoff. When precipitation exceeds predefined thresholds and river runoff reaches dangerous levels, warnings are issued through speakers and sirens. The Thai landslide network has been very successful in achieving its top priority of saving lives. In addition, the volunteer training educated and sensitized the population about the dangers of landslides and their resulting risks.

In addition, Weathernews Inc. (2016) offers a resident-participatory "disaster mitigation project" for municipal services in Japan, where residents register and share various information about their respective districts, such as water levels, flooding, traffic conditions, and lifeline status, and use information from other residents to assess the situation and act. Based on the information provided by the residents, local governments can assess the local situation on a map and consider response measures. There are also examples of residents participating in observations of slopes and spring water levels in relation to landslide disasters. Similar examples can be seen in the framework of cross-national information sharing by companies, such as Netatmo in France and Weather Link by Davis Instruments Corporation in the United States.

These initiatives have also been discussed from a broad CS perspective. Tipaldo and Allamano (2017) illustrate the recent discussion of CS with a case study of citizen participation in rainfall observations. They organized CS projects for participatory rainfall monitoring (Table 1) and pointed out that all efforts were based on the traditional CS paradigm, a process of rain monitoring initiated primarily by the sender and concerned residents to serve as data collectors. Subsequently, it is pointed out that while many studies on science, technology, and society exist in a wide variety of fields, CS-focused interdisciplinary approaches remain in the minority, and there is a lack of research on the social mechanisms that drive non-specialists (i.e., citizens). Similarly, in the natural sciences, weather forecasting requires continuous observations at very different scales while highlighting the problems that need to be addressed in establishing and maintaining networks (Muller et al. 2013).

Thus, given the diversity of cases and the immaturity of citizen participation in local observation as CS, an IoT–COS discussion is essential for its maturation.

3 Methodology

The present study aimed to identify examples of IoT observations related to disasters, with a particular focus on heavy rainfall disasters, and to evaluate these observations from both the aspects of natural and social sciences. Specifically, what elements are being observed and for what purpose in terms of natural science, and what are the social applications of these observations in terms of social science? By organizing these elements, we clarify the existing situation and challenges for IoT–COS, which aims to build an environment for the regional utilization of IoT technology.

The research methodology will not cover all cases but will identify target cases for discussion. In particular, this type of research needs to understand the actual situation on the ground and not just the superficial situation. This is because it is difficult to properly ascertain the field's viewpoint using only superficial information. For this reason, cases of IoT observations in which the authors were involved were selected as targets. The characteristics of these cases are summarized in terms of both risk assessment (natural science) and utilization in the community (social science). Based on these results, we discuss the characteristics and issues of the social deployment of local observations using IoT technology.

4 Results

4.1 Extraction of Cases

First, as target cases, we extracted observation cases familiar to local communities using IoT technology in Japan, as shown in Table 1. In all cases, the targets of use are primarily governments, districts, and residents and are not independently conducted by residents.

Table 1. Examples of targeted local observations

No	Target area	Observation items (target disasters)	Purpose of Observation	Subject of use
1	Shirutani River basin, Ise City, Mie Prefecture	River water level (flood)	Assessing river conditions	Not specified (prefecture, city, districts, resident)
2	Ichiuda area, Ise City, Mie Prefecture	River level (flood)	Assessing river conditions	Not specified (prefecture, city, community, resident)
3	Tadewara District, Fukuchiyama City, Kyoto Prefecture	Inundation occurrence (internal flooding)	Notification of internal flooding	Cities, community, residents
4	Kakimoto District, Fukuchiyama City, Kyoto	Valley water level (mudslide)	Monitoring changes in valley water conditions	City, community, resident
5	Araki District, Fukuchiyama City, Kyoto Prefecture	Slope inclination (landslide)	Assessing changes in slope	City, community, resident
6	Taisho District, Shimanto Town, Kochi Prefecture	Spring water amount (mudslide, landslide)	Assessing changes in the amount of spring water	Town, community, resident

Cases 1 and 2 are examples of government-led efforts to utilize the crisis management-type water-level gauge (Ministry of Land, Infrastructure, Transport and Tourism 2019) (Figs. 1 and 2). In Japan, water-level observations have been enhanced for large rivers. However, in recent years, the risk of flooding in small- and medium-sized rivers, such as tributary overflows, has increased, leading to discussions on developing inexpensive, power-saving, and maintenance-free water-level gauges that utilize IoT technology. Then, the "Innovative River Technology Project" was launched in 2016, and since its completion in 2017, 5,832 water-level gauges (as of August 22, 2021) have been installed throughout Japan as crisis management-type water-level gauges, the installation of which was promoted nationwide. However, while these gauges have been maintained, little progress has been made in discussing their utilization by communities and residents (Hinoda et al. 2022).

Case 3 is an IoT observation developed to report the occurrence of internal flooding due to poor drainage in a community frequently suffering from external flooding because such flooding has become more frequent with the improvement of river levees (Fig. 3). The system uses a mobile application to easily report the occurrence of flooding. Similar IoT observations are also being promoted by the national government, and installation is being promoted as a one-coin inundation sensor (Ministry of Land, Infrastructure, Transport and Tourism 2021).

Cases 4, 5, and 6 are examples of local observations related to landslide disasters using IoT technology, although the districts and observation items were different. Risk assessment is more difficult for landslide disasters than for floods and inland flooding. IoT observations are conducted in order to be used by communities and residents based on discussions at the research level, respectively. Case 4 is a district that experienced damage from a mudslide in the past, and water-level observations of water flowing from the valley were conducted at 10-min intervals to observe valley water conditions (Fig. 4). Case 5 is also a district that has experienced localized landslides in the past, and because some slopes remain without collapsing, observations of changes in slope inclination are being conducted at 10-min intervals using an inclinometer (Fig. 5) (Miyamoto et al. 2023). Case 6 is an example of assessing the risk of landslide disasters by observing the amount of spring water from the slope rather than valley water because there are several slopes in the district where spring water is generated. Observations are also made every 10 min (Fig. 6).

Fig. 1. Crisis management-type water-level gauge (Shirutani River basin, Ise City, Mie Prefecture)

Fig. 2. Crisis management-type water-level gauge (Ichiuda area, Ise City, Mie Prefecture)

4.2 Risk Assessment by Local Observation

For the extracted cases discussed in the previous section, we organized how risk assessment was discussed from the perspective of natural science at each site.

The crisis management-type water-level gauges in Cases 1 and 2 were subject to the risk of assessing flood inundation. As mentioned above, they primarily target small- and medium-sized rivers. From the viewpoint of achieving power conservation and understanding flooding, they are designed to start observations every 10 min after a certain water level (observation start level) is exceeded. This observation start level is set in

Fig. 3. Inundation detection sensor (Tadewara area, Fukuchiyama City, Kyoto Prefecture)

Fig. 4. Valley water sensor (Kakimoto district, Fukuchiyama City, Kyoto Prefecture)

Fig. 5. Slope inclination sensor (Araki district, Fukuchiyama City, Kyoto Prefecture)

Fig. 6. Spring water amount sensor (Taisho district, Shimanto Town, Kochi Prefecture)

consideration of the lead time required for evacuation actions, river channel characteristics, and the provisional water level at the location where the water-level gauge is installed (Ministry of Land, Infrastructure, Transport and Tourism, 2019). However, in an actual situation, sufficient lead time is not always available because of the characteristics of small- and medium-sized rivers, which fluctuate rapidly. Figures 7 and 8 show examples of the water-level changes in Cases 1 and 2, respectively. As shown in Fig. 7, there are water-level gauges whose peak times deviate from the precipitation by several hours, while there are also water-level gauges that show peak values within several 10-min intervals, as shown in Fig. 8. Thus, although the risk targets to be evaluated are the same, their response characteristics differ, and risk evaluation methods must be discussed according to the situation. As discussed below, Case 2 confirms the current situation, where it is difficult to use as a discussion of evacuation and is primarily used only as an understanding of the occurrence of administrative inundation.

Case 3 is a very simple instrument that only evaluates the occurrence or non-occurrence of inundation situations. This differs from the other cases in that it is not a quantitative observation but a qualitative one. As a risk assessment, it may only be used when the objective is clear.

Cases 4, 5, and 6 have one thing in common with the risk assessments: their correspondence with the actual occurrence of a disaster is unclear. In all of these cases, any anomaly is considered a possibility close to the occurrence of a disaster. However, it is not clear how long and to what degree of certainty they will occur. This is because the occurrence mechanism of landslides is highly uncertain as well as regional and localized. In this context, the risk assessment method is unique in that it is tailored to the characteristics of each district. In Cases 4 and 6, if there is no valley or spring water, it is difficult to conduct observations in the first place. Case 5 can be conducted anywhere on a slope, but there are no clear rules on which slope is appropriate, and it cannot be used to evaluate debris flows.

Although not the subject of this study, as discussed in previous studies, there are many cases in Japan where only rainfall observations have been conducted. The cases covered in this study involved observations tailored to the disaster risks faced by each district. However, these observations are based on precipitation. Precipitation observations can easily be conducted anywhere. Although these methods may not be as accurate, they have certain advantages.

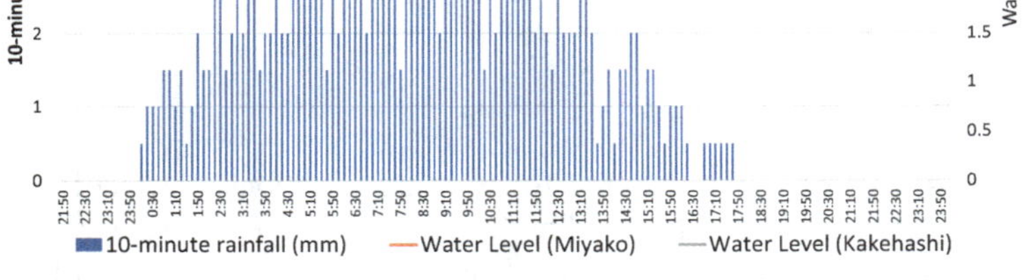

Fig. 7. An observation result in the Shirutani River (12 October 2019)

4.3 Use of Local Observation in the Community

Next, we summarize the utilization of local observations in the communities of target cases.

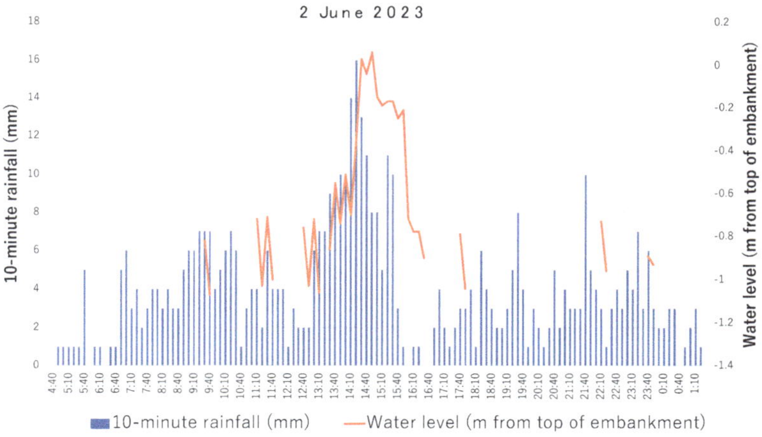

Fig. 8. An observation result in the Ichiuda River (2 June 2023)

For Case 1, Hinoda et al. (2022) found that only 4% of the 107 community associations in Ise City, Mie Prefecture, used crisis management-type water-level gauges, which is very low. The authors launched the "Shiru Tanigawa Minna de Observation Project" in 2021 to promote the use of water-level gauges in community disaster response and the sharing of water-level gauge information through a smartphone application. These efforts demonstrate that while familiar local observations have been enhanced, discussions on their utilization have not progressed. However, some characteristic stories have been identified in terms of social utilization. For example, the Kakehashi district participating in this project suffered flood damage in the past because of the overflow of the Shirutani River. However, residents of the Kakehashi district objected to the location of the crisis management-type water-level gauge. In this district, the Mukoseko River, which flows into the Shirutani River, inundates earlier than the Shirutani River. Evacuation by passenger cars is not possible unless the situation of the Mukoseko River is understood. Therefore, it is argued that a water-level gauge should be installed on the Mukoseko River. As a result of the government installing the gauges without discussing how they are utilized with the stakeholders, a situation has arisen in which supply and demand do not match (lack of a social utilization perspective). This story demonstrates that even in the case of utilization, how it is utilized and its positioning differs from district to district.

Similarly, in Case 2, the community did not use a crisis management-type water-level gauge. In the discussion on its utilization, it is being promoted to be used for chronological disaster responses (timeline disaster management). Drills such as the sounding of evacuation sirens using water-level gauges, have been conducted (in conjunction with the disaster response mechanism). However, important issues have arisen in the practical use of the system. The district is located in a mountainous area, and the distance between the mountains and villages is small. Therefore, when heavy rains occur, the river rises rapidly, and water-level gauges may not provide sufficient lead time (difficulty in community utilization). Consequently, crisis management-type water-level gauges have been installed to monitor the situation of the river on the administrative side, and the main purpose was to monitor the overflow of water. However, water-level gauges may not be used in communities in such cases. By accumulating data on rainfall and water levels, water overflow can be anticipated in advance based on rainfall forecasts. The problem lies in the fact that local observations are conducted without such discussions.

Case 3 does not have as many problems as the other cases in terms of its use in the community. This is because the function of local observation is simple, and the type of utilization is discussed and clarified in advance. In Tadewara district, based on past experiences of external flooding, there have been numerous discussions on mutual aid policies and evacuation sites within the district. This was because it was decided to use local observations as a catalyst for action within the district (to clarify the purpose of use). The equipment was installed after discussions between the community and the government. The fact that local observation is equipped with the necessary functions at the necessary locations, rather than because the equipment is simple, is considered the reason why there are few issues.

Case 4 is a district in which the community is small and has originally experienced disasters; however, because many residents are elderly, voluntary early evacuation is not being actively pursued. Under such circumstances, local observations are being conducted from the perspective of detecting abnormalities in valley waters and promoting evacuation (from the perspective of promoting disaster awareness). Because of this situation, the use of the system itself is maintained by the officers of the community association. A system has been established whereby the representatives view the valley water flow rates obtained from local observations on their smartphones, which are then used to call for evacuation to the surrounding residents.

The district in Case 5 was originally discussed for evacuation, not through local observation but through risk-level mesh information on landslide disasters provided by the Japan Meteorological Agency and the Kyoto Prefectural Government. Currently, landslide-related alerts in Japan are determined by the soil water index, which indicates the amount of moisture in the soil (Japan Meteorological Agency 2018). This information is available on a website, and as shown in Fig. 9, the government and residents determine index values, such as values from past landslide experiences, and the person in charge of checking the information in the district checks the status of the soil water index to call for evacuation and provide support. In this context, a complementary high-precision inclinometer is now being used to check the situation with the support of businesses (additional information to existing efforts). When the inclinometer records a certain rate of inclination, a notification is sent via e-mail, and the information is shared with a local LINE group.

Similar to Cases 4 and 5, Case 6 is a landslide-related initiative; however, its use is slightly different.

The Disaster Response Switch (DRSwitch) initiative was originally implemented in this district (Takenouchi 2020a). DRSwitch is a social mechanism that combines specialized disaster information with familiar local information to determine the timing of evacuation in advance and to respond to disasters accordingly. Unlike floods, in which the danger is clear from the water level, landslide disasters are superficially unclear in terms of their degree of danger. Therefore, the district in question used valley water

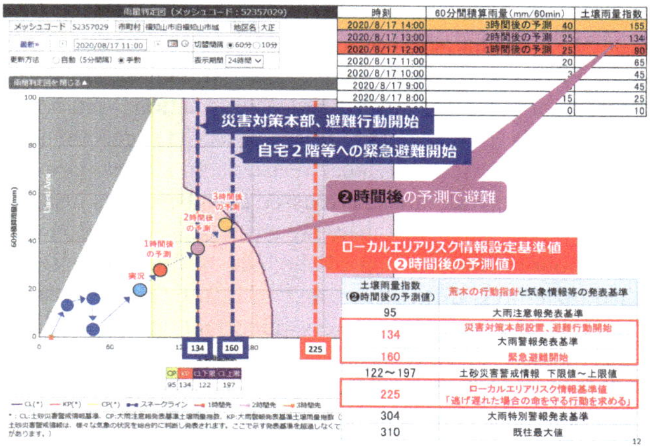

Fig. 9. Example of Soil Water Index Application (Fukuchiyama City, 2021) (Araki District, Fukuchiyama City, Kyoto Prefecture)

as an indicator, and the difference from normal times as a reference for the timing of evacuation. However, because this was a qualitative and unclear assessment of the situation, local residents demanded information that would enable them to directly assess the mountain situation. Therefore, after consulting with experts, local observations of the amount of water flowing into the mountain were initiated to understand the situation more objectively (improvement of uncertainty). Currently, real-time observation information is available on web pages, along with historical information.

5 Discussion

5.1 What Should Local Observations be like - Characteristics of Local Observations

In this study, the characteristics of local observation cases in Japan related to the authors were summarized from the viewpoints of natural and social sciences, respectively, as risk assessment and social application. The results are summarized in Table 2, which shows that local observations in the discussion on disasters are very diverse from both aspects.

First, from the aspect of risk assessment, some of them were evaluated as highly accurate information (Cases 1, 2, and 3) such as the case of water-level observation, which has existed for a long time, while others, such as slope and water inflow rate, are used to improve their accuracy (Cases 4, 5, and 6) such as the case of landslide disasters with high uncertainty. All these are still in the provision of objective information. However, some were identified as having a short lead time (Case 2) or as only observing the occurrence of a disaster (Case 3). These characteristics are closely related to the characteristics of the type of disaster. Observations and risk assessments are made according to these characteristics.

From the perspective of social application, as with risk assessment, some characteristics that are related to the characteristics of disaster types but differ from those of risk assessment, can be identified. In other words, some of them are used as a complement to existing disaster action rules and other disaster information, even if they are uncertain about risk assessments (Cases 4, 5, and 6). This is observed in cases where local observations are discussed in conjunction with local disaster action rules. In some cases, such as Case 3, the utilization of the system was discussed even before its installation, making it very simple and easy to use. However, there are cases (Cases 1 and 2) where, even when rules for disaster response actions are established, the positioning of the rules differs per region, and their utilization is difficult because of insufficient discussion prior to their establishment. In light of these points, a very important characteristic of the aspect of social utilization is that it is strongly influenced by local characteristics, as indicated by the term "local observation." If the purpose of the observation was to be utilized in the community, there could have been other options as a method of observation. However, this was not the case in the present study. The flooding sensor in Case 3 is unique in that it is the most appropriate one for the community rather than for other observations. As for the amount of spring water, the daily symbolic features of valley water from the mountains and springs that are also used in daily life have led to its utilization.

Thus, the characteristics of local observations can vary greatly depending on the current status of the risk assessment associated with the type of disaster and the local situation in social utilization.

5.2 How Should Local Observations be Conducted - Perspectives for Relationship-Building

In the previous section, we reviewed the points that should be considered when taking into account the characteristics of the local observations in each case study. Then, what should be the purpose of local observations in the first place? We would like to attempt an evaluation from a more meta-perspective. Such a discussion is necessary because, as shown in Cases 1 and 2, there is a concern that local observations with only superficial objectives cannot be truly utilized by local communities.

First, the typical purpose of local observation of heavy rainfall disasters is to understand the likelihood of disaster occurrence and its characteristics, similar to the perspective of risk assessment. Although there are cases of post-disaster observations that focus primarily on the aftermath of some types of disasters, such as earthquakes, observations that focus primarily on risk assessment are often seen in those related to meteorological disasters. In this way, risk assessment is discussed in most observations related to disaster prevention worldwide, although a few observations are made independently by individuals or communities; in most cases, they are established by specialized organizations. However, the achievement of the objectives centered on such a specialized institution is not as clear as Takenouchi (2020b) points out in Phase 3 of the "Outline of the relationship between the technical level of weather and hydrological information and residents' awareness of participatory information" that it creates a strong information-dependent relationship that does not lead to community response and action.

One issue that must be addressed is the need to create linkages and partnerships among the parties involved. The cases discussed in this study can be viewed as examples

Table 2. Characteristics of local observation from the perspective of both natural and social sciences

No	Target area	Target disaster	Risk assessment (natural science)	Social application (social science)
1	Shirutani River basin, Ise City, Mie Prefecture	Flooding	High accurate Sufficient lead time	Use of an action rule in each district Positioning differs from region to region
2	Ichiuda area, Ise City, Mie Prefecture	Flooding	High accurate Short lead time	Utilization of an action rule Difficult to utilize
3	Tadewara District, Fukuchiyama City, Kyoto Prefecture	Flooding	Highly accurate Whether or not a disaster will occur	Utilization in an action rule Easy to use
4	Kakimoto District, Fukuchiyama City, Kyoto	Mudslides	Uncertain	Used as a reference by some members
5	Araki District, Fukuchiyama City, Kyoto Prefecture	Landslide	Uncertain	Used as a supplement to other information
6	Taisho District, Shimanto Town, Kochi Prefecture	Mudslide, Land slide	Uncertain	Used as a supplement to an existing action rule

of communities trying to become involved in local observations. Even in cases where such perspectives are incorporated later, such as in Cases 1 and 2, collaboration with government agencies and river administrators is promoted in discussions on the use of local observations. That is, these cases are characterized by the fact that, through discussions on local observation, not only risk assessment according to the community but also relationship-building among the parties involved is established.

Traditionally, such discussions of collaboration across responsibility boundaries have not been fully intended in Japan due to the dominance of "self-help, mutual-help, and public-help" terms that describe the division of roles. However, the importance of overlapping disaster responses has been highlighted (e.g., Research Strategy Center in Hyogo Earthquake Memorial 21st Century Research Institute 2019, Sakamoto 2019). Similar to these discussions, they also indicated that "building local observations together" and "utilizing local observations together" are important objectives in promoting the social application of local observations. As the enhancement of local observations using IoT technology accelerates, this perspective of discussion can provide important suggestions.

6 Conclusion

In this study, local observation, which is expanding along with the improvement of IoT technology, was analyzed from the perspectives of natural science (risk assessment) and social science (social application) using case studies in Japan.

Although risk assessment depends on the type of disaster, social applications also differ from risk assessment. In addition, while the purpose of risk assessment is to assess risk, it is also important to build relationships among the parties involved.

The discussion in this study is based on the premise that communities are directly involved in local observations. It is, of course, possible to take initiatives that do not involve such local observations but only aim to utilize the information that is the product of such observations. However, unlike conventional low-resolution information, very familiar observation data installed in the community can reveal the situation of the residential area and can lead to the society of Phase 4 of Takenouchi (2020b), which points to the revival of the relationship between people and information.

An even greater variety of local observation efforts has been identified worldwide. By examining the increasing number of cases as the local observations increase, it is necessary to more carefully and actively discuss the social positioning of observation networks in next-generation disaster management.

References

Abon, C.C., Primo, C., David, C., Tabios, G.Q.: Community-based monitoring for flood early warning system: an example in central Bicol River basin Philippines. Disaster Prevent. Manage. **21**(1), 85–96 (2012). https://doi.org/10.1108/09653561211202728

Andrejevic, M., Burdon, M.: Defining the sensor society. Telev. New Media **16**(1), 19–36 (2014). https://doi.org/10.1177/1527476414541552

Chen, S.C., Wu, C.Y.: Debris flow disaster prevention and mitigation of non-structural strategies in Taiwan. J. Mountain Sci. **11**, 308–322 (2014). https://doi.org/10.1007/s11629-014-2987-3

Follett, R., Strezov, V.: An analysis of citizen science based research: usage and publication patterns. PLoS ONE **10**(11) (2015). https://doi.org/10.1371/journal.pone.0143687

Hinoda, K., Takenouchi, K., Tanaka, K., Matsuda, Y.: Usage status of the flood crisis management water level gauge and issues for utilization in the local communities. J. Jpn. Soc. Nat. Disaster Sci. **41**(S09), 125–140 (2022). https://doi.org/10.24762/jndsj.41.S09_125

IFRC: community early warning systems: guiding principles: International federation of red cross and red crescent societies (2012). https://www.ifrc.org/sites/default/files/CEWS-Guiding-Principles-EN.pdf. Accessed 23 Mar 2024

Japan meteorological agency: soil water index, runoff index, and surface water index (2018). https://www.jma.go.jp/jma/en/Activities/qmws_2018/Presentation/3.1/Soil%20Water%20Index,%20Runoff%20Index,%20and%20Surface%20Water%20Index_revised.pdf. Accessed 23 Mar 2024

Mercy corps and practical action: establishing community based early warning system: practitioner's handbook, mercy corps and practical action (2010). http://www.preventionweb.net/educational/view/19893. Accessed 23 Mar 2024

Ministry of internal affairs and communications: 100 examples of ICT regional revitalization (2018). https://www.soumu.go.jp/main_sosiki/joho_tsusin/top/local_support/ict/jirei/index.html. Accessed 23 Mar 2024

Ministry of land, infrastructure, transport and tourism: guidance for the installation of critical water level gauges (Draft) (2019). https://www.mlit.go.jp/river/shishin_guideline/kasen/pdf/kikikanri_tebiki.pdf. Accessed 23 Mar 2024

Ministry of land, infrastructure, transport and tourism: one-coin flooding sensor demonstration experiment (2021). https://www.mlit.go.jp/river/gijutsu/wankoinsensa/index.html. Accessed 23 Mar 2024

Miyamoto, Y., Oji, S., Fujitani, H., Yamori, K.: Development and trial of slope disaster prevention monitoring system with resident participation. Special Issue (Global Environment Engineering) Paper, **79**(27) (2023). (in Japanese). https://doi.org/10.2208/jscejj.23-27021

Muller, C.L., Chapman, L., Grimmond, C.S.B., Young, D.T., Cai, X.: Review sensors and the city: a review of urban meteorological networks. Int. J. Climatol. **33**, 1585–1600 (2013). https://doi.org/10.1002/joc.3678

Pocock, M.J.O., Chapman, D.S., Sheppard, L.J., Roy, H.E.: Choosing and using citizen science: A Guide to When and How to Use Citizen Science to Monitor Biodiversity and the Environment. Centre Ecology and Hydrology (2014)

Research strategy center in Hyogo earthquake memorial 21st century research institute: Research and survey department: research on the improvement of disaster preparedness of local communities – towards inclusive local disaster management. Research Survey Report (2019). (in Japanese)

Sakamoto, H.: 236th Sangyo seminar: analysis of Japanese Self-help, mutual-help, and public-help consciousness. Seminar Annual Report, pp. 97–107 (2019). (in Japanese)

Sasada, K., Yi-Tzu, L.Y.T., Sato, N.: Development of Non-structural measures toward dangerous area for debris flow disaster in the mountainous area of Taiwan and Implications for Japan. J. Japan Soc. Nat. Disaster Sci. **34**(3), 189–211 (2015). (in Japanese)

Savitri, E., Wahyuningrum, N., Nugroho, H., Sudiana, N., Indrajaya, Y., Christanto, N.: Identification of the rain gauge stations for the participatory flood and landslide mitigation in the Serayu river basin, Central Java. IOP Conf. Ser.: Earth Environ. Sci. **1109**, 012007 (2022). https://doi.org/10.1088/1755-1315/1109/1/012007

Schmidt-thomé, P., Tatong, T., Kunthasap, P., Wathanaprida, S.: Community based landslide risk mitigation in Thailand. Episodes J. Int. Geosci. **41**(4) (2018). https://doi.org/10.18814/epiiugs/2018/018017

Smith, P.J., Brown, S., Dugar, S.: Community-based early warning systems for flood risk mitigation in Nepal. Nat. Hazards Earth Syst. Sci. **17**, 423–437 (2017). https://doi.org/10.5194/nhess-17-423-2017

Starkey, E., Parkin, G., Birkinshaw, S., Large, A., Quinn, P., Gibson, C.: Demonstrating the value of community-based ('citizen science') observations for catchment modelling and characterisation. J. Hydro **548**, 801–817 (2017). https://doi.org/10.1016/j.jhydrol.2017.03.019

Takenouchi, K.: Incorporating public participation into landslide risk information and response: disaster response switch in the Taisho District of Shimanto-cho, Kochi. Japan. J. Integr. Disaster Risk Manage. **10**(1), 43–68 (2020a). https://doi.org/10.5595/001c.17362

Takenouchi, K.: Collaborative Community Weather Information, In: K. Yamori, ed. Disaster Risk Communication – A Challenge from a Social Psychological Perspective, pp. 65–81, Integrated Disaster Risk Management, Springer (2020b). https://doi.org/10.1007/978-981-13-2318-8_5

Takenouchi K., Takeuchi, M.: Stakeholders' involvement in constructing IoT-COS: utilization of IoT observations for Community Disaster Reduction through Community Disaster Prevention Plan. In: C+Bousai by the Japan Society of Community Disaster Management Plan, vol. 28, pp. 37–53 (2023). (in Japanese)

The business research company, IoT Global Market Report: Industrial IoT Market (By component: solution, services, platform; by end-use: manufacturing, energy & power, oil & gas, healthcare, logistics & transport, agriculture, others) - Global Industry Analysis, Size, Share, Growth,

Trends, Regional Outlook, and Forecast 2023–2032 (2024). https://www.fortunebusinessinsights.com/industry-reports/internet-of-things-iot-market-100307. Accessed 23 Mar 2024

Tipaldo, G., Allamano, P.: Citizen science and community-based rain monitoring initiatives: an interdisciplinary approach across sociology and water science. WIREs Water **4**, e1200 (2017). https://doi.org/10.1002/wat2.1200

United Nations: Sendai framework for disaster risk reduction 2015–2030 (2015). http://www.unisdr.org/we/coordinate/sendai-framework. Accessed 23 Mar 2024

Weathernews Inc. Disaster reduction and mitigation by participation and share in self- and mutual-helps, Japan (in Japanese) (2016). http://ictfss.nict.go.jp/ictfss-2016/1_unozawa.pdf. Accessed 23 Mar 2024

Disaster Preparedness and Risk Perception in Southern Italy: Context, Beliefs and Motivation in Messina 114 years After the Disaster

Irene Petraroli[1(✉)], Roger C. Baars[1], Marco Centorrino[2], and Antonino Baglio[2]

[1] Graduate School of Global Environmental Studies, Kyoto University, Kyoto, Japan
irene.annachiara.petraroli@gmail.com,
baars.rogercloud.6a@kyoto-u.ac.jp

[2] Dipartimento di Civilta' Antiche e Moderne, Messina University, Messina, Italy
{marco.centorrino,antonino.baglio}@unime.it

Abstract. More than 100 years have passed since 91% of the urban area of Messina was completely destroyed by the joint earthquake and tsunami in 1908. Although no survivor lives today, the issue of risk perception in the city of Messina, in Southern Italy, remains a topic of interest for its peculiar history with disasters and geological vulnerabilities. This study focused on the analysis of the risk perception (Priority 1 of the Sendai Framework 2015) of *Messinese* respondents, focusing on their personal characteristics and belief system to understand their level of interest and engagement in disaster preparedness. The analysis of the belief system, which included place attachment, risk perception, attitude, opinion of disaster preparedness, and trust in the authorities, gives an overview of the complexity of understanding risk perception in a local and cultural context. This study is intended to be the starting point of future discussion and qualitative studies.

Keywords: risk perception · disaster preparedness · liquidity · 1908 disaster · Italy

1 Introduction

It was during the UN World Conference on Disaster Risk Reduction (DRR) in Sendai, Japan (2015) that the importance of Understanding Risk was codified as the first of four priorities: 1. Understanding Risk, 2. Strengthening Governance, 3. Investing in DRR for Resilience and 4. Enhancing Disaster Preparedness. One of the reasons for this primacy is that understanding perceptions of risk is necessary to foster motivation and engagement in preparedness actions (Huurne and Gutteling, 2008; Gerhold, Wahl and Dombrowsky, 2019). Learning what motivates people to take preventive actions is crucial to resilient communities. In fact, it is from the people sharing about their motivations and worries that policy makers and researchers can identify and address their disaster vulnerabilities and empower them effectively to be more prepared.

H. Tatano et al. (Eds.): GSRIDRR 2023, DRRGADRIBS, pp. 497–513, 2025.
https://doi.org/10.1007/978-981-96-6609-6_33

Thus, this project falls into the category of DRR studies on the social aspects of vulnerability and preparedness, focusing on the links between local authorities, enforcers and the public (Spielman et al. 2020). This study looks at the non-expert population to uncover the reasons that promote or prevent interest in disaster preparedness measures. The objective is to investigate the impact of a variety of factors on motivation and engagement in disaster preparedness. This study contributes to the state of the art on disaster motivation and engagement by focusing on the joint impact of beliefs (Gero et al., 2021), attachment to place (Wallis, Fischer and Abrahamse, 2022), and socio-cultural context (Terpstra and Gutteling, 2008), which are traditionally analysed independently. Furthermore, this study uses the notion of "liquidity", developed by sociologist and philosopher Zygmunt Baumann on the issues of modernity, globalization, and consumerism, to give a possible interpretation of disaster risk perception results.

2 Holistic Approach

This study uses a holistic approach to measuring risk perception by combining literatures from disaster studies, human geography, and sociology. DRR and emergency studies focused on the importance of beliefs to understand risk perception, coping and preventive measures. A study conducted in Japan showed that optimism is an important resilience resource in disaster survivors (Gero et al., 2021). A study conducted in the USA showed the importance of political belief in abiding to COVID-19 restrictions rules. In other words, psychological, political, and religious beliefs have an impact on how way of acting and thinking. Human geography literature approached DRR studies from a different perspective, focusing on "place attachment", defined as the "emotional bonds between person and place" (Inalhan, Yang and Weber, 2021, p. 16). Place attachment has been found not only to affect individual's behaviour and interaction with the physical environments (ibid.), but also to have an important role in promoting place-based resilience (Wallis, Fischer and Abrahamse, 2022). Finally, another crucial piece of the puzzle that is understanding risk perception and promoting citizen participation lies in the relationship and perception of the local governments and relevant authorities (Terpstra and Gutteling, 2008).

These literatures were used to create a theoretical framework composed of three elements: context, belief system and engagement. Below is a graphic representation of the framework (Table 1). The article focuses on the statistically significant survey variables. In the discussion, the concept of "liquidity" will be used to give an additional qualitatively interpretation of the quantitative results.

2.1 The Case-Study: Messina and the Mezzogiorno

The city of Messina (2021 population 221,246), is the closest point to continental Italy, familiarly called "the boot" for its characteristic shape. The city of Messina was chosen for this study due to its exposure to seasonal hazards. Historically, Messina has been struggling with wildfires, landslides and floods, worsened by dry heat and strong Saharan winds (*scirocco).* Among the most devastating recent events we find the 2009 flash floods and landslides in Giampilieri, which caused 37 deaths (Aronica, Brigandí and Morey,

Table 1. Description of study model (elements, variables, methods)

Elements	Variables
Context	1. Demographic (gender, age, level of education, experience of disaster)
Belief system	2. Place attachment 3. Risk perception 4. Attitude 5. Disaster preparedness opinion 6. Trust in local authorities
Engagement	7. Disaster preparedness measures 8. Engagement and motivation

2012), and the wildfires in summer 2023, which consumed 48,990 hectares, 75% of which were in Sicily.[1]

In addition to the recent experience of natural hazards, to understand the context of Messina it is crucial to consider that the 91% of the urban area was destroyed by a compound earthquake-tsunami disaster in 1908. That disaster, although now distant in time, has remained in historical memory and is the only one among other tragic events that is annually commemorated, unlike for example the destructive bombings of the Second World War. The new generations, therefore, have personal knowledge of the 1908 disaster and collectively experience its consequences in Messina's urbanistic, cultural and behavioural traces. In fact, fear and precariousness, which characterised the disaster and the reconstruction, have remained in the minds of those who participated in the rebirth of the city and in the very foundation of the new Messina, being passed down to subsequent generations (Centorrino, 2010). Also, Messina is part of a bigger economic and socio-cultural discussion that involves Southern Italy, known as the "Mezzogiorno Issue". This term is an umbrella term to cover Southern Italy's complicated history with the National government, and the socio-economic progress vis-a-vis the "North", tracing back to the unification in 1864 (Morris, 1997; Angrist, Battistin and Vuri, 2017).

To take into account this background of continuous struggle against nature, the authorities and the difficult and often unequal social progress, the notion of "liquidity" is helpful. The term was introduced by Zygmunt Bauman to articulate the transition from the era of modernity to that of postmodernity (Bauman 2000 and 2005). For Bauman (2006), liquidity involves not only the concept of fear, but is an umbrella term to describe modern anxieties and uncertainties about threats - even if not real. The characteristic of these threats is that we cannot fully understand them and we are unable to find any remedies: the most fearsome fear is that without a clear direction and cause; fear without a reason, the threat that is glimpsed everywhere, but never shows itself clearly. The concept of "liquidity" is helpful in this case study because it resonates with the historical transition from the destruction of the city with the 1908 disaster and determined the process of social and political change that led to the current identity of the Messinese people.

[1] ISPRA https://www.isprambiente.gov.it/it/news/ecosistemi-forestali-ed-incendi-aggiornamento-1-settembre-2023. Accessed Sept. 4th, 2023.

2.2 Methods

To analyse the current perception of risk based on elements of human attitudes and behaviours, a mixed method was chosen. 72 quantitative surveys were collected between October and December 2022 in Messina. 65 surveys were collected from university students at the University of Messina (UniMe). To expand the respondents' pool, the students were then asked to share the survey with at least one family member. 20 paper questionnaires were also given at the local church "San Gabriele dell'Addolorata", 7 were returned.

The survey protocol included 7 variables and 35 questions (Table 1). The questions had the Likert scale response form 1 (Strongly Disagree) to 5 (Strongly Agree). To test knowledge and preparedness of disaster preparedness measures, a dichotomous response (yes/no) was given to six disaster measures: 1. Emergency number; 2. Sufficient knowledge about first aid; 3. Emergency kit (a backpack, bag); 4. Evacuation center; 5. Personal emergency contact; 6. Food and water stock (Table 8). The "don't know" responses were coded as null. The statistical analysis of the survey results focused on descriptive statistical data to identify trends, Pearson's correlations to identify relations between the variables.

3 Results and Discussion

The survey respondents were categorised based on their gender (Female = 1, Male = 0), age groups ((Young Adults (18~39 years old); Middle Age (40~65); Elderly (over 65 years old)), level of education (Primary = 1, High School = 2, University = 3, Masters and over = 4), and housing typology (apartment, independent house). In total, 27 women (38%) and 45 men (61%) responded to the survey, for an average age of 44 years old (19~77). With regards to the level of education, 61% of the respondents completed a high school degree, 4% a bachelor's degree and 35% a master or above. 75% of respondents lived in an apartment while 25% in an independent house. Also, the respondents were asked to grade their experience of disaster from 0 (no experience) to 5 (experienced a disaster more than 4 times). 38% of the respondents consider themselves never to have experienced a disaster (1 time = 22%; 2 times = 15%; 3 times = 6%; 4 times = 19%). The first analysis focused on the descriptive statistics of all variables, shown in Table 2.

3.1 Place Attachment

As for place attachment, the descriptive statistics show that the respondents have low sense of belonging to their neighborhood (Q3 – mean = 2.81, SD = 1.16; Q4 – mean = 2.22, SD = 1.25). On the other hand, the respondents show medium attachment to the city with regards to the pride of being Messinese (Q1– mean = 3.36, SD = 1.17) and the lowest mean score when it comes to missing the city (Q2– mean = 2.86, SD = 1.33). However, they also showed a strongly positive sense of attachment to the city's historical preservation (Q6 – mean = 4.46, SD = 1.03).

These results ground on the intrinsically complicated relationship that the Messinese people have with their city. During informal discussions, various respondents showed a

Table 2. Descriptive statistics of survey variables

	Descriptive Statistics					
		Valid	Missing	Mean	Std. Deviation	Variance
Place attachment	Q1 I am proud of living in Messina	72	0	3.361	1.166	1.361
	Q2 When I leave Messina for the holidays, I miss it	72	0	2.861	1.335	1.783
	Q3 I feel like I belong in my neighborhood	72	0	2.806	1.158	1.342
	Q4 I participate in events in my neighborhood	72	0	2.222	1.247	1.556
	Q5 Taking care of the house is important to me	72	0	4.375	0.83	0.688
	Q6 Preserving Messina's historical buildings is important to me	72	0	4.458	1.034	1.069
Risk perception	Q7 I am concerned a disaster will occur in Messina in the next 10 years	72	0	3.194	1.182	1.398
	Q8 I am concerned about disaster damage to my home	72	0	3.472	1.278	1.633
	Q9 I am concerned my close family and friends might be in danger in case of a disaster	72	0	4.194	1.096	1.201
	Q10 If there were a high chance of disaster, I would leave Messina	72	0	3.139	1.179	1.389
	Q11 I believe disasters can be accurately predicted by experts	72	0	3.403	0.944	0.892

(continued)

Table 2. (*continued*)

Attitude	Q12 I believe in the existence of superior beings who care about my existence"	70	2	3.643	1.642	2.697
	Q13 I feel embarassed when other people know I prepare for emergencies	72	0	1.5	0.751	0.563
	Q14 Having disaster preparedness measures makes me feel safer	72	0	3.819	0.969	0.939
	Q15 Big disasters like 1908 are inevitable, disaster preparedness does not help much	72	0	2.361	1.237	1.53
	Q16 In case of emergency, I would say people can be trusted	71	1	3.38	1.113	1.239
	Q17 I remain positive in uncertain times	72	0	3.597	1.159	1.343
Disaster preparedness opinion	Q18 I think there should be more disaster education in schools	72	0	4.639	0.678	0.459
	Q19 Emergency plans should be discussed more in neighborhoods and local communities	72	0	4.5	0.732	0.535
	Q20 First Aid is important	69	3	4.29	1.099	1.209
	Q21 Emergency Kit at home is important	69	3	3.667	1.28	1.637
	Q22 Meeting point is important	70	2	4.014	1.083	1.174

(*continued*)

Table 2. (*continued*)

	Q23 Food and water stock is important	69	3	3.797	1.158	1.341
	Q24 Having an emergency contact is important	72	0	4.431	0.747	0.558
Local authorities' role and trust	Q25 The local government should give information about how to be ready for emergencies	72	0	4.569	0.784	0.615
	Q26 I trust my local government to warn me in time to evacuate in case of emergency	71	1	2.789	1.094	1.198
	Q27 After a disaster, the government should tell me what to do	71	1	4.38	0.884	0.782
	Q28 After a disaster, I trust the government will take care of the recovery actions	72	0	2.264	1.113	1.239
Disaster preparedness measures	Q29 With my family, we discussed an emergency plan	72	0	2.375	1.326	1.759
	Q30 I believe the information we have about disaster preparedness is useful in case of emergency	71	1	2.845	1.25	1.561
	Q31 I think often about how to prepare for a disaster	72	0	2.111	1.157	1.34

(*continued*)

Table 2. (*continued*)

	Q32 If the need arises, I can easily find information about disaster	71	1	2.831	1.171	1.371
Motivation and engagement	Q33 I would participate in disaster preparedness events	71	1	3.437	1.216	1.478
	Q34 I don't have time to participate in disaster preparedness events	69	3	2.87	1.097	1.203
	Q35 I need more information about disasters	71	1	3.732	0.999	0.999

combination of dissatisfaction towards the city's many problems, predominantly about economic difficulties, traffic, and unregulated waste management, and a strong sense of pride in Messina's history and scenery, especially its signature seaside and port.

Place attachment is positively correlated higher family preparedness (Q29 - With my family, we discussed an emergency plan), confidence in available information (Q30 - I believe the information we received is useful in case of emergency), and thinking about disaster preparedness (Q31 - I think often about how to prepare for a disaster). Also, stronger sense of belonging to the neighborhood is positively correlated with self-efficacy in finding information if the need arises (Q32 - If the need arises, I can easily find information about disaster). Despite the positive impact of place attachment on efficacy and disaster awareness, place attachment variables are not directly correlated with engagement and motivation variables (Table 3).

3.2 Risk Perception

To measure perception of risk, different concerns were considered: the general perception of risk for a disaster within 10 years (Q7 – mean = 3.194, SD = 1.182), two questions to identify personal identification with risk regarding harm to one's home (Q8 – mean = 3.472, SD = 1.278) and harm to family and friends (Q9 – mean = 4.194, SD = 1.096), the disposition to leave Messina in case of high chance of disaster (Q10 – mean = 3.139, SD = 1.179), and whether disaster risk could be predicted accurately by experts (Q11 – mean = 3.403, SD = 0.944). The mean values shown above suggest that the respondents have overall a high perception of risk towards family and friends.

The literature suggests higher risk perception motivates interest in disaster preparedness measures and engagement in disaster preparedness actions (Huurne and Gutteling, 2008; Yong et al., 2017). In this survey, these are considered respectively in "Disaster Preparedness Measures" (Q29 to Q32) and "Engagement and Motivation" (Q33-Q35).

Table 3. Pearson's significant correlations of place attachment with disaster preparedness measures and engagement and motivation variables.

		Place attachment					
		Q1	Q2	Q3	Q4	Q5	Q6
Dis. Prep. Measures	Q29	−0.143 0.229	−0.010 0.934	−0.034 0.774	0.204 0.085	0.344** 0.003	0.376** 0.001
	Q30	0.291* 0.014	0.045 0.708	0.263* 0.026	0.306** 0.009	0.181 0.132	0.319** 0.007*
	Q31	0.272* 0.021	0.238* 0.044	−0.015 0.899	0.188 0.115	0.132 0.269	0.286 0.015
	Q32	0.040 0.741	−0.058 0.628	0.266* 0.025	0.292* 0.013	0.211 0.077	0.065 0.588
Engag. & Motivat	Q33	0.163 0.175	0.137 0.254	−0.028 0.815	−0.004 0.974	0.052 0.670	−0.016 0.895
	Q34	−0.122 0.320	0.011 0.928	−0.068 0.578	−0.002 0.988	0.054 0.659	5.547 × 10−4 0.996
	Q35	−0.169 0.158	0.095 0.428	−0.177 0.139	−0.062 0.605	0.034 0.781	0.053 0.663

This statement is supported by looking at the strong positive correlations between the risk perception, disaster preparedness measures, and motivation and engagement (Table 4).

As shown in the table, personal identification with risk (Q7-Q8-Q9) significantly correlate with disaster preparedness measures (Q29~Q32) and motivation and engagement (Q33~Q35). The concern about a disaster happening in Messina in 10 years (Q7) is positively correlated with thinking about one's personal preparedness (Q31 – r(5) = .48, p. = < .001) and the interest in participating in disaster preparedness events (Q33 – r(5) = .25, p. = < .039).

The concern about the damage to the house is also strongly correlated with thinking about one's preparedness (Q31– r(5) = .41, p = < .001) but also with not having time to participate in disaster events (Q34 – r(5) = .26, p = .032). This result highlights the importance of time constraints when informing the public about disaster preparedness. In fact, although respondents are aware of the risks of personal harm and damage in the house, this awareness does not push individuals to find the time to improve their household preparedness. Time constraints should be taken into account when developing and organizing disaster preparedness events. Finally, the concern about one's family and friends (Q9) has the highest number of significant positive correlations. Concern for family and friends is correlated with thinking about disaster preparedness often (Q31 – r(5) = .48, p = .016), interest in participating in disaster preparedness events (Q33 – r(5) = .24, p = 041) and the need to receive further information (Q35 – r(5) = .32, p = .006). This result suggests that the family unit is more relevant than the individual when considering disaster preparedness measures and information.

Table 4. Pearson's correlations values of Risk perception with Disaster Preparedness Measures and Engagement and Motivation. Significant correlations are shaded in grey

		Risk Perception				
		Q7	Q8	Q9	Q10	Q11
Dis. Prep. Measures	Q29	0.070 (0.561)	0.052 (0.665)	0.124 (0.301)	0.011 (0.925)	0.103 (0.391)
	Q30	−0.036 (0.769)	0.028 (0.819)	0.052 (0.666)	−0.003 (0.981)	0.032 (0.789)
	Q31	0.478*** (< .001)	0.412*** (< .001)	0.283* (0.016)	0.133 (0.265)	−0.042 (0.729)
	Q32	0.065 (0.589)	−0.077 (0.523)	0.048 (0.689)	0.035 (0.771)	0.020 (0.865)
Engag. & Motivat	Q33	0.246* (0.039)	0.190 (0.112)	0.244* (0.041)	−0.207 (0.084)	0.041 (0.737)
	Q34	0.044 (0.718)	0.258* (0.032)	0.055 (0.653)	−0.036 (0.770)	−0.178 (0.143)
	Q35	0.213 (0.074)	0.201 (0.092)	0.321** (0.006)	−0.219 (0.066)	0.080 (0.506)

3.3 Attitude

With regards to the attitude variable, the study shows that the respondents have a high belief in "superior beings who care about my existence" (Q12 – mean = 3.643, SD = 1.642). This is not surprising considering the general identification of Italians as Catholics and the national trend of stronger religious belief in Southern Italy. The Italian National Institute of Statistics (ISTAT) showed in religious service attendance by region in 2021 is 24% in Sicily, the highest of the whole Italian peninsula[2].

Religious belief does not seem to lead to a fatalistic understanding of disasters, and hence a negative opinion on disaster preparedness. The respondents overall feel safer when they have disaster preparedness measures (Q14 – mean = 3.819, SD = 0.969). The idea that religious beliefs do not necessarily lead to a fatalistic view of disasters is also showed by the low score of Q15 "big disasters like 1908 are inevitable, disaster preparedness does not help much" (mean = 2.361, SD = 1.237). In other words, religious beliefs are not detrimental to disaster awareness, rather, giving a religious meaning to life events, including disasters, can promote higher disaster preparedness through a more community-based view of life and the establishment of a social network which can support disaster mitigation (Sun, Deng and Qi, 2018). Similarly, in this case-study we can see that higher religious belief is correlated with higher participation in neighborhood events (Q4), higher attachment to the city's historical value (Q6), higher trust in other

[2] ISTAT "Aspetti della vita quotidiana: Practica religiosa – eta', titolo di studio" Available at: http://dati.istat.it/Index.aspx?QueryId=24346. Aaccessed Oct. 1, 2023).

people (Q16- mean = 3.380, SD = 1.113) and in the local authorities (Q26), and positive attitude (Q17 – mean = 3.597, SD = 1.159) (Table 5).

Table 5. Pearson's correlations of Q12 "Religious belief" with other variables

Variables	Statements	Pearson's r	p
Place attachment	Q4. I participate in events in my neighborhood	0.315**	0.008
	Q6. Preserving Messina's historical buildings is important to me	0.274*	0.022
Attitude	Q14. Having disaster preparedness measures makes me feel safer	0.281*	0.018
	Q17. I remain positive in uncertain times	0.294*	0.014
Trust in local authorities	Q26. I trust my local government to warn me in time to evacuate in case of emergency	0.317**	0.008

3.4 Opinion on Disaster Preparedness

As for the respondents' opinion on disaster preparedness, the descriptive statistics show there is a strong desire for more education in schools (Q18 – mean = 4.64, SD = .678); and to localise disaster preparedness in local associations and neighborhoods (Q19 – mean = 4.50, SD = .732). The respondents also ranked the importance of five disaster preparedness measures: First Aid, Emergency Kit (sch as emergency backpack), Meeting point, Food and water stock, and Emergency contact. Whilst all measures scored high in importance, the measures based on social connections (Q24 Emergency contact in the neighborhood – mean = 4.43, SD = .747), physical harm mitigation (Q20 First Aid – mean = 4.29, SD = .678) and outside preparedness (Q22 Meeting point – mean = 4.01, SD = 1.08) scored the highest. On the other hands, household preparedness measures, Food and water stock (Q23 – mean = 3.80, SD = 1.158) and an Emergency Kit (Q21– mean = 3.67, SD = 1.28), scored relatively lower compared to the other measures.

It is interesting to note that none of the disaster preparedness opinion is correlated with any disaster preparedness measures. Meaning that even though there are high perception of disaster preparedness measures, these do not translate in better preparedness. Also, even among the measures (Q20~Q24) only the emergency contact in the neighborhood (Q24) is correlated with desire to participate in disaster events (Q33 – r(5) = .26, p = .030) and need for more information (Q35 – r(5) = .23, p = .049), while the recognition of the importance of the other measures (First aid, Emergency Kit, Meeting point, Food and water stock) has no impact on disaster preparedness behaviours and actions (Table 6).

Table 6. Pearson's correlation of Disaster Preparedness Opinion with Disaster Preparedness Measures and Engagement and Motivation (significant correlations shaded in grey).

		Disaster Preparedness Opinion						
		Q18	Q19	Q20	Q21	Q22	Q23	Q24
Dis. Prep. Measures	Q29	0.137 0.251	0.210 0.076	−0.027 0.826	0.053 0.667	−0.054 0.656	0.044 0.722	0.034 0.778
	Q30	−0.032 0.794	0.056 0.645	0.085 0.486	−0.055 0.657	−0.051 0.678	0.010 0.935	0.088 0.464
	Q31	0.088 0.463	0.250 0.035	−0.038 0.757	0.033 0.789	−0.059 0.628	−0.026 0.833	0.204 0.085
	Q32	0.100 0.405	0.035 0.774	0.131 0.285	0.092 0.454	0.014 0.911	0.158 0.193	0.004 0.974
Engag. & Motivat	Q33	0.213 0.074	0.341** 0.004	0.047 0.703	0.196 0.106	0.039 0.749	−0.009 0.939	0.258* 0.030
	Q34	0.194 0.110	0.336** 0.005	−0.127 0.300	0.048 0.696	−0.083 0.496	−0.056 0.651	0.195 0.108
	Q35	0.274* 0.021	0.421*** < .001	−0.152 0.213	0.080 0.515	−0.050 0.684	−0.090 0.465	0.234* 0.049

3.5 Local Authorities

With regards to the role of local authorities, there are strong expectations that local institutions' should take on disaster preparedness strategies and information before a disaster (Q25 – mean = 4.57, SD = .784) and after an emergency (Q27 – mean = 4.38, SD = .884). However, the high level of reliance that the average respondent has on local authorities is paired with low trust in the same institutions to issue evacuation warnings in a timely manner (Q26 – mean = 2.79, SD = 1.094) and take care of the reconstruction process (Q28 – mean = 2.26, SD = 1.113).

Lack of trust in the institutions is a relevant component of disaster preparedness and evacuation behaviour (Ohtomo *et al.*, 2017). While the Pearson's correlation analysis shown no relationship between trust in the authorities and personal disaster preparedness measures (Q25~Q28), the two trust statements (Q26-Q28) are positively correlated with the belief on the efficacy of the information (Q30), supporting the notion that there is a relationship between higher trust in the authorities and higher "trust" in the information received (Table 7). Also, the correlation analysis tells us that people with higher expectations on local authorities before (Q25) and after a disaster (Q27) have stronger interest in disaster information (Q33). People with higher expectation on local authorities before a disaster also have higher desire to participate in disaster events (Q35) but less time to do so (Q34). In other words, people who belief that local authorities should take care of preventive information have less available time to attend disaster events suggesting that institutions should take care of what information the individual citizen has no time to obtain.

Table 7. Significant correlations of local authorities' variables with disaster preparedness measures and motivation and engagement

Local authorities' variables	Pearson's r	p
Q25 The local government should give information about how to be ready for emergencies		
Q33 I would participate in disaster preparedness events	0.381**	0.001
Q34 I don't have time to participate in disaster preparedness events	0.240*	0.047
Q35 I need more information about disasters	0.321**	0.006
Q26 I trust my local government to warn me in time to evacuate in case of emergency		
Q30 I believe the information we have about disaster preparedness is useful in case of emergency	0.300*	0.011
Q27 After a disaster, the government should tell me what to do		
Q33 I would participate in disaster preparedness events	0.292*	0.014
Q28 After a disaster, I trust the government will take care of the recovery actions		
Q30 I believe the information we have about disaster preparedness is useful in case of emergency	0.391***	< .001

* p < .05, ** p < .01, *** p < .001

3.6 Disaster Preparedness Measures

Disaster preparedness efficacy scored low average responses in all four components: discussion of an emergency plan with the family (Q29 – mean = 2.37, SD = 1.326), efficacy of received information (Q30 – mean = 2.84, SD = 1.25), thinking about personal disaster measures (Q31 – mean = 2.11, SD = 1.157), and self-efficacy in finding information if the need arises (Q32 – mean = 2.83, SD = 1.171). The lack of confidence and reliance on existing resources is confirmed by the self-assessment of six **disaster prevention measures** (Table 8). 5 of these measures' importance was measured as part of the disaster preparedness opinion (Q20–24). The results show that despite the high important of First aid (Q20), 81% of the respondents did not have sufficient information about First aid. Also, the results confirm that there is little interest and awareness about household preparedness (Q21, Q23), with high rate of people who did not have or know how to prepare c. emergency kits - emergency backpacks/bags (75%), and f. food and water stock (81%). The social and spatial measures scored the highest in terms of perceived importance (Q22, Q24) and self-assessment: a. emergency number to call (44% of people knew the number - 112), d. location of the closest meeting point (40%), and e. close emergency contact (40%).

The descriptive statistics of Q29~Q32 and of the individual measures shown above illustrate that relational measures (close emergency contact, evacuation shelter, request outside help) scored higher awareness and efficacy compared to practical household disaster preparedness measures, in particular in the provision and preparation of objects or tools in the house space, such as food, water, and emergency supplies. This finding suggests that relational preparedness is more important than individual preparedness in

Table 8. Self-assessment disaster prevention measures

Disaster prevention measures	Importance (Question, mean)	No	Si	Non so
a. Emergency number		38%	44%	15%
b. Sufficient knowledge about first aid	High (Q20, 4.29)	50%	18%	31%
c. Emergency kit (a backpack, bag)	Medium (Q21, 3.67)	72%	24%	3%
d. Meeting point	High (Q22, 4.01)	39%	40%	19%
e. Personal emergency contact	High (Q24, 4.43)	29%	40%	29%
f. Food and water stock	Medium (Q23, 3.80)	74%	19%	7%

the overall perception of the respondents. This idea is also confirmed by the high scores found in the analysis risk perception (Q9), attitude (Q16), and disaster preparedness opinion (Q18~Q24), and is supported by the cultural sense of community and family traditional of Southern Italy and especially presents in the Messinese post-disaster context (Dickie, 2008; Boatti, 2017). Furthermore, another reason why household preparedness is secondary to emergency preparedness lies in the model of disaster education. Disaster education is done prevalently in schools, and due to shortage of time it focuses mostly on effective evacuation and not on how to integrate daily changes to create resilience. Although there are examples of such household preparedness information (Consigli per la sicurezza, 2003), these are not efficient channels to share this information with the public. These cultural and disaster education values are reflected in these results.

3.7 Engagement and Motivation

Finally, the survey focused on the desire to participate in disaster preparedness event, the perception of time constraint and the desire for more information. The results indicate a positive disposition of the respondents. The descriptive analysis shows that on average, the respondents have high desire to participate in disaster preparedness events (Q33 – mean = 3.44, SD = 1.216), an overall a low perception of time constraints (Q34 – mean = 2.87, SD = 1.097), and needs more information about disasters (Q35 – mean = 3.72, SD = .999).

As mentioned before, disaster preparedness in Messina is almost exclusively done in school, targeting students from kindergarten to the university level. The high interest in participating in disaster preparedness events suggests that additional events targeting other age groups could be implemented. The current system also raises the question of a disparity in access to information. For example, those who leave high school to work (compulsory school age is 16 in Italy) have higher difficulties in accessing to free disaster training and are outside the system of compulsory disaster education. Secondly, although the average response for time constraint was low (Q34 – mean = 2.87, SD = 1.097), the issue of limited time, especially for the working people, should be taken into account. There are various possible solutions, such as allocating an amount of time to discuss disaster preparedness in the offices, and including disaster training events during other initiatives, such as national or local festivities. Finally, the higher desire for more information suggests that the content of the available disaster preparedness information could also be broadened from its current focus on the emergency evacuation to include more information about household preparedness and personal safety at home and in the workplace.

The concept of liquidity can be used to interpret this study's results from the sociological perspective. As mentioned in the introduction, the 20th century "new" Messina was structured around a search for solidity, as opposed to the "liquid" process of social and political change (Zygmunt Bauman 2000 and 2005). The current discussion on climate change, with its looming future threat and the experience of the 1908 disaster contribute to the creation of this "threatening feeling" that lasts until today. The study shows that fear does not disappear. It is approached, however, differently and the feedback on the importance of being prepared for disaster and confidence in prevention plans felt by the respondents can be framed as a way to deal with the desire for solidity while battling the fear of modernity. Furthermore, it is interesting to note that against the response to more "liquid" scenarios assumed by Bauman, in which individualism becomes antidote to such invisible threats, the respondents referred to a solidarity dimension to countertrend risks and uncertainty.

4 Conclusions and Further Research

In this study, we analysed the current risk perception of a selected group of the population of Messina through quantitative survey, focusing on three main elements: context, belief system, interest and engagement. There are three main takeaways from this preliminary quantitative study. Firstly, that high religious belief does oppose, but actually correlates with higher place attachment, higher confidence that preparedness makes a difference, and positive disposition towards helping and getting help from others. Secondly, the study showed that respondents are not confident in their own preparedness, and lack trust in the authorities that are responsible for information dissemination. Thirdly, the high place attachment to the home and the city suggests that including more household preparedness to make the home safer might be a possible solution to the current overfocus on emergency evacuation.

Ultimately, this study meant to show an overview of the relevant topics related to disaster risk awareness and interest in disaster preparedness in the case study of Messina, which is layered by its lessons learned from past disasters, current natural hazard challenges, cultural notions of family and community, as well as modern issues of trust in the local government, directly correlated with other municipal issues, such as waste and local economy. Furthermore, this study expanded on the liquidity argument of post-modern uncertainty by mentioning the struggles of combining perceptions of risk with individualism, emotional ties with place and relations, and the perceived constraints in taking actions.

While this method represents an interesting starting point for discussion, these results must be further explored through qualitative research on different groups within the respondents (i.e. women, disabled people, elderly etc.) as well as the interactional issues and concerns between local governments, enforcers (Firefighters, Civil Protection) and the public. Also important would be to further explore the implications of adapting and developing the "liquidity" concept in the growing field of climate-change uncertainty and risks.

References

Angrist, J.D., Battistin, E., Vuri, D.: In a small moment: class size and moral hazard in the italian mezzogiorno. Am. Econ. J. Appl. Econ. **9**(4), 216–249 (2017). https://doi.org/10.1257/app.20160267

Aronica, G.T., Brigandí, G., Morey, N.: Flash floods and debris flow in the city area of Messina, north-east part of Sicily, Italy in October 2009: the case of the Giampilieri catchment. Nat. Hazards Earth Syst. Sci. **12**(5), 1295–1309 (2012). https://doi.org/10.5194/nhess-12-1295-2012

Boatti, G.: La terra trema: Messina 28 dicembre 1908 : i trenta secondi che cambiarono l'Italia, non gli italiani. Milano: A. Mondadori (2017)

Consigli per la sicurezza Roma: Associazione Nazionale Vigili del Fuoco del Corpo Nazionale, Ministero dell'Interno (2003)

Dickie, J.: Una catastrofe patriottica: 1908, il terremoto di Messina. 1. ed. Roma: GLF editori Laterza (Storia e società) (2008)

Gerhold, L., Wahl, S., Dombrowsky, W.R.: Risk perception and emergency food preparedness in Germany. Int. J. Disaster Risk Reduction **37**, 101183 (2019). https://doi.org/10.1016/j.ijdrr.2019.101183

Gero, K., et al.: Dispositional optimism and disaster resilience: a natural experiment from the 2011 great east japan earthquake and tsunami. Soc. Sci. Med. **273**, 113777 (2021). https://doi.org/10.1016/j.socscimed.2021.113777

Huurne, E.T., Gutteling, J.: Information needs and risk perception as predictors of risk information seeking. J. Risk Res. **11**(7), 847–862 (2008). https://doi.org/10.1080/13669870701875750

Inalhan, G., Yang, E., Weber, C.: Place Attachment Theory. In: Appel-Meulenbroek, R., Danivska, V. (eds.) A Handbook of Theories on Designing Alignment between People and the Office Environment. 1st edn. London: Routledge, pp. 181–194 (2021). https://doi.org/10.1201/9781003128830-16

Morris, J.: Challenging meridionalismo. In: Lumley, R., Morris, J. (eds.) The New History of the Italian South: the Mezzogiorno Revisited, pp. 1–17. University of Exeter Press, Exeter (1997)

Ohtomo, S., et al.: The influences of residents' evacuation patterns in the 2016 kumamoto earthquake on public risk perceptions and trust toward authorities. J. Disaster Res. **12**(6), 1139–1150 (2017). https://doi.org/10.20965/jdr.2017.p1139

Scheier, M.F., Carver, C.S., Bridges, M.W.: Distinguishing optimism from neuroticism (and trait anxiety, self-mastery, and self-esteem): a reevaluation of the life orientation test. J. Personal. Soc. Psychol. **67**(6), 1063–1078 (1994). https://doi.org/10.1037/0022-3514.67.6.1063

Sun, L., Deng, Y., Qi, W.: Two impact pathways from religious belief to public disaster response: findings from a literature review. Int. J. Disaster Risk Reduction **27**, 588–595 (2018). https://doi.org/10.1016/j.ijdrr.2017.10.004

Terpstra, T., Gutteling, J.M.: Households perceived responsibilities in flood risk management in The Netherlands. Int. J. Water Resour. Dev. **24**(4), 555–565 (2008). https://doi.org/10.1080/07900620801923385

Wallis, A., Fischer, R., Abrahamse, W.: Place attachment and disaster preparedness: examining the role of place scale and preparedness type. Environ. Behav. **54**(3), 670–711 (2022). https://doi.org/10.1177/00139165211064196

Yong, A.G., et al.: Risk perception and disaster preparedness in immigrants and canadian-born adults: analysis of a national survey on similarities and differences: risk perception and disaster preparedness in immigrants and canadian-born adults. Risk Anal. **37**(12), 2321–2333 (2017). https://doi.org/10.1111/risa.12797

Immersive Training and Education Using Mixed Reality

Katarína Hollá[1(✉)], Samuel Kočkár[1], Justina Pluktaite[2], and Jozef Ristvej[1]

[1] University of Zilina, Zilina, Slovakia
{katarina.holla,samuel.kockar,jozef.ristvej}@uniza.sk
[2] Upper Secondary School of Chemistry Pardubice, Zilina, Slovakia
pluktaite@spsch.cz

Abstract. The training of the crisis management and emergency services in HE and VET is one of the critical success factors that must be addressed if an adequate prevention and response to be achieved. The development of the studying programme involves all levels of management and establishing preparedness is only part of the overall process. To ensure an adequate standard of teaching, an immersive training of students is required. As part of this process, simulation exercises should be carried out to test the competency of those likely to be involved in accident response.

For these purposes HE and VET are using different kind of tools to transfer the knowledge and decision-making. More recently, lives and education have further shifted as our world has become more digital (Reedy, A. 2021). The COVID-19 pandemic hastened applications of digital teaching and learning (Houlden, S. 2019). Therefore, educational institutions need to consider strategies to sustain learning and teaching even during challenging times. Teachers say, "We need to learn to do this better." Educational organizations have learned from recent times one valuable lesson: We have to move beyond webinars and toward true training experiences (Immersive Reality: Unleash Augmented, Virtual, and Mixed Reality 2023). No matter the delivery medium, in this article we know that learning is accomplished by ensuring content is relevant, applicable, and timely. A design approach that really works to accomplish these basics of students´ learning theory is virtual scenario-based training (VSBT), VR and AR which is not commonly used in HEIs and VETs. Trainers/lecturers create experiences (scenarios) in which learners can solve problems and interact with provided situations to learn new skills. In this article we are going to present results of the project FightARs dealing with the mixed reality and its implementation into the training and educational process.

1 Virtual Scenario Based Training

Virtual Scenario-Based Training (VSBT) effectively combines realistic practice with storytelling to enhance learning experiences. This approach is particularly valuable in fields like vocational education and training (VET) and universities, where decision-making and critical thinking are essential skills.

H. Tatano et al. (Eds.): GSRIDRR 2023, DRRGADRIBS, pp. 514–523, 2025.
https://doi.org/10.1007/978-981-96-6609-6_34

The Estonian Academy of Security Sciences (EASS) serves as a prime example of successful implementation, utilizing virtual scenarios for over a decade to train and assess competencies in rescue and police work. Their collaboration with other institutions highlights the benefits of shared learning and best practices in the deployment of VSBT.

Stella Polikarpus's research focuses on the crucial role of trainers in the VSBT implementation process, emphasizing that trainers who actively engage in creating and utilizing virtual scenarios often exhibit more positive attitudes towards these methods. This suggests that empowering trainers through involvement in VSBT design can lead to enhanced training outcomes and greater acceptance of innovative teaching approaches.

Overall, the findings indicate the importance of trainer involvement and attitudes in successfully implementing VSBT within organizations, potentially leading to more effective training programs in various fields. If you're interested in exploring specific applications or case studies further, please let me know!Also University of Zilina wanted to involve into education more effective decision-making and problem-solving in safety area and also to improve Technological Pedagogical Content Knowledge (TPACK) of teachers and trainers. In the organisational context, trainers have a central role in facilitating VSBT as they need to be aware of the capabilities of the technology need to choose or build these VS for training and eventually integrate these into their training approach. Therefore, in order to implement VSBT into an organisation, the role of trainers needs to be better understood. (Polikarpus, S. 2023).

These are mainly immersion technologies, which are (Fig. 1):

- Virtual Reality (hereafter VR)
- Augmented Reality (hereafter AR)
- Mixed Reality (hereafter MR)

Since the definition of mixed reality is not clear-cut and rather includes everything that between the purely real world and the virtual environment, the author is inclined to the view.

According to the main advantages of using VR and AR to enable processes and trainings innovations are mainly:

- It creates a standardised environment.
- Creates the illusion of a (yet) non-existent environment.
- Creates the illusion of a hard-to-access environment.
- Replicability of experiments.
- Recordability.
- Possibility of parallel data processing.
- Attractiveness of the technology.
- Cost-effectiveness.

We have been using so far virtual reality mainly in occupational safety and health education, augmented reality for mainly medical purposes and mixed reality for the crisis managers and emergency services (Fig. 1). In this article I want to discuss how we at the University of Žilina have implemented mixed reality scenarios in the teaching of dynamic-decision making and situational awareness and point out their advantages and disadvantages.

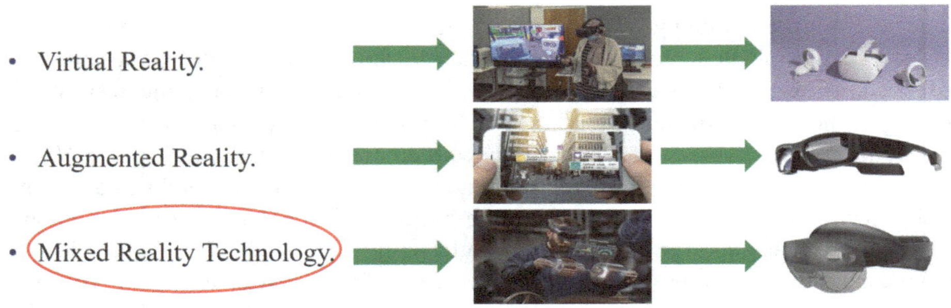

Fig. 1. VSBT in University of Zilina in safety area

2 Mixed Reality (MR)

Mixed reality (MR) is a hybrid technology that combines elements of both VR and AR. MR blends the real world with digital content in a way that allows users to interact with and manipulate virtual objects within their physical environment. (Immersive Reality: Unleash Augmented, Virtual, and Mixed Reality, 2023).

2.1 Project FightARs

FightARs was an international Erasmus + project that started in September 2020 and ended in April 2023. 6 partners from 6 European Union Member States participated in the project:

- SBG Dresden mbH (Germany);
- SCP Academy (Cyprus);
- Faculty of Security Engineering, University of Žilina (Slovakia);
- Secondary School of Chemistry Pardubice (Czech Republic - coordinator);
- Firefighters Training School - (Lithuania);
- Sisekaitseakadeemia - (Estonia).

The aim of this project was to identify the possibilities of using MR technology in the training of firefighters and crisis managers by means of concrete scenarios. The very aim of the FightARs project was to create a software program for training and education of firefighters, intervention commanders and crisis managers using innovative technologies of tomorrow or to obtain the necessary information for potential future creation of teaching materials with MR implementation in educational practice. MR education is future-focused, in which educators and technologies work together to provide trainees with the knowledge and skills. University of Zilina took also part in this project and implemented reszlts into education.

As a basis for assessing current and future training requirements, an online questionnaire survey was conducted in April 2021 among fire stations and educational institutions for the training of firefighters and crisis managers in the countries participating members of the FightARs project (Czech Republic, Slovak Republic, Republic of Estonia, Republic of Lithuania).Almost half of the 63 respondents already had experience with virtual reality (VR), but only about 10% had experience with glasses or applications in MR.

The dominant theme is the provision of technical and methodological - (procedural) skills through interaction with a digital object (hologram) or a digitally enriched real object in order to sufficiently exploit the potential of AR/MR. AR/MR is considered suitable for the fields of extinguishing burning cars, in terms of actuality electric or hybrid, or cutting persons out of a crashed car and subsequent rescue of injured persons. 80% of the respondents are interested in AR and VR applied in different training formats.

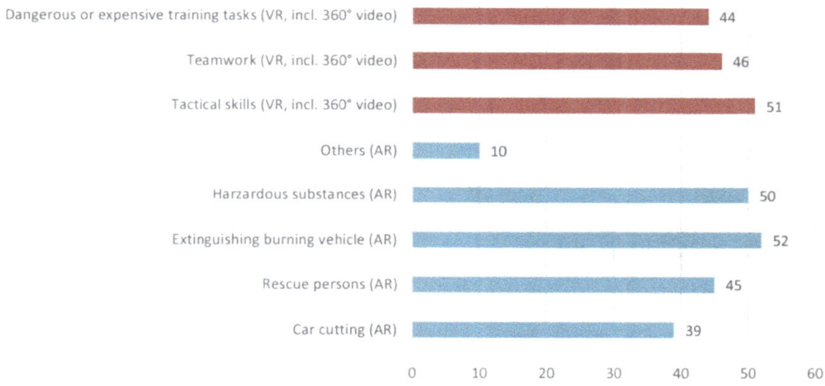

Fig. 2. Graph of preferred training and education scenarios with augmented/mixed (AR/MR) and virtual reality (VR) (Hofmann, J. 2022)

The interest of professional firefighters and teachers in educational institutions also showed that 80% are interested in AR and VR applied in different training formats. The use of 360° video is considered appropriate by 75% of respondents. The combination of on- and off-line received the highest approval with almost 80%, followed by hands-on workshops with about 60%. Purely online meetings were only considered useful by around 50% of respondents.

The following scenarios have been created within the FightARs project:

1. Deactivation of the electric car.
2. Stabilisation and cutting of the electric car while rescuing people.
3. Providing first aid for selected injuries in a car accident.
4. ADR hazardous substance spill from a tanker truck.
5. Complex scene of electric car and ADR tank truck crash with injured person.

Deactivation of the Electrical Car

The scenario is intended to provide the trainee with an introduction to the important components of an electric car in terms of their location in the body of the electric car. And also, information on the procedures for the correct safe shutdown - deactivation of the electric car, which is part of a road traffic accident. The theoretical part consists of a 3D hologram of an electric car with the possibility of displaying individual layers focusing

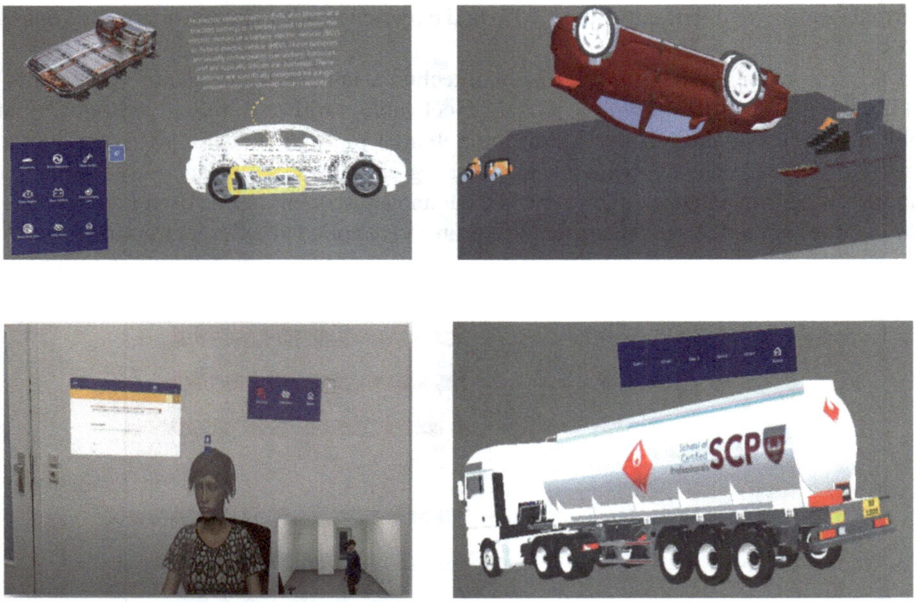

Fig. 3. Pictures from 4 scenarios within the FightARs application

on a specific component (high voltage battery, electric circuit wires, charging port and electric motor). In the practical part, it is possible to perform a training deactivation procedure on the 3D hologram by clicking on the correct sequence of deactivation.

Stabilization and Cutting of the Electrical Car While Rescuing People
The objective of this scenario is to familiarize trainees with specific stabilization techniques necessary to secure a crashed vehicle, preventing unwanted movement during interventions conducted by the emergency response units (ERU). Additionally, the scenario addresses critical locations on passenger vehicle bodywork where hydraulic spreaders and cutters are utilized; these locations are visually represented in the theoretical section via a digital twin, accompanied by detailed informational descriptions sourced from the Learning Management System (LMS). The theoretical component further elaborates on the significance and functionality of various devices and tools employed in the stabilization process. The practical portion of the training includes three variations of the positioning of the electric vehicle, as depicted in the accompanying figures. Trainees are tasked with executing the correct methodological procedures by engaging with 3D interactive holograms of the available equipment and tools within the designated scenario area.

Providing First Aid for the Selected Injuries in a Car Accident
In this scenario, trainees are introduced to the specific types of injuries that are most commonly associated with road traffic accidents. The theoretical component provides trainees with essential knowledge regarding the different types of injuries and the appropriate treatment methods, as well as the necessary resources and materials required for

effective intervention. Injuries are illustrated using 3D holograms of human figures, which are activated when trainees scan a designated QR code. The practical component of the training requires students to perform treatment procedures corresponding to each type of injury, applying these methods to the 3D holograms representing the injured individuals. Furthermore, this scenario is designed to be integrated with a physical training dummy or torso. Upon scanning the QR code affixed to the dummy, the HoloLens 2 device generates a digital overlay of the injured person in the form of a 3D hologram, enhancing the immersive learning experience for the trainees.

Within FightARs application there have been created several scenes:

- female, sitting, no injuries visible
- female, sitting, leg amputation
- female, laying down, no visible injuries
- female, sitting, head injury right of head

Hazardous Substances Spill from a Tanker Truck

The purpose of this scenario is to familiarize the trainees with the ADR tank truck carrying hazardous substances and the safety signs for the specific types of hazardous substances transported. The scenario contains within its variants 5 types of hazardous substances: Gasoline, Ammonia NH3, Chlorine Cl2, Sulphuric acid H2SO4, Liquefied Petroleum Gas LPG. Video presenting: how to use HoloLens 2 how to use QR codes with FightARs application Tip: Print all the needed QR codes in advance and mark on the paper what exactly are those for better logistics. Erasmus + FightARs: Best Practice Guide 11 Fig. 9 Hologram 3D tank ADR [2] For each substance, the theoretical part of the scenario contains the necessary information from the safety data sheets in connection with the activities carried out at the scene of the emergency to deal with the consequences or to treat the affected persons, as well as the methodical procedure for firefighters to deal with the presence of a hazardous substance in the event of accidents in road transport. Other information of no less importance is on the possibility of stopping the leak, decontamination and recovery for each type of hazardous substance concerned. The theoretical part is concluded with a short case study of simulating a leakage of the hazardous substance Chlorine by means of the software tool ALOHA on the D1 motorway section between the towns of Žilina and Bytča. In the practical part, the trainee has to complete a test in which his/her task is to answer correctly the questions related to the theory. [6] [2].

In the final complex scenario, the trainee is in the scene of a traffic accident and should apply all the knowledge learned. However, the final scenario itself is only in a state where it can only be used as training for "situation awareness; understudying" analysis and creating situational awareness and understanding. The experience is also supported by acoustic perceptions, where the individual phenomena present in the accident such as the leakage of a dangerous substance (hissing, spilling), the sound of the truck engine, the surroundings but also the moaning of the injured person are "immersive" to the greatest extent possible.

One of the 5 loads and one of the three accident/collision types can be selected when simulating the situation (Table 1). Thus, in case of exercises we are having a total of

15 different exam options. Table 1 Case and Load descriptions in Accident scene for FightARs application.

Table 1. All 5 loads within the complex scene scenario

Scenario no	Case no in app	Dynamics in the FightARs application	Load no	Load CMR	Substance (SDS)
1	1	Smoke from car	5	Load 5 CMR	LPG
2	1	Smoke from car	1	Load 1 CMR	Gasoline
3	2	Nothing dynamic	2	Load 2 CMR	Ammonia
4	3	Yellow puddle	4	Load 4 CMR	H_2SO_4
5	3	Yellow puddle	3	Load 3 CMR	Chlorine

Fig. 4. P Complex scene for the testing

Overall Fightars Application Training Content - LMS

Overall partners have developed FightARs application that has several sections that covers above mentioned chapters – scenarios.

Everybody can join developed course textual content on project website Learning Management System – LMS (piloting partners languages are going to be redirected to specific partners MOODLE systems): http://fight-ar.com/courses/

Main content is in English.

Other piloting partners languages (Estonian, Lithuanian, Czech and Slovak) are available as well.

What are the steps to join the FightARs Course?

1. Go to the website: http://fight-ar.com/courses/
2. Register or Log in to your account.
3. Choose FightARs course and start.

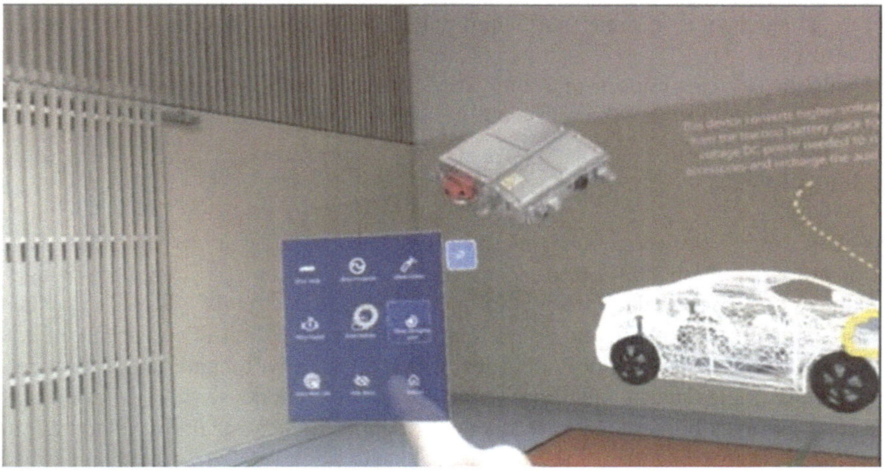

Fig. 5. YouTube video: FightARs application introduction [https://youtu.be/CWBk5PYkKDA]

All scenarios are used on University of Zilina in education within the subject Risks of Industrial Accidents. During last four years we have already identified advantages and disadvantages of Mixed reality. You can read them bellow.

Advantages of MR:

- The FightARs application is the only one of its kind that is currently looking at the potential usability of the HoloLens 2 device with digital twins training firefighters, crisis managers and students.
- The FightARs application is programmed in the Unity 3D cross-platform game engine.
- Positive feedback from test subjects on the concept of training through holograms and scenarios.
- Using the "Dynamics 365 Remote Assist" tool, trainees can be navigated in real time around the scene by drawing holographically in space.
- Wide range of variability for testing - heavy reliance on setting precise objectives and tasks to be fulfilled by the scenarios.
- Ability to recognize objects and anchor holograms to physical objects and then display layers with marked parts or components.
- The application can be controlled by voice commands.

Disadvantages of MR:

- The potential of the project results has been affected by the global pandemic of COVID-19.
- The HoloLens 2 can only be used effectively for a maximum of approximately 50 min at a time, after which the battery runs out.
- HoloLens 2 tends to overheat sometimes after 30 min of use.
- Some people may react neurologically - feel dizzy or have a burning sensation in their eyes.

- Using the device in areas with minimal lighting causes the sensors to become confused.
- Using the device outdoors on a sunny day - bright or indoors under a stronger source of artificial light makes the user environment difficult to see.

3 Conclusion

The choice of innovative methods in education depends on the approach of the individual or primarily the institution to technology and innovation. Defining today's age as dynamic with a great deal of information and knowledge is essentially a well-established phrase, which has its justification. It is important to receive the right information at the right time, but it is equally important to deal with it effectively. Technological advances in sensor technologies such as spatial position recognition or depth sensing, combined with powerful graphics processors, have opened up opportunities for the development and subsequent commercial sale of affordable but also dimensionally (hardware) affordable augmented (AR) and mixed reality (MR) devices. In terms of market availability, this is a relatively new technology whose potential for effective use is still being explored. This is evidenced by the number of scientific articles in bibliographic and citation databases dealing with this topic with a view to effectively applying this technology to various sectors or human activities. The activity at FightARs project piloting organisations, where the activity was developed in the search of possibilities of implementation of AR/MR scenarios, also viewed virtual reality solutions in teaching. The role of this training is not to replace conventional training and educational procedures on the contrary it is to form as the name AR implies an expansion of the possibilities and variety of information and sensations to be received as well as to try out methodologies in scenarios that would be costly or dangerous to simulate on a polygon. The FightARs has provided valuable insights into the development of scenarios for such training as well as information from surveys that there is interest in this type of training in practice. In conclusion, therefore, it can be assessed that AR/MR has potential in practical training and the issues raised by the FightARs project are not insurmountable and will be taken forward by the SAFAR project as starting points for further resolution. The project was approved for the funding.

References

Reedy, A., Pfitzner, D., Rook, L., et al.: Responding to the COVID-19 emergency: student and academic staff perceptions of academic integrity in the transition to online exams at three Australian universities. Int. J. Educ. Integr. **17**, 9 (2021). https://doi.org/10.1007/s40979-021-00075-9

Houlden, S., Veletsianos, G.: A posthumanist critique of flexible online learning and its "anytime anyplace" claims. British J. Educ. Technol. **50**(3), 1–14 (2019). https://doi.org/10.1111/bjet.12779

Immersive Reality: Unleash Augmented, Virtual, and Mixed Reality. https://www.learningguild.com/articles/immersive-reality-unleash-augmented-virtual--and-mixed-reality-/

The Application Of Scenario-Based Learning In VR Training Programs: 3 Use Case Examples (2023). https://elearningindustry.com/application-of-scenario-based-learning-in-vr-training-programs

Polikarpus, S., Luik, P., Poom-Valickis, K., et al.: The role of trainers in implementing virtual simulation-based training: effects on attitude and TPACK knowledge. Vocat. Learn. **16**, 459–486 (2023). https://doi.org/10.1007/s12186-023-09322-1

Responding to the COVID-19 emergency: student and academic staff perceptions of academic integrity in the transition to online exams at three Australian universities. https://edintegrity.biomedcentral.com/articles/https://doi.org/10.1007/s40979-021-00075-9

Hofmann, J. Pedagogical guidelines FightARs (2022). http://fight-ar.com/ [Accessed 15 April 2023]

SAFAR - Situational Awareness Training of Firefighters within an immersive XR training site. Pardubice: Střední průmyslová škola chemická Pardubice: SPŠCH

Kočkár, S.: Faculty of Security Engineering. University of Zilina (2022)

MODULE: DEACTIVATION, FightARs Course. http://fight-ar.com/course/fightars/ [Accessed 16 April 2023]

MODULE: CUTTING, FightARs Course. http://fightar.com/course/fightars/ [Accessed 16 April 2023]

MODULE: FIRST AID, FightARs Course http://fightar.com/course/fightars/ [Accessed 16 April 2023]

MODULE: HAZARDOUS SUBSTANCES, FightARs Course. http://fight-ar.com/course/fightars/ [Accessed 16 April 2023]

Resolutions of the 6th Global Summit of GADRI: Engaging Sciences with Action

Final Resolution of the 6th Global Summit of the Global Alliance of Disaster Research Institutes

Andrew Collins(✉)

DDN Northumbria University, Newcastle upon Tyne, UK
andrew.collins@northumbria.ac.uk

Abstract. The member institutes of the Global Alliance of Disaster Research Institutes (GADRI) fully recognise the importance of the United Nations Office for Disaster Risk Reduction (UNISDR) Sendai Framework for Disaster Risk Reduction 2015–2030 Mid Term Review.

1 The Summit Affirms That

1.1 Traditional approaches to risk management are being overwhelmed by the increasing complexity of systemic risk and its cascading impacts. The findings of the Midterm Review of the Sendai Framework, Global Assessment Report 2022 and the work of various GADRI members points to the need for radical changes to strengthen risk-informed decision-making.

1.2 The intensity and impact of the recent earthquakes in Turkey are a reminder of unresolved high levels of exposure and vulnerability to large scale rapid onset hazards occurring against a backdrop of global instability.

1.3 Climate change is projected to increase and intensify extreme weather events and associated disasters. Climate change is adversely impacting the health of humans, animals and entire ecosystems; the climate crisis is a health crisis. Meanwhile, the potential to adapt to climate change is not limitless.

1.4 Disaster research, policy and practice should be inclusive and equitable. Adopting an intersectional approach will ensure inclusivity and equity in practice. Gender should be understood as relational. Research into gender inequalities and inclusivity need to be central concerns of forums such as GADRI and researchers should strive to collect and analyse gender disaggregated data.

1.5 To be able to provide resilience-informed decision support that provides socially equitable solutions whole city models need to combine approaches from different disciplines crossing traditional boundaries; essentially combining physics-based and data-driven models.

Towards GADRI Objectives of Achieving a Sustainable Disaster-Resilient World, held at the Disaster Prevention Research Institute (DPRI), Kyoto University, Kyoto, Japan from 15 to 17th March 2023.

H. Tatano et al. (Eds.): GSRIDRR 2023, DRRGADRIBS, pp. 527–532, 2025.
https://doi.org/10.1007/978-981-96-6609-6_35

1.6 Science should provide information to support policy planning for avoiding extreme short- and long-term emergencies including through further investment in and interpretation of the role of early warning and action.

1.7 In recovery there is political will and financial support for bold action to address awareness, incentives, and change in regulations. Recovery planning is essential to achieve transformative DRR.

1.8 The emergence of polycrisis demands a more coherent, integrated, interdisciplinary and inclusive strategy to deal with multiple hazards at the same time. The disaster management agencies need better decision-making tools to deal with conflicting values, multiple trade-offs and complex causal structures. To be legitimate and socially accepted these judgments must be made in a transparent and inclusive manner.

1.9 Enhanced situational awareness for crisis management requires better early warning for crises based on anticipation of their impacts. Science needs to enable acting on those risks by providing local, regional and national authorities with knowledge, data and tools to help them to act on resilience building, risk reduction and climate change adaptation.

1.10 Theory is a 'road map' in disaster risk reduction but, as the modern world is changing very rapidly, we need to update theory to take account of very dynamic conditions.

1.11 In order to save more lives and reduce damage in disasters, we need to adopt a more rigorous, scientific approach to emergency planning and management.

1.12 Even in larger research institutes, it is difficult to gather sufficient researchers to cope with various types of disaster events, but cooperative systems over several institutes greatly helps to solve this problem.

1.13 Multi-nation pathways should work together to confront climate related hazard exposure and vulnerability through co-funded projects that facilitate mechanisms for researchers to share experiences to be able to expand research impact.

2 The GADRI Committees on Networking, Science and Technology Roadmap, Institutional Capacity Building, Data and Information Sharing and Advocacy report the Following

2.1 Networking

2.1.1 GADRI should create a Global Disaster Researcher Directory: voluntary compilation of individuals engaged in disaster research, channeled through GADRI member institutes, perhaps building on Orcid IDs.

2.1.2 GADRI should develop a strong social media presence (Facebook, Instagram, Twitter, Linked-in, YouTube). Each platform requires a different approach. YouTube allows the use of GADRI lectures and more detailed lengthier presentations appropriate to academic audiences and can be foundational for all five platforms.

2.1.3 GADRI should create GADRI Fellow and Young Researcher Programs: This would be based on (a) Nominations of accomplished senior disaster researchers, from which GADRI Fellows will be selected. GADRI will extend invitation for Fellows to visit/lecture disaster research institutes on a case-by-case basis at no financial burden to GADRI. (b) A Young Researcher program to foster exchange of early career researchers.

2.2 Science and Technology Roadmap

2.2.1 The future vision and directions of Science and Technology activities should be shared among members of GADRI and hopefully influence policy and systems transformation towards disaster resilient and sustainable societies.

2.2.2. The science and technology community will develop and strengthen three key functions of including knowledge integration, capacity integration and process integration.

Knowledge integration can be promoted by integrating the knowledge of natural and social sciences and humanities using well-organized observation, modelling, and data and information systems based on the Open Science policy.

Capacity integration can be fostered by integrating "facilitators" to work as catalysts capable of providing expert advice based on a broad range of scientific and indigenous knowledge in the local context.

Process integration can be operated by establishing cross-sectoral frameworks at local, national, regional, and global levels to link cutting-edge science beyond disciplines with on-site decision-making and action using an "end-to-end approach".

2.3 Institutional Capacity Building

2.3.1 GADRI should place enhanced focus on assisting member institutions with sharing approaches for overcoming challenges to institutional capacity building, focusing on some targeted and concrete challenges that many or most of us face. These include, for our higher education institutes, overcoming the challenges of disciplinary silos and creating new opportunities for faculty and students for interdisciplinary research, educational and service activities.

2.4 Data and Information Sharing

2.4.1 GADRI is committed to promoting "action data", "bridging gaps to knowledge sharing" and developing an "active database".

2.4.2 The GADRI Committee is acutely aware of data issues flagged in the Sendai Framework MTR under Priority One (understanding risk). The GADRI Committee is committed to continuing in many ways the work of the Global STAG working group on data.

2.4.3 The committee emphasised the importance of the various elements of FAIR data: Findable, Accessible, Interoperable and Reusable.

Inputs at the Summit brought influential examples such as the following

2.4.4 Databases and their sustainability and the use of big data and high-performance computing (DesignSafe) supporting engineering research and publishing workflows.

2.4.5 The GADRI database project that intends to regularly scrape URLs from members and analyse the pages with keywords, so that the data becomes searchable.

2.4.6 Local level data from narratives on risk from marginalized communities in particular children, which responds to the Sendai objective to understand risk for the most vulnerable people. In this context, action-oriented data collection methods have included through drawings, art and stories, which work well for health-related risks and assessing the mitigating impact of wellbeing.

Priorities for this GADRI committee to consider include:

2.4.7 Good implementation of the FAIR principles by all its members, thinking of the sustainability of databases. E.g. Identification of platforms or databases that can help make GADRI data FAIR. However, existing initiatives should be considered for their fitness for purpose. In particular, the UNDRR's own initiative on Risk Information Exchange. The design of the GADRI database must be flexible, able to collect quantitative and qualitative data, and developed by data scientists in collaboration with domain experts.

2.4.8 Collecting microdata from its members is an opportunity to develop a database of narratives about the most vulnerable people. There are issues of accessibility and a need for curation of a data policy and governance of its members, and recommendations to all DRR actors, including on data principles in times of crises.

2.4.9 Reusable data means that members are encouraged to monitor the ongoing use of data that supports the purpose of DRR.

2.5 Advocacy

2.5.1 The Committee emphasises that science advocacy needs to be cross-cutting, from grassroots to policy formulation. To do that, researchers need to co-design solutions along with civil society as well as top policy makers. The processes of co-design, co-development, and co-delivery of advocacy messages are important.

2.5.2 Media plays an important role in science advocacy, where risk communication to the people and communities as well as to the decision makers can be done effectively. Using diverse media, from conventional broadcasting to social media to community radio, and key change agents who can plan the interface of researcher and communities, customized science advocacy can be effective.

3 Panel discussions on challenges for action by GADRI reported the following:

3.1 Big Science for DRR: Large-Scale Experiments

There are significant challenges of large-scale experimental facilities for DRR. The panel identified the following actions for ongoing attention:

3.1.1 Acknowledgement of the importance of sharing facilities and existing experimental datasets with a systematic format / digital foundation.

3.1.2 Emphasise that benchmark tests using results of large-scale experiments will be beneficial to the community to understand the capability of numerical models/simulation codes.

3.1.3 Strengthen the tie between large-scale experiments and numerical modelling.

3.1.4 Enhance the use of new technologies to lower the costs of observation in the field and structures as well as experimental facilities.

3.1.5 Explore the use of large-scale experiments for disaster education to students and public, and not limit experimentation to scientific use.

3.2 Sustainable DRR Integrating Climate Action, SDGs, Field DRR; and Data Sharing Experience

The panel emphasised;

3.2.1 The importance of long-term in-situ field observations and data accumulation for climate change monitoring in climate change vulnerable regions. Sharing such observational data will be useful not only for climate change research but also for regional collaborations and cooperation.

3.2.2 That impact-based forecasting and projection is important to build community awareness of disaster mitigation and climate change adaptation.

3.2.3 That future projections for hazard, vulnerability and exposure will help societies both campaign for sustainable risk financing and prepare for situations where financing is not available.

3.2.4 That alignments of targets of different frameworks (SDG, SFDRR, CC) will help to address real world problems with strategic interventions.

3.2.5 The importance of a multi-disciplinary approach.

3.3 Gender and Inclusivity in DRR Policy and Practice

The session provided a space where gender inequalities and inclusivity in DRR policy and practice were discussed and concluded by emphasising the following:

3.3.1 Gender does not mean just women. It includes men, women, and gender and sexual minorities and all these categories have multiple identities. Intersectionality helps us understand intersecting identities of people in disaster settings and, with their participation, helps us identify appropriate interventions.

3.3.2 That following this session participants are asked to engage their home disaster research institutes on the need to listen, research and lead on the implementation of the Sendai Framework for Action in a gender aware way.

3.3.2 That recommendations arising from the discussion session are proposed for consideration by the GADRI Board and the UNDRR Sendai Framework mid-term review.

The group resolved that

3.3.4 Tackling inequalities at the root should be a priority for implementing the Sendai Framework for Action and gender should be understood as relational, impacting all, including men, women, children and gender and sexual minorities.

3.3.5 Adopting an intersectional approach to DRR policy will ensure inclusivity and equity in practice.

3.3.6 In order to build back better, it is important to understand structural inequality. A gender perspective helps us understand the everyday experiences of people and their lived reality.

3.4 Putting Health into Disaster Risk Reduction and Recovery

3.4.1 The session affirmed the broad and critical scope of putting health into DRR and recovery, including through a Japan based case study presented in the session that emphasised local level self-help.

3.4.2 GADRI members with these interests are encouraged to consider joining the WHO Health Emergency and Disaster Risk Management Research Network.

3.4.3 In the interests of developing opportunities for ongoing interactions in the health emergency and disaster risk management domains, a forthcoming joint meeting between GADRI and WHO Kobe H-EDRM will reflect an understanding that GADRI will incorporate an increased focus on health and well-being.

3.5 Youth and DRR

The session emphasised that

3.5.1 Increasing the engagement of youth and young professionals, and the networks they have created, is of vital importance to enhancing the concrete impact of our DRR work.

3.5.2 This engagement can and should be linked to innovations in our higher education programs, through placing a greater emphasis on multidisciplinary, multisectoral, multinational models of integrating empirically based DRR-related projects into our curriculum, across the majors.

3.5.3 Funding agencies and the private sector, at local, national and multinational levels, should prioritize providing support to existing and emerging youth organizations.

4 In response to these points recorded from the Summit, GADRI resolves to:

Bring all the above affirmations, findings and outcomes to its Board of Directors as GADRI resolutions for ongoing attention.

Closing Session

Closing Banquet of the 6th Global Summit of the Global Alliance of Disaster Research Institutes (GADRI)

Eiichi Nakakita(✉)

Disaster Prevention Research Institute (DPRI), Kyoto University, Kyoto, Japan

Abstract. The Sixth Global Summit of the Global Alliance of Disaster Research Institutes (GADRI) celebrated the successful conclusion of the conference with a closing banquet held at the Rihga Royal Hotel, Kyoto, Japan on 17 March 2023. Prof. Kyoko Inagaki, Vice President, Kyoto University, Japan; and Mr. Daisaku Kadokawa, Kyoto City Mayor, Japan were among the dignitaries who greeted the participants. There was a unique performance of the taiko-drums by the Kyoto Tachibana High School students.

Keyword: Midterm Review of the Sendai Framework for Disaster Risk Reduction; keynote speakers; SDGs; actionable and practical ideas

Closing Remarks by Prof. Eiichi Nakakita, Director, Disaster Prevention Research Institute (DPRI), Kyoto University
Friday, 17 March 2023

Distinguished Guests, Ladies and Gentlemen, Colleagues.

I would like to start my remarks by saying - congratulations! to all of you, – and express my gratitude to all speakers, and panelists for enlightening us, and sharing with us, their expertise, and visions during the past two and half days. My deepest appreciation also goes to all participants, presenters, supporters and everyone involved from the planning, organization, and implementation of the 6th Global Summit of GADRI.

There was a fascinating line-up of keynote speakers – experienced experts in diverse backgrounds – who had us captivated with their speeches, broaden our scope, and make us want to re-evaluate our research to do more for our communities, countries and regions.

Just to mention a few examples:

We are very happy to welcome Ms. Mami Mizutori from UNDRR. She appreciated the pre-summit survey conducted by GADRI. She presented some of the key findings from the recently completed Midterm Review of the Sendai Framework for Disaster Risk Reduction and highlights from the UN Global Assessment Report 2022. She mentioned that in the 2019 GAR, declared that the era of hazards-by-hazards risk reduction is over and that risk is systemic and it crosses borders. They are not siloed but the approach to risk management is very much siloed. And that conventional risk management and crisis responses system are not working. Failure to plan is a plan to fail. She thanked us for being a community that cares for the resilience of our people

H. Tatano et al. (Eds.): GSRIDRR 2023, DRRGADRIBS, pp. 535–545, 2025.
https://doi.org/10.1007/978-981-96-6609-6

and encouraged us to market resilience in a way that we can ask people to build their resilience to have a stable future.

We are glad to have Prof. Mustafa Erdik from Boğaziçi University who shared with us firsthand and detailed information about the tragic two-large earthquakes that hit southern region of Turkey just nine hours apart. He informed us that what is needed is a retrofit campaign to improve the earthquake performance of buildings that does not require tenants to vacate their apartments; and a revision for earthquake performance objectives for retrofit.

Prof. Masahide Kimoto stated that – science should contribute to people's action. Science can contribute and should provide information to support policy planning and make citizen life better.

While Prof. Paul Kovacs – presenting on recovery planning – stated – we have the potential to do better within GADRI because of the science and knowledge that we are generating. The recovery conversation is a big opportunity to be bold, be transformative, to take our science and push for things in recovery that are hard to do the rest of the time.

Prof. Virginia Murray, stated that the climate crisis is definitely a health crisis. It is certainly growing. With many examples of recently published literature sources, she made us aware and understand some of the risks of extreme events and highly translated into impacts and disasters.

Digital Twinning of Communities for Disaster Risk Reduction - Prof. John van de Lindt – discussed how to improve resilience – a tool to measure resilience of a community to improve it – and improving resilience through communication, decision, and actionable strategies is the goal.

Africa is one of the regions most impacted by climate change. Dr. Chipo Mudavanhu concentrated on climate change and disaster risks in the context of sustainable development in Southern Africa. The past 10 years experienced a lot of cyclones which are increasing in intensify and frequency. Talking about challenges, she stated a several factors - that when the information is packaged, it has to ensure that the end user can understand and make appropriate action with it; there are inadequate budgetary allocations – more resources are put towards disaster response rather than mitigation and preparedness and the importance of community involvement in the decision-making processes.

Panel discussion sessions brought another group of scholars from diverse background. Group I-A had a café-style discussion session.

In Group IC – The Prof. Reini Wirahadikusumah, Rector, Institute of Technology Bandung (ITB), Indonesia delivered an online keynote speech to kick-off the session.

These messages coupled with the presentations by all other panelists from the academia, government, media and the UN, have given us much food for thought. Undoubtedly, will further enhance our perceptions on disaster risk reduction and resilience, and will greatly contribute to stimulate our work and collaborations as called in all-of-society approach.

I would like to thank all panel group chairs and co-chairs for their roles in organizing their respective session with much care and diligence to deliver the results we saw in the outcome report.

I would also like to thank the session chairs, who, too, did an excellent job supporting each plenary and discussion session, and keeping everything lively and on track. Most of all, I thank you, the participants. Without your contributions, inputs and ideas, the discussion sessions and this summit would not have achieved the results noted in the final outcome document. Our expectations from the conference have exceeded our lofty goals.

Now, it is up to all of us to take the outcomes confirmed in the conference resolution document forward to the future. We will, once again, have the chance to evaluate the results of our research activities and how best we were able to promote the implementations of the goals and targets of the Sendai Framework Agenda in two years' time.

I am convinced that the contacts and collaborations formed here are useful and beneficial; and they will continue in to the future to form fruitful research and institutional collaborations.

I would like to end my remarks by stating that it is a great privilege for us, the DPRI, Kyoto University and Kyoto University, to host the GADRI. While supporting GADRI, it also gives us the chance to increase our visibility among the international community as an International Research Institute of Excellence in the field of disaster risk reduction and resilience to disasters. Together, we can strengthen collaborations and build-on our common objectives to create a sustainable disaster-resilient world.

I enjoyed interacting with you during the past three days, educated myself with much inspiring and stimulating discussions.

I wish you all the very best in your endeavours, and wish you future prosperity. And, a safe journey back home.

I now officially close the 6th Global Summit of GADRI.

Thank you.

Eiichi Nakakita
Director
Disaster Prevention Research Institute (DPRI)
Kyoto University, Japan

Fig. 1. Group photo with the Taiko Drum Performers of the Kyoto Tachibana High School

Closing Banquet of the 6th Global Summit of GADRI: Towards GADRI Objectives of Achieving a Sustainable Disaster-Resilient World

Greetings at the Closing Banquet

Fig. 2. Greetings delivered by Prof. Kyoko Inagaki

Greetings by Prof. Kyoko Inagaki, Executive Vice-President Kyoto University
Good evening everyone.

As introduced, I am Kyoko Inagaki, Kyoto University's executive vice-president for gender equality, international affairs, and external affairs.

Allow me to begin by applauding all of the speakers and panelists here, as well as those who joined us online, for helping to make the 6th Global Summit of GADRI such a great success.

During the past three days, the active contributions from all participants have enabled us to explore a very wide range of approaches to achieving a more disaster-resilient world, and I know that the groundwork and connections made during this summit will resonate through future collaboration for a long time to come.

I know that President Minato also mentioned this in his welcome address on Wednesday, but as Kyoto University's executive in charge of gender equality and diversity, I was personally very pleased to see a significant number of women among the summit's participants, as well as a session dedicated to gender and inclusivity in DRR policy.

Goal 5 of the UN Sustainable Development Goals is the achievement of gender equality and empowerment for all women and girls. We are now only seven years away from the goal's target year of 2030, but we still have a long way to go.

It is sometimes said that "disasters do not discriminate," so it is appropriate and encouraging to see an organization like GADRI taking the lead in raising awareness of gender and diversity issues among its members.

Likewise, GADRI's efforts to actively encourage the involvement of the younger generation of researchers and professionals is also laudable. The young are our future, and they have the energy and potential to bring positive change. It is vital that they are provided with a platform where their voices can be heard in an open-minded environment, and where they too can contribute, collaborate, and learn.

As President Minato said, diversity and inclusivity are key focus areas for Kyoto University at the moment, and the rising awareness of those issues among our collaborating organizations like GADRI will help us all to move forward into an era of more inclusive, more diverse, and consequently more effective endeavors.

In addition to gender equality, I also serve as Kyoto University's executive in charge of international affairs. In that capacity, I am always delighted to see the university's active involvement in international collaborative events. Some of our scholars also recently attended the 3rd World Bosai Forum, which was held in Sendai earlier this month, and we are also involved in the UN 2023 Water Conference, which will be held in New York next week. We are particularly happy, of course, to have been able to play such a central role in the organization and hosting of this Global Summit of GADRI.

The aim of this summit was to deliver actionable and practical ideas that can be carried out by all communities. I am looking forward to receiving a copy of the summit's outcome document, and reading about the achievements of the past three days.

Before I conclude, I would like to take this opportunity to thank the organizers of the summit, who worked tirelessly amidst the many challenges and obstacles caused by the coronavirus pandemic to make the event a success. Your efforts are greatly appreciated.

In closing, I wish you all the very best in your endeavors, and for those who are travelling, I wish you a safe journey home.

Now, let us relax and enjoy our evening. Thank you for your attention.

Fig. 3. Greetings by Mr. Daisaku Kadokawa, Mayor, Kyoto City, Japan

第６回世界防災研究所連合サミット及び第３回世界防災研究所連合総会懇親会

市長挨拶骨子（令和５年３月１７日（金）午後５時３０分～）

１　主催者・来賓への謝辞等

世界防災研究所連合（ＧＡＤＲＩ（ガドリ））事務局長　多々納（たたの）　裕一（ひろかず）氏（京都大学防災研究所教授）

２　本題、話題【逐次通訳有】

・　世界各国の災害研究・防災研究機関が連携し、災害リスクの軽減とレジリエンスの向上を目指す世界防災研究所連合。開催に御尽力の京都大学防

災研究所教授の多々納　裕一事務局長はじめ関係者の皆様に、改めて深く敬意を表する。【通訳】

・　今回のサミットでは、世界の 30 以上の国と地域から 160 名を超える防災の専門家の皆様が参加され、3 日間にわたり、活発な議論が行われた。この会議の御成功をお慶び申し上げる。【通訳】

＜過去の開催状況＞

・サミットは、宇治市で開催
・総会及び懇親会は、京都市で開催（今回で本市開催は３回目）
・市長が総会及び懇親会に参加されるのは、２０１９年の第２回の時以来

・　先月（２／６）、トルコ及びシリアで発生した大地震により、多くの死傷者が生じたことに、大変心を痛めている。亡くなられた方々に哀悼の意を表するとともに、被災された方々にお見舞い申し上げる。京都市でも、被災地の支援に役立てるよう、市内 73 施設でいち早く救援金の受付を開始し、市民の皆様から温かい善意が寄せられているところ。引き続き、可能な限りの支援を行ってまいる。【通訳】

＜2023 年トルコ・シリア地震に対する救援金＞

・実施期間：令和５年２月１０日（金）～令和５年５月３１日（水）

・受付場所：市役所本庁舎、分庁舎正面入口、各区役所・支所地域力推進室、事業所等

・　さて、京都は、1200 年を越える歴史を有するまち。その長い歴史の中で、幾度も地震や　大火、争乱等の危機に直面。しかし、その都度、市民の皆様が力を合わせ、困難を乗り越えて、今日まで発展を続けてきた。【通訳】

・　そんな「レジリエンス」が息づくまち京都で行われた本サミットが、世界各地の防災力の向上にとっての大きな一歩となることを念じている。【通訳】

・　さて、この総会・懇親会でも、先ほどの太鼓演奏はじめ、日本の文化を体験できる催しがいくつも御用意されている。「日本の文化の都」である京都で、その魅力を感じていただきながら、世界各地からお越しの皆様が、この後、大いに交流を深められ、友情を温めることで、世界防災研究所連合がますます発展されることを、心から祈念する。【通訳】

Message by Taiko Drum Performers of the Kyoto Tachibana High School, Japan.

Prior to the performance of the traditional taiko drums, the students of the Kyoto Tachibana High School share the message they would like to convey through their music.

Fig. 4. After the greetings, Mr. Kadokawa took time to greet all participants at the Closing Banquet

こんにちは。京都橘高校です。

私たちの学校は水と緑に囲まれた歴史の町、京都伏見の地にあります。1902 年に京都女子手芸学校として創立し、今年度で 120 周年を迎えた中高一貫校です。多くの生徒が勉強と両立して部活動に励んでいます。たくさんある部活動の中で、吹奏楽部、女子バレーボール部、男子サッカー部、陸上競技部、そして私たち太鼓部が強化指定クラブに指定されており、全国の舞台で活躍しています。

私たち太鼓部は中高合同で、「心はひとつ」を合言葉に日々活動しています。

これから「若い世代の持つ平和な未来への希望」について話します。

「平和な未来」という言葉から私たちは何を思い浮かべるでしょう。私たちの「平和な未来」を脅かすものはたくさんあります。ロシアによるウクライナ侵攻、1978年から断続的に続いているアフガニスタン紛争と、世界で今起こっている武力紛争の数は56ほどにものぼり、その惨状を目にしない日はありません。しかし、平和を脅かすものは何も戦争や紛争だけではありません。自然現象による災害、たとえば地震、火山、気象、土砂による被害、それによって私たちの命が奪われ、平和な生活が失われてしまうことも多いのです。そこで、「平和な未来」に向けて私たちにできることは何かを考えてみたいと思います。

この日本でも毎年さまざまな自然災害が起きています。阪神淡路大震災では、建物の倒壊や火災で多くの人の命と生活が奪われました。そして、東日本大震災では人々が津波にのみ込まれ、多くの人が命を失いました。これらの辛く悲しい過去からわたしたちが学び続けることこそが「平和な未来への希望」につながるのではないかと私は考えます。自然災害は戦争などと同じように、大切な命や何気ない生活を奪います。しかし、そのことから私たち人間は逃れることはできません。そうであれば、自然災害を未然に防ぐためにメカニズムを解明する、起きてしまった災害とどう向き合い、どのように対処するかを考えるなど、私たちひとりひとりが防災ということに対して常に意識をし、考え続けることが大切なのではないでしょうか。私たちにできることは限られているかもしれません。しかし、私はまず戦争や自然災害のことを知り、今世界で起こっている問題に対してできることはないのかを考えて行動していくべきだと思います。これからの世界が平和であるためにできることは何なのか。他人事ではなく自分のことだと捉えて考えることで私たちの生きていく未来は少しずつ変わるのではないでしょうか。私はこれからの世界そして未来が明るくなっていくことを望みます。

これから演奏します曲は「屋台囃子」です。それではお聞きください。

Hello. This is Kyoto Tachibana High School. Our school is located in Fushimi, Kyoto, a historic city surrounded by water and greenery. Founded in 1902 as Kyoto Women's Handicraft School, we are now a combined junior and senior high school celebrating our 120th anniversary this year. Many of our students participate in extracurricular activities while balancing their studies. Among the many clubs, our brass band, girls' volleyball team, boys' soccer team, track and field team, and our taiko drumming club are designated as strengthening clubs and are active on national stages.

Our taiko club practices together with both junior and senior high school students under the slogan of "One Heart". We will now talk about "hope for a peaceful future held by the younger generation." What comes to mind when we hear the words "peaceful future"? There are many things that threaten our "peaceful future." The number of armed conflicts occurring in the world, such as the Russian invasion of Ukraine and the ongoing conflict in Afghanistan that has been happening intermittently since 1978, is as many as 56, and there is not a day where we don't see the horrific state of these conflicts. However, what threatens peace is not just wars and conflicts. Natural disasters, such as earthquakes, volcanoes, weather, and landslides, also take away our

Fig. 5. Taiko drums arranged for the performance

Fig. 6. Message read by the students of the Kyoto Tachibana High School

Fig. 7. Taiko Drums performance by the Kyoto Tachibana High School Students

precious lives and peaceful lifestyles. Therefore, we want to think about what we can do for a "peaceful future."

Various natural disasters occur every year in Japan as well. In the Great Hanshin-Awaji Earthquake, many people lost their lives and livelihoods due to collapsed buildings and fires. And in the Great East Japan Earthquake, many people lost their lives as they were swallowed by the tsunami. I believe that continuing to learn from these painful and sad experiences is what leads us to "hope for a peaceful future." Natural disasters, like wars, take away important lives and ordinary lifestyles. However, we humans cannot escape from this fact. If so, I think it is important for each and every one of us to always be aware of disaster prevention, such as understanding the mechanisms of natural disasters and thinking about how to deal with them when they occur. We may have limited abilities, but I think we should first learn about wars and natural disasters, think about what we can do for the problems happening in the world today, and take action. What can we do to make the world peaceful in the future? By thinking about it as our own problem and not just someone else's, our future will gradually change. I hope that the world and the future will become brighter from now on.

The song we will perform next is "Yatai Bayashi." Please listen.

Author Index

H. Tatano et al. (Eds.): GSRIDRR 2023, DRRGADRIBS, pp. 547–548, 2025.
https://doi.org/10.1007/978-981-96-6609-6